AF719411

Current Research in Parasitic Weeds

Current Research in Parasitic Weeds

Editor: Smith Torres

MURPHY & MOORE
www.murphy-moorepublishing.com

Murphy & Moore Publishing,
1 Rockefeller Plaza,
New York City, NY 10020, USA

Visit us on the World Wide Web at:
www.murphy-moorepublishing.com

© Murphy & Moore Publishing, 2022

This book contains information obtained from authentic and highly regarded sources. Copyright for all individual chapters remain with the respective authors as indicated. All chapters are published with permission under the Creative Commons Attribution License or equivalent. A wide variety of references are listed. Permission and sources are indicated; for detailed attributions, please refer to the permissions page and list of contributors. Reasonable efforts have been made to publish reliable data and information, but the authors, editors and publisher cannot assume any responsibility for the validity of all materials or the consequences of their use.

ISBN: 978-1-63987-147-6 (Hardback)

Trademark Notice: Registered trademark of products or corporate names are used only for explanation and identification without intent to infringe.

Cataloging-in-Publication Data

Current research in parasitic weeds / edited by Smith Torres.
p. cm.
Includes bibliographical references and index.
ISBN 978-1-63987-147-6
1. Weeds. 2. Parasitic plants--Control. 3. Weeds--Control. 4. Plant parasites. I. Torres, Smith.
SB611 .C87 2022
632.5--dc23

Table of Contents

Preface

The purpose of the book is to provide a glimpse into the dynamics and to present opinions and studies of some of the scientists engaged in the development of new ideas in the field from very different standpoints. This book will prove useful to students and researchers owing to its high content quality.

Parasitic weeds are those plants which have lost their autotrophic nature during the course of evolution. According to the dependency on host plant, parasitic weeds can be classified into partial and total parasites. Partial parasites only require the minerals from the host and can prepare food on their own. Total parasitic weeds are completely dependent on host for their nutrition. On the basis of their attack site, parasitic weeds can be further classified into stem parasites and root parasites. Transplanting, crop rotation and intercropping techniques are a few ways of managing parasitic weeds. This book explores all the important aspects of parasitic weeds in the present day scenario. The current research on parasitic weeds are glanced at and their applications as well as ramifications are looked at in detail. The extensive content of this book provides the readers with a thorough understanding of the subject.

At the end, I would like to appreciate all the efforts made by the authors in completing their chapters professionally. I express my deepest gratitude to all of them for contributing to this book by sharing their valuable works. A special thanks to my family and friends for their constant support in this journey.

Editor

1

Increased Virulence in Sunflower Broomrape (*Orobanche cumana* Wallr.) Populations from Southern Spain is Associated with Greater Genetic Diversity

Edited by:
Monica Fernandez-Aparicio, Institut National de la Recherche Agronomique, France

Reviewed by:
Pascal Labrousse, Université de Limoges, France
Shaoqing Tang, Guangxi Normal University, China
Birgit Gemeinholzer, Justus Liebig University Giessen, Germany

***Correspondence:**
Leonardo Velasco
lvelasco@ias.csic.es

†Present address:
Jebri Malek, Laboratory of Molecular Genetics, Immunology and Biotechnology, Faculty of Science of Tunis, Tunis El Manar, Tunisia

Alberto Martín-Sanz[1], Jebri Malek[2†], José M. Fernández-Martínez[2], Begoña Pérez-Vich[2] and Leonardo Velasco[2]*

[1] Pioneer Hi-Bred Agro Servicios Spain SL, Sevilla, Spain, [2] Institute for Sustainable Agriculture, Consejo Superior de Investigaciones Científicas, Córdoba, Spain

Orobanche cumana Wallr. (sunflower broomrape) is a holoparasitic weed that infects roots of sunflower in large areas of Europe and Asia. Two distant *O. cumana* gene pools have been identified in Spain, one in Cuenca province in the Center and another one in the Guadalquivir Valley in the South. Race F has been hypothesized to have arisen by separate mutational events in both gene pools. In the Guadalquivir Valley, race F spread in the middle 1990's to become predominant and contained so far with race F hybrids. Recently, enhanced virulent populations of *O. cumana* have been observed in commercial fields parasitizing race F resistant hybrids. From them, we collected four independent populations and conducted virulence and SSR marker-based genetic diversity analysis. Virulence essays confirmed that the four populations studied can parasitize most of the race F resistant hybrids tested, but they cannot parasitize the differential inbred lines DEB-2, carrying resistance to race F and G, and P-96, resistant to F but susceptible to races G from other countries. Accordingly, the new populations have been classified as race G_{GV} to distinguish them from other races G. Cluster analysis with a set of populations from the two Spanish gene pools and from other areas, mainly Eastern Europe, confirmed that race G_{GV} populations maintain close genetic relatedness with the Guadalquivir Valley gene pool. This suggested that increased virulence was not caused by new introductions from other countries. Genetic diversity parameters revealed that the four populations had much greater genetic diversity than conventional populations of the same area, containing only alleles present in the Guadalquivir Valley and Cuenca gene pools. The results suggested that increased virulence may have resulted from admixture of populations from the Guadalquivir Valley and Cuenca followed by recombination of avirulence genes.

Keywords: sunflower broomrape, *Orobanche cumana*, new race, virulence, genetic diversity, gene pools, genetic recombination

INTRODUCTION

Orobanche cumana Wallr. (sunflower broomrape) is a holoparasitic plant species with a restricted range of hosts both in the wild, where it mainly parasitizes *Artemisia* spp., as well as in agricultural fields, where it only grows on sunflower (Fernández-Martínez et al., 2015). The parasitic interaction between sunflower and *O. cumana* generally follows a gene for gene model, with resistance in sunflower (Vrânceanu et al., 1980) and avirulence in *O. cumana* (Rodríguez-Ojeda et al., 2013b) controlled by dominant alleles at single loci. Nonetheless, more complex genetic control of resistance to *O. cumana* has been also reported in some sunflower resistant sources, including two dominant genes (Domínguez, 1996), one dominant and one recessive gene (Akhtouch et al., 2002; Akhtouch et al., 2016), one dominant and one modifying gene (Velasco et al., 2007), one recessive gene (Imerovski et al., 2016), two recessive genes (Rodríguez-Ojeda et al., 2001; Akhtouch et al., 2002), or polygenic genetic control (Labrousse et al., 2004).

The general occurrence of a gene for gene interaction between sunflower and *O. cumana* and the associated development of physiological races of the parasite is an exception in parasitic systems involving *Orobanche* spp., which are in general under quantitative or horizontal genetic control (Pérez-Vich et al., 2013). Vrânceanu et al. (1980) reported the existence of five races of *O. cumana* named as A to E, controlled by resistance genes *Or1* to *Or5*. New populations overcoming *Or5* resistance and named as race F were identified from the middle 1990's in most of the areas infested by *O. cumana*, namely Spain (Alonso et al., 1996), Romania (Pacureanu-Joita et al., 2004), Turkey (Kaya et al., 2004), Bulgaria (Shindrova, 2006), Ukraine (Burlov and Burlov, 2010), and Russia (Antonova et al., 2013). Nowadays, increasingly virulent populations classified as races G and H are becoming predominant in countries around the Black Sea (Kaya, 2014).

O. cumana has been traditionally considered as one of the few exceptions of the genus that are self-pollinating, which has been based on its flower morphology, with small lower lips that do not facilitate the action of big pollinators such as bees and bumblebees (Satovic et al., 2009), and the structure of its populations, characterized by low intra-population and large inter-population genetic variation (Gagne et al., 1998). However, experimental research using a mutant line lacking anthocyanin pigmentation evidenced the existence of a percentage of cross fertilization in this species of up to 40% under the conditions of the experiments, in which small insects were identified as pollinating agents (Rodríguez-Ojeda et al., 2013a).

Orobanche cumana is not present in the wild in Spain, but exclusively found in agricultural fields parasitizing sunflower (Pujadas-Salvà and Velasco, 2000). Recent studies have identified two well separated gene pools, one in Cuenca province in Central Spain and another one in the Guadalquivir Valley in southern Spain (Pineda-Martos et al., 2013; Molinero-Ruiz et al., 2014). The study of Pineda-Martos et al. (2013), conducted on 50 populations collected from both areas of Spain, reported very low intra-population and inter-population genetic diversity within each gene pool, which was hypothesized to be caused by a founder effect in separate introductions. Even so, greater genetic diversity was detected in a small number of populations, in which the presence of individuals from both gene pools and heterozygotes resulting from their hybridization were identified. Interestingly, both gene pools included populations classified as race E and race F that showed very high genetic similarity, suggesting that race F probably arose from punctual mutations within each gene pool (Pineda-Martos et al., 2013). To avoid misunderstanding, we are naming race F from the Guadalquivir Valley as F_{GV}.

The Guadalquivir Valley in southern Spain is one of the main areas of sunflower cultivation in this country. *O. cumana* race F_{GV} appeared in this area in the middle 1990's and spread rapidly to become predominant until the present day. In 2014, small spots of *O. cumana* plants parasitizing sunflower hybrids resistant to race F_{GV} were observed in several fields. The objectives of this research were (i) to evaluate the virulence of these populations on a set of differential lines and hybrids; (ii) to analyze their genetic relatedness to local and foreign populations in order to test whether they resulted from new introductions; and (iii) to determine the genetic diversity of the populations.

MATERIALS AND METHODS

Orobanche cumana Populations

Spots of *O. cumana* plants growing on sunflower hybrids with complete resistance to race F_{GV} were identified in 2014 in Sevilla Province, in the Guadalquivir Valley. From them, seeds of mature *O. cumana* plants were collected in four independent sunflower fields at three distant locations in July 2014. Populations BR-24 (36° 58′ 47.9″ N; 5°54′ 57.1″ W) and BR-27 (36° 59′ 34.0″ N; 5° 54′ 49.2″ W) were collected in Las Cabezas de San Juan, population BR-25 (37° 27′ 59.4″ N; 5° 40′ 24.2″ W) was collected in Carmona, and population BR-28 (37° 24′ 25.8″ N; 5° 14′ 47.6″ W) was collected in Écija, in the four cases on commercial sunflower hybrids resistant to race F_{GV}. The reason for not informing on the names of the hybrids on which the populations were collected is given below. *O. cumana* plants were found scattered in patches of no more than 15 m of diameter. The longest distance between the fields where the populations were collected was 98 km between BR-24 and BR-28, while the shortest distance was 1,6 km between BR-24 and BR-27. Seeds of a conventional race F population of the Guadalquivir Valley (F_{GV}) named as SP were used as a control in the phenotypic evaluations. Seeds of SP population were collected in 2001 in Écija (Sevilla) and the population is multiplied periodically on cultivar NR5, resistant to races A to E and susceptible to race F_{GV}. Six *O. cumana* populations from several Eastern European countries, all of them classified as race G on the basis of their ability to parasitize race F resistant hybrids, were used for comparative phenotypic evaluation of the alleged new race G population from the Guadalquivir Valley. They were collected in Ialomita County, Romania (G_{RO}), Thrace region of Greece (G_{GR}), Edirne area of Turkey (G_{TK}), southern area of Bulgaria (G_{BU}), Rostov area of Russia (G_{RU}), and Lugansk area of Ukraine (G_{UK}).

Twenty *O. cumana* populations from Spain and other countries were used as a reference to compare the molecular genetic similarity between the four populations under study with those from other geographical areas. The selection of the populations did not pretend to be an exhaustive representation of the main sunflower production areas, but just a set of geographically diverse populations to test the hypothesis on whether the new populations with increased virulence evolved in the Guadalquivir Valley area or, contrarily, whether they might have resulted from new introductions. The set of reference populations included four populations from the Guadalquivir Valley: CO-02, SE-10 (Pineda-Martos et al., 2013), EK-23 (Rodríguez-Ojeda et al., 2013b), and SP; four populations from Cuenca province in Central Spain: IASCum-4, CU-05, CU-07, CU-12 (Pineda-Martos et al., 2014b); one population from Serbia (Boro-9); one population from Romania (Boro-10); one population from Israel (Boro-11); two populations from Turkey (Boro-14, Boro-15); and seven populations from Bulgaria (Boro-16 to Boro-22). In all cases seeds were collected in the respective areas and tissue from 20 individual plants growing on sunflower susceptible line B117 was collected for DNA extraction. Equal amounts of DNA of the plants from each population were pooled and used as a template for PCR amplification.

The study of intrapopulation diversity was based on the four populations under study (BR-24, BR-25, BR-27, BR-28). Two conventional populations of the Guadalquivir Valley (SE-10) and Cuenca (IASCum-4) were used as a reference for both gene pools based on a previous study (Pineda-Martos et al., 2013). DNA from 12 randomly selected individual plants per population was used as a template for PCR amplification.

Sunflower Lines and Hybrids

Four differential inbred lines, two commercial hybrids susceptible to race F_{GV}, and 13 commercial hybrids with complete resistance to race F_{GV} were used in the experiments conducted to evaluate the virulence of the *O. cumana* populations. The inbred lines were B117, susceptible to all tested *O. cumana* races; NR5, resistant to Guadalquivir Valley's race E; P-96, resistant to race F_{GV} and susceptible to races G from other countries; and DEB-2, resistant to both race F_{GV} and races G. Commercial hybrids susceptible to race F_{GV} were 63D82 and P64LE19 (Pioneer). Commercial hybrids resistant to race F_{GV} were named as Hybrid 1 through Hybrid 13. All of them are hybrids widely cultivated in the Guadalquivir Valley area for several years with complete resistance to race F_{GV}. The names of the hybrids are not provided because the objective of the research was not to conduct a survey on resistance of commercial hybrids to a new race, but just to ensure that the new populations under study can be considered as a new race. Accordingly, the hybrids were not selected following the criterion of representativeness, and some seed companies may be overrepresented and others not included in the study, which might result in an unbalanced impact on the commercial interests of seed companies.

Phenotypic Evaluation

Virulence of the four populations BR-24, BR-25, BR-27, and BR-28 together with SP, G_{RO}, G_{GR}, G_{TK}, G_{BU}, G_{RU}, and G_{UK} used as a control were evaluated on four differential lines (B117, NR5, P-96, DEB-2) and six hybrids (63D82, P64LE19, Hybrid 1, Hybrid 2, Hybrid 3, Hybrid 4). Because of the limitation in the amount of *O. cumana* seed available, the experiments were conducted on multi-pot trays with a pot volume of 40 cm^3. Twenty plants of each line or hybrid, separated into two replications of ten plants each, were evaluated for each *O. cumana* population. An additional experiment based on a single replication of 10 plants was conducted to test the virulence of population BR-28, for which additional seeds were available, on 14 hybrids: 13 commercial hybrids resistant to race F_{GV} and the control 63D82, susceptible to race F_{GV}.

In all experiments, the soil consisted in a mixture of sand and peat (1:1 by vol). The soil was inoculated with *O. cumana* seeds at an approximate concentration of 0.28 mg of seeds per g of soil. This roughly corresponds to 185 seeds per g of soil or 4,000 seeds per pot. The soil mixture containing the *O. cumana* seeds was carefully mixed to obtain a homogeneously infested substrate. Sunflower seeds were germinated in moistened filter paper and then planted in the pots. The plants were grown in a growth chamber at 25/20°C (day/night) with a 16-h photoperiod, and photon flux density of 300 $\mu mol\ m^{-2}\ s^{-1}$.

Phenotypic evaluation of virulence was conducted by determining the level of *O. cumana* incidence (percentage of sunflower plants sustaining the growth of *O. cumana* shoots), number of *O. cumana* attachments per sunflower plant, and the *O. cumana* stage of development classified as: T1 = small nodules with diameter less than 2 mm; T2 = nodules with diameter greater than 2 mm and stem differentiation not yet apparent; T3 = nodules typically between 5 and 10 mm in which stem differentiation can be clearly observed or even the stem has grown several centimeters. In this paper we are only reporting the sum of nodules at T2 and T3 stages.

Tissue Collection, DNA Extraction, and SSR Analysis

Individual *O. cumana* plants showing well developed stems (T3 stage) were collected on sunflower line B117 for each population. The plants were frozen at −80°C, lyophilized and ground to a fine powder. DNA was extracted from individual plants using a modified version of the protocol described in Pérez-Vich et al. (2004). For the study of interpopulation diversity, equal amounts of DNA of 24 *O. cumana* plants from each population were pooled and used as a template for PCR amplification. For the study of intrapopulation diversity, DNA from individual plants was used as a template for PCR amplification. Microsatellite analyses were carried out as described in Pineda-Martos et al. (2014b) using 20 high-quality, polymorphic SSR primer pairs reported in that work: Ocum-003, Ocum-023, Ocum-052, Ocum-059, Ocum-063, Ocum-070, Ocum-074, Ocum-075, Ocum-081, Ocum-087, Ocum-091, Ocum-092, Ocum-108, Ocum-122, Ocum-141, Ocum-151, Ocum-160, Ocum-174, Ocum-196, and Ocum-197. Amplification products were analyzed by gel electrophoresis using 3% Metaphor agarose (BMA, Rockland, ME, USA) in 1x TBE buffer and SaveView Nucleic Acid Stain (NBS Biologicals Ltd., Huntingdon, UK) and visualized under UV light. A 100 bp DNA ladder (Solis BioDyne, Tartu, Estonia) was used as a standard molecular weight marker to get an approximate size of DNA fragments. Bands were scored manually with the aid of Quantity One® 1-D Analysis Software (Bio-Rad Laboratories Inc., Hercules, CA, USA). Amplified fragments were scored for the presence (1) or absence (0) of homologous bands and compiled into a binary data matrix or for their estimated molecular weight and compiled into a codominant data matrix.

Analysis of Inter-population Similarities

A dissimilarity matrix was created using DICE dissimilarity index, from which cluster analysis was conducted using the UPGMA (Unweighted Pair Group Method with Arithmetic Mean) method of NTSYSpc ver. 2.21q (Applied Biostatistics, Inc, Port Jefferson, NY, USA). The cophenetic correlation coefficient was calculated to test the goodness of fit of the dendrogram to the original dissimilarity matrix.

Genetic Diversity Analysis

The percentage of polymorphic loci (P), the average observed number of alleles (Na), the number of different alleles with a frequency ≥5% (Na ≥5%), the number of effective alleles (Ne), the number of private alleles unique to a single population (Npa), the observed and expected heterozygosity (Ho and He), the Shannon's diversity index (I), and the fixation index (Fis) were calculated for all loci at each population using GenAlEx ver. 6.5. Pairwise genetic distances between populations were calculated as the genetic distance coefficient GST using GenAlEx ver. 6.5 using 1000 random permutations to assess significance. The matrix of GST pairwise distances was used as input for a principal coordinates analysis (PCoA).

The average number of pairwise differences between individuals within each population was calculated as a measure of intrapopulation genetic diversity. Analysis of molecular variance (AMOVA) was conducted on the distance matrix to separate total variance into variance attributable to differences between individuals within a population and variance attributable to differences between populations. In both cases, Arlequin ver. 3.5.2.2 (L. Excoffier, CMPG, University of Bern, Switzerland) was used.

RESULTS

Phenotypic Evaluation of Virulence

The four alleged race G *O. cumana* populations BR-24, BR-25, BR-27, and BR-28 showed a similar pattern of virulence, resulting in 100% of parasitized plants and a high number of attachments per plant (9.2 to 25.8) on the race F_{GV} susceptible genotypes B117, 63D82, NR5, and P64LE19, high level of incidence (75 to 100%) and moderate number of attachments per plant (3.2 to 10.1) on Hybrid 1 and Hybrid 3, moderate level of incidence (20 to 50%) and low number of attachments per plant (1.3 to 2.0) on Hybrid 2, low level of incidence (0 to 15%) and low number of attachments per plant (1.0 to 2.0) on Hybrid 4, and no parasitization on lines P96 and DEB2 (**Table 1**). Population SP, a conventional race F_{GV} population used as a control, only parasitized on the four race F susceptible genotypes, with 100% incidence and between 10.8 and 22.2 attachments per

TABLE 1 | Percentage of susceptible plants (%S) and average number of *Orobanche cumana* attachments in susceptible plants (Xs) in ten sunflower lines and hybrids evaluated with *O. cumana* populations BR-24, BR-25, BR-27, BR-28, and SP (race F_{GV}) from the Guadalquivir Valley, and populations classified as race G from Romania (G_{RO}), Greece (G_{GR}), Turkey (G_{TK}), Bulgaria (G_{BU}), Russia (G_{RU}), and Ukraine (G_{UK}).

	Orobanche cumana populations																					
	BR-24		*BR-25*		*BR-27*		*BR-28*		*SP*		G_{RO}		G_{GR}		G_{TK}		G_{BU}		G_{RU}		G_{UK}	
Line/Hybrid[a]	**%S**	**Xs**	**%S**	**Xs**	**%S**	**Xs**	**%S**	**Xs**	**%S**	**Xs**	**%S**	**Xs**	**%S**	**Xs**	**%S**	**Xs**	**%S**	**Xs**	**%S**	**Xs**	**%S**	**Xs**
B117	100	25,8	100	12,7	100	20,3	100	11,3	100	22,2	100	5,1	100	8,5	100	14,0	100	10,5	100	9,2	100	11,0
63D82	100	14,8	100	10,1	100	13,1	100	8,0	100	11,8	100	5,1	100	12,4	100	12,2	100	14,0	100	15,0	100	10,5
NR5	100	16,8	100	15,2	100	9,2	100	7,7	100	10,8	100	5,7	100	12,2	100	14,5	100	16,4	100	14,7	100	11,8
P64LE19	100	15,8	100	11,3	100	13,3	100	7,6	100	15,7	100	8,0	100	11,2	100	14,8	100	14,8	100	15,0	100	15,0
P96	0	0,0	0	0,0	0	0,0	0	0,0	0	0,0	50	1,4	100	4,8	100	7,8	50,0	4,0	90	3,0	90	4,4
DEB2	0	0,0	0	0,0	0	0,0	0	0,0	0	0,0	0	0,0	0	0,0	0	0,0	0	0,0	0	0,0	0	0,0
Hybrid 1	95	7,8	85	5,4	100	10,1	75	5,7	0	0,0	100	3,2	100	2,4	100	7,3	0,0	0,0	90	1,8	100	3,3
Hybrid 2	30	1,5	20	2,0	50	1,3	30	1,3	0	0,0	80	2,0	0	0,0	80	2,0	0,0	0,0	40	1,5	30	1,0
Hybrid 3	85	4,4	75	3,5	100	6,2	90	3,2	0	0,0	100	2,8	40	2,5	100	11,6	0,0	0,0	70	3,2	100	3,7
Hybrid 4	5	2,0	0	0,0	10	2,0	15	1,0	0	0,0	30	1,0	0	0,0	30	1,6	0,0	0,0	0	0,0	70	1,0

The number of plants evaluated per line/hybrid and population was 20 in all cases.
[a]*Previous information indicated that B117 and 63D82 are susceptible to all O. cumana races, NR5 and P64LE19 are resistant to race E, P96 is resistant to races E and F_{GV}, DEB2 is resistant to races E, F_{GV}, and G_{TK}, and Hybrids 1 to 4 are resistant to races E and F_{GV}.*

plant, similar to the populations under study (**Table 1**). Race G populations from Romania (G_{RO}), Greece (G_{GR}), Turkey (G_{TK}), Bulgaria (G_{BU}), Russia (G_{RU}), and Ukraine (G_{UK}) parasitized in all cases line P96, with incidence between 50% (G_{RO} and G_{BU}) and 100% (G_{GR} and G_{TK}), which marked a significant difference between the four populations from the Guadalquivir Valley and race G populations from Eastern Europe. None of the race G populations parasitized on DEB2 line. The six race G populations from Eastern Europe showed some differences between them, from G_{BU} that failed to parasitize the four race F resistant hybrids to G_{RO}, G_{TK}, and G_{UK} that parasitized all of them (**Table 1**).

A second experiment, focused on the evaluation of population BR-28 with a set of 12 commercial hybrids resistant to race F_{GV}, revealed a degree of incidence between 56 and 100% and average number of attachments per plant between 1.4 and 12.8, compared to 100% incidence and 30.0 attachments per plant in the susceptible control (**Table 2**).

Interpopulation Relatedness

Cluster analysis revealed that the four populations with increased virulence (BR-24, BR-25, BR-27, and BR-28) were genetically related to conventional populations of the Guadalquivir Valley (**Figure 1**). The analysis, with a high cophenetic correlation coefficient ($r = 0.86$, $P < 0.01$), resulted in two main clusters, one formed by the eight populations of the Guadalquivir Valley and one population of Bulgaria at a greater distance, and a second cluster including the populations from Cuenca together with all other populations from Israel, Serbia, Romania, Turkey, and Bulgaria.

Genetic Diversity of the New Populations

Compared with the population of the Guadalquivir Valley used as a reference (SE-10, race F), the four populations collected on race F_{GV} resistant hybrids showed considerably greater genetic diversity (**Figure 2**). The four populations showed greater values of all the genetic parameters evaluated, except the number of private alleles (**Table 3**). For example, the percentage of polymorphic loci ranged from 43.75 in BR-25 and BR-28 to 87.50 in BR-24, compared to zero in SE-10, while the average number of pairwise differences between individuals of a population ranged from 2.05 in BR-28 to 4.59 in BR-24, compared to zero in SE-10. The AMOVA confirmed the existence of intra-population diversity, which accounted for 75.6% of the total variance (**Table 4**). All polymorphic loci contained two alleles, the one found in populations of the Guadalquivir Valley and the one found in populations of Cuenca (**Figure 3**). No other alleles were present in the populations.

TABLE 2 | Percentage of susceptible plants (%S) and average number of *O. cumana* attachments in susceptible plants (Xs) in 12 sunflower hybrids resistant to *O. cumana* race F_{GV} and the susceptible control 63D82 evaluated with *O. cumana* population BR-28.

Hybrid	n	%S	Xs
Hybrid 1	6	100	11,7
Hybrid 2	9	89	3,0
Hybrid 3	10	100	12,8
Hybrid 5	9	56	2,6
Hybrid 6	10	100	5,8
Hybrid 7	10	100	4,8
Hybrid 8	10	100	2,9
Hybrid 9	10	70	1,4
Hybrid 10	5	100	11,6
Hybrid 11	10	100	5,4
Hybrid 12	10	100	5,5
Hybrid 13	9	100	7,0
63D82	10	100	30,0

DISCUSSION

Four populations of *O. cumana*, collected on race F_{GV} resistant hybrids in the Guadalquivir Valley area of Spain, were confirmed to have increased virulence as compared with conventional populations in that area. The four populations showed similar level of virulence and parasitized most of the race F_{GV} resistant genotypes evaluated, with level of incidence of 100% or close to it in most cases, with the exception of two lines (P96, DEB2) that were fully resistant and some hybrids that showed moderate or low level of incidence and degree of attack. The four populations from the Guadalquivir Valley clearly differed from race G populations from Eastern Europe for their inability to parasitize P96 line, while the reaction on race F resistant hybrids followed different patterns in the race G populations from Eastern Europe. Currently, race classification of *O. cumana* populations is based on sets of differential lines. However, there is no universal set of differential lines, as it is for example the case for downy mildew in this crop (Tourvieille de Labrouhe et al., 2000). Differential lines used for *O. cumana* are specific to individual seed companies or research groups, particularly for races above E (Fernández-Martínez et al., 2012). For populations of the Guadalquivir Valley, the line NR5 derived from the differential line P1380 (Vrânceanu et al., 1980) has been used as differential between races E and F_{GV}, and line P-96 (Fernández-Martínez et al., 2004) as differential line with race F_{GV} resistance. The line P-96 was selected as differential line because it was completely resistant to race F_{GV} and to the race F predominant in Central Spain, and susceptible to the first reported race G, which was identified in the Edirne area of Turkey around 2000 (Kaya et al., 2004). P-96 line has been also reported to be susceptible to the race G present in Russia (Antonova et al., 2013) and the Black Sea area of Romania (Hladni et al., 2014). According to the results of the present study, it is also susceptible to race G populations from other areas. Therefore it becomes evident that the virulence of the new populations of the Guadalquivir Valley area is different to races G populations from Eastern Europe. We are accordingly naming the new populations as a local race G of the Guadalquivir Valley (G_{GV}) to differentiate its level of virulence from the currently predominant race F_{GV} and from races G described for other areas such as Turkey (Kaya et al., 2004), Bulgaria (Shindrova and Penchev, 2012), Romania (Pacureanu-Joita et al., 2009), Russia (Antonova et al., 2013), China (Shi et al., 2015), Greece, and Ukraine (present study). This study also concluded that the inbred line P-96 is not valid

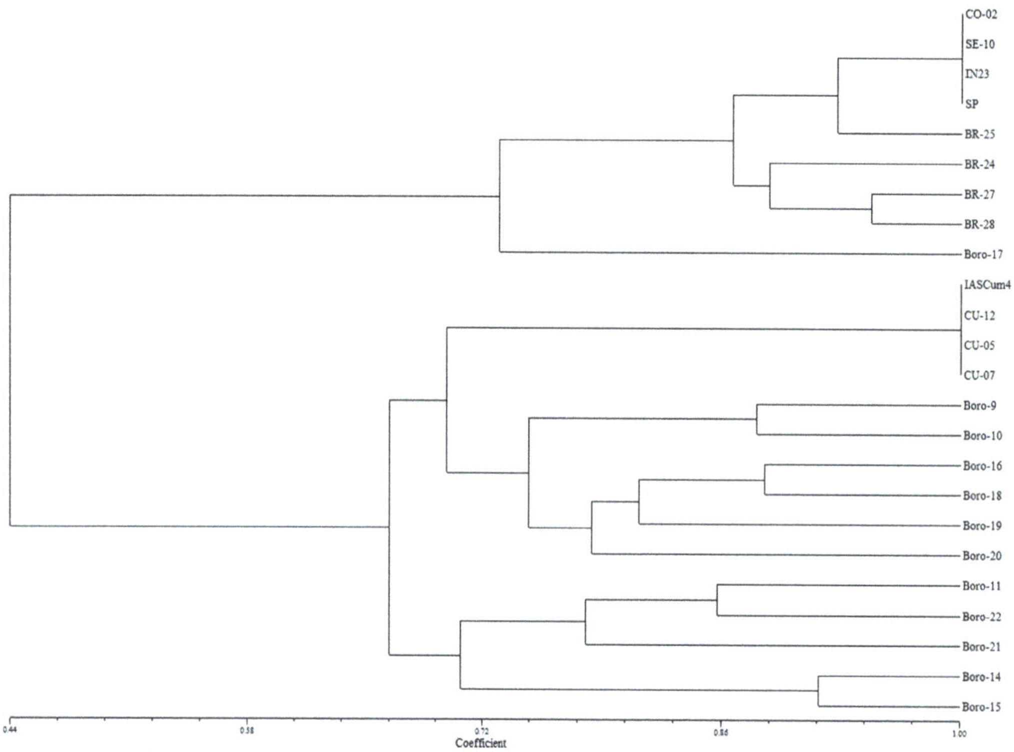

FIGURE 1 | Tree plot of the four populations under study (BR-24, BR-25, BR-27, BR-28) together with reference populations from the Guadalquivir Valley (CO-02, SE-10, EK-23, SP), Central Spain (IASCum-4, CU-05, CU-07, CU-12), Serbia (Boro-9), Romania (Boro-10), Israel (Boro-11), Turkey (Boro-14, Boro-15), and Bulgaria (Boro-16 to Boro-22).

as differential line between races F_{GV} and G_{GV}, as it is resistant to both groups of populations. Research is ongoing to identify and release a public line to differentiate between both races.

In a previous study, Pineda-Martos et al. (2013) identified the presence of two genetically distant pools of *O. cumana* in Spain, one in the Guadalquivir Valley in southern Spain and another one in Cuenca province in Central Spain. The study identified extremely low interpopulation and intrapopulation genetic diversity within each gene pool, but also the existence of a few populations with greater genetic diversity that resulted from admixture between populations of both gene pools and subsequent hybridization. In the present study, the four populations classified as race G_{GV} were genetically related to the Guadalquivir Valley gene pool when compared with a set of populations of the two Spanish gene pools and from other areas, mainly Eastern Europe. This suggested that race G_{GV} populations have evolved *in situ* and do not correspond to new introductions from other areas.

Race G_{GV} populations were characterized by greater genetic diversity than the conventional populations of the Guadalquivir Valley (Pineda-Martos et al., 2013). The later study included 30 populations from the Guadalquivir Valley area, mainly belonging to races E and F_{GV}, and even populations previous to race E. The low genetic diversity between populations and within populations was attributed to a founder effect, taking into account that *O. cumana* is not present in the wild in Spain (Pujadas-Salvà and Velasco, 2000). The high genetic similarity between race E and race F_{GV} populations suggested that race F_{GV} evolved *in situ* by punctual mutation from race E (Pineda-Martos et al., 2013). This was confirmed by Rodríguez-Ojeda et al. (2013b), who found segregation at a single locus in segregating populations from crosses between

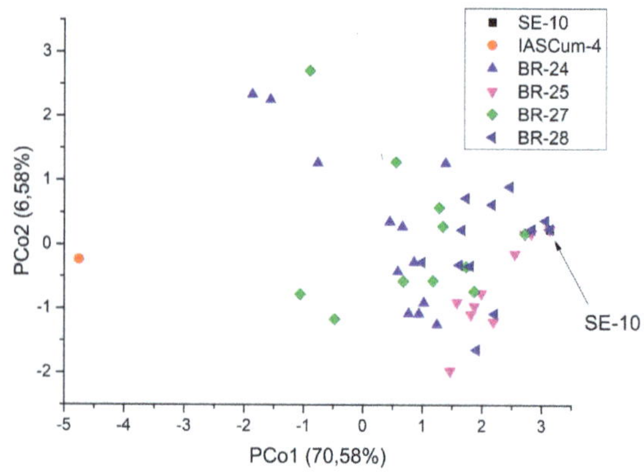

FIGURE 2 | Principal coordinates analysis of four *Orobanche cumana* populations (BR-24, BR-25, BR-27, BR-28) from the Guadalquivir Valley with increased virulence, and reference populations from the Guadalquivir Valley (SE-10) and Central Spain (IASCum-4).

TABLE 3 | Genetic diversity parameters of four *O. cumana* populations (BR-24, BR-25, BR-27, and BR-28) from the Guadalquivir Valley with increased virulence, and a reference population from the Guadalquivir Valley (SE-10).

Population	*P*	Na (±SE)	Na ≥5% (±SE)	Ne (±SE)	Npa (±SE)	H_0 (±SE)	He (±SE)	I (±SE)	Pairwise differences
BR-24	87.50	1.88 ± 0.09	1.88 ± 0.09	1.65 ± 0.10	0.00 ± 0.00	0.19 ± 0.05	0.35 ± 0.05	0.51 ± 0.06	4.59
BR-25	43.75	1.44 ± 0.13	1.44 ± 0.13	1.26 ± 0.11	0.00 ± 0.00	0.05 ± 0.02	0.14 ± 0.05	0.21 ± 0.07	2.26
BR-27	68.75	1.69 ± 0.12	1.69 ± 0.12	1.49 ± 0.11	0.00 ± 0.00	0.05 ± 0.02	0.27 ± 0.05	0.39 ± 0.08	3.67
BR-28	43.75	1.44 ± 0.13	1.44 ± 0.13	1.25 ± 0.08	0.00 ± 0.00	0.05 ± 0.02	0.15 ± 0.05	0.23 ± 0.07	2.05
SE-10	0.00	1.00 ± 0.00	1.00 ± 0.00	1.00 ± 0.00	0.00 ± 0.00	0.00 ± 0.00	0.00 ± 0.00	0.00 ± 0.00	0.00

P, percentage of polymorphic loci; Na, average observed allele number, Na >5%, number of different alleles with a frequency ≥5%; Ne, number of effective alleles; Npa, number of private alleles unique to a single population; Ho, observed heterozygosity; He, expected heterozygosity; I, Shannon's diversity index; Pairwise differences: average number of pairwise differences between individuals within each population.

TABLE 4 | Analysis of molecular variance (AMOVA) in a set of four populations of *O. cumana* with increased virulence collected in the Guadalquivir Valley area (BR-24, BR-25, BR-27, and BR-28).

Source of variation	Sum of squares	variance components	% Variation	*P* value
Among populations	36.52	0.47	24.42	<0.01
Within populations	127.29	1.45	75.58	<0.01

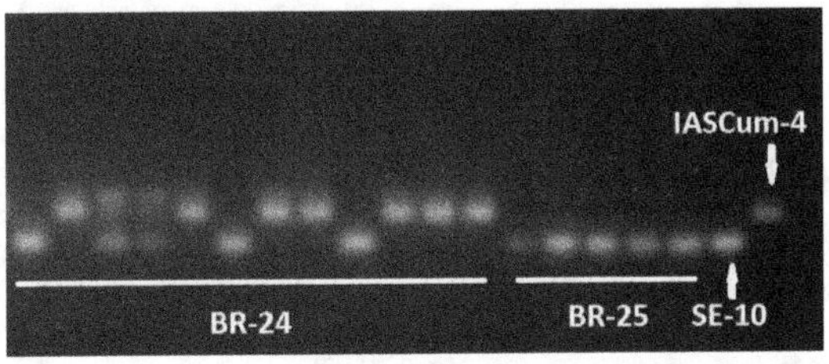

FIGURE 3 | Detail of an electrophoresis gel showing allelic variation at the Ocum-197 SSR locus in 12 single plants of *O. cumana* population BR-24, 5 single plants of population BR-25, and the reference populations from the Guadalquivir Valley (SE-10) and Central Spain (IASCum-4).

both races from the same gene pool. Conversely, genetic differentiation of race G_{GV} populations from the main gene pool of the Guadalquivir Valley suggests alternative mechanisms of race evolution. Genetic diversity parameters of the populations included in the present research are comparable to those reported by Pineda-Martos et al. (2013) in a set of populations with increased genetic diversity, with He values between 0.15 and 0.35 in this study compared with values between 0.12 and 0.49 in the previous one. Similarly to the study of Pineda-Martos et al. (2013), we have observed that the alleles present in polymorphic markers in the populations with increased virulence are those from the Guadalquivir Valley and Cuenca, with no new alleles being detected. Accordingly, we hypothesize that increased virulence in the four populations evaluated in the present research may have resulted from genetic recombination between avirulence genes present in both gene pools. Molinero-Ruiz et al. (2014) reported differences for virulence between race F populations from the Guadalquivir Valley and Cuenca province. Admixture between populations, hybridization and recombination of avirulence genes has been identified as an important mechanism to create increasing virulent races in sunflower downy mildew (Ahmed et al., 2012). In *Orobanche* spp., the occurrence of gene flow between populations has been previously documented in *O. minor* (Thorogood et al., 2009) and *O. cumana* (Pineda-Martos et al., 2014a).

The previous study of Pineda-Martos et al. (2013) included two populations with increased genetic diversity from the Guadalquivir Valley area, SE-02 and SE-07. The former was collected in 1991 and classified as race below E while the latter was collected in 1999 and classified as race E. The presence of alleles from the gene pool of Cuenca province was documented in both populations, although in that study greater genetic diversity was not associated with increased virulence. The reason for this could be that both gene pools did not differ for virulence alleles before the appearance of race F, although this has to be confirmed. The different situation in the populations collected in 2014 suggests that both gene pools may differ for race F virulence alleles and that their genetic recombination may be considered as a plausible mechanism underlying the recent appearance of the new race G_{GV}.

AUTHOR CONTRIBUTIONS

LV, JF-M and AM-S conceived the work and planned the experiments. AM-S, LV, and JF-M conducted the phenotypic evaluations. BP-V coordinated the molecular markers research and conducted statistical analyses. JM conducted molecular markers analyses. AM-S and LV wrote the draft of the manuscript. All authors read it critically and contributed to the discussion.

FUNDING

The research has been partially supported by DuPont Pioneer and research project AGL2014-53886-P of the Spanish Ministry of Economy and Innovation, co-funded with EU FEDER Funds.

ACKNOWLEDGMENTS

The authors thank Dr. Yalçin Kaya (Trakya University, Edirne, Turkey), Dr. Rossitza Batchvarova (AgroBio Institute, Sofia, Bulgaria), Dr. Kiril Stoyanov (Agricultural University of Plovdiv, Bulgaria), Dr. Maria Pacureanu-Joita (National Agricultural Research and Development Institute, Fundulea, Romania), Dr. Daniel M. Joel (Newe-Ya'ar Research Center, Ramat-Yishay, Israel), and Prof. Dragan Škorić (Serbian Academy of Sciences and Arts, Novi Sad, Serbia) for their contribution in the collection of *O. cumana* seed populations.

REFERENCES

Ahmed, S., Tourvieille de Labrouhe, D., and Delmotte, F. (2012). Emerging virulence arising from hybridisation facilitated by multiple introductions of the sunflower downy mildew pathogen *Plasmopara halstedii*. *Fungal Genet. Biol.* 49, 847–855. doi: 10.1016/j.fgb.2012.06.012

Akhtouch, B., del Moral, L., Leon, A., Velasco, L., Fernández-Martínez, J. M., and Pérez-Vich, B. (2016). Genetic study of recessive broomrape resistance in sunflower. *Euphytica* 1–10. doi: 10.1007/s10681-016-1652-z

Akhtouch, B., Muñoz-Ruz, J., Melero-Vara, J., Fernández-Martínez, J., and Domínguez, J. (2002). Inheritance of resistance to race F of broomrape in sunflower lines of different origins. *Plant Breed.* 121, 266–268. doi: 10.1046/j.1439-0523.2002.00701.x

Alonso, L. C., Fernández-Escobar, J., López, G., Rodríguez-Ojeda, M. I., and Sallago, F. (1996). "New highly virulent sunflower broomrape (*Orobanche cernua* Loefl.) pathotype in Spain," in *Proceedings of the Sixth International Parasitic Weed Symposium: Advances in Parasitic Plant Research*, eds M. T. Moreno, J. I. Cubero, D. Berner, D. Joel, L. J. Musselman, and C. Parker (Sevilla: Junta de Andalucia), 639–644. doi: 10.1046/j.1439-0523.2002.00701.x

Antonova, T. S., Araslanova, N. M., Strelnikov, E. A., Ramazanova, S. A., Guchetl, S. Z., and Chelyustnikova, T. A. (2013). Distribution of highly virulent races of sunflower broomrape (*Orobanche cumana Wallr.*) in the Southern regions of the Russian Federation. *Russ. Agric. Sci.* 39, 46–50. doi: 10.3103/S1068367413010023

Burlov, V., and Burlov, V. (2010). Breeding of sunflower resistant to new races of broomrape (*Orobanche cumana* Wallr.). *Helia* 53, 165–172. doi: 10.2298/HEL1053165B

Domínguez, J. (1996). R-41, a sunflower restorer inbred line, carrying two genes for resistance against a highly virulent Spanish population of *Orobanche cernua*. *Plant Breed.* 115, 203–204. doi: 10.1111/j.1439-0523.1996.tb00904.x

Fernández-Martínez, J. M., Pérez-Vich, B., Akhtouch, B., Velasco, L., Muñoz-Ruz, J., Melero-Vara, J. M., et al. (2004). Registration of four sunflower germplasms resistant to race F of broomrape. *Crop Sci.* 44, 1033–1034. doi: 10.2135/cropsci2004.1033

Fernández-Martínez, J. M., Pérez-Vich, B., and Velasco, L. (2015). "Sunflower broomrape (*Orobanche cumana* Wallr.)," in *Sunflower Oilseed: Chemistry, Production, Processing and Utilization*, eds E. Martínez-Force, N. T. Dunford, and J. J. Salas (Champaign, IL: AOCS Press), 129–156. doi: 10.1016/b978-1-893997-94-3.50011-8

Fernández-Martínez, J. M., Velasco, L., and Pérez-Vich, B. (2012). Progress in research on breeding for resistance to sunflower broomrape. *Helia* 57, 47–56. doi: 10.2298/HEL1257047F

Gagne, G., Roeckel-Drevet, P., Grezes-Besset, B., Shindrova, P., Ivanov, P., Grand-Ravel, C., et al. (1998). Study of the variability and evolution of *Orobanche cumana* populations infesting sunflower in different European countries. *Theor. Appl. Genet.* 96, 1216–1222. doi: 10.1007/s001220050859

Hladni, N., Dedić, B., Jocić, S., Marinković, R., and Mikli̇č, V. (2014). "Screening resistance of new NS sunflower hybrids to broomrape," in *Proceedings of the 3rd International Symposium on Broomrape (Orobanche spp.) in Sunflower, Córdoba, Spain, June 3–14, 2014* (Paris: International Sunflower Association), 207–212.

Imerovski, I., Dimitrijević, A., Miladinović, D., Dediæ, B., Jocić, S., Koèiš Tubić, N., et al. (2016). Mapping of a new gene for resistance to broomrape races higher than F. *Euphytica* 1–9. doi: 10.1007/s10681-015-1597-7

Kaya, Y. (2014). "Current situation of sunflower broomrape around the world," in *Proceedings of the 3rd International Symposium on Broomrape (Orobanche spp.) in Sunflower, Córdoba, Spain, June 3-14, 2014* (Paris: International Sunflower Association), 9–18.

Kaya, Y., Evci, G., Pekcan, V., and Gucer, T. (2004). Determining new broomrape-infested areas, resistant lines and hybrids in Trakya region in Turkey. *Helia* 40, 211–218. doi: 10.2298/hel0440211k

Labrousse, P., Arnaud, M. C., Griveau, Y., Fer, A., and Thalouarn, P. (2004). Analysis of resistance criteria of sunflower recombined inbred lines against *Orobanche cumana* Wallr. *Crop Prot.* 23, 407–413. doi: 10.1016/j.cropro.2003.09.013

Molinero-Ruiz, M. L., García-Carneros, A. B., Collado-Romero, M., Raranciuc, S., Domínguez, J., and Melero-Vara, J. M. (2014). Pathogenic and molecular diversity in highly virulent populations of the parasitic weed *Orobanche cumana* from Europe. *Weed Res.* 54, 87–96. doi: 10.1111/wre.12056

Pacureanu-Joita, M., Raranciuc, S., Sava, E., Stanciu, D., and Nastase, D. (2009). Virulence and aggressiveness of sunflower broomrape (*Orobanche cumana* Wallr.) populations in Romania. *Helia* 51, 111–118. doi: 10.2298/hel0951111p

Pacureanu-Joita, M., Veronesi, C., Raranciuc, S., and Stanciu, D. (2004). "Parasite-host plant interaction of *Orobanche cumana* Wallr. (*Orobanche cernua* Loefl) with *Helianthus annuus*," in *Proceedings of the 16th International Sunflower Conference, Fargo, ND, USA, Aug 29–Sept 2, 2004* (Paris: International Sunflower Association), 171–177.

Pérez-Vich, B., Akhtouch, B., Knapp, S. J., León, A. J., Velasco, L., Fernández-Martínez, J. M., et al. (2004). Quantitative trait loci for broomrape (*Orobanche cumana* Wallr.) resistance in sunflower. *Theor. Appl. Genet.* 109, 92–102. doi: 10.1007/s00122-004-1599-7

Pérez-Vich, B., Velasco, L., Rich, P. J., and Ejeta, G. (2013). "Marker-assisted and physiology-based breeding for resistance to root parasitic Orobanchaceae," in *Parasitic Orobanchaceae. Parasitic Mechanisms and Control Strategies*, eds D. M. Joel, J. Gressel, and L. J. Musselman (New York, NY: Springer), 369–391.

Pineda-Martos, R., Pujadas-Salvà, A. J., Fernández-Martínez, J. M., Stoyanov, K., Velasco, L., and Pérez-Vich, B. (2014a). The genetic structure of wild *Orobanche cumana* Wallr. (Orobanchaceae) populations in eastern Bulgaria reflects introgressions from weedy populations. *Sci. World J.* 2014, 15. doi: 10.1155/2014/150432

Pineda-Martos, R., Velasco, L., Fernández-Escobar, J., Fernández-Martínez, J. M., and Pérez-Vich, B. (2013). Genetic diversity of sunflower broomrape (*Orobanche cumana*) populations from Spain. *Weed Res.* 53, 279–289. doi: 10.1111/wre.12022

Pineda-Martos, R., Velasco, L., and Pérez-Vich, B. (2014b). Identification, characterization, and discriminatory power of microsatellite markers in the parasitic weed *Orobanche cumana*. *Weed Res.* 54, 120–132. doi: 10.1111/wre.12062

Pujadas-Salvà, A. J., and Velasco, L. (2000). Comparative studies on *Orobanche cernua* L. and O. Cumana Wallr. (Orobanchaceae) in the Iberian Peninsula. *Bot. J. Linn. Soc.* 134, 513–527. doi: 10.1006/bojl.2000.0346

Rodríguez-Ojeda, M. I., Fernández-Escobar, J., and Alonso, L. C. (2001). "Sunflower inbred line (KI-374) carrying two recessive genes for resistance against a highly virulent Spanish population of *Orobanche cernua* Loefl. / O.Cumana Wallr. race F," in *Proceedings of the 7th International Parasitic Weed Symposium*, eds L. J. Muselman, C. Parker, and J. A. C. Verkleij (Nantes: Faculté des Sciences, Université de Nantes), 208–211.

Rodríguez-Ojeda, M. I., Fernández-Martínez, J. M., Velasco, L., and Pérez-Vich, B. (2013a). Extent of cross-fertilization in *Orobanche cumana* Wallr. *Biol. Plant.* 57, 559–562. doi: 10.1007/s10535-012-0301-1

Rodríguez-Ojeda, M. I., Pineda-Martos, R., Alonso, L. C., Fernández-Escobar, J., Fernández-Martínez, J. M., Pérez-Vich, B., et al. (2013b). A dominant avirulence gene in *Orobanche cumana* triggers Or5 resistance in sunflower. *Weed Res.* 53, 322–327. doi: 10.1111/wre.12034

Satovic, Z., Joel, D. M., Rubiales, D., Cubero, J. I., and Román, B. (2009). Population genetics in weedy species of *Orobanche*. *Aust. Plant Pathol.* 38, 228–234. doi: 10.1071/AP08100

Shi, B. X., Chen, G. H., Zhang, Z. J., Hao, J. J., Jing, L., Zhou, H. Y., et al. (2015). First report of race composition and distribution of sunflower broomrape, *Orobanche cumana*, in China. *Plant Dis.* 99, 291. doi: 10.1094/PDIS-07-14-0721-PDN

Shindrova, P. (2006). Broomrape (*Orobanche cumana* Wallr.) in Bulgaria. Distribution and race composition. *Helia* 44, 111–120. doi: 10.2298/hel0644111s

Shindrova, P., and Penchev, E. (2012). Race composition and distribution of broomrape (*Orobanche cumana* Wallr.) in Bulgaria during 2007-2011. *Helia* 57, 87–94. doi: 10.2298/HEL1257087S

Thorogood, C. J., Rumsey, F. J., Harris, S. A., and Hiscock, J. (2009). Gene flow between alien and native races of the holoparasitic angiosperm Orobanche minor (Orobanchaceae). *Plant Syst. Evol.* 282, 31–42. doi: 10.1007/s00606-009-0204-6

Tourvieille de Labrouhe, D., Gulya, T. J., Masirevic, S., Penaud, A., Rashid, K. Y., and Virányi, F. (2000). "New nomenclature of races of *Plasmopara halstedii* (sunflower downy mildew)," in *Proceedings of the 15th International Sunflower Conference, Toulouse, France, June 12–15, 2000*, Vol. 2 (Paris: International Sunflower Association), 61–66.

Velasco, L., Pérez-Vich, B., Jan, C. C., and Fernández-Martínez, J. M. (2007). Inheritance of resistance to broomrape (*Orobanche cumana* Wallr.) race f in a sunflower line derived from wild sunflower species. *Plant Breed.* 126, 67–71. doi: 10.1111/j.1439-0523.2006.01278.x

Vrânceanu, A. V., Tudor, V. A., Stoenescu, F. M., and Pirvu, N. (1980). "Virulence groups of *Orobanche cumana* Wallr. differential hosts and resistance sources and genes in sunflower," in *Proceedings of the 9th International Sunflower Conference, Torremolinos, Spain, July 8–13, 1980* (Paris: International Sunflower Association), 74–80.

Conflict of Interest Statement: The manuscript is the result of public-private research collaboration, between the company Dupont Pioneer and the Institute for Sustainable Agriculture (IAS), which belongs to the Spanish National Scientific Research Council. Research conducted at the company has been funded by the own company, whereas research conducted at IAS has been funded by research project AGL2014-53886-P of the Spanish Ministry for Economy and Innovation, with involvement of EU FEDER Funds. Collaboration has been based on interchange of ideas and plant material, which led to the design of experiments of interest for both parties and the preparation of the present manuscript to inform the scientific community on the results of our experiments. For the sake of the experiments, sunflower commercial hybrids from seed companies other than Dupont Pioneer were included in the research. Both Dupont Pioneer and IAS consider they would have a conflict of interest if the names of the commercial hybrids and their corresponding companies were reported together with their performance under artificial infestation with an alleged new race of sunflower broomrape that is being reported for the first time in this manuscript. Since the objective of the research was not to conduct a survey on resistance of commercial hybrids to a new race, but just to ensure that the new populations under study can be considered as a new race, the hybrids were not selected following the criterion of representativeness, and some seed companies may be overrepresented and others not included in the study, which might result in an unbalanced impact on the commercial interests of seed companies. For this reason, the names of the hybrids and their corresponding companies have been omitted in the manuscript.

2

Striga Biocontrol on a Toothpick: A Readily Deployable and Inexpensive Method for Smallholder Farmers

Edited by:
Maurizio Vurro,
National Research Council, Italy

Reviewed by:
Grama Nanjappa Dhanapal,
University of Agricultural Sciences,
Bangalore, India
Alan Kemball Watson,
McGill University, Canada

***Correspondence:**
David C. Sands
davidsands41@yahoo.com

Henry S. Nzioki[1], Florence Oyosi[2], Cindy E. Morris[3,4], Eylul Kaya[4], Alice L. Pilgeram[4], Claire S. Baker[5] and David C. Sands[4]*

[1] Kenya Agriculture and Livestock Research Organization, Machakos, Kenya, [2] Liberty Initiators Network, Maseno, Kenya, [3] Plant Pathology, INRA-PACA, Avignon, France, [4] Department of Plant Sciences and Plant Pathology, Montana State University, Bozeman, MT, USA, [5] Biotech Investments, Bozeman, MT, USA

Striga hermonthica (witchweed) is a parasitic weed that attacks and significantly reduces the yields of maize, sorghum, millet, and sugarcane throughout sub-Saharan Africa. Low cost management methods such as hand weeding, short crop rotations, trap cropping, or conventional biocontrol have not been effective. Likewise, *Striga*-tolerant or herbicide-resistant maize cultivars are higher yielding, but are often beyond the economic means of sustenance farmers. The fungal pathogen, *Fusarium oxysporum* f.sp. *strigae*, has been the object of numerous studies to develop *Striga* biocontrol. Under experimental conditions this pathogen can reduce the incidence of *Striga* infestation but field use is not extensive, perhaps because it has not been sufficiently effective in restoring crop yield and reducing the soil *Striga* seed bank. Here we brought together Kenyan and US crop scientists with smallholder farmers to develop and validate an effective biocontrol strategy for management of *Striga* on smallholder farms. Key components of this research project were the following: (1) Development of a two-step method of fungal delivery, including laboratory coating of primary inoculum on toothpicks, followed by on-farm production of secondary field inoculum in boiled rice enabling delivery of vigorous, fresh inoculum directly to the seedbed; (2) Training of smallholder farmers (85% women), to produce the biocontrol agent and incorporate it into their maize plantings in *Striga*-infested soils and collect agronomic data. The field tests expanded from 30 smallholder farmers to a two-season, 500-farmer plot trial including paired plus and minus biocontrol plots with fertilizer and hybrid seed in both plots and; (3) Concerted selection of variants of the pathogen identified for enhanced virulence, as has been demonstrated in other host parasite systems were employed here on *Striga* via pathogen excretion of the amino acids L-leucine and L-tyrosine that are toxic to *Striga* but innocuous to maize. This overall strategy resulted in an average of >50% increased maize yield in the March to June rains season and >40% in the September to December rains season. Integration of this enhanced plant pathogen to *Striga* management in maize can significantly increase the maize yield of smallholder farmers in Kenya.

Keywords: *Striga hermonthica*, *Fusarium oxysporum*, Kenya, maize, witchweed, biocontrol, amino acid, toothpick

INTRODUCTION

Striga hermonthica (witchweed) is a parasitic weed that attacks cereal crops and significantly reduces yields. Countries with nascent infestation of *Striga* only 25 years ago now have heavy infestations of Striga resulting in significant losses of crop yield, adversely affecting about 300 million people in sub-Saharan Africa. Up to 50 million hectares of croplands in Africa show varying degrees of *Striga* infestation (Ejeta and Gressel, 2007). The most severe impact is made by *S. hermonthica* on maize (Atera et al., 2013). In some locations and years, *Striga* infestation results in total crop failure (Sauerborn, 1991; Abbasher et al., 1998) and abandonment of fields (Debrah, 1994). In Western Kenya alone, *Striga* has infested over 217,000 ha of crop land, resulting in maize losses of 182,227 tons per year valued at 53 million USD (Atera et al., 2013). Recently, *Striga* was reported in the highlands of Kenya[1]. Land use intensification and increasing cereal mono-cropping combined with its quick adaptation to new climatic conditions has exacerbated the *S. hermonthica* problem in sub-Saharan Africa. *Striga* thrives in challenged conditions, including drought and nutrient deficient soils, making it even more detrimental to all farmers but especially smallholder farmers (Ejeta and Gressel, 2007).

Conventional weed management methods including hand pulling emerged *Striga* stalks after they had already damaged the crop, catch/trap crops, crop rotation, and use of tolerant varieties are reported to be not reliably effective against *Striga* (Ogborn, 1987; Atera et al., 2013). There are ongoing efforts to develop maize varieties that are resistant to striga (Badu-Apraku and Yallow, 2009; Venne et al., 2009; Badu-Apraku and Akinwale, 2015) or that are resistant to herbicides that could then be used to control *Striga* (Kanampiu et al., 2003; Ransom et al., 2012). *Striga* tolerant maize varieties generally yield better than susceptible varieties in *Striga*-infested environments; however, their yield is still reduced relative to the yield from non-infested sites (Badu-Apraku and Akinwale, 2015). Such use of hybrid maize germplasm is an effective option for farmers if affordable and available. Farmers need affordable and effective strategies for *Striga* management that can be readily integrated into their production practices (Berner et al., 1996; Ejeta and Gressel, 2007; Hassan et al., 2009). Non-governmental organizations (NGOs) focusing on farm inputs and on education have steered away from investments in farms with *Striga* due to the increased risk of crop failure (Guerena, personal communication 2016; Seward, personal communication 2016), except in cases where they have committed to *Striga* management evaluations, as with the push pull approach with the *Striga* suppressing plant *Desmodium* (Khan et al., 2002).

Biological control of weeds is a promising field with some successes, even though there is an additional complication with biological control of parasitic plants: care must be taken to not harm the host plant. The herbicide industry has confronted the same complication – how to differentially kill one plant and not another. *Fusarium oxysporum* f. sp. *strigae* has a potential for biocontrol of *Striga* (Ciotola et al., 1995, 2000; Savard et al., 1997; Marley et al., 1999; Venne et al., 2009; Ndambi et al., 2011; Avedi et al., 2014). As a rule, *Fusarium,* mainly *F. oxysporum,* isolated from *S. hermonthica* selectively attack *Striga* spp. (Abbasher and Sauerborn, 1992; Ciotola et al., 1995; Savard et al., 1997; Abbasher et al., 1998; Marley et al., 1999; Amalfitano et al., 2002; Marley and Shebayan, 2005; Elzein and Kroschel, 2006; Schaube et al., 2006; Venne et al., 2009; Watson, 2013). An isolate from Mali in combination with fertilizer inputs successfully prevented *Striga* emergence and resulted in 400% increase of sorghum dry matter (Ciotola et al., 1995). The growth of crop species (sorghum, pearl millet, maize, rice, fonio, cotton, groundnut, cowpea, and okra) was unaffected by this pathogen. However, recent host range studies have shown that some Solanaceous plants are susceptible to *F. oxysporum* isolates used for biocontrol of *Striga* thus they should not be intercropped with *Striga*-host crops where *F. oxysporum* has been applied to control *Striga* (Zarafi et al., 2014). While wild type strains of *F. oxysporum* have been reported to reduce the incidence of *Striga* infestation, their use in the field has not yet been widespread or practical. Perhaps they are not effective enough in the manner applied to restore crop yield or to significantly reduce the soil *Striga* seed bank. Whatever the underlying reasons, biocontrol of *Striga* is not yet used as a routine means of management of this weed and no products are widely available for African smallholder farmers.

The objectives of the work reported here were to validate under realistic field conditions a practical, effective, and inexpensive *Striga* biological control technology that can enable smallholders to greatly improve their economic outlook. Biocontrol agents portend good prospects for effective integration into subsistence farming in particular because of their applicability across a range of crops and varieties, and the minimal economic input. However, there have been several obstacles preventing the widespread acceptance of weed biocontrol practices: (i) cost and availability to the farmer; (ii) sufficient early control of the target weed before it can damage the crop; (iii) reliability of results from season to season; and (iv) greater yields per input of labor. By focusing on these obstacles we developed a labor saving biocontrol of *Striga* technology for rural Kenya and tested it for three seasons in research trials and validated it for two seasons with 500 smallholder farmers. We considered the options of carefully optimizing each variable of our preliminary studies or proceeding to the here-described large scale field testing of our biocontrol strategy in paired plot tests on 500 smallholder farmers' fields based on preliminary results of the performance of the biocontrol agent. We chose the latter to provide rationale to justify further optimization research.

MATERIALS AND METHODS

Fungal Strains: Isolation and Preservation of *Fusarium oxysporum*

Fusarium oxysporum was isolated from wilted diseased *Striga* plants in Western Kenya and identified as described by Nelson et al. (1983). The virulence and host-specificity of the isolates was confirmed in glass house and field studies. The fungal cultures were maintained on PDA (potato dextrose agar (Difco,

[1] http://striga.aatf-africa.org

Detroit, MI, USA)) slants or on sterile wooden toothpicks. Colonized toothpicks were produced by inoculating a plate of PDA with a single culture and incubating for 24–72 h. Sterile toothpicks were then placed directly on the expanding colonies and incubated for a further 72 h. Colonized toothpicks were aseptically removed from the plate and dried in sterile, open glass vials or paper envelopes in a laminar flow hood (24–72 h). The vials or envelopes were then sealed for long-term storage at room temperature (>years). Culture viability and purity after storage was determined by placing a single toothpick on a plate of PDA and incubating for 24–72 h.

Selection of Strains for Enhanced Virulence to *Striga*

Determination of Amino Acid Sensitivity of *S. hermonthica* and Maize in Pot Tests

Amino acid toxicity has long been observed in plants. 'Frenching' disease of tobacco, first described in the early 1700's, is caused by high amounts of isoleucine excreted in the rhizosphere by bacteria (Steinberg, 1952). This amino acid disrupts plant growth and development because of the tight regulation of free amino acid levels in plants (Sands et al., 2003; Sands and Pilgeram, 2009; Pilgeram and Sands, 2010). The amino acids that inhibit *S. hermonthica* but do not inhibit maize were identified using an adaptation of the protocol described by Sands and Pilgeram (2009) in glass house pot tests at KALRO (Kenya Agricultural & Livestock Research Organization) in Kibos. Ten grams of *Striga* seeds that had undergone a dormancy period of more than 4 months were thoroughly mixed with 5 kg of sand. Pots (36 cm diameter) were filled with *Striga*-free black cotton soil. Each pot was infested with *Striga* by mixing one tablespoon of *Striga* -sand mixture (about 20 g) into the soil layer 2 cm below the surface and mixed with the soil. Each pot was then sown with four seeds of a pesticide-free local maize cultivar (cv. Rachari) placed ~2 cm below the soil surface. Five amino-acids (L-lysine, L-leucine, L-tyrosine, L-tryptophan, and L-threonine) were evaluated (5 pots/amino acid). Following planting, pots were watered daily with a solution containing 1 mM of the given amino acid. The control treatment was watered with plain tap water. Watering with amino acids or plain tap water continued until 10 weeks after planting. The trial continued until 15 weeks after planting during which time, both maize and *Striga* had reached maturity. At the end of the trial the dry *Striga* biomass and dry maize biomass were determined.

Selection of Amino Acid-Overproducing Variants of *Fusarium oxysporum*

After identifying which amino acids both inhibited *Striga* and did not affect maize (previous step), we selected variants of pathogenic *F. oxysporum* that excreted the selected amino acids. Selection of amino acid-overproducing lines of microorganisms is a relatively straightforward process (Sands and Hankin, 1974; Tiourebaev et al., 2001). The *F. oxysporum* strains were grown in minimal CATSUP media [Czapek's glucose salts medium (Difco Laboratories, Detroit]) supplemented with thiamine (4 mg/L), proline (2100 mg L^{-1}), uracil (20 mg L^{-1}), and a commercial vitamin mixture (e.g., Sesame Street Brand, or similar) at 37°C. Amino acid overproducers were selected by plating a suspension of ~10^4 conidia per dish onto CATSUP agar (2%) supplemented with increasing concentrations (10–1000 mg L^{-1}/ml) of an amino acid analog (Sigma–Aldrich, St. Louis). The analogs were: L-3-fluoro-tyrosine for selection of tyrosine excretors, lucinol, or L-norvaline for selection of leucine excretors, and L-seleno-methionine for methionine over-producers. The vast majority of colonies that survived and grew on media supplemented with the toxic analog excreted the desired amino acid which presumably diluted the analog. For example, a colony that grows on media supplemented with 3-fluoro-tyrosine is likely excreting tyrosine. Any colonies growing on media supplemented with an amino acid analog were screened for excretion of the target amino acid using specific amino acid assay media (Difco Laboratories, Detroit) amended with a biological indicator of amino acid excretion [*Pediococcus cerevisiae* (ATTC 4023)]. The relative level of amino acid excretion of selected lines in CATSUP broth was further quantified using HPLC mass spectroscopy. In addition to selecting *Fusarium* variants that excreted amino acids that were toxic to *Striga*, we selected variants of *F. oxysporum* f.sp. *strigae* that excreted the amino acid methionine using the analog seleno-methionine. We chose methionine because it is known to stimulate germination of the *Striga* soil seed bank (Primrose, 1976; Logan and Stewart, 1991) and could enhance the effectiveness of mycelial invasion of the soil seed bank.

Preparation of T14 Field Inoculum

Preparation of Primary Inoculum

Three different amino acid-overproducing variants of *F. oxysporum* were individually cultured on PDA. After the cultures grew for 3–4 days, sterile wooden toothpicks were placed onto the culture. The fungi grew, ramifying into the toothpicks. The toothpicks were removed from the plate and aseptically dried in a laminar flow hood. Once fully dried, the trio of toothpicks was stored together in sealed, sterile drinking straws and transported to the farmers. This 1:1:1 ratio of a trio of toothpicks with variants of *F. oxysporum* f. sp. *strigae* is referred to as Foxy T14.

Development and Preparation of Field Inoculum

Kenyan smallholders were given a straw containing the three fungus-carrying toothpicks to inoculate cooked and cooled pearled rice, providing fresh on-farm inoculum of the fungi. In an alcohol-swabbed disinfected plastic container with a lid, the on-farm inoculum was prepared by room temperature incubation of the toothpicks in pearled rice. Although other grains were also tested as substrates, the pearled rice consistently supported abundant mycelial growth if kept unopened and shaken twice daily for a 3-day period. We determined that a period of 3 days of incubation at room temperature was optimal to avoid depletion of the carbohydrates in the inoculum as well as risk of external contamination. The Foxy T14 field inoculum contained all three variant strains of *F. oxysporum* that each excreted tyrosine, leucine and/or methionine.

Paired-Plot Field Trials on 500 Smallholder Farms

After first successfully testing the method on 50 farms (data not shown), we embarked on a large scale field trial. In 2014 we evaluated Foxy T14 on over 500 smallholder farms near Maseno, Kenya. The 500 trial farms were in a 127 km^2 region bounded by latitudes 0.059°N and 0.057°S and longitudes 34.588°E and 34.503°E. A survey was first conducted on all of the farms to determine that *Striga* was present and was an apparent constraint to maize production Training of the farmers to deploy Foxy T14 and collect data was coordinated by the association called Liberty Initiators Network (LIN) based in Maseno. Twenty seven LIN farmers, who had previously collaborated with us to evaluate the technology on KALRO research farms, were further trained as implementers to instruct 500 smallholder farmers to establish paired 4.5 m × 10 m plots on their farms. On each farm the Foxy T14 treatment was compared to a standardized farmer-practice. Both treatments used the same hybrid maize, fertilizer, and plot management. LIN purchased maize seed Western Seed cultivar Western 302, untreated with fungicide and fertilizer in bulk for distribution to LIN members. At planting, DAP fertilizer was applied at a rate of 175 kg/ha (80.5 kg P_20_5) ha^{-1}. After thinning, CAN fertilizer was top dressed at a rate of 150 kg/ha (40.5 kg N ha^{-1}). The amount of fertilizer applied to each hill was approximately 4.4 and 5.3 g of DAP and CAN, respectively. The Foxy T14 treatment was applied as taught to the smallholder farmers by the implementers. As maize is planted by hand by these smallholder farmers, approximately 2.5 g of inoculated rice was placed in each planting hole along with fertilizer and 2 kernels of maize seed (**Supplementary Figure S1**). Each hill was later thinned to a single plant. *Striga* plants were not culled from neither the farmer practice plots nor the T14 plots and were recorded along with dry grain yield at harvest. Manure was applied at 22 g/hill (10 tons per hectare), far below the KALRO's recommended rate of 100 tons per hectare to have an effect on Striga. A few smallholders did not fertilize the paired plots equally due to an inadequate supply of manure. Their yield data were excluded from analysis. The LIN implementers collaborated with the farmers to collect biweekly readings of the number of emerged and wilting *Striga* plants. The trial was conducted during the March - June rains season and repeated during the September–December rains season.

Verification of the Absence of *Fusarium* Toxin Production by the T14 Biocontrol Lines of *F. oxysporum* fsp *strigae*

The three enhanced Foxy T14 lines and a wild type line were grown to exhaustion on boiled rice. The colonized rice cultures were tested at The Mycotoxin Laboratory at Virginia Polytechnic University (Niki McMaster and David Schmale, III) using certified GC-Mass Spectroscopy detection methods for five of the common *Fusarium* elicited toxins: deoxynivalenol (DON), 3-acetyldeoxynivalenol (3-ADON), 15-acetyldeoxynivalenol (15-ADON), nivalenol (NIV), and zearalenone (ZEA). None were detected in the Foxy 14 lines nor from a wildtype strain of *F. oxysporum* f. sp. *strigae* also from Kenya. In a separate assay procedure [Veratox-Fumonisin test 5/10, Neogen Corp. Lansing Michigan, Limit of detection: 0.2 ppm (determined by the mean average of 10 fumonisin free samples plus 2 standard deviations) and the limit of quantitation: 0.5 ppm (described as the lowest concentration point on the calibration curve that this test can reliably detect fumonisin)], they tested for the presence of fumonisins (Khatibi et al., 2014). A reference sample of dried distiller grains was concurrently analyzed as a positive control.

RESULTS

Screening Amino Acid Sensitivity of *Striga* and Maize Reveals an Amino Acid Cocktail for Virulence Enhancement of *F. oxysporum*

None of the tested amino acids were inhibitory to the maize. In all the treatments, *Striga* emerged 5 weeks after planting. There were apparent trends but with no significant differences ($P > 0.05$) in the number of emerged *Striga* plants, the number of wilting *Striga* plants, or flower/seed capsule formation (data not shown). The treatments varied significantly in *Striga* dry weight (**Figure 1**). Tyrosine resulted in the lowest dry *Striga* weight followed by leucine and lysine. The *Striga* dry weight was highest in the control and the tryptophan treatments. There were no significant differences in maize stover dry weight among all the treatments (data not shown). Based on these results, tyrosine and leucine were the amino acids chosen for virulence enhancement studies. We also selected for methionine over-production in our biocontrol mix (Foxy T14) because methionine is converted into ethylene by soil microbes and ethylene stimulates *Striga* seed germination (Primrose, 1976; Logan and Stewart, 1991) which would render the *Striga* more susceptible to *F. oxysporum.*

Based on the amino acid sensitivity studies, we selected variants of the wild type isolate (Ken Foxy 1/KSM1) that

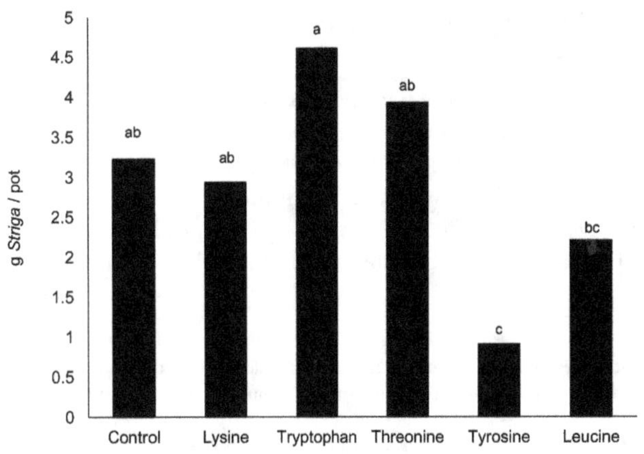

FIGURE 1 | Effect of an excess of amino acid on *Striga* biomass 15 weeks after planting (WAP) in *Striga* infested maize in pots at KALRO, Kibos. Means followed by the same letter are not significantly different at $P = 0.05$.

overproduced tyrosine, leucine, and or methionine. We pooled three strains (Leu2a, Z6a and Z5a) that in combination produce excess leucine, methionine, and tyrosine as shown by GC-Mass Spectroscopy analysis of CATSUP broth grown cultures of *F. oxysporum* (**Table 1**). The amounts of production of these amino acids in **Table 1** are only relative to a control wild type strain in *in vitro* culture and are not necessarily indicative of what might be produced by the three strains in the soil or plant tissue.

Significant and Consistent Yield Increases were Achieved in On-Farm Testing of Foxy T14

The implementers each trained approximately 20 participating farmers on secondary inoculum production and Foxy T14 application. The implementers monitored the fields and collected all the data. Data on 500 *Striga* infested farms were obtained for both growing seasons with almost all the farmers participating in both seasons. Most (99.6%) of the farmers had equal or greater yield in their Foxy T14 plots relative to yield in their comparable farmer-practice plots without Foxy T14 (**Figure 2**). The average maize yield in the March–June rains season was increased by 56.5% in Foxy T14 plots relative to the farmer-practice plots ($p < 0.0001$, pair-wise *t*-test; **Figure 2**). Approximately one third of the farmers doubled their yield in this test (**Figure 2**). Overall, yield in the September–December rains season was reduced by drought but we still observed an average increase of yield of 42% ($t < 0.0001$; **Figure 2**). Typical of maize smallholder farmers in Western Kenya, 85% of the farmers were women. There was no significant linkage between gender and yield.

Striga stalk emergence was reduced in 80% of long season Foxy T14 plots and in 92% of the short season Foxy T14 plots (**Figure 3**). Maize yield increases were positively correlated with inherent yield capacity of fields, negatively with above-ground *Striga* density, and positively with the magnitude of reduction of above-ground *Striga* (**Supplementary Figure S2**).

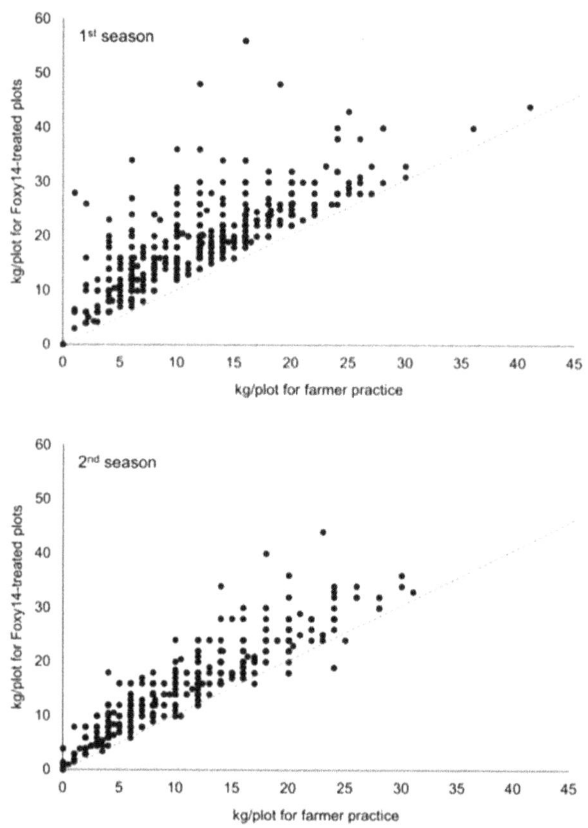

FIGURE 2 | Maize yield (2014) in the long rainy-season (first season) and in the short rainy season (second season) for Foxy T14 plots relative to maize yield in the corresponding farmer-practice plots (i.e., control). The average maize yield in the long rainy season increased by 56.5%. Average yield improvement of short rainy season was 42%. Numerous values are super-imposed. The light gray dotted line indicates yields for Foxy T14 plots that are equal (1:1) to that in farmer-practice plots.

T14 Biocontrol Lines of *F. oxysporum* Do Not Produce Detectable Levels of Fusarial Toxins

No traces of any of the five common *Fusarium* produced toxins [deoxynivalenol (DON), 3-acetyldeoxynivalenol (3-ADON), 15-acetyldeoxynivalenol (15-ADON), nivalenol (NIV) and zearalenone (ZEA)] were detected in the Foxy 14 lines nor from a wildtype strain of *F. oxysporum* f. sp. *strigae* also from Kenya. Additionally, Niki McMaster and David Schmale, conducted a separate analysis, in which additional toxins (fumonisins) were also not detected in any of the samples.

TABLE 1 | Relative enhancement of amino acid excretion in minimal medium (% of total amino acids excreted minus excretion by wildtype).

Amino acid	Strain Leu2a	Strain Z6a	Strain Z5a
Tyrosine	0	0.3	0
Leucine	0.3	4.0	0.2
Methionine	0.1	0.2	0.1

Foxy T14 inoculum was a 1:1:1 combination of the three strains below.

DISCUSSION

Our work clearly shows that deployment in the field of actively growing strains of *F. oxysporum* pathogenic to *S. hermonthica* and selected for over-excretion of amino acids that are inhibitory to this parasitic weed can significantly increase the yields for smallholder farmers under real production conditions in Kenya. The marked effectiveness of this method provides justification for more in depth molecular studies to assess the precise dynamics of amino acid excretion *in situ* and the extent to which it harms *Striga* development. The complexity and expense of such studies could have hindered by several years the transfer of this technology to the field. In this light, we recognize that, although we have demonstrated the effectiveness of the Foxy14 for biocontrol of *Striga*, the precise mechanisms of its effectiveness remain to be fully clarified.

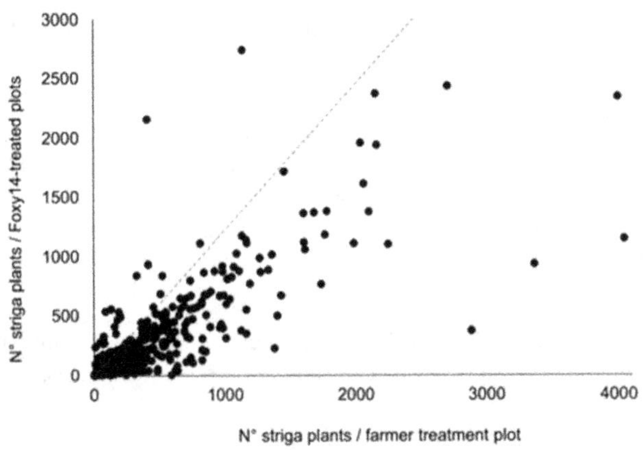

FIGURE 3 | Number of emerged *Striga* plants in the long rainy -season (first season) for Foxy T14 plots relative to the number in the corresponding Farmer-practice plots (i.e., control) at harvest (14 weeks after planting). The light gray dotted line indicates *Striga* counts for Foxy T14 plots that are equal (1:1) to those in Farmer-practice plots. The number of *Striga* plants was reduced in 80% of long rainy season Foxy T14 plots.

These results are significant for several reasons. Firstly, yield increase was obtained under all conditions tested including different production seasons and different yield potentials of farms and farmers. Secondly, the technology was readily deployable in that it was correctly implemented by 500 smallholder farmers and is compatible with their economic limitations. Most biocontrol agents are produced in large facilities and then packaged for long term storage for later farm distribution and not optimal for delivery of fresh rapidly growing inoculum. These biocontrol agents are not efficacious in the case in sub-Saharan Africa and Asia where operating farms of 10 hectares or less are frequently vulnerable points in supply chains. The Foxy T14 on-farm inoculum method is ideal for sustainable deployment of biocontrol of *Striga*. Thirdly, by aiding farmers to reliably achieve yields that allow for economic success, this biocontrol technology will better allow smallholder farmers to invest in inputs such as high quality seed and fertilizers.

The use of various forma speciales of *F. oxysporum* for biological control of weeds and in particular of parasitic weeds such as *Striga* has faced numerous obstacles: (1) Plant pathogenic fungi, especially host specific types, are seldom sufficiently lethal to reduce their target weed to the "knockdown" levels that synthetic herbicides can give. (2) Parasitic weeds are especially difficult to control because the control action must not harm the host crop. Fungi have pathological traits including toxin production and hormone actions that can potentially affect the host crop or the farmer. (3) A single *Striga* plant can produce up to 500,000 seeds in a season and these seeds can survive in the soil seed bank for over a decade (Berner et al., 1995). (4) The traditional delivery systems for biological agents involve field inoculum being manufactured in large central facilities and distributed to farms. The inoculum in such formulations has been stabilized (dried) to maximize shelf life and transportation durability. Such biocontrol agents are essentially dormant when they are applied to the soil and take considerable lag time and optimal conditions to ensure maximum viability. They must quickly encounter and infect a host or die from lack of nutrition. By placing the fungus in rice, it has a source of nutrition for a prolonged period, and allows long distance mycelial growth, increasing the likelihood of encountering a germinated *Striga* seed. Once the scientific development costs of centrally produced commercial biocontrol agents are factored into the product cost, they are beyond the means of the primary clients, smallholder farmers. (5) Biological controls, to be truly effective need to be integrated into overall crop management strategies which may include input investments of labor, soil amendments and quality seed. These investments are not justified if *Striga* is present. (6) Biocontrol agents, even if they are isolated locally, face expensive requirements involved in registration of pesticides, including toxicological, environmental, and efficacy studies. (7) Biocontrol is not a silver bullet that will miraculously maximize crop yield, but it can act as an organizing driver for communities to further impact their wellbeing. This was observed by their continued participation of the women in the Liberty Initiators Network.

Once a problematic weed is suppressed, crop yield is still influenced by the quality of seed, soil fertility, and water availability. When the above obstacles are overcome, there is still the challenge of getting the technology into the hands of the smallholder farmer who is simultaneously battling poor nutrition, disease, grain storage problems, illiteracy and lack of education, gender inequality and poverty. The successful biocontrol of *Striga* can have a strong impact when integrated into a rural community, not only for its direct economic impact, but it also strongly contributes to womens' empowerment, rural nutrition, and education for children. With effective *Striga* management, smallholder farmers can reduce the risk of crop failure and enabling them to improve their outlook by adopting a series of agronomic practices. Effective *Striga* management clearly and concretely contributes to sustainable food security for these farmers.

The method of biocontrol of weeds described here is also compatible with farming systems based on agroecology. The amino acid excretion approach to possible virulence enhancement does not involve directly amplifying plant hormones or phytotoxins. Since selections of amino acid excretion are commonly done by industry to produce large amounts of amino acids for animal feed and even for human nutritional supplementation, the choice of this enhancement method was based on safety considerations relative to the selection of toxin-producing strains. Numerous countries are attempting to reduce the use of and exposure to chemical pesticides, however, replacement strategies are not emerging quickly. The approach demonstrated here of selecting for enhanced virulent biocontrol agents may be applicable to many weeds in that amino acid sensitivity of plants is common, as are host-specific pathogens that could be enhanced for virulence in this manner. We should note that there are rare cases where phenylalanine, an essential amino acid, can have negative effects on humans.

The virulence enhancement and extensive field trials of Foxy T14 led to a replicable technology that provides consistent reduction of *Striga* paired with increased maize yield on smallholder farms. In the hands of farmers, this technology increased maize yield by 56.5% in the long rains season and by 42% in the short rains season. After limiting the effect of *Striga* in the paired plot experiments, there were still great two–fourfold differences in yield among the 500 farms. These differences can probably be attributed to multiple factors including farmer agronomic knowledge, planting date, drought, flooding, and soil nutrient deficiency. We speculate that soil nutrient deficiency, particularly of trace minerals not addressed with the DAP and CAN input used in the paired plot trials, is a principal compounding factor. Trace element deficiency could not only impact crop yields but, as a consequence of a mineral deficient diet, could also have a negative impact on smallholder health with increased incidence of malaria and cognitive disorders.

CONCLUSION

Field deployment of strains of *F. oxysporum,* inherently pathogenic to *S. hermonthica* and then selected for their excretion of certain amino acids, could significantly increase maize yield of smallholder farmers under real production conditions in Kenya. The results are significant for several reasons. Significant yield increases were obtained under all conditions tested including different production seasons and different yield potentials of farms and farmers. The technology was easily and eagerly implemented by 500 smallholder farmers. This technology is inexpensive to manufacture, making it an affordable investment for smallholder farmers. The toothpick primary inoculum system is stable, inherently inexpensive, easily distributed, and it was embraced as part of the on-farm culture where cooking is done daily. By aiding farmers to curtail *Striga* and reliably achieve yields, this biocontrol technology could enable smallholder farmers to purchase inputs such as high quality seed and fertilizers, understanding that their investment will not be rendered futile by *Striga* infestation.

Endemic pathogens may be best adapted to local conditions, and they may present less risk from unforeseen side effects if toxigenic strains of *Fusarium* are avoided. This technology of strain enhancement relies on concerted conventional selection of mutations in amino acid biosynthesis pathways. We chose to develop a three component enhancement of virulence to increase the field effectiveness of the biocontrol treatment and to hopefully delay resistance as is common in weeds when exposed to single herbicides. We expect that this inherently accessible and sustainable low input technology can be incorporated into integrated pest management systems, promoting reduced herbicide and toxin exposure, and, through increased yield income, allow for optimization of plant germplasm diversity and quality.

We envision that this biological control of *Striga* could be adopted in each country where there are *Striga* infestations.

AUTHOR CONTRIBUTIONS

HN, FO, and DS designed the study. HN and FO collected the data. HN, DS, CM, EK, AP, and CB analyzed the data. HN, DS, CM, AP, and CB interpreted the data. HN, DS, CM, AP, and CB drafted the manuscript and critically revised the manuscript. All authors read and approved the final version of the manuscript.

FUNDING

The authors are deeply grateful to the Starfish Foundation of San Diego, California for ongoing financial support founded by Dr. John Sands (1939–2010), to whose memory this manuscript is dedicated, The 500-farm field trials were generously funded by a grant to DS from the Bill and Melinda Gates Foundation: Grand Challenges Exploration, "Biocontrol of *Striga* at the Village Level to Save Labor" (OPP-1097731).

ACKNOWLEDGMENTS

We thank George Oriyo and Denis Omanje for assistance in fieldwork and Lydia Anderson for initial lab excretion selection. We are grateful that Virginia Polytech University (Niki McMaster and David Schmale, III) conducted mycotoxin analysis.

FIGURE S1 | Smallholder farmers apply the inoculated rice while planting.

FIGURE S2 | Trial plots: left plot used hybrid seed, fertilizer, and manure. Middle plot used Foxy T14, hybrid seed, fertilizer and manure. Right third plot used farmer seed saved from a previous harvest.

REFERENCES

Abbasher, A. A., Hess, D. E., and Sauerborn, J. (1998). Fungal pathogens for biological control of *Striga hermonthica* on sorghum and pearl millet in West Africa. *Afr. Crop Sci. J.* 6, 179–188. doi: 10.4314/acsj.v6i2.27814

Abbasher, A. A., and Sauerborn, J. (1992). *Fusarium nygamai*, a potential bioherbicide for *Striga hermonthica* control in sorghum. *Biol. Control* 2, 291–295. doi: 10.1016/1049-9644(92)90021-5

Amalfitano, C. R., Pengue, A., Andolfi, M., Vurro, M. C., Zonno, X. X., and Evidente, D. A. (2002). HPLC analysis of fusaric acid, 9, 10-dehydrofusaric acid and their methyl esters, toxic metabolites produced by weed pathogenic *Fusarium* species. *Phytochem. Anal.* 13, 277–282. doi: 10.1002/pca.648

Atera, A. E., Ishii, T., Onyango, J. C., Itoh, K., and Azuma, T. (2013). *Striga* infestation in Kenya: status, distribution and management options. *Sustain. Agric. Res.* 2, 99–108. doi: 10.5539/sar.v2n2p99

Avedi, E. K., Ochieno, D. M., Ajanga, S., Wanyama, C., Wainwright, H., Elzein, A., et al. (2014). *Fusarium oxysporum* f. sp. strigae strain Foxy 2 did not achieve biological control of *Striga hermonthica* parasitizing maize in Western Kenya. *Biol. Control* 77, 7–14. doi: 10.1016/j.biocontrol.2014.05.012

Badu-Apraku, B., and Akinwale, R. O. (2015). Cultivar evaluation and trait analysis of tropical early maturing maize under *Striga*-infested and *Striga*-free environments. *Field Crops Res.* 121, 186–194. doi: 10.1016/j.fcr.2010.12.011

Badu-Apraku, B., and Yallow, C. G. (2009). Registration of *Striga*–resistant and drought-tolerant tropical early maize populations TZE-W pop DT STR C4 and TZE-Y pop DT STR C4. *J. Plant Regist.* 3, 86–90. doi: 10.3198/jpr2008.06.0356crg

Berner, D. K., Alabi, M. O., Di-Umba, U., and Ikie, F. O. (1996) "Proposed integrated control program for *Striga hermonthica* in Africa," in *Advances in Parasitic Plant Research, Proceedings 6th Parasitic Weeds Symposium 1996*, eds M. T. Moreno, J. I. Cubero, D. K. Berner, D. Joel, L. Musselman, and C. Parker (Cordoba: Junta de Andaluciia, General Directorate of Agricultural Research), 817–825.

Berner, D. K., Kling, J. G., and Singh, B. B. (1995). *Striga* research and control – a perspective from Africa. *Plant Dis.* 79, 652–660.

Ciotola, M., Ditommaso, A., and Watson, A. K. (2000). Chlamydospore production, inoculation methods and pathogenicity of *Fusarium oxysporum* M12-4A, a biocontrol for *Striga hermonthica. Biocontrol Sci. Technol.* 10, 129–145. doi: 10.1080/09583150029279

Ciotola, M., Watson, A. K., and Hallet, S. G. (1995). Discovery of an isolate of *Fusarium oxysporum* with potential to control *Striga hermonthica* in West Africa. *Weed Res.* 35, 303–307. doi: 10.1111/j.1365-3180.1995.tb01793.x

Debrah, S. K. (1994). Socio-economic constraints to the adoption of weed control techniques: the case of *Striga* control in the West African Semi-Arid Tropics. *Int. J. Pest. Manag.* 40, 153–158. doi: 10.1080/09670879409371874

Ejeta, G., and Gressel, J. (2007). *Integrating New Technologies for Striga Control: Towards Ending the Witch-Hunt.* Singapore: World Scientific Publishing Co. doi: 10.1142/6470

Elzein, A., and Kroschel, J. (2006). Host range studies of *Fusarium oxysporum* "Foxy 2": an evidence for a new forma specialis and its implications for *Striga* control. *J. Plant Dis. Prot.* 20, 875–887.

Hassan, M. M., Gani, M. E. S. A., and Babiker, A. G. E. T. (2009). Management of *Striga hermonthica* in sorghum using soil rhizosphere bacteria and host plant resistance. *Int. J. Agric. Biol.* 11, 367–373.

Kanampiu, F. K., Kabambe, V., Massawe, C., Jasi, L., Ransom, J. K., Friesen, D., et al. (2003). Multisite, multi-season field tests demonstrate that herbicide seed-coating herbicide-resistance maize controls *Striga* spp. and increases yields. *Crop Prot.* 22, 697–670. doi: 10.1016/S0261-2194(03)00007-3

Khan, Z. R., Hassanali, A., Overholt, W., Khamis, T. M., Hooper, A. M., Pickett, J. A., et al. (2002). Control of witchweed *Striga hermonthica* by intercropping with *Desmodium* spp., and the mechanism defined as allelopathic. *J. Chem. Ecol.* 28, 1871–1885. doi: 10.1023/A:1020525521180

Khatibi, P. A., McMaster, N., Musser, R., and Lii, D. G. (2014). Survey of mycotoxins in corn distillers' dried grains with solubles from seventy-eight ethanol plants in twelve states in the U.S. in 2011. *Toxins (Basel)* 6, 1155–1168. doi: 10.3390/toxins6041155

Logan, D. C., and Stewart, G. R. (1991). Role of ethylene in the germination of the hemiparasite *Striga hermonthica. Plant Physiol.* 97, 1435–1438. doi: 10.1104/pp.97.4.1435

Marley, P. S., Ahmed, S. M., Shebayan, J. A. Y., and Lagoke, S. T. O. (1999). Isolation of *Fusarium oxysporum* with potential for biocontrol of the witch weed (*Striga hermonthica*).in the Nigerian savanna. *Biocontrol Sci. Technol.* 9, 159–163. doi: 10.1080/09583159929749

Marley, P. S., and Shebayan, J. A. Y. (2005). Field assessment of *Fusarium oxysporum* based mycoherbicide for control of *Striga hermonthica* in Nigeria. *Biocontrol* 50, 398–399. doi: 10.1007/s10526-004-0461-9

Ndambi, B., Cadish, G., Elzein, A., and Heller, A. (2011). Colonization and control of *Striga hermonthica* by *Fusarium oxysporum* f.sp. strigae, a mycoherbicide component: an anatomical study. *Biol. Control* 58, 149–159. doi: 10.1016/j.biocontrol.2011.04.015

Nelson, P. E., Toussoun, T. A., and Marasas, W. F. O. (1983). *Fusarium species: An illustrated Manual for Identification.* Pennsylvania, PA: Pennsylvania State University Press.

Ogborn, J. E. (1987). "*Striga* control under peasant farming conditions," in *Parasitic Weeds in Agriculture: Striga*, Vol. 1, ed. L. J. Musselman (Boca Raton, FL: CRC Press), 145–158.

Pilgeram, A. L., and Sands, D. C. (2010). "Bioherbicides," in *Industrial Applications: The Mycota*, 2 Edn, ed. M. Hofrichter (Berlin: Springer Verlag), 395–405.

Primrose, S. B. (1976). Ethylene-forming bacterial from soil and water. *J. Gen. Microbiol.* 97, 343–346.

Ransom, J., Kanampiu, F., Gressel, J., De Groote, H., Burnet, M., and Odhiambo, G. (2012). Herbicides applied to the seed of imidazolinone-resistant maize as a *Striga* control option for small-scale African farmers. *Weed Sci.* 60, 283–289. doi: 10.1614/WS-D-11-00060.1

Sands, D. C., and Hankin, L. (1974). Selecting lysine-excreting mutants of lactobacilli for use in food and feed enrichment. *Appl. Microbiol.* 28, 523–524.

Sands, D. C., and Pilgeram, A. L. (2009). Methods for selecting hypervirulent biocontrol agents of weeds: why and how. *Pest Manag. Sci.* 65, 581–587. doi: 10.1002/ps.1739

Sands, D. C., Pilgeram, A. L., Zidack, N. K., Jacobsen, B. J., and Tiourebaev, K. S. (2003). "Enhancing the efficacy of bioherbicides," in *Pseudomonas syringae and Related Pathogens*, eds N. S. Iacobellis, A. Collmer, S. Hutcheson, J. Mansfield, C. E. Morris, J. Murillo, et al. (Dordrecht: Kluwer Academic Publishers), 431–441.

Savard, M. E., Miller, J. D., Ciotola, M., and Watson, A. K. (1997). Secondary metabolites produced by a strain of *Fusarium oxysporum* used for *Striga* control in West Africa. *Biocontrol Sci. Technol.* 7, 61–64. doi: 10.1080/09583159731045

Schaube, B., Marley, P. S., Elzein, A. E. M., and Kroschel, J. (2006). Field evaluation of an integrated *Striga hermonthica* management in Sub-Saharan Africa: synergy between *Striga*-mycoherbicides (biocontrol).and sorghum and maize resistant varieties. *J. Plant Dis. Prot.* 20, 691–699.

Steinberg, R. A. (1952). Frenching symptoms produced in *Nicotiana tabacum* and *Nicotiana rustica* with optical isomers of isoleucine and leucine and with *Bacillus cereus* toxin. *Plant Physiol.* 27, 302–308. doi: 10.1104/pp.27.2.302

Sauerborn, J. (1991). "The economic importance of Phytoparasites *Orobanche* and *Striga*," in *Proceedings of the 5th International Symposium on Parasitc Weeds*, eds J. K. Ransom, L. J. Musselman, A. D. Worsham, and C. Parker (Nairobi: CIMMYT), 137–143.

Tiourebaev, K. S., Semenchenko, G. V., Dolgovskaya, M., McCarthy, M. K., Anderson, T. W., Carsten, L. D., et al. (2001). Biological control of infestations of ditchweed (*Cannabis sativa*).with *Fusarium oxysporum* f.sp. cannabina in Kazakhstan. *Biocontrol Sci. Technol.* 11, 535–540. doi: 10.1080/09583150120067562

Venne, J., Beed, F., Avocanhy, A., and Watson, A. (2009). Integrating *Fusarium oxysporum* f. sp. strigae into cereal cropping systems in Africa. *Pest Manag. Sci.* 65, 572–580. doi: 10.1002/ps.1741

Watson, A. K. (2013). "Biocontrol," in *Parasitic Orobanchaceae, Parasitic Mechanisms and Control Strategies*, eds D. M. Joel, J. Gressel, and L. J. Musselman, (Heidelberg: Springer), 469–497. doi: 10.1007/978-3-642-38146-1_26

Zarafi, A. B., Elzein, A., Abdulkadir, D. I., Beed, F., and Akinola, O. M. (2014). Host range studies of *Fusarium oxysporum* f. sp. strigae meant for the biological control of *Striga hermonthica* on maize and sorghum. *Arch. Phytopathol. Plant Prot.* 48, 1–9. doi: 10.1080/03235408.2014.880580

Conflict of Interest Statement: The authors declare that the research was conducted in the absence of any commercial or financial relationships that could be construed as a potential conflict of interest.

3

The Influence of the Plant Growth Regulator Maleic Hydrazide on Egyptian Broomrape Early Developmental Stages and its Control Efficacy in Tomato under Greenhouse and Field Conditions

Edited by:
Monica Fernandez-Aparicio,
Institut National de la Recherche Agronomique (INRA), France

Reviewed by:
Lytton John Musselman,
Old Dominion University, USA
Simier Philippe,
Nantes University, France
Bruno Chauvel,
Institut National de la Recherche Agronomique (INRA), France
Mohamed Kharrat,
Institut National de Recherche Agronomique de Tunis, Tunisia
Ricardo Alcántara-de la Cruz,
Federal University of Viçosa, Brazil
Joel Ransom,
North Dakota State University, USA

***Correspondence:**
Joseph Hershenhorn
josephhe@volcani.agri.gov.il

*Ariel Venezian[1], Evgenia Dor[2], Guy Achdari[2], Dina Plakhine[2], Evgeny Smirnov[2] and Joseph Hershenhorn[2]**

[1] Efal Agri Ltd., Netanya, Israel, [2] Department of Plant Phytopathology and Weed Research, Newe Ya'ar Research Center, Agricultural Research Organization, Ramat Yishai, Israel

Broomrapes (*Phelipanche* spp. and *Orobanche* spp.) are holoparasitic plants that cause tremendous losses of agricultural crops worldwide. Broomrape control is extremely difficult and only amino acid biosynthesis-inhibiting herbicides present an acceptable control level. It is expected that broomrape resistance to these herbicides is not long in coming. Our objective was to develop a broomrape control system in tomato (*Solanum lycopersicum* L.) based on the plant growth regulator maleic hydrazide (MH). Petri-dish and polyethylene-bag system experiments revealed that MH has a slight inhibitory effect on *Phelipanche aegyptiaca* seed germination but is a potent inhibitor of the first stages of parasitism, namely attachment and the tubercle stage. MH phytotoxicity toward tomato and its *P. aegyptiaca*-control efficacy were tested in greenhouse experiments. MH was applied at 25, 50, 75, 150, 300, and 600 g a.i. ha^{-1} to tomato foliage grown in *P. aegyptiaca*-infested soil at 200 growing degree days (GDD) and again at 400 GDD. The treatments had no influence on tomato foliage or root dry weight. The total number of *P. aegyptiaca* attachments counted on the roots of the treated plants was significantly lower at 75 g a.i. ha^{-1} and also at higher MH rates. *Phelipanche aegyptiaca* biomass was close to zero at rates of 150, 300, and 600 g a.i. ha^{-1} MH. Field experiments were conducted to optimize the rate, timing and number of MH applications. Two application sequences gave superior results, both with five split applications applied at 100, 200, 400, 700, and 1000 GDD: (a) constant rate of 400 g a.i. ha^{-1}; (b) first two applications at 270 g a.i. ha^{-1} and the next three applications at 540 g a.i. ha^{-1}. Based on the results of this study, MH was registered for use in Israel in 2013 with the specified protocol and today, it is widely used by most Israeli tomato growers.

Keywords: ***Solanum lycopersicum*** **L.,** ***Phelipanche aegyptiaca*****, parasitic weeds, chemical control,** ***Orobanche***

INTRODUCTION

Broomrapes (*Phelipanche* spp. and *Orobanche* spp.) are holoparasitic plants that parasitize many dicotyledonous crops, causing tremendous losses in crop yield and quality worldwide (Joel et al., 2007). The parasite produces tiny seeds (0.2–0.3 mm) that remain viable in the soil for many years and germinate in response to substances—mostly are strigolactones—secreted from plant roots (Xie et al., 2010). The germinating seed produces a radicle that grows and reaches the root surface and produces a haustorium (the attachment stage). Intrusive haustorium cells penetrate the host root via enzymatic activity and physical force and form connections between the vascular systems of the host and parasite. The parasite develops into a tubercle that grows underground on the host root surface, sucking all of its nutrient needs from the host. Toward the end of its life cycle, the parasite develops flowering shoots that emerge aboveground, producing hundreds of thousands of seeds per plant (Parker and Riches, 1993).

Several methods were suggested for broomrape management that may be classified to resistant varieties, cultural, physical, biological, and chemical control by herbicides (Dhanapal et al., 1996; Ali, 2007). Resistant varieties and chemical control are probably the most commonly used commercially. Effective chemical control of broomrape is extremely difficult to achieve because these plants are anatomically and physiologically connected to the host. In many cases, when a broomrape sensitive crop is grown for the first time, the farmer is unaware of the parasite's presence or infestation level in the field until broomrape shoots appear aboveground, at which point it is too late because most of the damage has already been done. In addition, control of aboveground shoots is much more difficult than at early developmental stages (Eizenberg et al., 2013). For effective control with systemic herbicides, the host should be highly selective to the herbicide without reducing the latter's phytotoxicity as it passes through the host conductive tissues toward the roots (Joel et al., 2007). Broomrapes are chlorophyll-lacking parasites. Therefore photosynthesis-inhibiting herbicides cannot be used because there is no target site. To date, only amino acid biosynthesis-inhibiting herbicides have been used successfully to control broomrapes at a commercial level. These include imidazolinones, sulfonylureas, and glyphosate (Eizenberg et al., 2013). Imidazolinones and sulfonylureas inhibits acetolactate synthase (ALS), also known as acetohydroxy acid synthase (EC 4.1.3.18), a key enzyme in the biosynthesis of the branched-chain amino acids leucine, isoleucine and valine - HRAC group B (Herbicide Resistance Action Committee [HRAC], 2010). Glyphosate inhibits the enzyme 5-enolpyruvylshikimate-3-phosphate synthase (EPSPS; EC 2.5.1.19), which is essential for the synthesis of the aromatic amino acids phenylalanine, tryptophan and tyrosine - HRAC group G (Schönbrunn et al., 2001; Duggleby et al., 2008; Herbicide Resistance Action Committee [HRAC], 2010).

Phelipanche aegyptiaca is a devastating parasite on tomato (*Solanum lycopersicum* L., synonyms *Lycopersicon esculentum* Mill.), that endangers the existence of the processing tomato industry in Israel, among other countries (Hershenhorn et al., 2009). Sulfosulfuron (SS) has been registered in Israel for *P. aegyptiaca* control in tomato. Its usage involves three split applications with precise timing, each application followed by upper irrigation (Hershenhorn et al., 2009). The herbicide is not active when applied on tomato foliage, probably due to detoxification by a P450-type oxidase enzyme (Buker et al., 2004). To be effective, the herbicide should be washed into the soil so that it can establish direct contact with the germinating *P. aegyptiaca* seeds (Eizenberg et al., 2004). This method of control has several drawbacks. Any deviation from the prescribed application timing may drastically reduce control efficacy (Hershenhorn et al., 2009). Upper irrigation systems are no longer in use for tomato growth as all fields are drip-irrigated, so the farmer must invest in expensive irrigation equipment such as moving pivots. In addition, SS has a long residual effect in the soil, jeopardizing proper crop rotation. The upper irrigation, which enhances disease development, the expensive equipment needed, the precise timing and the long residual effect of SS in the soil considerably limit the number of fields in which this method is implemented in Israel.

Maleic hydrazide (MH; 1,2-dihydro-3,6-pyridazinedione) has been extensively used as a commercial systemic plant growth regulator and as herbicide since its introduction in 1949 (Schoene and Hoffman, 1949). After application to foliage, MH is freely translocated in plants to meristematic tissues, with mobility in both phloem and xylem (Meyer et al., 1987). Its mode of action in plants is not clear, although several hypotheses have been proposed and investigated, such as inhibition of cell division by mitotic disruption. Others have suggested that MH acts as an anti-auxin, anti-gibberellin or regulator of auxin metabolism and other plant growth regulators (Hoffman and Parups, 1964). Some carcinogenic effects of MH in mice, rats and cultured human lymphocytes raised a concern of its risks to man (Swietlińska and Žuk, 1978; De Marco et al., 1995; Ribas et al., 1996). MH is registered in the United States to control sprouting of potatoes and onions, suckers in tobacco, and growth of weeds, grasses, and trees in and along lawns, turf, ornamental plants, non-bearing citrus, utility and highway rights-of-way, airports and industrial land (Environmental Protection Agency [EPA], 1994). MH is also registered in Europe (except Austria, Denmark, Finland), Canada, Brazil, Argentina, China, Dominican Republic, Israel, Malaysia, South Africa, and other countries as well for sprout suppression on onion, shallot, garlic, carrots, and sprout suppression and control of volunteers on potatoes (Boyd, 2006; Coresta, 2014; European Food Safety Authority [EFSA], 2016). Previous attempts to use MH for broomrape control were reported in watermelon for *P. aegyptiaca* (Prokudina, 1976) and *P. aegyptiaca* and *P. ramosa* in tobacco (Evtushenko et al., 1973a; Danko, 1980; Lolas, 1986, 1994; Zhelev, 1988; Covarelli, 2002; Brault-Hernandez and Mornet, 2007) with moderate to high control efficacy. MH was also considered for dodder (*Cuscuta* spp.) control on sugar beet (Evtushenko et al., 1973b).

The limited number of effective herbicides for broomrape control and their extremely narrow range of modes of action, make it hard to match an herbicide to a crop plant for successful broomrape control. Glyphosate and ALS inhibitors are among the most widely used herbicides in the world, resulting in the appearance of glyphosate- and ALS-herbicide-resistant weed populations (Heap, 2014). This phenomenon greatly increases the chances of developing sulfonylurea-, imidazolinone-, and glyphosate-resistant broomrape races, because they are plants with high level of genetic flexibility, especially in areas where the herbicides are heavily used (Hershenhorn et al., 2009). Therefore, there is an urgent need to find new agents that effectively control broomrape by combining new modes of action with ease of use. MH's mode of action differs from ALS and EPSPS inhibition and it has no residual soil effect.

The objectives of this work were: to study the effect of MH on early developmental stages of *P. aegyptiaca* and its translocation within tomato roots; and to optimize its ability to control of this parasitic weed in tomato cultivation under greenhouse and field conditions.

MATERIALS AND METHODS

Plant Material

Processing tomato (*Solanum lycopersicom*) cv. M82 seeds were obtained from Tarsis Agricultural Chemicals Ltd. (Petah Tikva, Israel). *P. aegyptiaca* seeds were collected in 2015 from Egyptian broomrape inflorescences parasitizing tomato grown in Kibbutz Mevo Hama, in the Golan Heights, northern Israel (32°73′67″N, 35°65′52″E). The inflorescences were left to dry at 23–35°C for 2 months. The seeds were separated with a 300-mesh size sieve (50 micron) and stored in the dark at 4°C until use.

Petri-Dish Experiment

The experiments was conducted according to Hershenhorn et al. (1997). *P. aegyptiaca* seeds were surface-sterilized in 70% ethanol solution for 1.5 min, transferred to 1% hypochlorite solution containing two drops of Tween 20 per 200 mL of water for 12 min, and washed three times with sterile water. The test was conducted in 45-mm-diameter Petri dishes, into which 8-mm-diameter GF/A glass microfiber filter paper disks (GFFP, Whatman International, Kent, UK) were placed on top of two layers of filter paper. Approximately 50 *P. aegyptiaca* seeds were placed on each disk. The Petri dishes were kept at 25°C for 5 days for preconditioning after which excess water was removed from the disks by blotting on a paper towel and they were transferred to new Petri dishes. Then 250 μL of the active isomer of GR24 (a synthetic strigol analog kindly provided by Prof. Yoneyama, Weed Science Center, Utsunomiya University, Utsunomiya, Japan) at a concentration of 10^{-6} M was added to each Petri dish. After 2 or 14 days the experiments were initiated by adding 30 μL of MH solution at concentrations of 1.4, 2.1, 2.8, 5.6, 11.2, or 22.4 mM to each Petri dish. Petri dishes receiving 30 μL water served as controls. Germ tube length of the radicle emerged from the germinating seeds was measured under a stereoscopic microscope at ×30 magnification 7 days after application (DAA) of water or MH. The Petri dishes were kept during the experiments in a growth chamber at 25°C in the dark. Each Petri dish contained five disks with three Petri dishes per treatment.

Polyethylene-Bag (PEB) Experiments

Phelipanche aegyptiaca seeds were surface-sterilized as described above and dispersed on a 14 cm × 20 cm glass microfiber filter paper (70 seeds cm^{-2}), equal to 20 000 seeds per 15 cm × 25 cm PEB into which the filter paper was inserted. A window (~12 cm × 20 cm) was cut on one side of the bag face. The cuts were taped over with masking tape. A 1-month-old tomato seedling was planted in the PEB through the upper opening, with the root system scattered over the filter paper containing the broomrape seeds (Dor et al., 2007). Half strength Hoagland solution was used to supplement the plant's growth in the PEB (Hoagland and Arnon, 1950). The bags were placed vertically in a paper folder covered on all sides with an opaque black polyethylene box to prevent light penetration into the root zone. The boxes were kept in a growth chamber at 25°C with 16 h/8 h day/night conditions, and a light intensity of 77 μEi m^{-2} s^{-1}. Five milliliter of 10^{-5} M GR24 were applied 10 days after planting (DAP). Broomrape attachments appeared on the tomato roots 1 week after application (WAA) of GR. GR24 was used to get uniform and synchronized seed germination. At this stage, the herbicides 0.5 μM imazapic (IM), 10 μM SS, and 75, 150, 300, and 600 μM MH were injected onto the filter paper inside the PEBs. In a second PEB experiment, the same previous doses of IM, SS, and MH were applied at the tubercle stage, 2 WAA of GR24. Observations were carried out 7 and 14 DAA. The number of live and necrotic attachments on the surface of the filter paper in the PEBs was counted and the number of tubercles smaller and bigger than 3 mm was recorded under a stereoscopic microscope at ×30 magnification. Each treatment included five replicates (PEBs), each containing one tomato plant.

In the "double"-PEB system experiment, the plant's roots were divided between two bags designated A – bag containing roots to which herbicide was applied, and B – bag containing the non-treated portion of the plant's roots. The number of healthy and necrotic attachments was counted 14 DAA in bag A and B. Each treatment included five replicates (five double PEBs), each containing one tomato plant. Five milliliter of 10^{-5} M GR24 were applied 10 DAP. The herbicides IM (0.5 μM), SS (10 μM), and MH (600 μM) were injected onto the filter paper inside the bag A.

Greenhouse Experiment

Tomato seeds were germinated in styrofoam growing trays and 4 weeks after sowing, were transferred to 2.0-L pots (one plant per pot) using soil from a field at Newe Ya'ar Research Center, Israel (medium-heavy clay-loam soil containing, on a dry wt. basis, 55% clay, 23% silt, 20% sand, 2% organic matter, pH 7.1), closely resemble soil characteristics of the field experimental plots except for a somewhat higher send content. Osmocote fertilizer (Osmocote®, Scotts Miracle-Gro, Marysville, OH, USA) at a concentration of 0.6% (w/v), and *P. aegyptiaca* seeds at a concentration of 15 mg seed kg^{-1} soil, (~2250 seeds kg^{-1}) were added to the soil. These components were mixed to homogeneity in a cement mixer for 10 min. Control pots did not contain *P. aegyptiaca* seeds. Soil temperatures in each experiment were recorded hourly throughout the experiment with temperature data logger (HOBO data logger, Onset Computer Corporation, Bourne, MA, USA) buried at one of the pots at 5 cm depth. The pots were placed in a greenhouse and drip-irrigated as needed. The experiments were arranged in a completely randomized design with four replications (pots) per treatment. MH at rates of 25, 50, 75, 150, 300, and 600 g a.i. ha^{-1} was applied twice at 200 (15 DAP) and 400 (30 DAP) growing degree days (GDD) on tomato foliage at 200 L ha^{-1} using a motorized sprayer equipped with a Tee Jet 8001E nozzle operated at a pressure of 300 kPa. Temperatures were converted to GDD as follows: $GDD = \Sigma\ (T_{mean} - T_{base})$, where T_{mean} is the mean daily soil temperature (°C) and T_{base} is the base temperature (°C). In this study, we used Dong et al. (2008) T_{base} value for tomato of 10°C.

The experiment was terminated when broomrape inflorescences in the non-treated control pots started to develop seeds, about 7 WAA of the herbicide. The roots were gently washed out of the pot and the number of broomrapes, and fresh broomrape and root biomass were determined. Tomato plants were harvested and foliage fresh and dry weights (dried for 72 h at 72°C) were determined at the end of the experiment, 80 DAP.

Field Experiments

Details of field experiments conducted between 2010 and 2012 at five locations in Israel (Ein Harod Ihud - three experiments, and one experiment each in Hulata, Mevo Hama, Kibbutz Mesilot and Havat Eden) are given in **Table 1**. All sites were plowed, cultivated and leveled, with 1.93-m-wide raised beds. Tomato plants were mechanically transplanted in two rows, 40 cm apart with 35 cm between plants in each row, and watered with the "Ra'am-Netafim" drip-irrigation system (Netafim Irrigation Equipment & Drip Systems, Kibbutz Hatzerim, Israel), placed between the two rows and comprised of a 17-mm diameter tube with drippers every 30–50 cm, each emitting 3.5 L h^{-1}. Soil temperatures in each experiment were recorded hourly throughout the growing season with temperature data loggers buried at 5-cm depth in the middle of the experimental plot. The experimental plots were 2 m wide and 10 m long, replicated four times per treatment in a randomized block design. Pesticide and herbicide treatments for non-parasitic weeds were applied following the recommendations of the Israeli Extension Service. MH was applied using a backpack sprayer equipped with a 2-m spraying boom with Tee Jet 11015 nozzles (Spraying Systems Co., Wheaton, IL, USA) calibrated to deliver 200 L ha^{-1} at 300 kPa (Echo SHR210, Echo Ltd., Lake Zurich, IL, USA). MH application through the drip-irrigation system was conducted with a Fertilizing pump connected to the drip-irrigation pipes delivering 10 L h^{-1} and was conducted during the last hour of the irrigation period. Non-treated plots served as controls. MH application rates are given on **Tables 2–4**. In all experiments, tomato plants were harvested manually from a 5-m length of bed in the middle of each plot between late July and late September according to fruit and field conditions. *Phelipanche aegyptiaca* shoots were counted several times during the growing season.

Statistical Analysis

The Petri dish, PEB, greenhouse and field experiments were conducted twice except the experiment for the tomato yield as influenced by five split applications of MH at various rates which was conducted in three locations. The results were subjected to ANOVA using JMP software version 5.0 (SAS Institute Inc., Cary, NC, USA). To meet the assumption on ANOVA, percentage data were arcsine-transformed before analysis. On the graphs, back-transformed means are presented. The results were compared

TABLE 1 | Details of field experiments conducted between 2010 and 2012 in five locations in Israel.

Location		**Ein Harod Ihud**		**Hulata**	**Mevo Hama**	**Kibbutz Mesilot**	**Havat Eden**
					Northern Israel		
Region		**Jezreel Valley**		**Hula Valley**	**Golan Heights**	**Beit She'an Valley**	
Coordinates		**32°56′0.31″N, 35°39′0.17″E**		**33°05′0.29″N, 35°60′0.92″E**	**32°73′67″N, 35°65′52″E**	**32°49′74″N, 35°47′47″E**	**32°28′28″N, 35°29′28″E**
Year	**2010**	**2011**	**2012**	**2010**	**2011**	**2012**	**2012**
Soil characteristics		Medium-heavy clay loam soil		Medium clay loam soil	Medium-heavy clay loam soil	Medium-heavy clay loam soil	Medium-heavy clay loam soil
Clay (%)		57.0		46.1	60.5	48.3	51.0
Silt (%)		20.7		30.2	24.8	26.9	24.0
Sand (%)		13.1		11.7	6.7	12.4	9.0
$CaCO_3$ (%)		8.9		9.1	7.1	10.5	14.8
Organic matter (%)		1.0		2.9	0.9	1.9	1.2
pH		7.1		7.3	7.1	7.4	7.1
Variety	5811	5811	LRT 3715	8892	9205	5811	2549
Planting date (dd.mm.)	10.03	15.03	28.02	11.05	02.05	23.02	22.02
Harvest date (dd.mm.)	05.07	04.07	05.07	–	10.08	18.06	14.06

using Tukey–Kramer Honest Significant Difference (HSD) test ($P < 0.05$). The two experiments conducted at Ein Harod Ihud in 2011 and 2012 were compared using Fisher's t-test, which showed homogeneity of variances, and therefore data were combined.

RESULTS

Petri-Dish Experiments

The germination rate of *P. aegyptiaca* seeds after preconditioning and GR24 application was 95%, with over 95% of the germinated seeds producing germ tube longer than 3 mm 9 days after GR24 (or 7 days after MH) application to the Petri dishes. Germination rate was not influenced by the lower MH application rates of 1.4, 2.1, 2.8, and 5.6 mM, but it was significantly reduced at 11.2 mM, and even more drastically at 22.4 mM. At 1.4 mM, MH had no influence on seed germ tube length, but higher concentrations reduced their length in a concentration-dependent manner up to 5.6 mM. Almost all of the germinated seeds found in the Petri dishes treated with MH at 5.6 and all the germ tube of the germinated seeds at 11.2 and 22.4 mM were shorter than 3 mm (**Figure 1**).

PEB System Experiments

Imazapic (0.5 μM), SS (10 μM), and MH (75, 150, 300, and 600 μM) applied to PEBs 2 DAA of GR24 almost completely prevented the formation of *P. aegyptiaca* attachments on the roots when counted 7 days later. This situation did not change 14 DAA of the herbicides except for 75 μM MH, where 12 new attachments were found on the roots as compared to 46 in the control PEBs. Although 75 μM MH was still significantly reducing the number of new attachments 14 DAA (by 75% as compared to the control PEBs), it was significantly less effective than the higher concentrations of MH (150, 300, and 600 μM) and IM and SS at 0.5 and 10 μM, respectively, which kept the number of new attachments close to zero (**Figure 2**).

In the second set of PEB experiments, MH, SS and IM were applied at the tubercle stage, 2 WAA of GR24. There was no difference in the number of tubercles at the time of application irrespective of their size. However, 14 DAA of herbicides, although there were still no differences in the number of tubercles smaller than 3 mm (0 in the herbicide-treated plants and only 3 in the control plants), there was a significantly lower number of tubercles bigger than 3 mm in the treated PEBs as compared to the controls (**Figure 3**).

In the double-PEB system experiments, 2 WAA of herbicides, the following counts were obtained. In the non-treated control bags, there were 9 healthy tubercles and no necrotic ones in bag A and about 14 and 2 healthy and necrotic tubercles, respectively (90% healthy tubercles) in bag B. No healthy tubercles could be found in the herbicide-treated A bags and a high proportion of necrotic tubercles were found in the non-treated B bags, ranging from 92, 76, and 73% in the MH, SS, and IM bags, respectively (**Figure 4**).

Greenhouse Experiments

Preliminary experiments were conducted in pots to determine MH phytotoxicity when applied on tomato foliage or directly into the pot soil. Results indicated that 1, 2, 3, or 4 split MH applications on the foliage or in the soil starting at 200 GDD with 200 GDD intervals at rates of 300, 450, or 600 g a.i. ha^{-1} had no effect on the plant's visual appearance, height or weight

TABLE 2 | *Phelipanche aegyptiaca* control and tomato yield as influenced by five split applications of maleic hydrazide (MH) at various rates at Ein Harod Ihud and Hulata, 2010.

Treatment MH g a.i. ha^{-1} × no. of applications	*P. aegyptiaca* shoots per m^2				Yield kg m^{-2}
	Ein Harod Ihud		Hulata		Ein Harod Ihud
	Days after planting				
	75	85	63	76	
Control	12.4a	31.2a	3.8a	5.2a	6.0a
67.5 × 5	6.4ab	13.7a	2.7ab	2.7ab	6.7a
135 × 5	6.0ab	11.8ab	3.9a	5.2a	6.4a
270 × 5	3.8b	4.6b	0.7b	0.7b	7.6b
540 × 5	0.0c	0.0c	0.2b	0.2b	7.8b
67.5, 125, 192.5, 270, 540	6.4ab	6.6ab	0.5b	1.6ab	7.1ab

In both experiments, MH was applied five times, 19 [197 growing degree days (GDD)], 32 (381 GDD), 46 (620 GDD), 60 (770 GDD), and 73 (980 GDD) days after planting (DAP) in Ein Harod Ihud and 21 (185 GDD), 35 (390 GDD), 49 (602 GDD), 64 (789 GDD), and 77 (990 GDD) DAP in Hulata. The data for each experiment were compared separately using Tukey–Kramer honest significant difference (HSD) test (P < 0.05) with JAMP program. Different letters indicate significant differences between treatments.

TABLE 3 | *Phelipanche aegyptiaca* control and tomato yield as influenced by one, two, three, or four split applications of MH at a constant rate of 540 g a.i. ha^{-1} at Ein Harod Ihud in 2011 and 2012 and at Mevo Hama in 2011.

Treatment	*P. aegyptiaca* shoots per m^2			
MH g a.i. ha^{-1} × no. of applications	Ein Harod Ihud	Mevo Hama	Yield kg m^{-2}	
	Days after planting		Ein Harod Ihud	Mevo Hama
	105	84	2011 and 2012	2011
	2011 and 2012	2011		
Control	28.9a	18.3a	7.4a	4.6a
540	22.5b	12.1b	8.6b	4.6a
540 × 2	11.7b	6.5c	9.0b	3.6a
540 × 3	1.1c	0.1d	9.3b	4.5a
540 × 4	1.0c	0.0d	9.1b	4.7a

Maleic hydrazide was applied at 51 (355 GDD), 70 (548 GDD), 86 (740 GDD), and 103 (1000 GDD) DAP in Ein Harod Ihud and at 29 (400 GDD), 36 (627 GDD), 49 (817 GDD), and 77 (1020 GDD) DAP in Mevo Hama. The two experiments in Ein Harod Ihud (2011, 2012) were compared by Fisher's t-test which showed homogeneity of variances, and therefore data were combined. The data for each experiment were compared separately using Tukey–Kramer HSD test (P < 0.05) with JAMP program. Different letters indicate significant differences between treatments.

TABLE 4 | *Phelipanche aegyptiaca* control and tomato yield as influenced by five split foliar applications of MH or three split applications through the drip-irrigation system at Kibbutz Mesilot and Havat Eden, 2012.

Time of application[a]								
100	200	400	700	1000	*P. aegyptiaca* shoots per m^2		Yield kg m^{-2}	
Days after planting					Days after planting		Mesilot	Havat Eden
24	38	53	73	90	Mesilot	Havat Eden		
Treatment MH g a.i. ha^{-1}					106	109		
Control	25.0a	6.2a	2.8a	10.0a				
270	270	540	540	540	0.1c	0.0b	8.7b	11.3b
400	400	400	400	400	0.9c	0.1b	7.1b	11.3b
	2700	2700	2700[b]		11.4b	2.3ab	5.4ab	8.2c
	2700	5200	7900[b]		9.0b	2.4ab	5.6ab	8.4c

Data for each experiment were compared separately using Tukey–Kramer HSD test ($P < 0.05$) with JAMP program. Different letters indicate significant differences between treatments. [a]Growing degree days. [b]Applied through the drip-irrigation system.

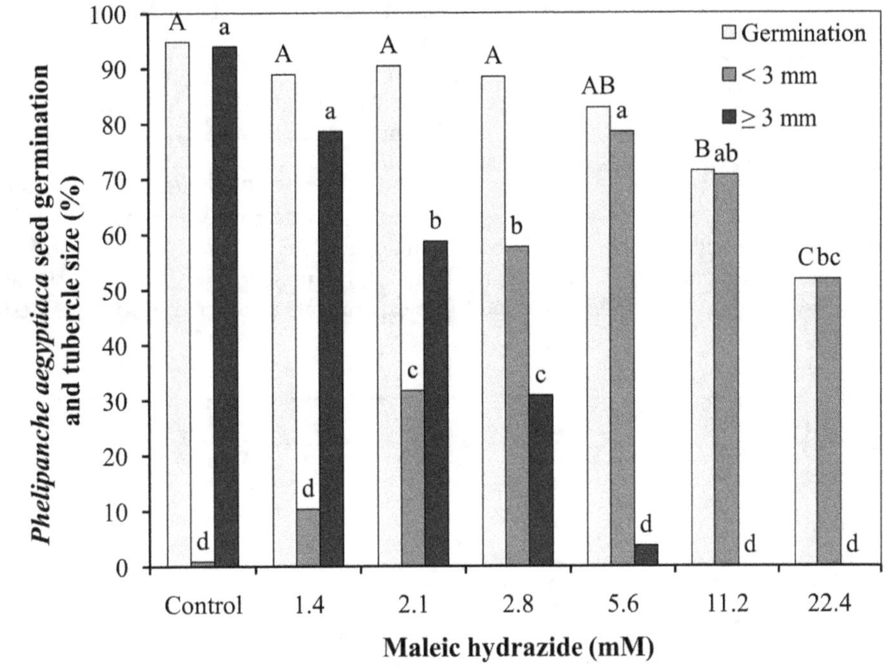

FIGURE 1 | Effect of maleic hydrazide (MH) on *Phelipanche aegyptiaca* seed germination in Petri dishes. Experiments were initiated 2 days after the addition of 10^{-6} M GR24 by adding 30 μL water (control plates) or 30 μL MH solution at 1.4, 2.1, 2.8, 5.6, 11.2, or 22.4 mM (treated plates). Radical length of the germinating seeds was measured 7 days after water or MH applications. The results were compared using Tukey–Kramer honest significant difference (HSD) test ($P < 0.05$) with JAMP program. Uppercase letters indicate significant differences in *P. aegyptiaca* seed germination between treatments. Lowercase letters indicate significant differences in the number of tubercles shorter than 3 mm between treatments or in the number of tubercles longer than 3 mm between treatments. To meet the assumption on ANOVA, percentage data were arcsine-transformed before analysis. On the graphs, back-transformed means are presented.

or on fruit number or weight. MH applied at the indicated rates and timing directly to *P. aegyptiaca*-infested soil had no effect on the number of parasitic shoots or the time of their aboveground emergence. Foliar MH application experiments were repeated with tomato grown in *P. aegyptiaca*-infested soil treated at 200 GDD and again at 400 GDD with 25, 50, 75, 150, 300, or 600 g a.i ha^{-1} MH. The treatments had no influence on tomato foliage or root dry weight as compared to the non-treated plants. The total number of *P. aegyptiaca* counted on the roots of the treated plants was significantly lower at 75 g a.i. ha^{-1} MH and higher. Nevertheless, *P. aegyptiaca* biomass was significantly reduced at the lowest MH rate of 25 g a.i. ha^{-1} and further reduced, to close to 0, at the higher rates of 150, 300, and 600 g a.i. ha^{-1} MH (**Figure 5**).

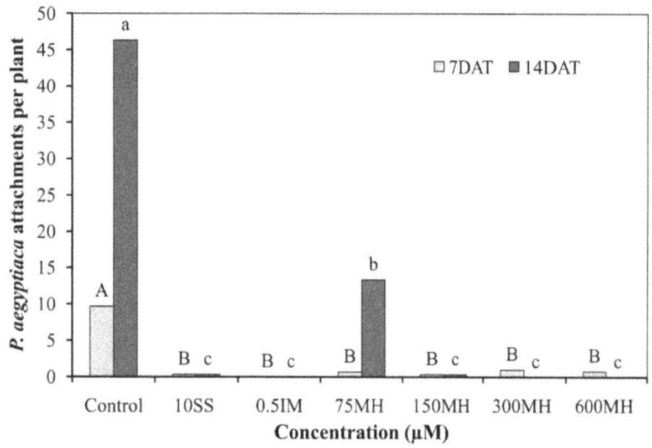

FIGURE 2 | Effect of MH and the herbicides sulfosulfuron (SS) and imazapic (IM) on *P. aegyptiaca* in the polyethylene-bag (PEB) system. MH (75, 150, 300, and 600 μM), SS (10 μM), and IM (0.5 μM) were applied 2 weeks after the addition of 10^{-5} M GR24. The number of *P. aegyptiaca* attachments was counted 7 and 14 days after herbicide application (DAT). Data from each observation date were compared separately using Tukey–Kramer HSD test ($P < 0.05$) with JAMP program. Different letters indicate significant differences between treatments. Uppercase letters represent observations made after 7 days and lowercase letters, observations made after 14 days.

Field Experiments

Ein Harod Ihud and Hulata, 2010

The number of *P. aegyptiaca* shoots counted in the Ein Harod Ihud and Hulata experiments was significantly lower than in the control plots at the two highest MH rates (270 and 540 g a.i. ha^{-1}) at the 75 and 63, and 85 and 76 DAP in Ein Harod Ihud and Hulata, respectively (**Table 2**). The tomato yield at Ein Harod Ihud was significantly higher than that in the control plots. No yield was harvested in Hulata due to pest attack on the last 2 weeks of the growth season. The number of broomrape was significantly lower at the plots treated with five MH applications of 270 or 540 or increasing application rates of 67.5, 125, 192.5, 270, and 540 g a.i. ha^{-1} at 63 DAP at Hulata but at the next observation, about 2 weeks later, only 5 applications of 270 or 540 g a.i. ha^{-1} MH caused a significant reduction in the number of *P. aegyptiaca* shoots between this treatment and the control.

Ein Harod Ihud 2011, 2012 and Kibbutz Mevo Hama 2011

At 540 g a.i. ha^{-1}, MH was enough to significantly reduce the number of *P. aegyptiaca* shoots per m^2 at both locations. Two applications at the same rate did not improve the control at Ein Harod Ihud but it did at Mevo Hama. Three and four split applications of the same rate further improved the control efficacy, with almost no *P. aegyptiaca* shoots at Mevo Hama and only 1 shoot m^{-2} at Ein Harod Ihud. The harvested yield in the Ein Harod Ihud experiments was significantly higher for all treatment applications, whereas at Mevo Hama, there was no detectable change in yield between the control and treated plots for any treatment (**Table 3**).

Kibbutz Mesilot and Havat Eden, 2012

Maleic hydrazide applied 5 times × 400 g a.i. ha^{-1} or in five split applications of 2 × 270 + 3 × 540 g a.i. ha^{-1} significantly reduced the number of *P. aegyptiaca* shoots to close to zero in the experiments conducted in Kibbutz Mesilot and Havat Eden

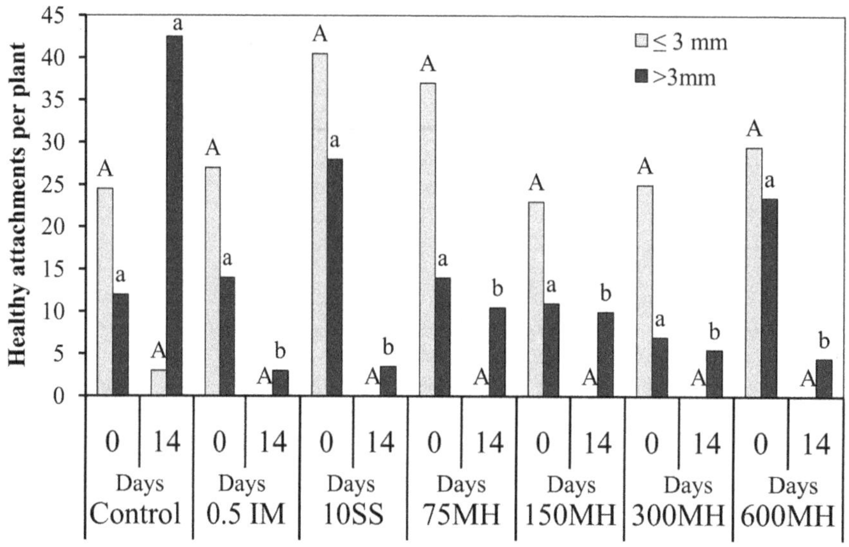

FIGURE 3 | Effect of MH and the herbicides SS and IM on *P. aegyptiaca* in the PEB system. MH (75, 150, 300, and 600 μM), SS (10 μM), and IM (0.5 μM) were applied 2 weeks after the addition 10^{-5} M GR24. The number of *P. aegyptiaca* attachments was counted at the time of application (0) and 14 days later. Data for each observation date were compared separately by LSD on the basis of the Tukey-Kramer Honestly Significant Difference test ($P < 0.05$). Different letters indicate significant differences between treatments. Uppercase letters indicate significant differences in the number of tubercles shorter than 3 mm between treatments for the observation at time 0, and between treatments for the observation at 14 days. Lowercase letters indicate significant differences in the number of tubercles longer than 3 mm between treatments for the observation at time 0, and between treatments for the observation at 14 days.

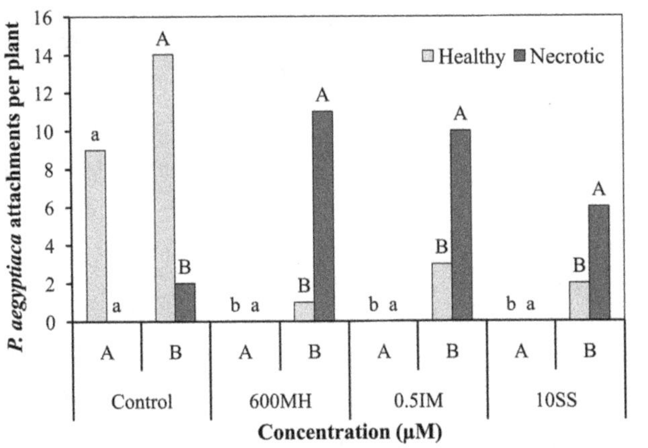

FIGURE 4 | Effect of MH, SS and IM on mortality of *P. aegyptiaca* attachments on tomato roots 2 weeks after their application in the double-polyethylene-bag system. Each bag contained half of the tomato root system (Bag A – herbicide-treated PEB; bag B – non-treated PEB). Five ml of 10^{-5} M GR24 were applied to both bags 10 days after planting. MH, SS, and IM were applied only to bag A 1 week after the addition of GR24. Data from MH-treated and non-treated PEBs were compared separately using Tukey–Kramer HSD test ($P < 0.05$) with JAMP program. Different letters indicate significant differences between treatments. Uppercase letters refer to comparison of MH-treated PEBs, lowercase letters to the non-treated PEBs.

in 2012. As a result, the yield was increased in both treatments and at both locations by an average of 1.3–5.9 kg m^{-2}. Three split MH applications through the drip-irrigation system of 2700 g a.i. ha^{-1} (total of 8100 g a.i. ha^{-1}) or three split applications at increasing rates of 2700, 5200, and 7900 g a.i. ha^{-1} (total of 15 800 g a.i. ha^{-1}) reduced the number of *P. aegyptiaca* shoots significantly in Kibbutz Mesilot as compared to the control, but was less effective than five foliar split applications of 400 g a.i. ha^{-1} or five foliar split applications of 270 × 2 + 3 × 540 g a.i. ha^{-1}. The treatments through the drip-irrigation system at Havat Eden were not effective and did not reduce *P. aegyptiaca* infection. The treatments through the drip-irrigation system had no effect on the yield at Kibbutz Mesilot and had a negative effect on the yield at Havat Eden (**Table 4**).

DISCUSSION

Maleic hydrazide has never been used for broomrape control in tomato, and its ability to control the parasite has thus never been compared to other relevant herbicides such as SS and IM (Eizenberg et al., 2013). The Petri-dish experiments were used to investigate MH's effect on *P. aegyptiaca* seeds at early developmental stages, i.e., seed germination and germ tube development and elongation, before reaching the host root. MH demonstrated limited capacity to prevent seed germination. Only the two highest MH concentrations, 11.2 and 22.4 mM, were able to reduce the seed-germination percentage by 25 and 45%, respectively, as compared to the untreated controls. Lattice seed germination was reduced by 93% in response to 15 mM MH, (Haber and White, 1960), and Grant and Harney (1960) did not find any inhibitory effect on tomato seed germination at 10 mM MH, while *Albizia lebbeck* L. seeds responded in a positive germination rate to increase of MH concentration (Tomar, 2008). These results demonstrate the variation in seed sensitivity of different plant species to MH. Using dormant lettuce and y-irradiated wheat seeds, Haber and White (1960) concluded that MH inhibits cell division but has no effect on cell expansion. Similar experiments were conducted by Hershenhorn et al. (1998) with sulfonylurea herbicides. Chlorsulfuron, bensulfuron, primisulfuron, and thifensulfuron under similar conditions reduced seed germination rate by 17, 18, 25, and 19% at a concentration of 10 μM, which is 1000–2000 times lower than the MH concentrations used in our experiments. The same concentration, 10 μM, of seven phytotoxins—fusarenon X, nivalenol, deoxynivalenol, T-2 toxin, HT-2 toxin, diacetoxyscirpenol, and neosolaniol produced by *Fusarium* species caused 100% inhibition of *Phelipanche ramosa* seed germination (Zonno and Vurro, 2002).

Maleic hydrazide was much more efficient at preventing germ tube elongation. The two highest MH concentrations reduced germ tube length of all treated germinating seeds to less than 3 mm. Hershenhorn et al. (1998) reported that the sulfonylurea herbicides chlorsulfuron, bensulfuron, nicosulfuron, triasulfuron and thifensulfuron reduced germ tube length by ~50% at a concentration of 10 μM. The fact that MH and the sulfonylurea herbicides were less effective at preventing seed germination and much more efficient at inhibiting seed germ tube elongation is not surprising, because most herbicides are active at the post-seed-germination stages rather than during initial seed germination phase (Cathey, 1964; Parka and Soper, 1977).

A PEB system was used to study the ability of MH, as compared to SS and IM, to prevent or inhibit the initial parasitism stages of *P. aegyptiaca*, namely, penetration and establishment in the host root tissues. In the first set of experiments, MH, SS, and IM were applied 2 days after the roots were exposed to GR24. The number of tubercles that formed on the control roots 7 DAA was high whereas in the treated plants, almost no tubercles formed, with no differences between the treatments. These results indicated very high control efficacy when the herbicides were applied shortly after the *P. aegyptiaca* seeds were triggered to germinate. These and the Petri-dish results suggest that the main MH inhibition mechanism is located at the penetration and establishment stages, because MH's effect on the germination process in the Petri-dish experiments was rather limited. The observation made 14 DAA revealed no change in the treatments' efficacy; except for 75 μM MH which, although somewhat less effective, still reduced tubercle number by 72% as compared to the control plants. To investigate the effect of the herbicides on a later stage of *P. aegyptiaca* parasitism, namely after tubercle formation, they were applied 2 WAA of GR24, when tubercles were already formed and were present on the roots. There were no significant differences between the treatments and the control plants in the number of healthy tubercles or their size at the time of herbicide application. However, all treatments significantly reduced the number of healthy tubercles and their size 2 WAA of herbicides. All treatments were equally effective, even the lower MH rate of 75 μM. This set of experiments demonstrated that MH can

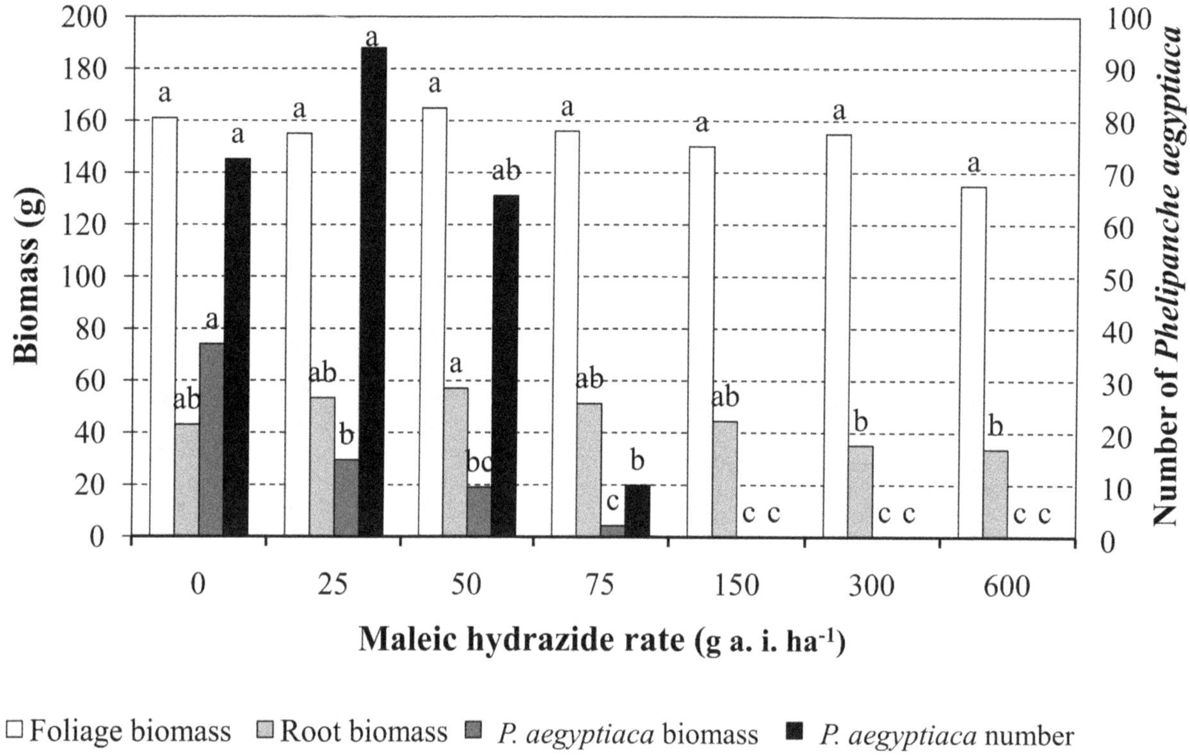

FIGURE 5 | The effect of MH applied to tomato foliage at 200 growing degree days (GDD) and 400 GDD on tomato foliage and root biomass and *P. aegyptiaca* number and biomass in pot experiments. The experiments were terminated when broomrape inflorescences in the non-treated control pots began to develop seeds, about 7 weeks after the last herbicide application. Data for each parameter (tomato foliage and root biomass and *P. aegyptiaca* number and biomass) were compared separately using Tukey–Kramer HSD test ($P < 0.05$) with JAMP program. Different letters indicate significant differences between treatments.

control *P. aegyptiaca* at the tubercle stage and is as effective as SS and IM, albeit at higher rates.

Translocation and movement of MH in the root system could enhance control efficacy if the compound flows, without being detoxified, into the *P. aegyptiaca* attachments, which serve as a strong sink (Fernández-Aparicio et al., 2016). The MH would then accumulate in the attachments until a lethal dose is achieved. To investigate MH translocation and flow in the roots, we used the double-PEB system. All attachments in the treated PEB were dead 2 WAA of herbicides regardless of the herbicide used—MH, SS, or IM. As expected, these results were in complete agreement with those obtained when using a PEB containing the whole plant root system. In the adjacent non-treated PEBs containing the other portion of the plant's root system, 93, 77, and 75% of the attachments were dead in the MH, IM, and SS treatments, respectively. These results suggest that MH moves through the root system without being degraded to non-toxic metabolites. Uptake from solution of MH by barley roots and its translocation to shoots is pH dependent. Uptake and translocation was higher with the decrease of the solution pH probably by passive diffusion of the undissociated form of the acid. Translocation to shoots was approximately proportional to the chemical concentrations in the roots (Briggs et al., 1987). Radioactive labeled MH applied to leaves of *Zebrina pendula*, *Tradescantia fluminensis* and barley seedlings moved relatively free in both phloem and xylem (Crafts and Yamaguchi, 1958). MH absorption from *Ricinus communis* petioles was rather slow but was readily phloem-translocated from the mature leaves allowing appreciable quantities to reach the roots and the nutrient solution (Rigitano et al., 1987). MH absorption by foliar application, translocation to roots and between roots without being detoxified suggests an increase chances for successful broomrape control under greenhouse and field conditions.

Greenhouse experiments in pots revealed that MH up to 600 g a.i. ha^{-1} applied twice, at 200 and 400 GDD, has no negative effect on tomato foliage or root biomass, demonstrating that MH is very safe for use on tomato plants. Further applications were not conducted and fruit number and weight were not determined as the pot volume became the main limiting factor for plant growth. The total number of *P. aegyptiaca* counted on the roots was significantly lower at application rates of 75 g a.i. ha^{-1} or more as compared to the non-treated plants, and *P. aegyptiaca* biomass was reduced significantly at 25 g a.i. ha^{-1}, the lowest rate tested. Those results prompted us to test MH control of *P. aegyptiaca* under field conditions.

Over 30 field experiments were conducted during 2010–2014 to optimize the timing, number and rates for maximal *P. aegyptiaca* control and tomato safety, as well as to minimize the total amount of MH applied. MH is registered in Israel for the prevention of onion and potato sprouting at rates of 2.2 and 2.5 kg ha^{-1}, respectively. Our goal was to limit the total rate used for *P. aegyptiaca* control to this range, in order to facilitate MH registration for its new purpose—*P. aegyptiaca* control in tomato.

The first set of field experiments was aimed at detecting the most effective rate under a constant number of applications. We used five split applications following the application scheme established for *P. aegyptiaca* control in tomato with SS (Hershenhorn et al., 1998, 2009; Eizenberg et al., 2004, 2013; Ephrath et al., 2012), which dictates three split applications of sulfonylurea herbicides at 200 GDD intervals starting at 200 GDD, followed by two applications of IM, for a total of five split applications. A significant reduction of *P. aegyptiaca* shoots in the 2010 experiments at Ein Harod Ihud and Hulata was achieved with 270 g a.i. ha^{-1} MH and a further reduction with 540 g a.i. ha^{-1}, a total of 1.35 and 2.7 kg a.i. ha^{-1} of MH, respectively. Gradually increasing the rates along the applications had no influence on *P. aegyptiaca* shoot number in Ein Harod Ihud but significantly decreased their number in Hulata at the early observation of 65 DAP; this effect then diminished and disappeared by 76 DAP. A positive correlation was found between *P. aegyptiaca* control and yield in Ein Harod Ihud. The yield was significantly higher with the treatments that were also effective at reducing the number of *P. aegyptiaca* shoots. Based on the results of these experiments, we further examined the minimal number of applications needed with 540 g a.i. ha^{-1}, the most successful rate in the 2010 experiments.

We conducted three experiments aimed at testing the efficacy of an increased number of applications under a constant MH rate of 540 g a.i. ha^{-1}. The results (**Table 3**) indicated that even one or two MH applications significantly reduce the number of *P. aegyptiaca* shoots. This was true for the two Ein Harod Ihud experiments whose results were combined. At Mevo Hama, two applications were more efficient than one application, which also reduced the number of *P. aegyptiaca* shoots significantly as compared to the non-treated control plots. In all three experiments, three and four MH applications gave superior results, with no aboveground *P. aegyptiaca* shoots in Mevo Hama and one aboveground *P. aegyptiaca* shoot per m^2 in the Ein Harod Ihud experiments. The higher control efficacy achieved in Mevo Hama may be explained by the lower *P. aegyptiaca* infestation level in that field, with 18 shoot m^{-2} in the non-treated plots as opposed to about 30 shoot m^{-2} in the Ein Harod Ihud control plots. Although the infestation level in Mevo Hama was lower and the *P. aegyptiaca* control efficacy higher than at Ein Harod Ihud, the yield—irrespective of the control efficacy—was lower. This was probably a result of pest and disease outbreaks in the field toward the end of the growing season. All treatments in Ein Harod Ihud improved the yield regardless of the number of applications.

Maleic hydrazide at a rate of 700 g a.i. ha^{-1} applied on tobacco once at 40 DAP reduced the number of *P. ramosa* plants by 60–80% depending on the tobacco cultivar. The same rate applied twice at 40 and 60 DAP reduced the number of *P. ramosa* plants by 80–90% and led to a considerable yield increase (Lolas, 1986). Lower rate of 4500 + 4100 g a.i. ha^{-1} applied at 40 and 60 or 50 and 70 DAP resulted only in a moderate control (Lolas, 1994). Those results are in agreement with the results obtained in the present study were increased MH rates improved the control efficacy. However, Lubanov (1973) reported complete broomrape control with one MH application of 450 g a.i. ha^{-1} on tobacco at the bud formation stage. This complete control may be explained by the fact that the field was previously disinfected with metam-sodium. Similar results were also reported by Jelev (1988) were MH was applied once to tobacco 40 DAP. MH at 2000 g a.i. ha^{-1} applied once on tobacco foliage 55 DAP prevented completely the appearance of *P. ramosa* plants above ground but the same rate applied 20 days later had no effect on the number of *P. ramosa* plants as compared to the control plots (Covarelli, 2002) emphasizing the importance of the application timing. The main parameter dictating broomrape parasitism process and as a consequence herbicide application is the accumulated heat expressed as GDD (Joel et al., 2007; Hershenhorn et al., 2009; Eizenberg et al., 2013). Comparison to our results cannot be conducted without available GDD data. However, it should be noted that MH early application had phytotoxic effect on tobacco plants. In our experiments one MH application of 540 g a.i. ha^{-1}, only 25% of the rate used by Covarelli (2002), 20 DAP reduced the number of *P. aegyptiaca* plants by 12% in the two experiments conducted at 2011 and 2012 at Ein Harod Ihud and by 34% in Mevo Hama without any phytotoxic effects. The host plant roots grow and penetrate in to deeper soil layers during the growing season and stimulate new broomrape seeds to germinate. Therefore repeated MH applications are needed to effectively control the parasite during the whole growing season. It may be hypothesized that complete control of the parasite by one MH application may occur if MH stays at an effective lethal concentration level in the tobacco roots during all the growing season. These contrasting results demonstrate the problematic of comparing experiments conducted in different crops under different conditions. Differences in physiology and morphology of the host plant species, soil type, environmental conditions such as temperature and rainfall, irrigation regime and equipment, growth conditions and as a result plant status and stress, broomrape species and sampling methods, spraying equipment and MH formulation and surfactants, infestation level and cultivation practices are only some of the factors which influence control efficacy.

In irrigated water melon a reduction of the infestation level by 67–96% was achieved with MH at 5000 g a.i. ha^{-1} repeated twice at 10–20 days (Prokudina, 1976). Such high MH rates were not tested in our experiments. In our preliminary experiments one MH application of 1100 g a.i. ha^{-1} caused a significant damage to tomato plants (curly and silvery leaves, flower drops and as a result extensive vegetative growth) and therefore were not continued to be tested (Sarkar et al., 2014).

In the last set of experiments conducted at Kibbutz Mesilot and Havat Eden in 2012 (**Table 4**), we tried two different options to reduce the total MH amount applied to the field during the growing season. Since we had found, in preliminary pot experiments, that moving the first application to 100 GDD has no negative effect on the tomato plant, we reduced the rate of the two first applications to 270 g a.i. ha^{-1} and started the first application at 100 GDD. The second option was a constant rate of 400 g a.i. ha^{-1}, also starting at 100 GDD. As MH may be absorbed by roots and was proven in our double-PEB experiments to flow through the root system to *P. aegyptiaca* attachments, we applied a very high amount of MH through the drip-irrigation system. Five foliar applications, regardless of rate, had excellent control efficacy, almost completely preventing the appearance of *P. aegyptiaca* shoots aboveground. This was especially impressive at Kibbutz Mesilot as the infestation level was high. MH applied three times through the drip-irrigation system had no effect on the number of *P. aegyptiaca* shoots in Havat Eden, and three split applications of 2700, 5200, and 7900 g a.i. ha^{-1} reduced shoot numbers significantly as compared to the non-treated plots but were not as effective as the foliar treatments. The effective control of the foliar applications resulted in a significant yield increase in both locations. The drip-irrigation applications had no effect on yield at Kibbutz Mesilot and a negative effect at Havat Eden. The inefficiency of MH control when applied through the soil, even at the very large amounts applied in our experiments, may be explained by the soil characteristics. Although MH is readily absorbed by roots, it rapidly disappear form soil (Levi and Crafts, 1952; Hoffman et al., 1962; Helweg, 1974). There are several soil components that enhance MH adsorption and inactivation, such as clay, specific surface area and pH, but not organic matter content (Hermosin et al., 1987) The end result is less herbicide available to the plant roots and therefore to *P. aegyptiaca*.

Based on these results and many other field and observation trials, MH was registered in March 2013 for *P. aegyptiaca* control in processing tomato fields in Israel under the two protocols presented in **Table 4**. Since its registration, MH has been widely used by Israeli tomato growers with excellent control results, crop safety and yield increases.

AUTHOR CONTRIBUTIONS

AV contributed to the conception of the work, conducted the experiments and analyzed the data, ED conducted the statistical analysis, contributed to data interpretation and edited the manuscript, GA conducted the experiments, DP conducted the experiments and contributed to the design of the work, ES conducted the experiments, JH planned the study, analyzed and interpreted the data, drafted the manuscript and ultimately approved the version to be published.

ACKNOWLEDGMENTS

This study received financial assistance from The Israeli National Executive Processing Tomato Committee and Efal Agri Ltd. Israel.

REFERENCES

Ali, R. (2007). Conventional and biotechnological approaches for control of parasitic weeds. *In Vitro Cell. Dev. Biol. Plant* 43, 304–317. doi: 10.3109/07388551.2012.743502

Boyd, D. R. (2006). *The Food We Eat: An International Comparison of Pesticide Regulations. 2006*. Vancouver, BC: David Suzuki Foundation.

Brault-Hernandez, M., and Mornet, F. (2007). "*Orobanche ramosa* on tobacco in France: extension, biology and control methods," in *Proceedings of the CORESTA Meeting, Agronomy/Phytopathology*, Krakow, 28.

Briggs, G. G., Rigitano, R. L. O., and Bromilow, R. H. (1987). Physico-chemical factors affecting uptake by roots and translocation to shoots of weak acids in barley. *Pest Manag. Sci.* 19, 101–112. doi: 10.1002/ps.2780190203

Buker, R. S., Rathinasabapathi, B., Stall, W. M., MacDonald, G., and Olson, S. M. (2004). Physiological basis for differential tolerance of tomato and pepper to rimsulfuron and halosulfuron: site of action study. *Weed Sci.* 52, 201–205. doi: 10.1614/WS-03-046R

Cathey, H. M. (1964). Physiology of growth retarding chemicals. *Annu. Rev. Plant Physiol.* 15, 271–302. doi: 10.1146/annurev.pp.15.060164.001415

Coresta (2014). *Maleic Hydrazide*. Available at: https://www.coresta.org/sites/default/files/events/2014Quebec-ListOfPapers.pdf

Covarelli, L. (2002). Studies on the control of broomrape (*Orobanche ramosa* L.) in Virginia tobacco (*Nicotiana tabacum* L.). *Beiträge Tabakforschung Int.* 20, 77–81.

Crafts, A. S., and Yamaguchi, S. (1958). Comparative tests on the uptake and distribution of labeled herbicides by *Zebrina pendula* and *Tradescantia fluminensis*. *Hilgardia* 27, 421–454. doi: 10.3733/hilg.v27n16p421

Danko, J. (1980). The effectiveness of chemicals for the control of broomrape on tobacco. *Ochrana Rostlin* 16, 61–65.

De Marco, A., De Simone, C., Raglione, M., and Lorenzoni, P. (1995). Influence of soil characteristics on the clastogenic activity of maleic hydrazide in root tips of *Vicia faba*. *Mutat. Res.* 344, 5–12. doi: 10.1016/0165-1218(95)90033-0

Dhanapal, G. N., Struik, P. C., Udayakumar, M., and Timmfrmans, P. C. J. M. (1996). Management of broomrape (Orobanche spp.) — a review. *J. Agron. Crop Sci.* 175, 335–359. doi: 10.1111/j.1439-037X.1996.tb00479.x

Dong, Q. X., Louarn, G., Wang, Y. M., Barczi, J. F., and De Reffye, P. (2008). Does the structure-function model GREENLAB deal with crop phenotypic plasticity induced by plant spacing? A case study on tomato. *Ann. Bot.* 101, 1195–1206. doi: 10.1093/aob/mcm317

Dor, E., Evidente, A., Amalfitano, C., Agrelly, D., and Hershenhorn, J. (2007). The influence of growth conditions on biomass, toxins and pathogenicity of fungus *Fusarium oxysporum* f. sp. *orthoceras*, a potential agent for broomrape biocontrol. *Weed Res.* 47, 345–352. doi: 10.1111/j.1365-3180.2007.00567.x

Duggleby, R. G., McCourt, J. A., and Guddat, L. W. (2008). Structure and mechanism of inhibition of plant acetohydroxyacid synthase. *Plant Physiol. Biochem.* 46, 309–324. doi: 10.1016/j.plaphy.2007.12.004

Eizenberg, H., Goldwasser, Y., Golan, S., Plakhine, D., and Hershenhorn, J. (2004). Egyptian broomrape (*Orobanche aegyptiaca* Pers.) control in tomato with sulfonylurea herbicides—greenhouse studies. *Weed Technol.* 18, 490–496. doi: 10.1614/WT-03-023R3

Eizenberg, H., Hershenhorn, J., Ephrath, H. J., and Kamampiu, F. (2013). "Chemical control," in *Parasitic Orobanchaceae. Parasitic mechanisms and Control Strategies*, eds D. M. Joel, J. Gressel, and L. J. Musselman (Heidelberg: Springer), 415–428.

Environmental Protection Agency [EPA] (1994). *Fact Sheet: Reregistration Eligibility Decision: (RED): Maleic Hydrazide*. Washington, DC: Environmental Protection Agency.

Ephrath, J. E., Hershenhorn, J., Achdari, G., Bringer, S., and Eizenberg, H. (2012). Use of logistic equation for detection of the initial parasitism phase of Egyptian broomrape (*Phelipanche aegyptiaca*) in tomato. *Weed Sci.* 60, 57–63. doi: 10.1614/WS-D-11-00070.1

European Food Safety Authority [EFSA] (2016). Peer review of the pesticide risk assessment of the active substance maleic hydrazide. *EFSA J.* 14, 4492–4514.

Evtushenko, G. A., Belyaeva, A. V., and Kolov, O. V. (1973a). "The use of MH for controlling broomrape on tobacco," in *Vliyanie Fizioligeschi*, ed. G. A. Evtuchenko 53–59.

Evtushenko, G. A., Iskhakova, N. A., Chepkasova, A. P., and Trushnikhina, E. L. (1973b). "The effect of MH on the productivity of sugarbeet and dodder (in Russian)," in *Vliyane Fiziologicheski Aktivnykh Soedinenii na Obmen Veschetstv i Produktivnost' Rastenii*, ed. G. A. Evtushenko 3–12.

Fernández-Aparicio, M., Reboud, X., and Gibot-Leclerc, S. (2016). Broomrape weeds. Underground mechanisms of parasitism and associated strategies for their control: a review. *Front. Plant Sci.* 7:135. doi: 10.3389/fpls.2016.00135

Grant, W. F., and Harney, P. M. (1960). Cytogenetic effects of maleic hydrazide treatment of tomato seed. *Can. J. Genet. Cytol.* 2, 162–174. doi: 10.1139/g60-016

Haber, A. H., and White, J. D. (1960). Action of maleic hydrazide on dormancy, cell division, and cell expansion. *Plant Physiol.* 35, 495–499. doi: 10.1104/pp.35.4.495

Heap, I. (2014). "Herbicide resistant weeds," in *Integrated Pest Management*, eds D. Pimentel and P. Peshin (Dordrecht: Springer), 281–303. doi: 10.1007/978-94-007-7796-5_12

Helweg, A. (1974). Degradation of 14C-labelled maleic hydrazide in soil as influenced by sterilization, concentration and pretreatment. *Weed Res.* 15, 53–58. doi: 10.1111/j.1365-3180.1975.tb01096.x

Herbicide Resistance Action Committee [HRAC] (2010). *The World of Herbicides According to HRAC Classification on Mode of Action*. Available at: http://www.hracglobal.com/pages/home.aspx

Hermosin, M. C., Cornejo, J., and Perez Rodriguez, J. L. (1987). Adsorption and desorption of maleic hydrazide as a function of soil properties. *Soil Sci.* 144, 250–256. doi: 10.1097/00010694-198710000-00003

Hershenhorn, J., Eizenberg, H., Dor, E., Kapulnik, Y., and Goldwasser, Y. (2009). Broomrape management in tomato. *Weed Res.* 49, 34–47. doi: 10.1111/j.1365-3180.2009.00739.x

Hershenhorn, J., Goldwasser, Y., Plakhine, D., Ali, R., Blumenfeld, T., Bucsbaum, H., et al. (1998). *Orobanche aegyptiaca* control in tomato fields with sulfonylurea herbicides. *Weed Res.* 38, 343–349. doi: 10.1046/j.1365-3180.1998.00105.x

Hershenhorn, J., Plakhine, D., Goldwasser, Y., Westwood, J. H., Foy, C. L., and Kleifeld, Y. (1997). Effect of sulfonylurea herbicides on early development of Egyptian broomrape (*Orobanche aegyptiaca*) in tomato (*Lycopersicon esculentum*). *Weed Technol.* 12, 108–114.

Hoagland, D. R., and Arnon, D. I. (1950). The water-culture method for growing plants without soil. *Circular* 347, 1–32.

Hoffman, I., and Parups, E. V. (1964). "Mode of action of maleic hydrazide in relation to residues in crops and soils," in *Residue Reviews, Residues of Pesticides and Other Foreign Chemicals in Foods and Feeds*, Vol. 7, ed. F. A. Gunther (Berlin: Springer-Verlag), 74–96.

Hoffman, I., Parups, E. V., and Carson, R. B. (1962). Analysis of maleic hydrazide. *J. Agric. Food Chem.* 10, 43–45.

Jelev, N. (1988). A method of applying maleic hydrazide in the control of tobacco broomrape. *Rasteniev' Nauki* 15, 42–48.

Joel, M. D., Hershenhorn, J., Eizenberg, H., Aly, R., Ejeta, G., Rich, P. J., et al. (2007). Biology and management of weedy root parasites. *Hortic. Rev.* 33, 267–349. doi: 10.1002/9780470168011.ch4

Levi, E., and Crafts, A. S. (1952). Toxicity of maleic hydrazide in California soils. *Hilgardia* 21, 431–440. doi: 10.3733/hilg.v21n16p431

Lolas, P. C. (1986). Control of broomrape (*Orobanche ramosa* L.) in tobacco (*Nicotiana tabacum* L.). *Weed Sci.* 34, 427–430.

Lolas, P. C. (1994). Herbicides for control of broomrape (*Orobanche ramosa* L.) in tobacco (*Nicotiana tabacum* L.). *Weed Res.* 34, 205–209. doi: 10.1111/j.1365-3180.1994.tb01988.x

Lubanov, J. (1973). "Sur le probleme pose par les mauvais herbes parasites en Bulgarie," in *Proceedings of the First EWRC Symposium on Parasitic Weeds*, (Wageningen: Parasitic Weeds Research Group), 18–27.

Meyer, S. A., Sheets, T. J., and Seltmann, H. (1987). "Maleic hydrazide residues in tobacco and their toxicological implications," in *Review of Environmental Contamination and Toxicology*, Vol. 98, ed. G. W. Ware (New York, NY: Springer-Verlag), 43–61.. doi: 10.1007/978-1-4612-4700-5_2

Parka, S. J., and Soper, O. F. (1977). The physiology and mode of action of the dinitroaniline herbicides. *Weed Sci.* 25, 79–87.

Parker, C., and Riches, C. R. (1993). *Parasitic Weeds of the World: Biology and Control*. Wallingford, CT: CAB International.

Prokudina, F. B. (1976). Malic hydrazide against broomrape in watermelons. *Zashchita Rastenii* 21, 23–25.

Ribas, G., Surralles, J., Carbonell, E., Xamena, N., Creus, A., and Marcos, R. (1996). Genotoxicity of the herbicides alachlor and maleic hydrazide in cultured human lymphocytes. *Mutagenesis* 11, 221–227. doi: 10.1093/mutage/11.3.221

Rigitano, R. L. O., Bromilow, R. H., Briggs, G. G., and Chamberlain, K. (1987). Phloem translocation of weak acids in *Ricinus communis*. *Pest Manag. Sci.* 19, 113–133. doi: 10.1002/ps.2780190204

Sarkar, M. D., Kabir, K., Jahan, M. S., and Arefin, S. M. A. (2014). Performance of summer tomato in response to maleic hydrazide. *Int. J. Sci. Res. Pub.* 4, 556–558.

Schoene, L. D., and Hoffman, O. L. (1949). Maleic hydrazide, a unique growth regulant. *Science* 109, 588–590. doi: 10.1126/science.109.2841.588-a

Schönbrunn, E., Eschenburg, S., Shuttleworth, A. W., Schloss, J. V., Amrheini, N., and Evans, J. N. S. (2001). Interaction of the herbicide glyphosate with its target enzyme 5-enolpyruvylshikimate 3-phosphate synthase in atomic detail. *Proc. Natl. Acad. Sci. U.S.A.* 98, 1376–1380. doi: 10.1073/pnas.98.4.1376

Swietlińska, Z., and Žuk, J. (1978). Cytotoxic effects of maleic hydrazide. *Mutat. Res.* 55, 15–30. doi: 10.1016/0165-1110(78)90010-6

Tomar, A. (2008). Effect of salicylic acid, L-arginine monohydrochloride, gibberellic acid, maleic hydrazide and indole acetic acid on seed germination and seedling growth of *Albizzia lebbeck* (L.). *Plant Arch.* 8, 495–496.

Xie, X., Yoneyama, K., and Yoneyama, K. (2010). The strigolactone story. *Annu. Rev. Phytopathol.* 48, 93–117. doi: 10.1146/annurev-phyto-073009-114453

Zhelev, N. (1988). A method of applying maleic hydrazide in the control of tobacco broomrape. *Rasteniev Nauki* 25, 42–48.

Zonno, M. C., and Vurro, M. (2002). Inhibition of germination of *Orobanche ramosa* seeds by *Fusarium* toxins. *Phytoparasitica* 30, 519–524. doi: 10.1007/BF02979757

Conflict of Interest Statement: The authors declare that the research was conducted in the absence of any commercial or financial relationships that could be construed as a potential conflict of interest.

The reviewer BC and handling Editor declared their shared affiliation, and the handling Editor states that the process nevertheless met the standards of a fair and objective review.

4

Fluorescence Imaging in the Red and Far-Red Region during Growth of Sunflower Plantlets: Diagnosis of the Early Infection by the Parasite *Orobanche cumana*

Edited by:
Maurizio Vurro,
National Research Council – Institute of Sciences of Food Production, Italy

Reviewed by:
Grama Nanjappa Dhanapal,
University of Agricultural Sciences Bangalore, India
Dragana Miladinović,
Institute of Field and Vegetable Crops, Serbia

***Correspondence:**
Leire Molinero-Ruiz
leire.molinero@csic.es

*Carmen M. Ortiz-Bustos[1], María L. Pérez-Bueno[2], Matilde Barón[2] and Leire Molinero-Ruiz[1]**

[1] *Department of Crop Protection, Institute for Sustainable Agriculture – Spanish National Research Council, Cordoba, Spain,*
[2] *Estación Experimental del Zaidín, Spanish National Research Council, Granada, Spain*

Broomrape, caused by the root holoparasite *Orobanche cumana*, is the main biotic constraint to sunflower oil production worldwide. By the time broomrape emerges, most of the metabolic imbalance has been produced by *O. cumana* to sunflower plants. UV-induced multicolor fluorescence imaging (MCFI) provides information on the fluorescence emitted by chlorophyll (Chl) *a* of plants in the spectral bands with peaks near 680 nm (red, F680) and 740 nm (far-red, F740). In this work MCFI was extensively applied to sunflowers, either healthy or parasitized plants, for the first time. The distribution of red and far-red fluorescence was analyzed in healthy sunflower grown in pots under greenhouse conditions. Fluorescence patterns were analyzed across the leaf surface and throughout the plant by comparing the first four leaf pairs (LPs) between the second and fifth week of growth. Similar fluorescence patterns, with a delay of 3 or 4 days between them, were obtained for LPs of healthy sunflower, showing that red and far-red fluorescence varied with the developmental stage of the leaf. The use of F680 and F740 as indicators of sunflower infection by *O. cumana* during underground development stages of the parasite was also evaluated under similar experimental conditions. Early increases in F680 and F740 as well as decreases in F680/F740 were detected upon infection by *O. cumana.* Significant differences between inoculated and control plants depended on the LP that was considered at any time. Measurements of Chl contents and final total Chl content supported the results of MCFI, but they were less sensitive in differentiating healthy from inoculated plants. Sunflower infection was confirmed by the presence of broomrape nodules in the roots at the end of the experiment. The potential of MCFI in the red and far-red region for an early detection of *O. cumana* infection in sunflower was revealed. This technique might have a particular interest for early phenotyping in sunflower breeding programs. To our knowledge, this is the first work where the effect of a parasitic plant in its host is analyzed by means of fluorescence imaging in the red and far-red spectral regions.

Keywords: broomrape, chlorophyll content, *Helianthus annuus* L., leaf pair, pre-symptomatic detection

INTRODUCTION

Orobanche cumana Wallr. is a holoparasitic plant that parasitizes sunflower (*Helianthus annuus* L.) and limits crop yield in all the countries of Southern Europe and areas around the Black Sea where sunflowers are grown, as well as in North Africa, Israel, and China (Molinero-Ruiz et al., 2015). On average, sunflower seed losses caused by *O. cumana* (broomrape) can be quantified as being above 50% when susceptible hybrids are grown, and they frequently reach 100% in heavily infested fields (Domínguez, 1996; Jestin et al., 2014). Crop yield reductions due to infection by *O. cumana* depend on several factors, such as aggressiveness of the parasite, sunflower genotype, earliness of broomrape emergence, soil depth and, particularly, the soil infestation level or seed-bank in the soil (Molinero-Ruiz et al., 2009; Jestin et al., 2014). Germination of *O. cumana* seeds in the soil is stimulated by chemical compounds exuded in the rhizosphere by surrounding sunflower roots (Joel et al., 2011; Yoneyama et al., 2011; Raupp and Spring, 2013; Ueno et al., 2014). The seed produces a radicle that grows toward the host root, reaches it, and adheres to the host plant. *Orobanche* spp. obtain water and inorganic compounds from the host through a xylem to xylem contact, and, unlike the vast majority of parasitic plants, they also utilize the carbon fixed by the host plants, which is transferred through phloem continuity between host and parasite (Molinero-Ruiz et al., 2015). Penetration of the intrusive cells into the xylem triggers their division, leading to the formation of a tubercle, a callus-shaped bulge of parasite cells outside the root (Molinero-Ruiz et al., 2015). In *O. cumana*-sunflower interaction, the utilization of host photoassimilates by the parasite results in depletion of resources which are necessary for the growth of sunflower and for the optimal development of the seeds.

The complete disruption of the host vascular system of sunflower forced by the parasitism of *O. cumana*, begins during the initial weeks of the crop. In fact, and as early as 3 weeks after inoculation, a decrease in secondary metabolites is observed in the host upon infection by the parasite (Pérez-Bueno et al., 2014). After underground development of tubercles parasite stems emerge aboveground at around flowering-time of sunflower plants (Melero-Vara and Alonso, 1988). By that time most of the metabolic imbalance has been produced irreversibly by the parasite to sunflower. From that moment onward, the impact of *O. cumana* on the crop yield is evident through absent or small sized capitula, their low number and small seeds, and even plant death (Molinero-Ruiz et al., 2015). Because the parasite establishes itself in the host plant in underground stages, non-destructive methods for diagnosing that establishment are greatly needed. Additionally, diagnosis tools used as early as possible after plant sowing and parasite infection will have a wide range of applications in the search for efficient control approaches, such as genetic resistance or treatments with potential biocontrol agents.

Multicolor fluorescence imaging (MCFI) is a technique based on induction of fluorescence emission through UV excitation. It provides leaf images of the autofluorescence emitted by plants in four spectral bands peaking at around 440 nm (blue, F440), 520 nm (green, F520), 690 nm (red, F690), and 740 nm (far-red, F740; Buschmann and Lichtenthaler, 1998; Buschmann et al., 2000). The first two fluorescence signals (F440 and F520) are related to secondary metabolites (Harris and Hartley, 1976; Morales et al., 1996) while the other two (F680 and F740) are emitted by Chl *a* (Buschmann et al., 2000). Also of particular interest are the fluorescence ratios; the red/far-red (F680/F740) ratio was found to be a good indicator of stress-induced decreases in Chl content (Hák et al., 1990; Buschmann, 2007). Therefore, MCFI is considered to be a powerful tool for the research on plant physiology related to development (Lang et al., 1994; Cerovic et al., 1999; Subhash et al., 1999). Moreover, this imaging technique is especially valuable for detecting early symptoms in leaves as those caused by *Tobacco mosaic virus* in tobacco plants (Chaerle et al., 2007) and *Pseudomonas syringae* in bean plants (Pérez-Bueno et al., 2015).

On the other hand, the use of Chl meters for quantifying leaf Chl levels has become common in recent years. Unlike traditional leaf Chl content estimation techniques consisting mainly of pigments extraction by organic solvents followed by *in vitro* measurements with a spectrophotometer, Chl content meters allow to asses this leaf parameter in a fast and non-destructive way. Furthermore, previous studies using Chl meters have shown that they provide good estimates of leaf Chl content (Fanizza et al., 1991; Brito et al., 2011). Other useful applications of Chl meters are the estimation of foliar N concentration (Chang and Robison, 2003; Liu et al., 2012) and crop disease detection (Calderón et al., 2014).

The objectives of this study were to: (a) analyze the distribution of red and far-red fluorescence emission during early growth stages of healthy sunflower plants, (b) evaluate the use of fluorescence imaging in the red and far-red region as an indicator of sunflower infection by *O. cumana* during underground development stages of the parasite, and (c) compare fluorescence imaging in the red and far-red region with measurements of total leaf Chl content as indicator of early infection by *O. cumana*.

MATERIALS AND METHODS

Distribution of Red and Far-Red Fluorescence Emission in Healthy Sunflower Plantlets

Healthy Plant Material

Four sunflower seeds of the inbred line NR5 were surface-disinfected by soaking them in 20% household bleach (50 g of active chlorine per liter) for 5 min, then seeds were rinsed in deionised water for 5 min and kept in a dark incubator at 25°C at saturation humidity until radicles were 2–5 mm long. After germination, seeds were individually transplanted into small pots with 250 g of a soil mixture (sand:silt:peat moss 2:1:1, *V*). Plants were grown in a glasshouse at 12–22°C without additional lighting for 5 weeks and watered as needed. When plants were 2-weeks-old and once a week for the next 2 weeks of growth, they were fertilized with 15 ml/pot of a nutrient solution with N:P:K (7:5:6).

Multicolor Fluorescence Imaging

Multicolour fluorescence imaging was carried out using an Open FluorCamFC 800-O and analyzed using the Fluorcam7 software (Photon Systems Instruments, Brno, Czech Republic) according to Pérez-Bueno et al. (2015). Fluorescence images in red (F680) and far-red (F740) regions of the spectrum (which are emitted by chlorophyll *a*) were acquired sequentially after excitation with UV light (360 nm). Images corresponding to the F680/F740 ratio were calculated by the mentioned software. Fully developed leaf pairs (LPs) were measured from 2 weeks after transplanting. Thereafter, they were measured twice a week for 3 weeks until the end of the experiment. Measurements were always taken at the same time of day on attached and unshaded leaves.

The F680 and F740 emission by the leaf of healthy sunflower plants was studied by defining three different leaf areas (base, apex, and a region in the middle of the leaf). In addition, progress of chlorophyll fluorescence emission on the first four LPs of the plants was analyzed.

Statistical Analyses

The F680 and F740 emission in leaf areas of healthy sunflower was analyzed in a completely randomized design. Eight replicates were considered and one blade of an LP was regarded as a replicate. Significant differences in the intensity of F680 and F740 between leaf areas was calculated using analysis of variance (ANOVA). If significant effects were obtained, Fisher's protected LSD tests ($P = 0.05$) were used as the comparison procedure. In the case of the time course of the F680 and F740 in LP, means and standard errors of 8–12 replicates were considered. Both experiments were repeated twice and ANOVA analysis was conducted to determine significant differences between the whole set of data. As no significant differences were found, data were pooled.

Red and Far-Red Fluorescence Emission, and Leaf Chl Content in Sunflower Plantlets upon Inoculation with *O. cumana*

Orobanche cumana Inoculation and Sunflower Plant Growth

Five sunflower seeds (replications) of the inbred line NR5, which is the differential for race F (Molinero-Ruiz et al., 2015), were grown in the greenhouse for 5 weeks under the same conditions as those previously described and used as controls. In the case of inoculated plants, five NR5 seedlings were individually transplanted into a soil mixture uniformly infested with 10 mg of *O. cumana* seeds of LPA13 population previously characterized as race F (García-Carneros et al., 2014). From then on, the experiment was carried out under the same conditions described in the section on healthy sunflower plantlets.

At the end of the experiment, sunflower plants were uprooted and their roots carefully cleaned to remove soil, washed and air-dried. Nodules of *O. cumana* in each plant were counted using a stereo microscope and weighed with a precision balance.

Measurements of the Total Leaf Chlorophyll Content

The total content on Chl was determined by an Chl content meter (CL-01 Hansatech Instruments, Ltd., Kings Lynn, UK), which provides a rapid and direct way to determine the relative content of Chl in leaves. Measurements were taken twice a week, from 2 and until 5 wai, in control and in inoculated plants. In each plant, one measurement (replicate) was made in the middle of all fully developed leaves, avoiding the main vein. For each time point, all LPs were individually analyzed. Between 12 and 20 replicates (measurements) were considered for each LP.

Spectrophotometric Chlorophyll Quantification

At the end of the experiment (5 wai), and before plants were uprooted, Chl *a* and Chl *b* content were determined spectrophotometrically for the second, third and fourth LP of four control and four inoculated plants. Two 1 cm diameter leaf disks were collected from each LP, weighted and immediately placed in vials and frozen in liquid nitrogen. Each sample was ground, the total pigments extracted with 4 ml of a solution of 80% acetone (v/v). The absorbance was measured at 470, 647 and 663 nm with a Shimadzu UV-1800 spectrophotometer (Shimadzu Corporation, Tokyo, Japan). Total Chl content and the Chl *a*/*b* ratio were calculated according to Lichtenthaler and Buschmann (2001).

Statistical Analyses

Sunflower plants were inoculated in two experiments which were set up as a factorial on a completely randomized design. Since ANOVA analysis did not detect any significant difference of data between them, the data were pooled. For each time point and LP, data of MCFI and Chl content included a maximum of 20 measurements. Sometimes, MCFI measurements were not taken on completely horizontal leaves resulting in a reduction to 14 replications for that particular time point. Similarly, if Chl content was determined on leaves that were too small to be clamped or had a very irregular surface, measurements were omitted and the replication number was reduced to 12. In any case, and for each time point, the same number of replications of control and inoculated plants was used in the statistical analyses. Significant differences in fluorescence parameters, broomrape incidence, total Chl content and Chl *a*/*b* ratio between inoculated and control plants were analyzed using the same statistical procedures as explained for the healthy sunflowers.

RESULTS

MCFI Distribution Pattern on Healthy Sunflower Plants

Figure 1 shows the average values of F680, F740 and their ratio for the first four LPs of healthy sunflower plants from the second to the fifth week of growth. Values of F680 and F740 of the first LP were significantly higher in the third week as compared to those recorded 1 week earlier. However, the F680/F740 ratio of the same LP showed a slight decrease in the third week. Concerning the second LP, no significant changes were detected in any of the fluorescence parameters analyzed.

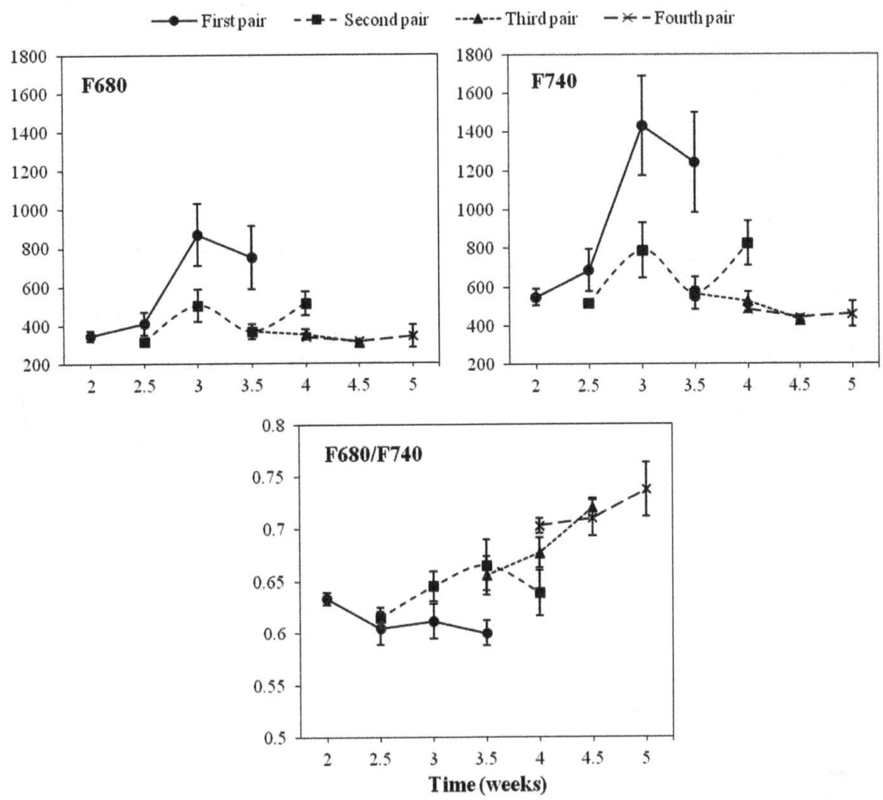

FIGURE 1 | Progress of chlorophyll fluorescence emission on the first, second, third and fourth pair of true leaves of a healthy sunflower plant. Error bars represent the standard error of the mean of 8–12 replications. Time is expressed as weeks after transplanting.

The third LP showed a significant increase in the F680/F740 ratio during the measurement period. Conversely, the other two variables decreased slightly throughout the same period. The three fluorescence parameters of the fourth LP remained constant during the measuring week (**Figure 1**).

The pattern of fluorescence across the leaves was found to be very homogeneous and therefore not dependent on measurement location on the leaf (data not shown). Consequently, a similar size area was selected in the middle of the leaf for the subsequent image analysis of healthy vs. *O. cumana* infected plants.

Alterations in MCFI Distribution Pattern on Sunflower Plants upon Infection with *O. cumana*

Figures 2 and **3** show the progress of F680 and F740, respectively in the first four LPs of sunflower plants inoculated with *O. cumana* compared with that of control plants.

The most informative parameters were those of the first LP from the second to the third week of growth and the second LP from the fourth week on. Between 2 and 3 wai, F680 and F740 of the first LP was significantly higher ($P < 0.05$) in the inoculated sunflowers as compared to the controls. In the same period of time, F680 and F740 signals in the second LP of inoculated plants were significantly higher than those of the controls ($P < 0.05$), as well as at 4 wai ($P < 0.05$). F680 and F740 signals of the third and fourth LP of inoculated plants were significantly higher than those of the controls at 4 wai in the third LP ($P < 0.05$) and half a week later in the fourth LP ($P < 0.05$).

Representative images of the two first LPs from control and inoculated sunflower are shown in **Figure 4**.

Figure 5 displays the values of the F680/F740 ratio in inoculated and control sunflower plants in the first four LPs as well as significant differences between treatments. A significant effect ($P < 0.05$) of the inoculation by *O. cumana* was observed in F680/F740 of the second LP, being lower in inoculated plants at 2.5 and 3 wai ($P < 0.05$), and higher during the rest of the measurements ($P < 0.05$). Significant differences ($P < 0.05$) of F680/F740 between treatments were obtained in the fourth LP (averages 0.73 and 0.71 for inoculated and control plants, respectively). Although no significant differences ($P > 0.05$) were obtained between inoculated and control plants when the F680/F740 ratio of the third LP was analyzed, similar trends were observed.

Chlorophyll Content in Sunflower Plants upon Infection with *O. cumana*

The time course of the total Chl content of the first four LPs (A to D) is presented in **Figure 6**. Differences between the Chl content of inoculated and control plants depended on the LP analyzed, being significantly low ($P < 0.05$) in the second LP of inoculated

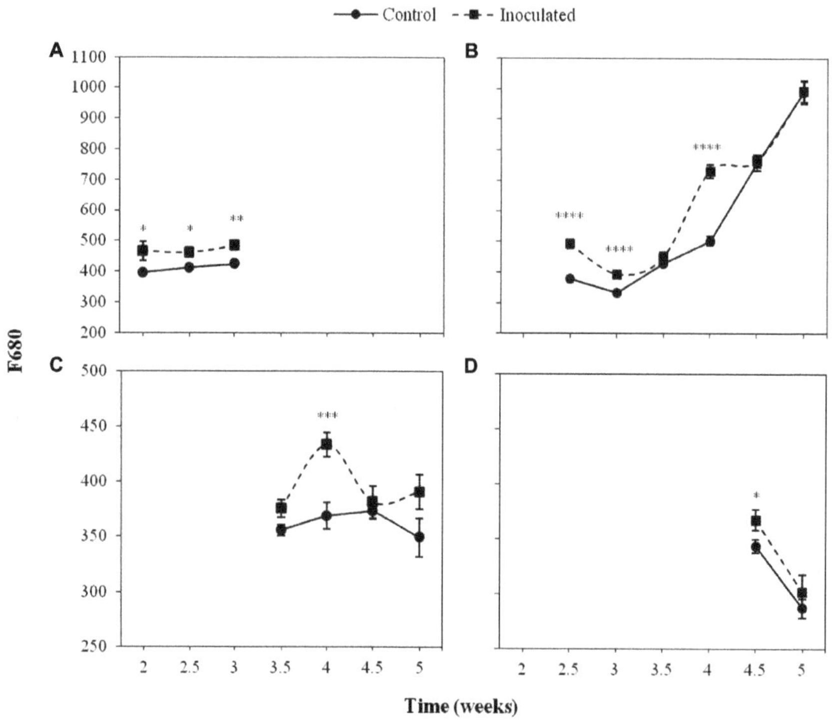

FIGURE 2 | Mean measurements of F680 of the first (A), second (B), third (C), and fourth (D) leaf pair (LP) of sunflower plants inoculated with *Orobanche cumana* and control plants. Vertical bars represent the standard error of the mean of 14–20 replications. Analyses of variance were conducted and asterisks indicate significant differences (*P < 0.05; **P < 0.01; ***P < 0.001; ****P < 0.0001).

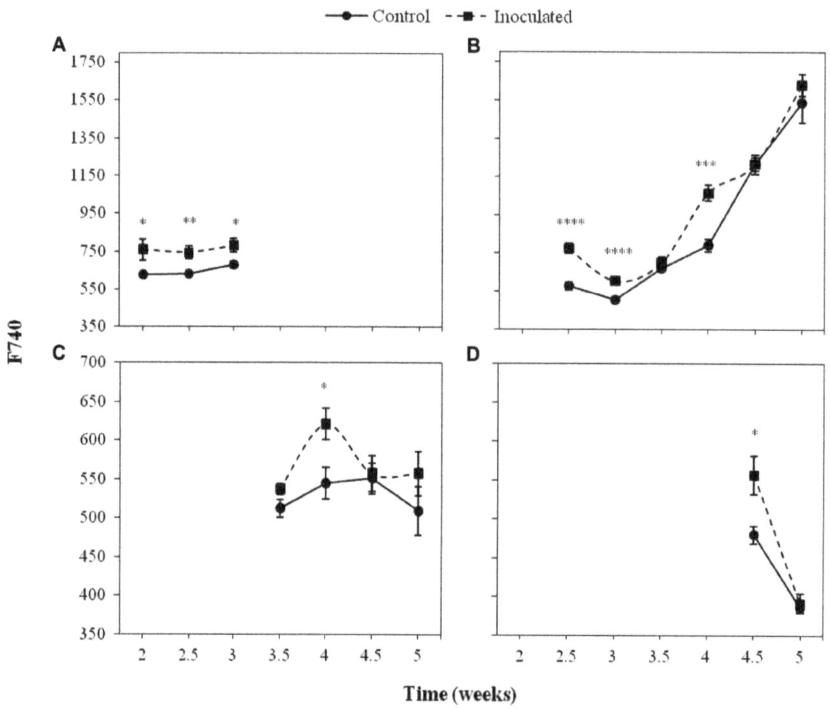

FIGURE 3 | Mean measurements of F740 of the first (A), second (B), third (C), and fourth (D) LP of sunflower plants inoculated with *O. cumana* and control plants. Vertical bars represent the standard error of the mean of 14–20 replications. Analyses of variance were conducted and asterisks indicate significant differences (*P < 0.05; **P < 0.01; ***P < 0.001; ****P < 0.0001).

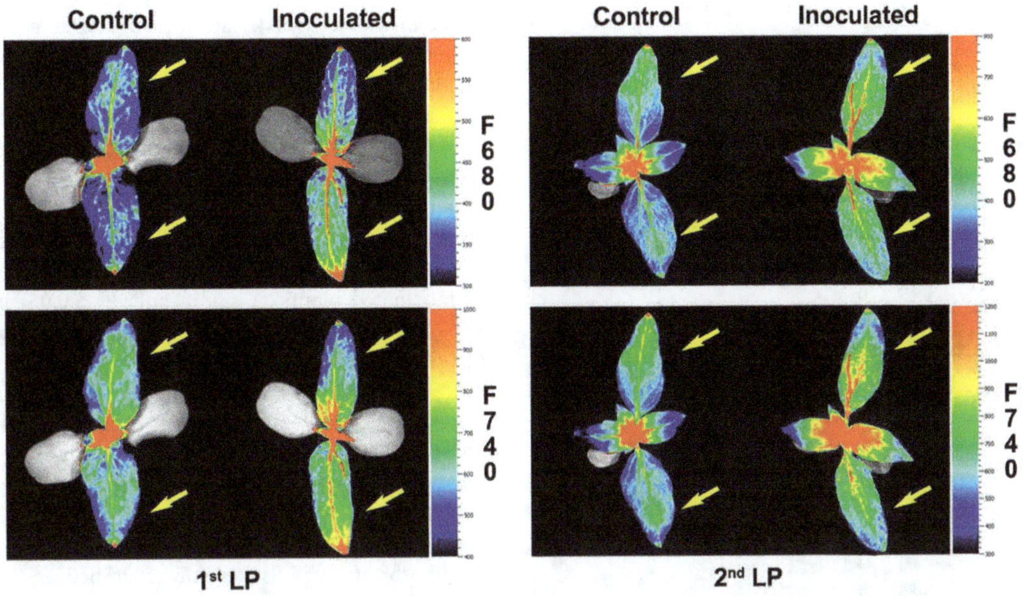

FIGURE 4 | Red fluorescence (F680) and far- red fluorescence (F740) images of the two first LP of sunflower plants inoculated with *O. cumana* and control plants at 2 and 2.5 growth weeks respectively. First LPs are marked with an arrow.

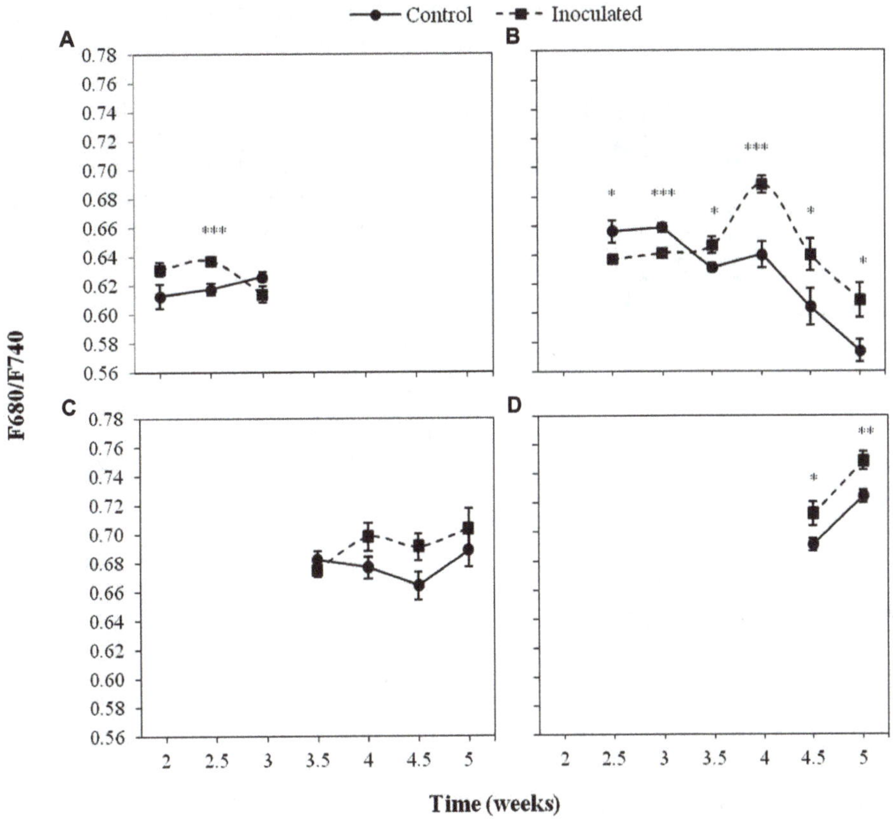

FIGURE 5 | Mean measurements of the F680/F740 ratio of the first (A), second (B), third (C), and fourth (D) LP of sunflower plants inoculated with *O. cumana* and control plants. Vertical bars represent the standard error of the mean of 14–20 replications. Analyses of variance were conducted and asterisks indicate significant differences (*$P < 0.05$; **$P < 0.01$; ***$P < 0.001$).

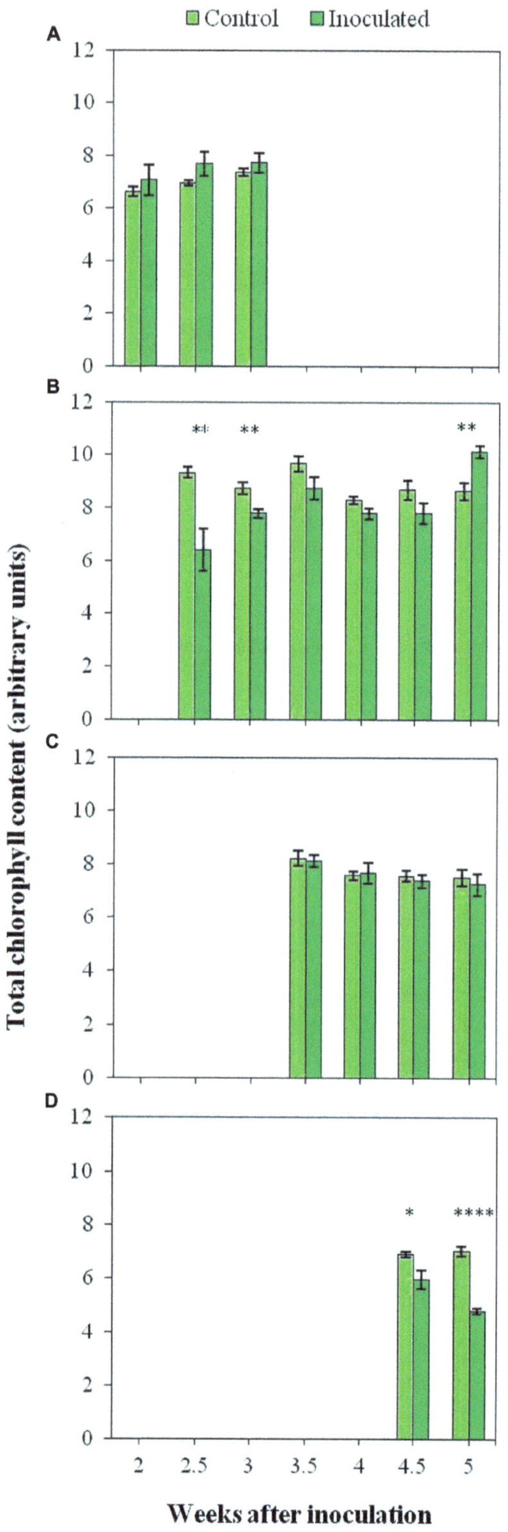

FIGURE 6 | Total chlorophyll content of the first (A), second (B), third (C), and fourth (D) LP of control plants and plants inoculated with *O. cumana*. Vertical bars represent the mean of 12–20 independent measurements with their standard error. Analyses of variance were conducted at $P < 0.05$ and asterisks indicate significant differences (*$P < 0.05$; **$P < 0.01$; ****$P < 0.0001$).

plants at nearly all time points (7.70 arbitrary units as compared to 8.94 in the controls). The measurements conducted on the fourth LP resulted also in significantly ($P < 0.05$) lower values of inoculated plants as compared to the controls (5.38 and 6.98 arbitrary units respectively). In the case of the first and the third LP, no significant ($P > 0.05$) differences in total Chl content were detected between inoculated and control plants.

Spectrophotometric measurements of Chl content are presented in **Supplementary Figure S1**, which shows the differences in the total Chl content of each LP between control and inoculated plants. A significant ($P < 0.05$) increase in the total Chl content of the second LP (**Supplementary Figure S1**) was observed in inoculated plants as compared to the controls (1659.90 and 1460.19 μg/g fresh weight, respectively). On the contrary, measurements of the fourth LP resulted in significantly ($P < 0.05$) lower total Chl content in inoculated plants than in control plants (1050.90 and 1639.31 μg/g fresh weight, respectively). No significant ($P > 0.05$) differences of total Chl content were detected in the third LP. Neither differences in the Chl a/b ratio of LPs were observed between inoculated and control plants (data not shown).

Determination of Final Degree of Sunflower Infection by *O. cumana*

When sunflower plants were inoculated with *O. cumana*, both inoculated and control plants developed four LPs at the end of the experiment (5 wai). No symptoms of parasite infection were observed in the inoculated plants throughout the experiment. Nevertheless, the presence of the parasitic plant in inoculated sunflowers was confirmed by the existence of tubercles in the roots at the end of the experiment. While no tubercles were observed in the controls, means of 24.4 tubercles/plant and 0.46 g tubercles/plant were obtained in inoculated sunflowers (data not shown).

DISCUSSION

During our study, the emission of red and far-red fluorescence in healthy sunflower depended on the LP analyzed. F680 and F740 nm showed the same trend in all the LPs analyzed, with higher F740 values compared to those of F680, probably due to the partial reabsorption of the red fluorescence emitted by leaf Chl *a* (Gitelson et al., 1998). Since F680/F740 was inversely correlated with the Chl content of leaves (Hák et al., 1990; Buschmann and Lichtenthaler, 1998), low values found in the first LP during early stages of the growth of sunflower, compared with the other three LPs under study, were indicative of higher Chl contents. A similar trend of low F680/F740 was observed in the case of the second LP, although with some delay in the time. In our study, the role of an LP as leader in Chl content was sequentially acquired by the one immediately above, and this happened approximately every 1–2 weeks. A different physiological activity between cotyledons and the first LP of sunflower, in terms of photosynthesis, has been reported by Jucknischke and Kutschera (1998). Also, differences in

Chl content and photosynthesis have been reported as being dependent on leaf position in sunflower (Connor and Sadras, 1992; Subrahmanyam and Rathore, 2004). The correlation between photosynthetic activity and Chl content in individual LPs upon infection with *O. cumana* should be investigated in the future.

Increases in both F680 and F740 upon infection of *O. cumana* were observed in all LPs analyzed compared to that of the healthy controls in agreement with previous results by our group (Pérez-Bueno et al., 2014). The evolution of F680/F740 throughout the time was similar to that of F680 and F740 in the same LP. Values of F680/F740 increased for the second to fourth LPs upon infection by *O. cumana* and at different times in each LP, which could be indicative of a decrease in the Chl content upon infection assuming a similarity in the results reported by Gitelson et al. (1999), where the ratio F700/F730 was presented as an indicator of Chl content. Similarly, to those observed in F680 and F740 of LP from healthy sunflowers, the decreases were detected in different LPs sequentially in the time. In inoculated plants, higher values of F680/F740 in the first LP until the third week after inoculation indicated lower Chl contents. From the third week on, a decrease in Chl content as compared to that of the controls occurred in the second LP of the plants. The same trend was observed in the third and fourth LPs when measurements of inoculated and control plants were compared later on in the time. The alteration in F680 and F740 due to *O. cumana* infection shows the usefulness of the MCFI as an early indicator of parasite infection, as was suggested in a previous work (Pérez-Bueno et al., 2014). Furthermore, our results showed that the best indicators of parasite infection were the measurements of F680, F740, and F680/F740 in the first and second LPs from the second to the third week of plant growth. These findings confirm previous results by our research group about early detection of parasite infection by means of fluorescence emission of the first LP of sunflower. Even 3 wai proposed by Pérez-Bueno et al. (2014) as the earliest time for discriminating infected from uninfected plants, was reduced in this work to 2 wai. Therefore, the non-destructive pre-symptomatic disease diagnosis might be feasible as early as 15 days after sunflower inoculation.

Low Chl contents in LP from infected plants at certain times were also shown by outputs of the portable Chl-meter, which showed significantly low Chl contents in the second LP 4.5 weeks after inoculation and in the fourth LP from that moment on. Our results also reveal the greater sensitivity and accuracy of MCFI in diagnosing sunflower infection by *O. cumana* as compared to the direct measurements of Chl in intact leaves, because trends observed when using the hand-held Chl-meter were statistically significant according to red and far-red Chl fluorescence emissions. The positive correlation between outputs of portable Chl-meters and Chl fluorescence measurements has been reported by other authors (Netto et al., 2002, 2005; Nyi et al., 2012; Hu and Fang, 2013). In our work, F680 and F740 in parasitized plants were in agreement with direct measurements of Chl, particularly when the second and fourth LPs were considered. In addition, the spectrophotometric measurements of Chl content at 5 wai provided further evidence that, upon parasite infection, the Chl content in sunflower leaves decreased. The positive correlation between outputs of portable Chl-meters and photosynthetic pigments extracted from leaves has been reported by many authors (Fanizza et al., 1991; Netto et al., 2005; Brito et al., 2011). Finally, the greater sensitivity of MCFI compared to measurements of Chl content might be related to the fact that fluorescence intensity is not only dependent on the Chl content, but also on other factors such as leaf architecture or chloroplast organization (Cerovic et al., 1999). The effect of parasite infection on them might be explored in the future.

Early studies detected a decrease in Chl content in leaves of *Trifolium repens* upon infection by *O. minor*, although no effect on the host photosynthesis was observed (Dale and Press, 1998). Some authors, such as Hibberd et al. (1999) and Watling and Press (2001), suggested that the high competitiveness of holoparasites as a source forces an increase in host photosynthesis. On the contrary, other authors reported low rates of photosynthetic activity of the host upon infection by the holoparasite, such as in the case of tomato plants infected by the broomrape *O. ramosa* (Mauromicale et al., 2008) or *Mikania micrantha* infected by *Cuscuta campestris* (Shen et al., 2007). It is widely known that Chl fluorescence is an excellent approach to photosynthetic efficiency, which can directly or indirectly reflect the impact of biotic and abiotic stresses on plants. Hence, a number of studies have used chlorophyll fluorescence imaging to examine the impact of plant pathogens, i.e., virus, bacteria, and fungi, on host photosynthesis, as excellently reviewed by Rolfe and Scholes (2010). In future research, imaging measurements of Chl fluorescence kinetics in sunflower upon infection by *O. cumana* might explore on the plant photosynthesis alteration as a consequence of the holoparasite infection.

Generally, MCFI has been applied to the study of local symptoms in leaves caused by infections of airborne pathogens (Chaerle et al., 2007; Konanz et al., 2014; Pérez-Bueno et al., 2015). However, symptoms in leaves can as well be a consequence of an infection that occurred far beyond those organs, such as roots in the case of soilborne pathogens (Granum et al., 2015). In the latter, the responses observed in leaves clearly relate to plant defense mechanisms and the alteration of the host physiology due to the infection, since any direct damage in them can be discarded. Up to now, plant infections by soilborne pathogens have only been addressed by means of MCFI in Granum et al. (2015). Rousseau et al. (2015) have recently proved the potential of Chl fluorescence to detect the effect of *Phelipanche ramosa* on *Arabidopsis thaliana*. However, to the best of our knowledge, this is the first time that the impact of a root parasite on its plant host is examined by using fluorescence imaging in the red and far-red region. This inexpensive, non-destructive technic has proved to be an effective diagnostic tool in the early detection of sunflower plants parasitized by *O. cumana.* Moreover, the parameters F680 and F740 detect the parasite infection as early as 2 weeks after inoculation, and this might have a particular interest for early phenotyping in sunflower breeding programs.

AUTHOR CONTRIBUTIONS

LM-R, MP-B, and MB: conceived and designed the experiments. CO-B, LM-R, and MP-B: conducted experiments. CO-B and LM-R analyzed data and interpreted the results. MP-B: mounted images. LM-R and MB: contributed materials, equipment, and analysis tools. CO-B and LM-R wrote the manuscript and all the authors reviewed it and approved the final version.

ACKNOWLEDGMENTS

Financial support for this research was provided by "Consejería de Economía, Innovación y Ciencia" of the Andalusian Government (P12-AGR1281 to LM-R and P12-AGR370 to MB) and the European Social Fund, and RECUPERA 2020 (grant number 20134R060 to MB) and FEDER Funds.

FIGURE S1 | Total chlorophyll content for the second, third and fourth LP of sunflower plants inoculated with *O. cumana* and control plants. Vertical bars represent the average of 8 independent measurements with their standard error. Analyses of variance at $P < 0.05$ between inoculated and control plants were conducted and asterisks indicate significant differences (*$P < 0.05$; ***$P < 0.001$).

REFERENCES

Brito, G. G., Sofiatti, V., Brandão, Z. N., Silva, V. B., Silva, F. M., and Silva, D. A. (2011). Non-destructive analysis of photosynthetic pigments in cotton plants. *Acta Scie. Agro.* 33, 671–678. doi: 10.4025/actasciagron.v33i4.10926

Buschmann, C. (2007). Variability and application of the chlorophyll fluorescence emission ratio red/far-red of leaves. *Photosynth. Res.* 92, 261–271. doi: 10.1007/s11120-007-9187-8

Buschmann, C., Langsdorf, G., and Lichtenthaler, H. K. (2000). Imaging of the blue, green, and red fluorescence emission of plants: an overview. *Photosynthetica* 38, 483–491. doi: 10.1023/A:1012440903014

Buschmann, C., and Lichtenthaler, H. K. (1998). Principles and characteristics of multicolour fluorescence imaging of plants. *J. Plant Physiol.* 152, 297–314. doi: 10.1016/S0176-1617(98)80144-2

Calderón, R., Lucena, C., Trapero-Casas, J. L., Zarco-Tejada, P. J., and Navas-Cortés, J. A. (2014). Soil temperature determines the reaction of olive cultivars to *Verticillium dahliae* Pathotypes. *PLoS ONE* 9: e110664. doi: 10.1371/journal.pone.0110664

Cerovic, Z. G., Samson, G., Morales, F., Tremblay, N., and Moya, I. (1999). Ultraviolet-induced fluorescence for plant monitoring: present state and prospects. *Agronomie* 19, 543–578. doi: 10.1051/agro:19990701

Chaerle, L., Lenk, S., Hagenbeek, D., Buschmann, C., and Van Der Straeten, D. (2007). Multicolor fluorescence imaging for early detection of the hypersensitive reaction to tobacco mosaic virus. *J. Plant Physiol.* 164, 253–262. doi: 10.1016/j.jplph.2006.01.011

Chang, S. X., and Robison, D. J. (2003). Nondestructive and rapid estimation of hardwood foliar nitrogen status using the SPAD-502 chlorophyll meter. *For. Ecol. Manage.* 181, 331–338. doi: 10.1016/S0378-1127(03)00004-5

Connor, D. J., and Sadras, V. O. (1992). Physiology of yield expression in sunflower. *Field Crop Res.* 30, 333–389. doi: 10.1016/0378-4290(92)90006-U

Dale, H., and Press, M. C. (1998). Elevated atmospheric CO2 influences the interaction between the parasitic angiosperm *Orobanche minor* and its host *Trifolium repens*. *New Phytol.* 65–73. doi: 10.1046/j.1469-8137.1998.00247.x

Domínguez, J. (1996). "Estimating effects on yield and other agronomic parameters in sunflower hybrids infested with the new races of sunflower broomrape," in: *Symposium I: Disease tolerance in Sunflower*, ed. A. Pouzet (Int. Sunflower Assoc: Paris), 118–123.

Fanizza, G., Ricciardi, L., and Bagnulo, C. (1991). Leaf greenness measurements to evaluate water stressed genotypes in *Vitis vinifera*. *Euphytica* 55, 27–31. doi: 10.1007/BF00022556

García-Carneros, A. B., Dedic, B., Miladinovic, D., and Molinero-Ruiz, L. (2014). "Pathogenic comparison of highly virulent *O. cumana* affecting sunflower in Moldova, the South of Russian Federation, Serbia and Spain," in *Proceedings of the Third International Symposium of Orobanche spp. of Sunflower*, ed. International Sunflower Association (Paris: International Sunflower Association), 173–177.

Gitelson, A. A., Buschmann, C., and Lichtenthaler, H. K. (1998). Leaf chlorophyll fluorescence corrected for re-absorption by means of absorption and reflectance measurements. *J. Plant Physiol.* 152, 283–296. doi: 10.1016/S0176-1617(98)80143-0

Gitelson, A. A., Buschmann, C., and Lichtenthaler, H. K. (1999). The chlorophyll fluorescence ratio F 735/F 700 as an accurate measure of the chlorophyll content in plants. *Remote Sens. Environ.* 69, 296–302. doi: 10.1016/S0034-4257(99)00023-1

Granum, E., Pérez-Bueno, M. L., Calderón, C. E., Ramos, C., de Vicente, A., Cazorla, F. M., et al. (2015). Metabolic responses of avocado plants to stress induced by *Rosellinia necatrix analysed by fluorescence and thermal imaging. Eur. J. Plant Pathol.* 142, 625–632. doi: 10.1007/s10658-015-0640-9

Hák, R., Lichtenthaler, H. K., and Rinderle, U. (1990). Decrease of the chlorophyll fluorescence ratio F690/F730 during greening and development of leaves. *Radiat. Environ. Biophys.* 29, 329–336. doi: 10.1007/BF01210413

Harris, P. J., and Hartley, R. D. (1976). Detection of bound ferulic acid in the cell walls of the Gramineae by ultraviolet fluorescence microscopy. *Nature* 259, 508–510. doi: 10.1038/259508a0

Hibberd, J. M., Quick, W. P., Press, M. C., Scholes, J. D., and Jeschke, W. D. (1999). Solute fluxes from tobacco to the parasitic angiosperm *Orobanche cernua* and the influence of infection on host carbon and nitrogen relations. *Plant Cell Environ.* 22, 937–947. doi: 10.1046/j.1365-3040.1999.00462.x

Hu, H., and Fang, S. (2013). "Relation of leaf image, chlorophyll fluorescence, reflectance and SPAD in rice and barley," in *Proceedings of the Second International Conference on Agro-Geoinformatics (Agro-Geoinformatics)* (Rome: IEEE), 28–31.

Jestin, C., Lecomte, V., and Duroueix, F. (2014). "Current situation of sunflower broomrape in France," in *Current Situation of Sunflower Broomrape Around the World, Proceedings of the Third International Symposium on Broomrape* (Orobanche spp.) *in Sunflower*, ed. International Sunflower Association (Paris: International Sunflower Association), 28–31.

Joel, D. M., Chaudhuri, S. K., Plakhine, D., Ziadna, H., and Steffens, J. C. (2011). Dehydrocostus lactone is exuded from sunflower roots and stimulates germination of the root parasite *Orobanche cumana*. *Phytochemistry* 72, 624–634. doi: 10.1016/j.phytochem.2011.01.037

Jucknischke, A., and Kutschera, U. (1998). The role of the cotyledons and primary leaves during seedling establishment in sunflower. *J. Plant Physiol.* 153, 700–705. doi: 10.1016/S0176-1617(98)80223-X

Konanz, S., Kocsányi, L., and Buschmann, C. (2014). Advanced multi-color fluorescence imaging system for detection of biotic and abiotic stresses in leaves. *Agriculture* 4, 79–95. doi: 10.3390/agriculture4020079

Lang, M., Lichtenthaler, H. K., Sowinska, M., Summ, P., and Heisel, F. (1994). Blue, green and red fluorescence signatures and images of tobacco leaves. *Bot. Acta* 107, 230–236. doi: 10.1111/j.1438-8677.1994.tb00790.x

Lichtenthaler, H. K., and Buschmann, C. (2001). Chlorophylls and carotenoids: measurement and characterization by UV-VIS spectroscopy. *Curr. Protoc. Food Anal. Chem.* (Suppl. 1), F4.3.1–F4.3.8. doi: 10.1002/0471142913.faf0403s01

Liu, Z. A., Yang, J. P., and Yang, Z. C. (2012). Using a chlorophyll meter to estimate tea leaf chlorophyll and nitrogen contents. *J. Soil Sci. Plant Nutr.* 12, 339–348. doi: 10.4067/S0718-95162012000200013

Mauromicale, G., Monaco, A. L., and Longo, A. M. (2008). Effect of branched broomrape (*Orobanche ramosa*) infection on the growth and photosynthesis of tomato. *Weed Sci.* 56, 574–581. doi: 10.1614/WS-07-147.1

Melero-Vara, J. M., and Alonso, L. C. (1988). "Las Enfermedades del Girasol," in *Enfermedades y Daños de Herbicidas en el Cultivo del Girasol*, ed. Koipesol (Madrid: Artes Gráficas EMA), 33–39.

Molinero-Ruiz, L., Delavault, P., Pérez-Vich, B., Pacureanu-Joita, M., Bulos, M., Altieri, E., et al. (2015). History of the race structure of *Orobanche cumana* and the breeding of sunflower for resistance to this parasitic weed: a review. *Span. J. Agric. Res.* 13: e10R01. doi: 10.5424/sjar/2015134-8080

Molinero-Ruiz, M. L., García-Ruiz, R., Melero-Vara, J. M., and Domínguez, J. (2009). *Orobanche cumana* race F: performance of resistant sunflower hybrids and aggressiveness of populations of the parasitic weed. *Weed Res.* 49, 469–478. doi: 10.1111/j.1365-3180.2009.00708.x

Morales, F., Cerovic, Z. G., and Moya, I. (1996). Time resolved blue-green fluorescence of sugar beet (*Beta vulgaris* L.) leaves: spectroscopic evidence for the presence of ferulic acid as the main fluorophore of the epidermis. *BBA Bioenergetics* 1273, 251–262. doi: 10.1016/0005-2728(95)00153-0

Netto, A. T., Campostrini, E., de Oliveira, J. G., and Bressan-Smith, R. E. (2005). Photosynthetic pigments, nitrogen, chlorophyll a fluorescence and SPAD-502 readings in coffee leaves. *Sci. Hortic.* 104, 199–209. doi: 10.1016/j.scienta.2004.08.013

Netto, A. T., Campostrini, E., de Oliveira, J. G., and Yamanishi, O. K. (2002). Portable chlorophyll meter for the quantification of photosynthetic pigments, nitrogen and the possible use for assessment of the photochemical process in *Carica papaya* L. *Braz. J. Plant Physiol.* 14, 203–210. doi: 10.1590/S1677-04202002000300004

Nyi, N., Sridokchan, W., Chai-arree, W., and Srinives, P. (2012). Nondestructive measurement of photosynthetic pigments and nitrogen status in *Jatropha* (*Jatropha curcas* L.) by chlorophyll meter. *Philipp. Agric. Sci.* 95, 83–89.

Pérez-Bueno, M. L., Barón, M., Garcia-Carneros, A. B., and Molinero-Ruiz, L. (2014). Diagnosis of the infection of sunflower by *Orobanche cumana* using multicolour fluorescence imaging. *Helia* 37, 161–174. doi: 10.(1515)/helia-2014-0015

Pérez-Bueno, M. L., Pineda, M., Díaz-Casado, E., and Barón, M. (2015). Spatial and temporal dynamics of primary and secondary metabolism in *Phaseolus vulgaris* challenged by *Pseudomonas syringae*. *Physiol. Plant.* 153, 161–174. doi: 10.1111/ppl.12237

Raupp, F. M., and Spring, O. (2013). New sesquiterpene lactones from sunflower root exudate as germination stimulants for *Orobanche cumana*. *J. Agric. Food Chem.* 61, 10481–10487. doi: 10.1021/jf402392e

Rolfe, S. A., and Scholes, J. D. (2010). Chlorophyll fluorescence imaging of plant-pathogen interactions. *Protoplasma* 247, 163–175. doi: 10.1007/s00709-010-0203-z

Rousseau, C., Hunault, G., Gaillard, S., Bourbeillon, J., Montiel, G., Simier, P., et al. (2015). Phenoplant: a web resource for the exploration of large chlorophyll fluorescence image datasets. *Plant Methods* 11, 24. doi: 10.1186/s13007-015-0068-4

Shen, H., Hong, L., Ye, W., Cao, H., and Wang, Z. (2007). The influence of the holoparasitic plant *Cuscuta campestris* on the growth and photosynthesis of its host *Mikania micrantha*. *J. Exp. Bot.* 58, 2929–2937. doi: 10.1093/jxb/erm168

Subhash, N., Wenzel, O., and Lichtenthaler, H. K. (1999). Changes in blue-green and chlorophyll fluorescence emission and fluorescence ratios during senescence of tobacco plants. *Remote Sens. Environ.* 69, 215–223. doi: 10.1016/S0034-4257(99)00029-2

Subrahmanyam, D., and Rathore, V. S. (2004). 14CO2 assimilation and 14C-photosynthate translocation in different leaves of sunflower. *Photosynthetica* 42, 313–316. doi: 10.1023/B:PHOT.0000040607.67963.38

Ueno, K., Furumoto, T., Umeda, S., Mizutani, M., Takikawa, H., Batchvarova, R., et al. (2014). Heliolactone, a non-sesquiterpene lactone germination stimulant for root parasitic weeds from sunflower. *Phytochemistry* 108, 122–128. doi: 10.1021/jf2024193

Watling, J. R., and Press, M. C. (2001). Impacts of infection by parasitic angiosperms on host photosynthesis. *Plant Biol.* 3, 244–250. doi: 10.1055/s-2001-15195

Yoneyama, K., Xie, X. N., Kisugi, T., Nomura, T., Sekimoto, H., Yokota, T., et al. (2011). Characterization of strigolactones exuded by Asteraceae plants. *Plant Growth Regul.* 65, 495–504. doi: 10.1007/s10725-011-9620-z

Conflict of Interest Statement: The authors declare that the research was conducted in the absence of any commercial or financial relationships that could be construed as a potential conflict of interest.

The reviewer DM declared a past co-authorship with one of the authors LM-R to the handling Editor, who ensured that the process met the standards of a fair and objective review.

5

Broomrape Weeds: Underground Mechanisms of Parasitism and Associated Strategies for their Control

Edited by:
Richard S. Winder,
Natural Resources Canada, Canada

Reviewed by:
Pascal Labrousse,
Université de Limoges, France
Tadao Asami,
University of Tokyo, Japan

***Correspondence:**
Mónica Fernández-Aparicio
monica.fernandez@dijon.inra.fr

Mónica Fernández-Aparicio[1]*, Xavier Reboud[1] and Stephanie Gibot-Leclerc[2]

[1] INRA, UMR 1347 Agroécologie, Dijon, France, [2] Agrosup-Dijon, UMR 1347 Agroécologie, Dijon, France

Broomrapes are plant-parasitic weeds which constitute one of the most difficult-to-control of all biotic constraints that affect crops in Mediterranean, central and eastern Europe, and Asia. Due to their physical and metabolic overlap with the crop, their underground parasitism, their achlorophyllous nature, and hardly destructible seed bank, broomrape weeds are usually not controlled by management strategies designed for non-parasitic weeds. Instead, broomrapes are in current state of intensification and spread due to lack of broomrape-specific control programs, unconscious introduction to new areas and may be decline of herbicide use and global warming to a lesser degree. We reviewed relevant facts about the biology and physiology of broomrape weeds and the major feasible control strategies. The points of vulnerability of some underground events, key for their parasitism such as crop-induced germination or haustorial development are reviewed as inhibition targets of the broomrape-crop association. Among the reviewed strategies are those aimed (1) to reduce broomrape seed bank viability, such as fumigation, herbigation, solarization and use of broomrape-specific pathogens; (2) diversion strategies to reduce the broomrape ability to timely detect the host such as those based on promotion of suicidal germination, on introduction of allelochemical interference, or on down-regulating host exudation of germination-inducing factors; (3) strategies to inhibit the capacity of the broomrape seedling to penetrate the crop and connect with the vascular system, such as biotic or abiotic inhibition of broomrape radicle growth and crop resistance to broomrape penetration either natural, genetically engineered or elicited by biotic- or abiotic-resistance-inducing agents; and (4) strategies acting once broomrape seedling has bridged its vascular system with that of the host, aimed to impede or to endure the parasitic sink such as those based on the delivery of herbicides via haustoria, use of resistant or tolerant varieties and implementation of cultural practices improving crop competitiveness.

Keywords: integrated pest management, *Orobanche*, *Phelipanche*, parasitism, germination, haustorium, plant recognition, seed bank

INTRODUCTION

The broomrapes are obligate plant-parasitic plants from the genera *Orobanche* and *Phelipanche* in the Orobanchaceae family (Bennett and Mathews, 2006; Tank et al., 2006; Joel, 2009). Due to their achlorophyllous nature, broomrapes are constrained to obtain their nutritional resources by feeding off other plants using the haustorium, an organ unique in parasitic plants through which the parasite diverts water and nutrients from the host (De Candolle, 1813; Kuijt, 1969; Musselman and Dickison, 1975; Westwood, 2013). The majority of broomrape species are botanical wonders parasitizing wild host plants in natural ecosystems. However, seven broomrape species, *Orobanche crenata, O. cernua, O. cumana, O. foetida, O. minor, Phelipanche aegyptiaca,* and *P. ramosa* have specialized on attacking crops causing trouble in agriculture along Mediterranean, central and eastern Europe, and Asia (Parker, 2009). The crops affected depend on the host range of the broomrape species considered but in general, those in the Asteraceae, Brassicaceae, Apiaceae, Fabaceae, or Solanaceae such as sunflower, oilseed rape, carrot, faba bean, or tomato among many others, sustain the major attacks (Parker and Riches, 1993). The damage induced in the crop by broomrape parasitism differs for each broomrape-host association. In general, parasitized crops suffer from reductions in total biomass at the greatest expense to the reproductive tissue (Barker et al., 1996; Manschadi et al., 1996; Lins et al., 2007). In some crops, the biomass loss equals to that accumulated by the parasite indicating that damage in the crop is directly attributed to the parasitic sink activity (Barker et al., 1996; Manschadi et al., 1996; Hibberd et al., 1998). However, in other broomrape-crop associations the damage induced by broomrape extends beyond assimilate diversion. In those cases, broomrape displays a pathogenic nature promoting disease in the crop mainly through negative effects on the crop photosynthetic machinery and hormonal balance (Stewart and Press, 1990; Mauromicale et al., 2008).

There are not figures based on rigorous data for the total area affected by broomrape weeds (Parker, 2009). Sauerborn (1991) estimated that 16 million ha in Mediterranean and West Asia regions risked being infested. In this regard, France is doing valuable work through the Technical Center for Oilseed Crops and Industrial Hemp, Terresinovia, where a nationwide survey of infested fields is actualized online on real time by the farmers with new cases emerging every year and recently toward new regions such as the French Centre region[1]. Several studies suggest that large areas of new territory are at risk of invasion by broomrape (Mohamed et al., 2006; Grenz and Sauerborn, 2007), and in fact, invasions in completely new regions are already emerging in countries such as Spain, UK, France, Algeria, Ethiopia, Egypt, Sudan (Reda, 2006; Babiker et al., 2007; Babiker, 2008; Rubiales et al., 2008; Abu-Irmaileh and Labrada, 2009; Parker, 2014).

Several factors contribute to the fact that broomrape weeds remain an uncontrolled agricultural problem. Control strategies designed for non-parasitic weeds such as cultural and chemical methods do not necessarily achieve the required level of control for broomrape due to its mixed traits as weed and as root parasite. Biological traits in broomrape such as achlorophyllous nature, underground parasitism, the physical connection and growth synchronization with the crop, and the exclusive uptake of resources via crop vascular system rather than from the soil make broomrape control a challenging agricultural task. In addition, the biological similarity between host and parasite characterizing broomrape-crop interactions is higher than in other plant pathosystems, which complicates the development of selective methods to control broomrape, without harmful effect in the crop from which it is feeding (Eizenberg et al., 2006; Hearne, 2009; Yoder and Scholes, 2010; Pérez-Vich et al., 2013).

Besides the difficulty of selectively controlling broomrape in the form of host-attached parasite, eradication of broomrape seed bank is extremely difficult due to prolific production of parasitic seeds, their easy dispersal, physiological dormancy, seed longevity, and germination synchronized with specialized range of host cultivation. New infestations can occur through the use of contaminated seeds or machinery and their prevention is essential. Despite of this fact, Seed Certification Services in some of the countries affected, do not include in their certification standards, inspection of crop seed samples for broomrape inoculum. Effective broomrape control should target the underground mechanisms of crop parasitism in order to meet both the short-term productivity expectations of the farmer and reduction of soil bank in the long run (**Figure 1**). Promising new control strategies have been investigated though the majority of them are under development or remain as prototypes to which farmers have not access. Although some examples of successful control do exist for some crops, the majority of commercially available control methods are either not fully effective or not applicable to many of the affected crops, especially in the case of low-input crops (Joel, 2000).

This paper reviews relevant facts about the biology of broomrape weeds, the key mechanisms they employ to attack crops and the control methods already developed or in development that directly target those mechanisms.

SPECIALIZED MECHANISMS IN BROOMRAPE WEEDS FOR A PARASITIC MODE OF LIFE

The evolution from autotrophic to heterotrophic mode of nutrition carried a reduction of the main broomrape vegetative organs toward vestigial versions, non-functional for autotrophy. Root system in mature broomrape plants is reduced to short adventitious parasitic roots with functions of anchorage and stabilization in the soil and their leaves are reduced to small achlorophyllous scales (Parker and Riches, 1993). In return they develop haustoria to feed off other plants (Kuijt, 1969; Musselman and Dickison, 1975). The haustorium is the key feature of plant parasitism which has evolved independently at least 11 times in angiosperms (Barkman et al., 2007; Westwood et al., 2012; Yang et al., 2015). Haustorium allows broomrape to attack crops by successive functions, first as host-adhesion

[1] http://www.terresinovia.fr/orobanche/carte.php

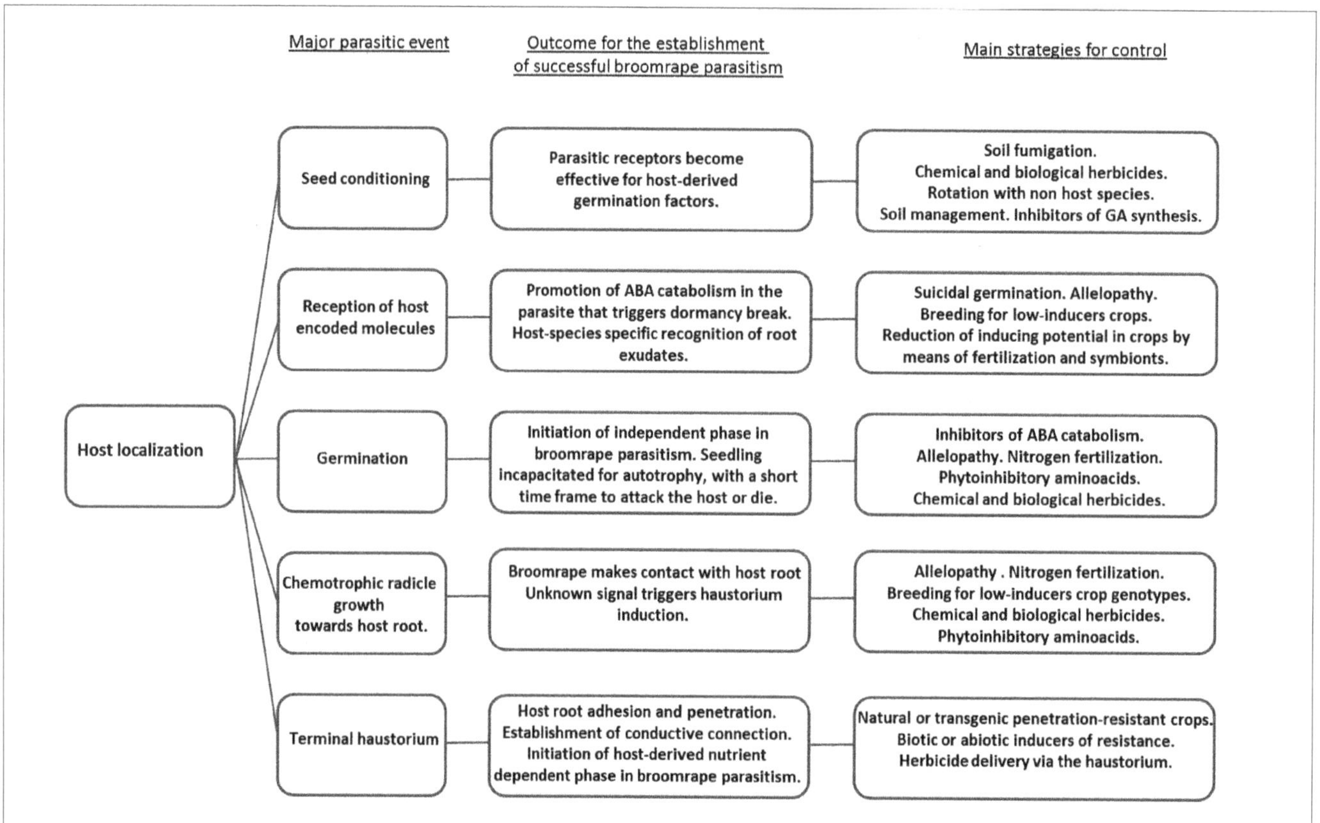

FIGURE 1 | Flowchart showing major underground parasitic events developed by broomrape weeds on susceptible crops and the control strategies that successfully target them.

organ, and subsequently as invasive organ toward host vascular system where finally establishes vascular continuity allowing the parasite to withdraw water and nutrients from the host (Riopel and Timko, 1995; Joel, 2013). The terminal haustorium develops at the apex of the seedling radicle upon host recognition (Musselman, 1980; Joel and Losner-Goshen, 1994). This is a short and delicate stage where the parasite either connects with the host or dies due to nutrient exhaustion. As a consequence of the high risk of establishment failure in the seedling, broomrapes have evolved germination strategies that "predict" establishment potential based on host chemodetection (Vaucher, 1823). In the following sections we describe the key developmental stages in the subterranean broomrape life cycle. These stages constitute sites of broomrape metabolism at which it is possible to design successful strategies to inhibit its sophisticated parasitism.

Host Localization

Seed Conditioning

Broomrape seed bank remains viable in the soil for many years until germination is triggered by the coincidence of several physical and chemical factors that are indicative of environmental conditions for successful seedling establishment: i.e., the nearby growth of a host plant in a physiological stage susceptible for broomrape invasion and subsequent parasitic reproductive growth (Linke and Saxena, 1991; López-Granados and García-Torres, 1996, 1999). Although hard seed coat has been described as dormancy mechanism in newly formed broomrape seeds (López-Granados and García-Torres, 1996), water uptake and imbibition are performed quickly by mature seeds through the micropyle without the need of scarification (Bar-Nun and Mayer, 1993; Joel et al., 2012). Broomrape seed bank presents annual cycles of non-deep physiological dormancy induced by seasonal changes in climatic conditions. In non-parasitic plants, physiological dormancy can be relieved through stratification but in the case of broomrape weeds, two consecutive processes are required to release dormancy: an environment-dependent first step of warm stratification called the conditioning phase, and a host-dependent second step of chemodetection. The first step of conditioning promotes in the parasitic seed receptors the required sensitivity for the second step of host detection (Musselman, 1980; Kebreab and Murdoch, 1999; Lechat et al., 2012, 2015; Murdoch and Kebreab, 2013).

The length and temperature required to promote seed conditioning depends on the broomrape species but are usually described under laboratory conditions in a range of 4–12 days at a temperature of 19–23°C, in dark and humid conditions (Kebreab and Murdoch, 1999; Gibot-Leclerc et al., 2004; Lechat et al., 2012). This treatment in the lab mimics the soil conditions in climatically suitable regions for broomrape such as Mediterranean non-irrigated agrosystems where the onset of warm and wet season coincides with the growth of juvenile stages of many annual crops (López-Granados and García-Torres, 1996; Grenz and Sauerborn, 2007). It is important for broomrape to initiate parasitism in young crops otherwise host reproductive organs in the rapid seed-filling stage will be able to endure a delayed parasitism by establishing a stronger competition with parasitic sinks (Manschadi et al., 1996; Fernández-Aparicio et al., 2009a, 2012a). Therefore broomrape seeds timely gain sensitivity for host chemodetection by means of conditioning (López-Granados and García-Torres, 1996). In absence of host detection the continuation of wet conditions allows broomrape seeds to enter again in deeper levels of dormancy, from which they will emerge upon the new onset of sequenced dry/wet seasons carrying new opportunities to encounter suitable hosts (Kebreab and Murdoch, 1999; López-Granados and García-Torres, 1999). Interestingly, experimentation carried out on broomrape species specialized on summer crops revealed their lower requirement for conditioning when compared with species specialized in winter annual crops highlighting the ecological adaptation of broomrape weeds to the cropping system in which they become specialized (Plakhine et al., 2009).

The metabolic activity of the seed conditioning in broomrape has been characterized in terms of patterns of respiration, synthesis and turnover of proteins, metabolism of nitrogen, carbohydrates and lipids and hormonal balance. Seed respiration patterns during conditioning indicate a strong activation of metabolism. Neither nitrogen nor lipid content change significantly during conditioning, while carbohydrate metabolism and protein synthesis seems to be crucial (Bar-Nun and Mayer, 1993, 2002; Mayer and Bar-Nun, 1994, 1997). It is well-established in autotrophic plants that abscisic acid (ABA) acts as a positive regulator of induction of seed dormancy and its maintenance and gibberelins (GAs) antagonizes with ABA, promoting dormancy release and subsequent germination (Finch-Savage and Leubner-Metzger, 2006). Accordingly, broomrape seed conditioning induces a decrease in ABA levels (Chae et al., 2004; Lechat et al., 2012) and GA synthesis (Joel et al., 1991; Zehhar et al., 2002). The reduction of ABA:GA ratio induced by stratification (conditioning) is enough to break dormancy and promote germination in dormant seeds of non-parasitic weeds but it is not enough for broomrape, which requires a further decrease in ABA levels induced by the activation of the ABA catabolic gene *PrCYP707A1* (Lechat et al., 2012). This gene remains silenced during conditioning phase and its activation occurs mediated by host-encoded germination stimulants, i.e., strigolactones, only after the conditioning phase is complete. Gain of host sensitivity in broomrape seeds at the end of the conditioning phase is mediated by demethylation of *PrCYP707A1* promoter. Besides the demethylation of *PrCYP707A1* promoter required for host-dependent *PrCYP707A1* expression, the high levels of global DNA demethylation observed at the end of conditioning period suggest that the epigenetic process occurring during the conditioning phase may be targeting other unknown molecules during conditioning. Whether the demethylation and host stimulation are independent or related processes remains to be clarified (Lechat et al., 2015).

Host-Induced Germination

The timing of germination is the most crucial event that obligated parasitic plants face along their life cycle (**Figure 2C**). Their absolute dependence on host-derived nutritive resources for successful seedling establishment and consequent growth makes necessary the synchronization of parasitic germination with the growth of its host. In order to achieve such synchrony they evolved mechanisms that release seed from dormancy triggering germination upon detection of specific molecules contained in host root exudates (Vaucher, 1823).

Broomrape species display high diversity with regard to their host range. Host specificity in broomrape species is usually indirectly related to the predictability of nutritive resources. The predictability of establishment on perennial hosts is high and therefore wild broomrape species feeding off perennial plants have narrow host ranges. On the contrary, weedy broomrape species are usually generalists attacking annual crops (Schneeweiss, 2007). The first mechanism involved in host specialization is displayed during broomrape germination and is mediated by the broomrape recognition of host root exudates in a species-specific manner. For each broomrape-crop association, broomrape germination potential is defined by the combination of both, the stimulatory capability of crop root exudates and the sensitivity of parasitic receptors to recognize specific forms of germination-inducing factors (Fernández-Aparicio et al., 2008a, 2009b, 2011).

The stimulatory capability of crop root exudates is defined by the qualitative and quantitative content of germination-inducing factors and varies across crop species and cultivars. In addition it also varies considerably in crops growing under different physiological status, growth stages and growing seasons, allowing broomrape to synchronize its germination with physiologically suitable hosts (López-Granados and García-Torres, 1996; Yoneyama et al., 2007a,b; Fernández-Aparicio et al., 2009b, 2014; Xie et al., 2010). Several classes of germination stimulants have been identified in root exudates such as strigolactones (Xie et al., 2010), peagol and peagoldione (Evidente et al., 2009), peapolyphenols A–C (Evidente et al., 2010), soyasapogenol B, *trans*-22-dehydrocampesterol (Evidente et al., 2011), dehydrocostus lactone (Joel et al., 2011), or isothyocyanates (Auger et al., 2012). The best studied group of germination-inducing factors are strigolactones, a group of terpenoid lactones. They are exuded by the crop to the rhizosphere under nutrient deficient conditions in order to promote symbiotic interactions (Akiyama et al., 2005). Parasitic plants eavesdrop the plant-to-symbiont communication to sense their hosts and germinate (Xie et al., 2010). Besides their role as extraorganismal signaling, recent research is uncovering new functions for strigolactones as plant hormone controlling crop development in response to the environment (Gomez-Roldan et al., 2008; Umehara et al., 2008). Post-germination development in broomrape could be probably regulated by their own broomrape-encoded strigolactones as it occurs in the close related parasite *Striga hermonthica* or in non-parasitic plants (Liu et al., 2014; Das et al., 2015). Although analytical chemistry methods have failed to detect strigolactones in parasitic plants (Liu et al., 2014), transcriptome

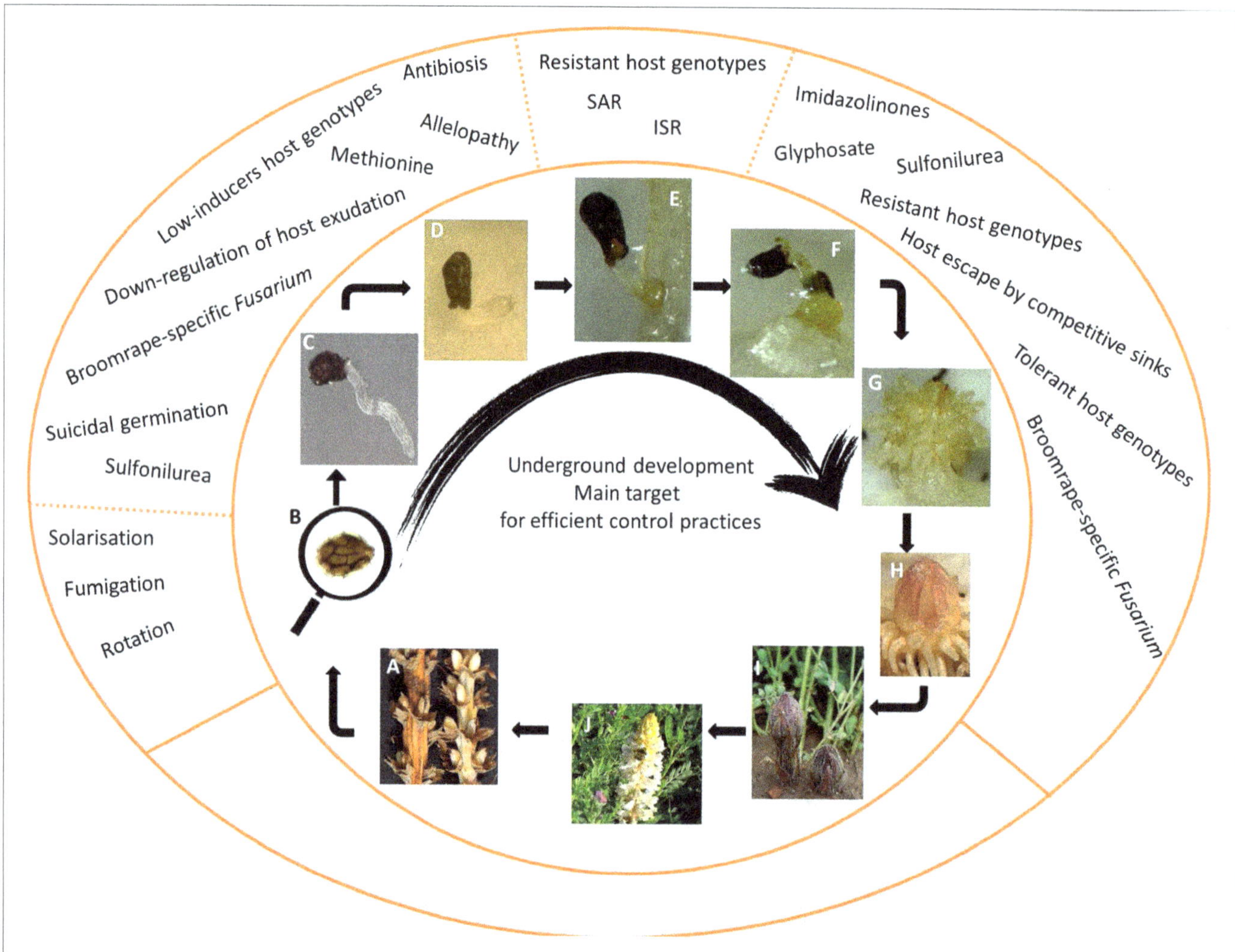

FIGURE 2 | Illustration of broomrape life stages and mechanisms of control. (A) Fructification and dehiscence of capsules containing mature seeds; **(B)** microscopic view of a seed (size ranging 0.2–2 mm) that undergoes sucessive dispersal, primary dormancy and annual release of secondary dormancy; **(C)** broomrape embryo does not develop morphologycaly identified cotyledons or shoot meristem and upon host-induced germination, only a radicle emerges from the seed with the function of searching and contacting the host root; **(D)** upon haustorial induction, the radicle stops elongating and a single terminal haustorium is differentiated. The first function of haustorium is as adhesion organ to host root surface mediated by a papillae cell layer; **(E)** adhesion to the root 3 days after germination induction; **(F)** upon vascular connection with the host, broomape initiates the development of the tubercle, the broomrape storage organ for host-derived nutrients. A swelling of the host root at the penetration point is also observed due the parasitic stimulation of host tissue proliferation; **(G)** tubercle develops a crown of adventicious roots; **(H)** tubercle differentiates apical shoot meristem (single shoot meristem for Orobanche species and several shoot meristems for Phelipanche species); **(I)** the underground shoot eventually emerges through the root surface; **(J)** flowering and pollination occur. Some broomrape species are outcrossers while others are self-pollinating. Reviewed in Joel et al. (2007).

sequencing reveals that all known strigolactone genes, both synthesis and perception are present in broomrapes with apparently full-length proteins (Péron et al., 2012; Das et al., 2015). The presence of strigolactone biosynthetic system in broomrapes raises the question on how the parasite performs diversified stimulant recognition in order to set the timing of germination. How broomrapes make the distinction not only between host-derived and their own-encoded strigolactones but also how they sense diversified strigolactone profiles in root exudates across species correlated with host ranges. Gene expression analysis could be indicating that parasitic plants down-regulate their synthesis of strigolactones at the end of conditioning period, and perhaps the creation of that internal deficit for broomrape-encoded strigolactones contributes to the broomrape sensitivity for external, host-derived strigolactones at the time of host detection (Das et al., 2015). In addition, the parasitic-specific receptor KAI2d that enables host detection in broomrapes has recently been identified. Multiple *KAI2d* genes across broomrape species genomes may allow diversified recognition of root exudates corresponding with suitable hosts (Conn et al., 2015). Additional mechanisms that could contribute to the selective action of host-derived strigolactones in broomrape germination could be (1) variations of molecular structure between host-derived and parasite-encoded strigolactones conferring different specificity for different biological functions or (2) different spatial localization inside the broomrape seed for functions of strigolactone detection and strigolactone synthesis (Das et al., 2015).

The requirement for germination-inducing factors in order to break dormancy in parasitic seeds are bypassed by ethylene or cytokinins (which promotes ethylene biosynthesis) in *Striga* sp. a close related parasitic weed genus, but these hormones are ineffective in promoting germination of broomrape weeds (Lieberman, 1979; Logan and Stewart, 1995; Berner et al., 1999; Joel, 2000; Toh et al., 2012). GA acts positively on germination in dormant non-parasitic species by counteracting ABA (Seo et al., 2009). However, exogenous application of GA alone is not sufficient to promote broomrape germination (Takeuchi et al., 1995; Chae et al., 2004) and strigolactone-mediated ABA catabolism in conditioned seeds is required to trigger germination (Lechat et al., 2012). A better understanding in the roles of major hormones in the process of broomrape germination would facilitate the design of feasible control strategies based on either inhibition of broomrape germination during crop cultivation or promotion of suicidal germination in the absence of the crop. In order to increase their applicability in low-input crops, the development of synthetic analogs of hormones would constitute a cheap alternative to natural bioregulators for seed bank control of weeds in general and parasitic weeds in particular.

Radicle Elongation and Haustorium Differentiation

The embryos in broomrapes have not morphologically identified cotyledons or shoot meristems and upon germination, only a radicle emerges through the seed coat with the only function of reaching and invading the host. The broomrape radicle shows no gravitropism and grows toward the host as a result of cell elongation. It has no root cap and does not develop procambium or conductive tissues (Joel and Losner-Goshen, 1994). Being deprived of the initiation of autotrophic mode of life, the growth of broomrape seedling toward the host is only sustained by water absorption and remobilization of reserve nutrients from the seed perisperm and endosperm (Joel, 2000; Joel et al., 2012). The seedling absorbs water both from the soil and from the seed endothelium, the later ensuring radicle development even in dry soil (Joel et al., 2012). The maximum radicle elongation is limited (1–5 mm) and its viability in the absence of host connection only last a few days after germination has been triggered (Veronesi et al., 2007). This spatial/temporal frame defines the maximum host-reaching distance for successful broomrape parasitism. Upon host detection, the broomrape radicle stops elongating and terminal haustorium is differentiated as an anchoring device. In this process, cellular expansion of the root meristem is redirected from longitudinal to radial and the root apex changes its form from conical to spherical. The external cell layer at the root tip differentiates into a papillate cell layer forming an adhesion epithelium (**Figure 2D**). The papillae form a crown around the apical cells that remain non-papillate but later will become intrusive cells with an essential function in the penetration process. This surface is covered by carbohydrate secretion that sticks the haustorium to the host surface. This structure is described as the external anchorage device of the pre-penetrated haustorium to the host surface (Joel and Losner-Goshen, 1994).

Close related parasitic plants of Orobanchaceae such as *Striga* and *Triphysaria* use host derived phenolic derivatives to induce haustorium differentiation (Riopel and Timko, 1995; Albrecht et al., 1999; Bandaranayake and Yoder, 2013). Hydrogen peroxide generated by parasitic radicles activates host peroxidases that catalyze the conversion of host cell walls into haustorium-inducing quinones (Keyes et al., 2000, 2007). Haustorium-inducing factors are structurally similar to allelopathic phytotoxins and gene expression of parasitic radicles exposed to haustorium-inducing factors is similar to that after radicle is exposed to phytotoxins (Tomilov et al., 2006). Parasitic plants probably evolved to recruit plant defense molecules as host recognition cues (Atsatt, 1977; Matvienko et al., 2001; Bandaranayake and Yoder, 2013). In broomrape species, the chemistry of host recognition for haustorium initiation remains uncharacterized. They have been traditionally considered the exception in parasitic Orobanchaceae that do not require host factors for haustorium initiation (Joel and Losner-Goshen, 1994; Bandaranayake and Yoder, 2013). However, results recently arisen from a molecule screening identified phytotoxins that induce the development of anchoring device in broomrape radicles (Cimmino et al., 2014, 2015). The significance of this structure in broomrape parasitism requires further investigation. A better understanding of the biochemistry of host recognition in broomrape will facilitate the generation of control strategies targeting the haustorium development.

Host Invasion

After host adhesion to host root surface the haustorium develops its invasive function of penetrating the host root (**Figure 2E**). The apical cells in the radicle apex develop into intrusive cells, which successively invade host root cortex, endodermis, and the central cylinder. During the host penetration process, broomrape does not dissolve the host cells in its way toward vascular cylinder. Mechanical force exerted by the haustorium development toward host vascular cylinder combined with enzymatic secretion promotes the separation of host cells without their lysis (Privat, 1960; Ben-Hod et al., 1993; Sholmer-Ilan, 1993; Singh and Singh, 1993; Antonova and Ter Borg, 1996; Bar-Nun et al., 1996; Losner-Goshen et al., 1998; Veronesi et al., 2005). Although broomrape pre-vascular connections benefits from host nutrients, the growth of broomrape in its way toward vascular cylinder is mainly sustained by consumption of seed reserves (Aber et al., 1983; Joel and Losner-Goshen, 1994; Joel, 2000). If the vascular connection is not successfully performed in few days the parasitic seedling dies of inanition and therefore quick invasion of the host is of advantage to avoid loss of viability. When resistant crops impose barriers to stop the parasitic development at this stage, broomrape exhausts and parasitism is quickly aborted. The chemical characteristics of the barriers of resistance to broomrape penetration have been extensively studied in Fabaceae crops (Pérez-de-Luque et al., 2009) and are reviewed in this article in Section "Resistant Crops to Broomrape Invasion."

Establishment of Vascular Continuity and Parasitic Sink

Shortly after host penetration and connection, the parasite begins its heterotrophic growth at the expense of host resources. Broomrape acts as a strong sink, depriving the host from water, mineral, and organic nutrients with the consequent negative impact on the growth of the host plant (Manschadi et al., 1996; Hibberd et al., 1998; Joel, 2000; Abbes et al., 2009). Few days after host vascular connection, the part of the broomrape seedling that remains outside the host root develops into a storage organ called tubercle. As the tubercle matures a crown of adventitious roots will emerge from this tubercle carrying capacity of developing lateral haustorial connections. Underground shoots will also develop from the tubercles that will eventually emerge through the soil surface leading into the development of reproductive organs (**Figures 2F–J**).

The transfer of nutrients from host to broomrape is performed through a continuous vascular system at the host-parasite interface. A continuous phloem system between broomrape and its host has been microscopically observed at the terminal haustoria. Sieve elements of both organisms are already interconnected by interspecific sieve pores at early stages of parasitism. It allows the parasite to quickly start tapping carbohydrates, amino acids, and organic acids from its host (Dörr and Kollmann, 1995; Nandula et al., 2000; Abbes et al., 2009). Although host phloem supplies the majority of nutrients including minerals, open xylem connections developed at the host-parasite interface allow additional mineral and water flow toward the parasite (Abbes et al., 2009; Westwood, 2013). These connections are probably developed from simultaneous differentiation of adjacent host and parasite cells to xylem elements (Dörr, 1997). The differentiation of xylem elements in the parasite are under the control of polar auxin transport (Harb et al., 2004; Bar-Nun et al., 2008). During the grafting between host and parasite, broomrape assumes the role of a root, orientating vascular tissues from the host shoot into itself (Bar-Nun et al., 2008). Acquisition of water is driven by a lower water potential in broomrape tissues (Ehleringer and Marshall, 1995). This is maintained by accumulation of solutes mainly potassium at higher concentrations than in the corresponding host tissues (Abbes et al., 2009). Regarding carbon assimilation broomrape takes it from the host phloem mainly in the form of sucrose (Aber et al., 1983; Hibberd et al., 1999). Once in the parasite system, sucrose is not accumulated but metabolized to other compounds. Cleavage of sucrose into glucose and fructose doubles the osmotic potential of the parasite. Sucrose is also metabolized to starch that is accumulated in the broomrape storage organ, the tubercle (Abbes et al., 2009; Draie et al., 2011).

Nitrogen metabolism remains largely unknown in broomrape. Activity of some nitrogen assimilating enzymes has been reported low in broomrapes. Nitrate reductase is not detectable (Lee and Stewart, 1978) and activity of glutamine synthetase is very low (McNally et al., 1983). Broomrape tubercles accumulate host-derived nitrogen in the form of either arginine or in the arginine and aspartate pair (Nandula et al., 2000; Abbes et al., 2009). Besides arginine and aspartate, other major forms of amino acids translocate from the host phloem but they are rapidly utilized by broomrape. Maintenance of relative low levels of those amino acids in tubercles either by low levels of synthetase activities (McNally et al., 1983) and or their rapid turnover of host-derived amino acids, establishes a decreasing concentration gradient that favors the unloading of amino acids into the parasite (Abbes et al., 2009). Despite the reports of broomrape inefficient machinery for nitrogen assimilation and broomrape dependence for host-derived organic forms of nitrogen demonstrated by the fact that broomrape growth is arrested when feeding on host cultivars with decreased amino acid-phloem levels (Abbes et al., 2009), inhibition of enzymes at the top of amino-acid biosynthetic pathway by means of either direct inhibitory action of herbicides (Gressel, 2009) or by feedback inhibition induced by amino-acid end-products (Vurro et al., 2006) are able to kill broomrape. This seems to indicate contribution of amino acid synthesis in broomrape mediated by broomrape-encoded enzymes although their identification and characterization remain unknown (Gressel, 2009; Eizenberg et al., 2012).

Understanding the key processes of host recognition, haustorium development and maturation and metabolic regulation of the parasitic sink allow virulence predictions and the design and implementation of highly calibrated, feasible, and durable control strategies leading to the arrest of broomrape parasitism minimizing simultaneously environmental impact and yield losses.

CONTROL STRATEGIES TARGETING UNDERGROUND BROOMRAPE STAGES

Successful broomrape control should target the underground broomrapes at their earlier life stages, prior attachment or as soon as it attach to the host, because of their highest vulnerability at those stages and the avoidance of yield loss in the current crop. The majority of strategies aimed to manage autotrophic weeds do not necessarily work for broomrapes and those that provide a degree of success for broomrape need to be optimized for each broomrape-crop species combination, local environmental conditions and broomrape population. Due to the high broomrape fecundity, long seed viability and for some weedy broomrape species, broad host range, the seed bank is easily replenished and long lasting. Therefore an integrated and sustained management strategy composed by several control methods acting at different broomrape life stages is highly recommended to keep away the broomrape weed problem in a durable manner (Kebreab and Murdoch, 2001). The following sections and **Table 1** review the major feasible control measures for broomrape control.

TABLE 1 | Methods for *Orobanche* and *Phelipanche* spp. control.

Technique		Broomrape stage targeted	Feasibility	Drawbacks or side effects
Cultural control				
Rotation including	Trap crops Catch crops Allelopathic crops	Seed germination Pre-attached seedling Young attachments	To avoid seed bank replenishing the rotation should avoid • Susceptible crops • Broomrape-susceptible non-parasitic weeds	• Miscalculation of optimal harvesting time for catch crops leads to broomrape multiplication • Avoidance of susceptible crops that in most cases are the most economically advantageous • Slow process requiring long term management perspective • Risk of widen broomrape host range
Intercropping susceptible crops with allelopathic species		Seed germination Pre-attached seedling	• Requires appropriate selection of sowing density and non-host component for each broomrape species • Requires greater management skills • Once optimized, rather easy to set up	• Yield uncertainty • Partial effect within a growing season
Fertilization	Nitrogen	Direct effect on seed germination and pre-attached stages Indirect effect on seed bank: negative regulation of host synthesis and exudation of germination-inducing factors	• Urea and ammonium but not nitrate forms inhibit broomrape seed germination and radicle elongation • Nitrogen regulates host exudation of germination-inducing factors in some species such as sorghum, lettuce, vetch, and wheat but not in important broomrape hosts such as clover, alfalfa, and tomato	• Environmental pollution • Potentially inappropriate Nitrogen supply for the crop species • Lower levels of symbiotic interactions in legume crops
	Phosphorus	Indirect effect on seed bank: negative regulation of host synthesis and exudation of germination-inducing factors	• At low or high pH the P solubility is reduced • Requires correct placement of the fertilizer both for easy access to crop root (P is immobile) and to avoid toxic effect to the crop seed (DAP-containing fertilizers produce free ammonia)	• Environmental pollution • Limited global resources. • Lower levels of symbiotic interactions in crops
Delayed sowing		Ungerminated seed bank (lower stimulatory capability in root exudates of mature crops growing late in the season) Young attachments (lower parasitic sink strength competing with rapid seed-filling stage of host fruits)	• Requires proper timing to obtain the best balance between inhibition of parasitic sink strength and crop productivity	• Shorter growing cycles not only reduces the amount of biomass partitioned to the parasite but also the overall crop productivity
Physical control				
Solarization		Ungerminated seed bank	• Requires high solar radiation	• Expensive • Plastic waste • For intercropping periods

(Continued)

TABLE 1 | Continued

Technique			Broomrape stage targeted	Feasibility	Drawbacks or side effects
Chemical control					
Soil fumigation	Methyl bromide		Ungerminated seed bank	• Banned by international agreement	• Environmental pollution • Expensive • Labor intensive
	Metham sodium Dazomet 1,3-dicloropropene		Ungerminated seed bank	• Tested as substitutes of methyl bromide but they are more expensive and less effective	• Environmental pollution • Expensive • Many active ingredients removed from the market for their unacceptable environmental profile under recent legislation • Labor intensive
Soil herbigation	Synthetic herbicides	Sulfonylurea	Seed germination Radicle elongation Young attachments	• Requires proper timing and application technology. • Requires a crop tolerant to sulfonylurea residues	• Environmental pollution. • High risk of appearance of sulfonylurea-resistance races.
	Inducers of suicidal broomrape germination	Synthetic analogs of strigolactones Fluridone GA agonists	Ungerminated seed bank	• Requires proper timing, application technology • Should be applied in absence of susceptible crops in order to avoid further increase of parasitism	• Though suicidal germination has been proved successful for *Striga* weed control in USA, it has not reached acceptable levels of seed bank reduction in broomrape
	Broomrape-specific phytotoxic amino acids	Methionine Lysine	Seed germination Radicle elongation Crop invasion	• Green and non-toxic • Requires proper timing, application technology, and tolerant crops • Commercially available at large scale as animal feed supplements • Mixed applications of two inhibitory amino acids can bring synergistic effect on weed control and reduced risk of resistance breakdown adaptation	• Soil pH alteration • Difficult application at the required broomrape toxic concentration (5 mM) • Technique under experimental development
	Toxins from microbial origin	Deoxynivalenol Diacetoxyscirpenol Fusarenon X HT-2 toxin Neosolaniol Nivalenol Roridin A T-2 toxin Verrucarrins A, B and M	Seed germination Radicle elongation	• Molecules of biological origin are biodegradable in comparison with synthetic herbicides • Requires proper timing and application technology • Requires drip-irrigation system for site-specific weed management and avoidance of toxin spread • Broomrape-specific toxins are preferred	• Lack of broomrape selectivity • Lack of knowledge on the physiological mechanism responsible for inhibitory action • Lack of knowledge on the persistence in the soil • Technique under experimental development • Lack of large scale production
Foliar application of systemic herbicides		Glyphosate Imidazolinones Sulfonylurea	Attached parasites (herbicide reaches the parasite via the haustorium)	• Requires proper timing and application technology • Requires crops with herbicide-resistance not based on metabolic degradation or inactivation • Where herbicide-resistance varieties are not available within a crop, repeated application at low doses are effective for young parasitic stages • Exact knowledge of the underground phenology related with thermal time is required to predict the best application timing	• Marginal crop selectivity • Environmental pollution • Injury of reproductive meristems in some crops • Sublethal concentrations of glyphosate suppress phytoalexin-based immunity in legumes toward pathogens • GM crop paradigm

(Continued)

TABLE 1 | Continued

			Technique		Broomrape stage targeted	Feasibility	Drawbacks or side effects
Biological control							
Broomrape-specific pests and pathogens	Fungi and bacteria	Wild forms of fungal and bacterial pathogens	Solid formulation	Wheat, corn or rice seeds Granular formulation (microbial agent+nutrients)	All stages from un-germinated seed bank to broomrape inflorescences	• Risk of low uniform distribution and control efficiency • Control efficiency can be increased by "multiple pathogen strategy" • Control efficiency can be increased with abiotic elicitors of resistance, e.g., BTH or sublethal herbicidal doses	• Allows only one application (at planting) • Moisture is limiting factor • Low uniform distribution and control efficiency • Regularity of the result can hardly be ascertained because of biotic interactions prevailing in the local conditions
			Microbigation		All stages from un-germinated seed bank to broomrape inflorescences	• Requires drip-irrigation system for site-specific weed management • Repeated applications of *Fusarium* suspensions provide moisture and optimization of concentration and timing	• Still under scientific evaluation. PGPR area to which this technique is related remains largely unexplored
		Enhanced-virulence pathogens	Hypervirulence transgenes	Enzymes that overcome broomrape defense mechanisms Broomrape-specific toxins	All stages from un-germinated seed bank to broomrape inflorescences	• Increased broomrape kill rate in comparison with their wild forms • Enhanced virulence should not alter specificity for broomrape weed • Spread of transgenic organisms could be avoided with the selection of asporogenic mutants	• Still under scientific evaluation. Techniques remain largely unexplored
			Pathogens overproducing and excreting phytotoxic amino acids				
	Insects		*Phytomyza orobanchia*		Avoids seed bank replenishing	• Broomrape-specific • Efficiency can be lowered by several agricultural practices and parasitoids • Efficient control requires multiple applications over several years • No artificial diet for rearing is available • Easiness to add to other agricultural controlling practices	• Low impact in a short term seed bank management • No impact in the productivity of the current crop
Symbiotic organisms			*Rhizobium leguminosarum* *Glomus mosseae*		Seed germination Radicle growth Crop invasion	• Requires optimal control of fertilization	• Strategy in its early developmental stage

(Continued)

TABLE 1 | Continued

	Technique	Broomrape stage targeted	Feasibility	Drawbacks or side effects
Host resistance				
Breeding for natural resistance	Broomrape-resistant varieties	Seed germination Host attachment Crop invasion	• Low cost of implementation once developed • Environmental compatible • Varieties with pyramiding gene resistance are preferred to avoid resistance breakdown by the parasite • Resistant varieties should be used in an integrated management strategy to avoid resistance breakdown by the parasite	• Not available for all crops • Risk of emergence of new virulent races
	Broomrape-tolerant varieties	Parasites attached to the host	• Feasible • Environmental compatible	• Not available for all crops
	Systemic herbicide-resistant varieties	Young parasitic seedling connected to host vascular system	• Feasible if herbicide resistance is not based on metabolic degradation or inactivation • Requires proper timing and application technology	• Not available for all crops • Risk of emergence of herbicide-resistant broomrape populations
Mutagenised crops	EMS-mutagenesis Fast-neutron-mutagenesis	Seed germination Host attachment Crop invasion	• Low cost of farm implementation of resistant varieties • Environmental compatible	• Risk of emergence of new virulent races
Transgenic resistance	Transgenes encoding for toxic products to broomrape	Young broomrapes	• Low cost of farm implementation of resistant varieties • Better results are obtained when using broomrape-responsive gene promoters	• Technique under experimental development • Public acceptance in some countries
	RNA interference	Young attached parasites (broomrape-specific RNAi reaches via the haustorium)	• Low cost • Environmental compatible • Requires previous identification of parasitic genes essential for broomrape virulence and metabolism • Resistant varieties should be used in an integrated management strategy to avoid resistance breakdown by the parasite	• Genetic redundancy may dilute silencing effect • Technique under experimental development
	Transgenic herbicide resistance	Attached parasites (herbicide reaches the parasite via the haustorium)	• Herbicide-resistance mechanism must be based in mechanism other than metabolic degradation or inactivation of herbicide by the transgenic crop • Low herbicide doses are required to kill the young parasite	• Food safety issues • Herbicide resistant-gene transfer to wild plants • Public acceptance in some countries

Management Strategies to Protect Crops from Detection by Broomrape Seeds

Due to the small size of the seeds and their inability to develop autotrophy, the establishment probability of broomrape seedlings is very low. Broomrapes counteract the high risk of failure in establishment on a host with highly evolved mechanisms of survival. Broomrape high fecundity, with thousands of seeds released per broomrape plant (**Figures 2A,B**), multiplies the chances of the next generation to encounter a host and achieve successful parasitism (Parker and Riches, 1993). In addition long lived seed banks under physiological dormancy ensure that germination will occur when a suitable host in its correct stage of development is present nearby (Rubiales et al., 2009b). Preventing the movement of parasitic seeds from infested to non-infested agricultural fields, by contaminated machinery or seed lots, is crucial (Panetta and Lawes, 2005). Once a field is infested, controlling the broomrape seed bank is very difficult due to its high resilience. A variety of methods have been developed to specifically neutralize broomrape pre-attached development though the majority of them are not commercially implemented because they are still at the stage of development or have not proved enough efficiency or applicability for large scale crops. These methods can be classified as cultural and physical, chemical, biological control, and use of host resistance (Rubiales et al., 2009b).

Cultural and Physical Control Practices

Fertilization can induce soil suppressiveness to initiation of broomrape parasitism. Application of phosphate or nitrogen to deficient soil reduces broomrape parasitism on clover and tomato (Southwood, 1971; Jain and Foy, 1992). Nutrients influence the crop-parasite pre-attached interaction in several ways. Direct toxic effects by urea and ammonium but not nitrate forms inhibit broomrape seed germination and radicle elongation (Jain and Foy, 1992; Abu-Irmaileh, 1994; van Hezewijk and Verkleij, 1996; Westwood and Foy, 1999). Accumulation of ammonium can be toxic to plants and its detoxification occurs via incorporation into organic compounds. The activity of glutamine synthetase in broomrape is very low and therefore carries a reduced broomrape ability to detoxify ammonium. Urea has no detrimental effects in plants but it is toxic to broomrape pre-attached stages probably exercised via ammonium after broomrape urease hydrolyses urea into ammonium. Nitrate is not toxic to broomrape as it lacks the ability to convert nitrate into ammonium (van Hezewijk and Verkleij, 1996). In addition to the toxic effects on broomrape seed and seedling, fertilization can protect crops from broomrape parasitism by means of down-regulating the crop synthesis and exudation of strigolactones, the most potent germination-inducing factors for broomrape. Phosphorous and nitrogen have been described to down regulate strigolactones exudation in some crop species (Yoneyama et al., 2007a,b, 2012). This effect may not be applicable to those broomrape species with preference for classes of germination-inducing factors other than strigolactones (Joel et al., 2011; Auger et al., 2012).

Intercropping systems cultivate simultaneously more than one species in close association to take agronomic advantage of biodiversity, competition, and complementarity between them. For broomrape control, this system seeks the simultaneous cultivation of susceptible host species with inhibitory species of broomrape parasitism. The release of phytochemicals by the roots of the allelopathic component in the intercrop inhibits the broomrape germination and/or radicle elongation toward the host component. Successful reduction of broomrape parasitism in the current crop is obtained by intercropping host species with inhibitory species of cereals, fenugreek, or berseem clover (Fernández-Aparicio et al., 2007, 2008b, 2010a). Careful selection of the non-host component in the intercrop is, however, required as some plant species can act as non-host facilitators and therefore increase the severity of broomrape infection in the host component (Gibot-Leclerc et al., 2013).

A rotation decreasing the frequency of host cultivation is one of the main ways that farmers deal with the broomrape-related problem. This prevents broomrape parasitism from taking place, maintaining the seed bank dormant and reducing the rate of seed bank replenishing. However, it is a long-term strategy due to the long viability of seed bank (Rubiales et al., 2009b), which requires at least a nine-course rotation in order to prevent broomrape seed bank increases (Grenz et al., 2005). Its efficacy for broomrape cultural control can be increased if the farmer includes trap and/or catch crops as components in the rotation (Rubiales et al., 2009b). The concept of trap crops refers to the cultivation of crop species whose root exudates exhibit high germination-inducing activity on broomrape seeds, but these species do not become infected because they are resistant to later stages of the parasitic process indirectly leading to the killing of the young broomrape seedlings due to the lack of proper host. The inductor potential of root exudates from a given species varies with the broomrape considered. Each broomrape species show specificity not only for root exudates in order to germinate but also for host species to invade and feed on, being the germination-stimulatory range usually broader than the actual host range (Fernández-Aparicio et al., 2009b). Potential trap crops have been suggested for broomrape weeds (Parker and Riches, 1993). For instance, root exudates of field pea induces high germination of the very destructive broomrape species *O. crenata, O. foetida, O. minor,* and *P. aegyptiaca,* however, it only becomes infected by *O. crenata* therefore pea may theoretically be a good trap crop against *O. foetida, O. minor,* and *P. aegyptiaca* but not for *O. crenata* infested field (Fernández-Aparicio and Rubiales, 2012). Many other interesting examples of trap crops emerged from a root exudates screening of important crops (Fernández-Aparicio et al., 2009b). The second possibility to increase rotation efficacy for broomrape control is to include catch crops, which are crops that also induce high broomrape germination but they are not resistant to it. On the contrary, they must be highly susceptible, as the farmer is the one with the role of stopping the parasitic process by harvesting the catch crop as a green vegetable before the parasite emerges. For instance, tori (*Brassica campestris* var. *toria*) when managed properly as a catch crop can result in up to a 30% reduction in the size of broomrape seed bank (Acharya et al., 2002).

Soil management affects the success of broomrape seeds in becoming established on the host and then the longevity of broomrape seed bank. Minimum tillage reduces the amount of viable seeds incorporated in the soil and then their capacity to reach the crop root system (Ghersa and Martinez-Ghersa, 2000; López-Bellido et al., 2009). The opposite agricultural practice deep-plowing, has been suggested to bring seeds of parasitic weeds to a depth with less oxygen availability and therefore a reduction in its germination capacity (Van Delft et al., 2000). According with pot experiments carried out in the tomato-*P. aegyptiaca* system, deep-plowing bringing the seeds to depth $\geq$ 12 cm will strongly reduce broomrape infection severity in terms of number of parasites, total parasitic biomass, delayed broomrape emergence and prevention of flower initiation and seed set (Eizenberg et al., 2007).

Solarization is a thermal soil disinfestation method that shows high efficiency reducing the viability of the broomrape seed bank along with other harmful organisms to crops such as plant-parasitic nematodes, disease causing microorganisms and non-parasitic weeds. This method consists in heating the soil through sun energy achieving temperatures above 45°C, by covering a wet soil with transparent polyethylene sheets for a period of 4–8 weeks during the warmest season (Katan, 1981; Mauro et al., 2015). Many beneficial organisms are either able to survive the solarization treatment or able to recolonize solarized soil (Sauerborn et al., 1989; Mauromicale et al., 2001). Not all areas infested by broomrape are suitable for efficient solarization. Hot air temperature and clear skies are required during the solarization period. Its high cost per surface unit makes this method not readily applicable at large scale (Joel, 2000). In addition, this technique generates a considerable amount of plastic waste but the emergence of new materials at low-cost, of biological origin and biodegradable may in the future reduce earth pollution with plastic debris derived from agriculture practices (Fernandez and Ingber, 2013).

Chemical Control of Seed Bank

Soil fumigation with methyl bromide has been proved one of the most effective methods to eradicate broomrape seed bank, but this chemical has been banned from use due to its toxic effects on the environment (Joel, 2000; Hershenhorn et al., 2009). Use of other soil sterilants such as metham sodium, dazomet, and 1,3-dichloropropene have shown different degrees of efficacy but their high cost, complex application and negative environmental effects have prevented their widespread use by farmers (Foy et al., 1989; Goldwasser et al., 1995; Hershenhorn et al., 2009) or conducted to the withdrawal of authorization, at least in some countries.

Soil herbigation (saturating the soil with herbicides such as sulfonylureas) effectively controls preattached stages of broomrapes (Hershenhorn et al., 2009) but is hardly compatible with other agricultural cropping practices as detrimental for many crop seedlings for several weeks or months. *In vitro* treatments of a large range of sulfonylurea herbicides inhibit broomrape germination and radicle elongation (Hershenhorn et al., 1998; Plakhine et al., 2001). Incorporation of sulfosulfuron and rimsulfuron directly to the soil provides successful control of preattached stages of broomrape weeds (Eizenberg et al., 2012).

Natural pesticides derived of microbial and plant origin are considered to be less harmful because they usually biodegrade quicker, resulting in less pollution-related problems. Several toxins have been identified with inhibitory activity on broomrape parasitism by interfering with broomrape germination and radicle elongation (Vurro et al., 2009; Fernández-Aparicio et al., 2013; Cimmino et al., 2014). Special interest arises from those metabolites with a favorable pattern of broomrape-specific effect (e.g., tenuazonic acid) and no described side-effect to other biosystems (Vurro et al., 2009). However, the efficacy of these molecules has been proved only in laboratory essays. Until now, difficulties of purification at industrial scale have hampered the field experimentation with such metabolites (Vurro et al., 2009) despite their interesting potential.

Certain amino acids strongly inhibit the early development of broomrape without phytotoxic effects in the host (Vurro et al., 2006). The amino acid approach to control weeds is inspired on the concept of frenching disease where amino acid end-product inhibits the activity of a controlling enzyme in the amino acid biosynthesis pathway (Vurro et al., 2006, 2009; Sands and Pilgeram, 2009). The effectiveness of amino acids as broomrape inhibitors has not been proved in real field conditions but field application of amino acids has been effective to manage other parasites such as plant-parasitic nematodes (Zhang et al., 2010). Among the amino acids producing the highest and most consistent inhibitory effects on broomrape germination and radicle elongation, some, such as methionine are being produced in large commercial scale as animal feed supplements. The use of those amino acids as pesticide is classified by the United States Environmental Protection Agency as innocuous to public and environment health (USEPA, 2004). Based on those conditions, methionine has the potential to be used as broomrape herbicide but it needs to be confirmed and its application adjusted to real field conditions.

Synthetic analogs of growth regulators can be successfully used to reduce parasitism by hampering the synchronization of the parasitic seed bank with the growth of the host. Inhibition of seed conditioning and subsequent germination mediated by inhibitors of GA synthesis reduces the receptivity of broomrape seeds to germination-inducing factors. In addition, inhibitors of ABA catabolism inhibit the germination-triggering effect of host-derived strigolactones. For example, soil application of uniconazole, a triazole that is commercially used for growth regulation has proved to reduce parasitism by inhibiting seed conditioning and subsequent germination (Joel, 2000; Zehhar et al., 2002; Song et al., 2005; Lechat et al., 2012).

Promotion of suicidal germination is the technique used to induce broomrape germination with synthetic molecules in the absence of a host to which broomrape can attach, a technique lethal for the parasite as the broomrape seedling is unable to acquire autotrophy. Therefore, it may be possible to achieve broomrape control by fooling the parasite with the delivering to the soil of synthetic analogs of the original host-derived germination-inducing factors such as strigolactones (Johnson et al., 1976). Direct application of strigolactones to the soil has been the subject of intense research. The first attempts to deplete parasitic weed seed bank was made by Johnson et al. (1976) by using the synthetic strigolactone analog GR7. However, instability of this compound, particularly at pH > 7.5, and lack of optimal formulations rendered this technique not applicable (Saghir, 1986; Babiker et al., 1987, 1988). Recent advances in this research area has led to new, more stable strigolactone analogs and optimization of field application protocols and formulations (Bhattacharya et al., 2009; Zwanenburg et al., 2009; Mwakaboko and Zwanenburg, 2011). Another strategy to induce suicidal germination of broomrape seed bank could be the use of gibberellin agonists. They elicit GA-like germination activity in dormant seeds of several autotrophic plant species (Suttle and Schreiner, 1982; Metzger, 1983), constituting a cheap alternative to natural bioregulators for weed seed bank control (Suttle, 1983). When they are applied *in vitro* to seeds of *P. ramosa* and *O. minor*, they bypass the effect of germination-inducing factors, promoting broomrape germination in absence of host or any germination stimulant (Cala et al., 2015).

Once broomrape germination has occurred, chemicals that reduce the growth of broomrape radicle reduce the chances of reaching the host and therefore parasitism. A novel metabolite, ryecyanatine-A recently isolated from rye (*Secale cereale* L.), presents potential for broomrape control by promoting rapid cessation of broomrape radicle growth and therefore inhibiting its ability to reach the host. In addition it promotes the development of a layer of papillae at the radicle apex in the absence of host contact, morphology that resembles the attachment organ (Joel and Losner-Goshen, 1994; Cimmino et al., 2015). Because the haustorial organ in broomrape radicle is terminal and its growth is not resumed unless it can immediately penetrate the host, cessation of radicle elongation and haustorial induction in the absence of a host is lethal to the parasite. Other interesting molecules that hamper the ability of broomrape radicle to reach the host have been recently discovered from different microbial and plant origins (Fernández-Aparicio et al., 2013; Cimmino et al., 2014).

Biological Control of Seed Bank with Living Organisms

Biological control of broomrape is based on the use of living organisms either by killing seed bank or interfering with its host-recognition ability. Assessment of pathogenicity or damages toward non-target plants has to be carefully assessed in order to avoid environmental risks. The efficient action of the biological control agent will depend on its ability to remain active over a large range of ecological conditions (Aly, 2007).

Phytophagous insects to prevent the build-up of broomrape seed bank

More than 40 insect herbivores from 22 families have been collected on broomrape plants but a majority of them are polyphagous without any specificity for broomrape species being some of them serious pests of important crops (Klein and Kroschel, 2002). *Phytomyza orobanchia* is reported to be broomrape-specific and its main action as biocontrol agent is by reduction of broomrape reproductive activity due to their feeding activity on ovules and young seeds. Often secondary infections by fungi cause early death of broomrape shoots or limit the development of flowers and ovules (Klein and Kroschel, 2002). The capacity of *P. orobanchia* to reduce broomrape populations is limited by cultural practices and antagonists (Klein and Kroschel, 2002; Aly, 2007).

Mycoherbicides attacking broomrape seeds and radicles

The broomrape seed bank efficiency to initiate parasitism can be reduced by incorporation to the soil of several pathogens able to infect preattached broomrape stages such as *Fusarium* sp. or *Ulocladium botrytis* (Müller-Stöver, 2001; Boari and Vurro, 2004; Dor and Hershenhorn, 2009). Like most seeds, broomrape seeds are resistant to rapid microbial degradation due to phenols located in its testa (Cezard, 1973). However, hyphae of specific pathogens are able to penetrate the seed coat of broomrape dormant seeds, dissolving the endosperm cell walls and metabolizing the cytoplasm. The advantage of this approach using fungi is that it can be used in absence of host cultivation (Thomas et al., 1999). The parasitic weed radicle that emerges from germinated seed and carries the attachment organ is also targeted by those mycoherbicides (Abbasher and Sauerborn, 1992). This approach would inhibit parasitism by killing the young seedling before it attaches to the host root. The control of broomrape by mycoherbicides does not so far provide the level of control required in highly infested soils (Aly, 2007). Novel approaches can increase broomrape control by fungi. A "multiple-pathogen strategy" in which two or more pathogens are combined has been proved successful for the control of broomrape causing a synergistic effect that can lead to 100% broomrape control (Dor and Hershenhorn, 2003; Müller-Stöver et al., 2005). Refined formulations and encapsulations of fungal propagules increase efficacy in biocontrol by reducing desiccation or microbial competition (Amsellem et al., 1999; Quimby et al., 1999; Kroschel et al., 2000; Müller-Stöver, 2001; Aybeke et al., 2015). Engineering of virulence-enhanced mycoherbicides is another approach of great interest. This approach is based on the selection of naturally occurring mutants that overproduce and excrete an enhanced amount of specific amino acid with broomrape inhibition properties on seed germination and radicle growth (Vurro et al., 2006; Sands and Pilgeram, 2009).

Bacteria as biocontrol agents

Pseudomonas aeruginosa, P. fluorescens, Bacillus atrophaeus, B. subtilis are promising biocontrol agents targeting the growth of broomrape radicles (Barghouthi and Salman, 2010). In addition to this direct effect, ethylene-producing bacteria such as *Pseudomonas syringae* pv. *glycinea* induce ethylene-mediated suicidal germination in *Striga* sp. (Berner et al., 1999; Ahonsi et al., 2003), a close relative of broomrapes, however, broomrape germination is not responsive to ethylene (Joel, 2000). *Azospirillum brasilense* is reported to inhibit broomrape radicle growth (Dadon et al., 2004).

Microbial interactions that interfere on broomrape ability to recognize its host

Broomrape seeds are less capable to recognize crop roots colonized by arbuscular mycorrhizal fungi, *Rhizobium leguminosarum* or *Azospirillum brasilense* due to change in the composition of the root exudates in colonized plants

(Dadon et al., 2004; Mabrouk et al., 2007a; Fernández-Aparicio et al., 2009c, 2010b; Louarn et al., 2012). A reduced content of broomrape germination-inducing factors in root exudates of mycorrhizal plants has been demonstrated (López-Ráez et al., 2011). Those interactions promote the broomrape seed bank remains dormant inhibiting the initiation of broomrape parasitism, and therefore its rates of seed bank replenishment.

Low-Inducers Crop Genotypes

Breeding for broomrape resistance stands out as the most economic, easy to adopt and environmentally friendly practice. Because parasitic weeds require host encoded molecules to stimulate the initiation of parasitism both at the level of seed germination and haustorium initiation, breeding for low-inducers genotypes of those processes are obvious targets for resistance (Yoder and Scholes, 2010). Sources of natural resistance based on low exudation of germination-inducing factors exist in legumes and sunflower and are highly effective in inhibiting broomrape weed parasitism (Labrousse et al., 2001, 2004; Rubiales et al., 2003b, 2009a; Pérez-de-Luque et al., 2005; Sillero et al., 2005; Abbes et al., 2010; Fernández-Aparicio et al., 2012b, 2014).

Sources of natural resistance based on reduced release of haustorium-inducing factors is a doubly interesting strategy to inhibit broomrape parasitism because not only it prevents broomrape parasitism in the current crop, but also it promotes the demise of the seed bank by promoting suicidal germination. Sources of low-inducers genotypes exist in crops species attacked by the close related parasitic weed *Striga* (Rich et al., 2004). It remains unknown whether host factors are required by broomrape radicle to initiate haustorium and consequently this strategy has not been fully explored.

Control Strategies Targeting Host Penetration

The major strategy that specifically impedes the broomrape capacity to penetrate the host once the radicle has made contact with host root, is based on the use of host resistance, either genetic resistance obtained by breeding (Pérez-de-Luque et al., 2009; Yoder and Scholes, 2010), or induced resistance by abiotic or biotic agents (Gonsior et al., 2004; Kusumoto et al., 2007).

Resistant Crops to Broomrape Invasion

Depending on the genetic background of the resistant host, the intrusive cells of broomrape seedling can be stopped at three different levels in their way of penetration through the root layers to achieve connection with the host vascular system. The first barriers are imposed at the cortex level with reinforced cell walls mediated by either protein cross-linking or with the deposition of metabolites such as suberin, or callose. In addition, accumulation of toxic phenolic compounds at the infection point can be observed in some resistant varieties. Resistance that occurs in the endodermis is mediated by lignification of endodermal and pericycle cell walls. Resistance that occurs in the central cylinder is related with accumulation of phenolic compounds in the surrounding tissues and nearby xylem vessels inducing a toxic release near the parasite impeding vascular connection (Pérez-de-Luque et al., 2009). However, selecting for high phenolic varieties is likely to induce many other side changes altering agronomic performance.

Lack of knowledge in the molecular regulation of the host-parasite interaction during crop invasion has impeded the development of varieties carrying transgenes with capacity to inhibit broomrape penetration. As alternative, transgenic resistant crops have been engineered with broomrape-inducible expression of toxins specifically targeting the penetrating broomrape seedling. This strategy to abort broomrape invasion requires regulating the toxin production with promoters specifically induced around the site of *Orobanche* penetration such as the *HMG2* promoter, ensuring correct delivery of the toxic effect to the broomrape penetrating seedling and overall low concentration of the toxin in the rhizosphere. As a consequence the crop is protected from broomrape invasion (Joel and Portnoy, 1998; Westwood et al., 1998; Hamamouch et al., 2005; Aly et al., 2006).

Abiotic Inducers of Resistance

Induced disease resistance mediated by endogenous salicylic acid (SA) also described as systemic acquired resistance (SAR) induces hypersensitive responses in many plant species against fungal, bacterial and viral diseases. SA promotes resistance to broomrape. Abiotic inducers of SAR thus represent an innovative approach to control broomrape parasitism. Benzo-(1,2,3)-thiadiazole-7-carbothioic acid *S*-methyl ester (BTH) acts as a functional analog of SA and activates defense responses in susceptible hosts leading to lignification of the endodermis and a consequent inhibition to up to 98% broomrape parasitism (Gonsior et al., 2004; Pérez-de-Luque et al., 2004; Kusumoto et al., 2007). Commercially available as Bion®, field doses of 0.8 kg ha^{-1} are recommended to inhibit *P. ramosa* parasitism in hemp and tobacco (Gonsior et al., 2004), crops for which resistant varieties are not available. In other pathosystems, amino acids such as D-L-β-amino-*n*-butyric acid or L-methionine induce resistance in crop plants against pathogen attack. This resistance is coordinated with the expression of genes encoding for pathogenesis-related proteins (Sarosh et al., 2005; Hasabi et al., 2014). The ability of L-methionine to stop the entrance of broomrape intrusive cells into the host-root layers has not been studied. However, when Vurro et al. (2006) applied L-methionine in pots to tomato roots the number of broomrape seedlings that successfully developed parasitism was highly reduced. Though, the effect of L-methionine on internal crop resistance was not studied and requires further investigation. If this effect is confirmed, L-methionine use to elicit resistance to broomrape in susceptible crops could be a straightforward strategy either by direct applications of this amino acid in the soil as explained in Section "Control Strategies Targeting Host Penetration" or delivered by overproducing and excreting microorganisms as explained in Section

"Strategies to Control Underground Broomrapes Acting after Establishment."

Biotic Inducers of Resistance

Biotic inducers of systemic resistance have also proved being successful against broomrape parasitism under experimental conditions. *Rhizobium leguminosarum* induces defense mechanisms based on elevated induction of the phenylpropanoid pathway conferring mechanical and chemical barriers to the parasite penetration (Mabrouk et al., 2007a,b,c, 2010). Although the effect of jasmonic-acid-dependent induced systemic resistance (ISR) against parasitic plants is less clear (Kusumoto et al., 2007; Hiraoka et al., 2009; Yoder and Scholes, 2010), strains of *Pseudomonas* sp. inducers of ISR (Gozzo, 2003) and commercially available as Proradix® can reduce broomrape parasitism by 80% in susceptible cultivars of hemp and tobacco without phytotoxic effect on the crop (Gonsior et al., 2004). Based on the results obtained in their greenhouse experiments, these authors recommended field doses of 1.6 kg ha^{-1} for crop densities of 32,000 tobacco plants ha^{-1}. Unfortunately this technique represents another example of highly promising broomrape control strategy that has never been validated in field experiments.

Strategies to Control Underground Broomrapes Acting after Establishment

Once broomrape has established connection with the vascular system of its hosts, broomrape management should be performed quickly to abort at earlier stages the strong parasitic sink for nutrients and water. Major feasible strategies for controlling broomrape and gain productivity in the current crop are those based on cultural practices that promote host scape to parasitic damage by improving host sink competitiveness, selective chemical control of the parasite via the haustorium, and host resistance based in physical, chemical barriers and physiological incompatibility. Some of the strategies discussed in previous sections such as biological control maintain their control action at post-attachment stages and will not be discussed again in this section.

Cultural Control of Post-attached Parasites

Delayed sowing date is a traditional method that can show high degree of success on inhibiting parasitism if implemented correctly (López-Granados and García-Torres, 1996; Rubiales et al., 2003a; Pérez-de-Luque et al., 2004; Grenz et al., 2005). This technique promotes the host plant to fulfill its required thermal time to flower in a shorter number of days, making the grain filling period shorter. The host reproductive sinks compete earlier and stronger against the parasitic sink and in consequence less nutritive resources are allocated to the parasite (Manschadi et al., 1996). However, the overall productivity of the host-parasite system is also reduced due to the shorter growing period being detrimental for crop yield. Therefore, decisions on the date of sowing has to be well-adjusted in order to balance the loss of productivity due to shorter growing period with gain of productivity due to reduced parasitism. Delaying sowing date has, however, a general drawback by reducing yield potential under normal development so that plant breeding program tend generally to favor long lasting cultivars with early sowing dates.

Besides date of sowing, nutrient management can promote both tolerance and increased resistance in crops to broomrape parasitism (Parker, 2009; Labrousse et al., 2010). Broomrape attack is more severe on crops growing in low fertility soils. Besides the effects of fertilization management on pre-attached broomrape stages described in previous sections, high soil fertility can induce crops to endure broomrape parasitism by helping the host to maintain a favorable osmotic potential that reduces the parasitic sink strength (Gworgwor and Weber, 1991).

Chemical Control of Post-attached Parasites

Four broomrape features define the post-attachment herbicidal strategy in comparison with non-parasitic weeds. First, broomrape weeds are achlorophyllous and therefore those herbicides that target photosynthetic process, e.g., triazines or substituted urease [C group in the Herbicide Resistance Action Committee (HRAC) classification], will have only limited effect on broomrapes. Second, broomrape weed exerts their damage underground right after attachment and therefore, contact herbicides applied after broomrape emergence, e.g., 2,4-D, had no effect on limiting yield loss in the current crop. Third, broomrape underground attachments do not take herbicides from the soil but only systemically from the host and therefore, this strategy is limited to systemic herbicides applied to herbicide-resistant crop varieties that do not metabolize the herbicide into inactive forms. And four, despite reports on broomrape inefficient machinery for nitrogen assimilation, and on amino acid fluxes from the host phloem to the parasite, herbicides inhibiting amino acid biosynthesis in the parasite via suppressive action on broomrape-encoded acetolactate synthase (ALS) and enolpyruvylshikimate phosphate synthase (EPSPS) enzymes are able to kill broomrape.

Current chemical control of post-attached broomrape life stages is mainly achieved with foliar applications of systemic herbicides inhibiting ALS (imidazolinones, sulfonylureas) or EPSPS (glyphosate) to the leaves of crop varieties carrying target-site resistances to those herbicides to avoid direct injury of their metabolism. The target-site herbicide-resistance is based on a modification of the enzyme in such a way that it binds to its normal substrate in the amino acid biosynthesis pathway but not to the herbicide. This kind of resistance is more interesting than other mechanisms of resistance that usually involve translocation and enhanced metabolism, resulting in lower herbicide concentration in the sap of the host plant. With target-site resistance, the herbicide translocates unmetabolised to the underground broomrape via the haustorium inflicting its suppressive action in the parasite (Gressel, 2009).

Target-site resistances have been successfully developed in crops either by classical breeding such as sunflower, by screening mutagenized crop populations such as the case of oilseed rape or by transgenic techniques such as tomato, tobacco, carrots, and oilseed rape (Joel et al., 1995; Aviv et al., 2002; Slavov et al., 2005; Tan et al., 2005). Such target-site resistance is also available in other broomrape-susceptible crops but remains to be tested and registered to control broomrape. Engineered host crops harboring herbicide-resistance transgenes have not yet been commercialized for broomrape management (Gressel, 2009[2]).

An alternative to the selective use of herbicides when target-site resistance is not available for a specific crop is the touchy use of repeated applications of non-selective herbicidal doses to promote sublethal effects for the crop but lethal effects to the initial stages of post-attached parasitism (Foy et al., 1989). This strategy requires a careful calibration of doses and timing depending on the host crop and underground phenology of broomrape determined by local conditions and crop (Hershenhorn et al., 1998, 2009; Eizenberg et al., 2006). Systemic translocation of nanoencapsulated herbicides could improve this herbicidal approach (Pérez-de-Luque and Rubiales, 2009).

Resistant and Tolerant Cultivars

Death of the young broomrape tubercles shortly after nutritive flow initiation has been observed in cultivars carrying post-haustorial resistance in the form of growth arrest and necrosis of young tubercles. Several mechanisms underlying this resistance have been described at this stage such as production of gel-like substances within host vessels blocking the transfer of nutrients, host-encoded toxic-compounds delivered into the parasitic tissue though the vascular system and hormonal incompatibility that leads to abnormal haustorial maturation with scarce vascular connections (Fernández-Aparicio et al., 2008c; Pérez-de-Luque et al., 2008, 2009). Those mechanisms kill the broomrape either by inducing toxic effects or by starving the parasite.

Tolerant varieties are able to endure infection with minor losses on productivity. Crops that reach their seed filling period earlier than broomrape initiates its underground bud development are able to restrict parasitic sink and endure parasitic damage (Manschadi et al., 1996; Grenz et al., 2005; Fernández-Aparicio et al., 2009a, 2012a). In addition, some modifications of host biochemistry have been described in tolerant crops inducing low performance of the parasite when attached. High osmotic potential in roots and drop in amino acid levels in the phloem has been reported in tolerant varieties of faba bean in response to broomrape parasitism. The consequent reduced flux of water and nutrients toward the parasite, low utilization of host-derived sucrose and lower levels of soluble proteins limits the parasitic sink strength and yield losses due to broomrape parasitism (Abbes et al., 2009).

CONCLUSION

If left uncontrolled during one or a few seasons, broomrape weeds build a hardly destructible seed bank in agricultural soils that further renovates at a rate of millions of seeds per ha each year a susceptible crop is infested. In addition, their mixed traits of weed and underground pathogen, make their control tricky. Even the easiest method of control, herbicides, requires broomrape specific-optimization for each cropping system to target the most vulnerable broomrape life stage, the young attachments while preserving the crop. It seems more and more obvious that a single strategy has low probability to control broomrapes. Instead an integrated control program including a battery of broomrape-specific measurements is preferable.

We have seen that several opportunities to stop the cycle of the parasite have been explored. The development of the solutions has usually not been conducted to their end so that many potential ways of controlling broomrape are not on the market. What we have often seen is that the solution has to propose a modification that makes the parasitic life cycle unfit to that of the crop. Still, as the parasite is synchronized on the crop development this means in some cases that the change disfavoring the parasite could also limit the maximum potential yield for the crop. As a consequence, except when deeply infested, the farmer (and thus the market) will not retain a solution that has economical negative drawbacks. This may well-explain why some several decades of parasitic plant research have not end up with satisfying and largely available tools for controlling this parasitic plant.

The advances yielded as intense research made connects the major critical steps of the life cycle of *Orobanche,* the external factors influencing it either through molecular dialog between the parasite and the crop or the soil and climatic environmental conditions naturally opens the way toward the potential effect of the cropping system in limiting broomrape parasitism: choice of the crop, timing, plant protection, soil perturbation, fertilization, etc. This lead us to build the list of the major feasible components that a model designed to quantify the effects of cropping systems on pest dynamics should include for specific broomrape control. Such a model would be a valuable tool to synthesize knowledge on broomrape life-cycle, to design and test management strategies and better predict the variability in effects observed for a given environment and set of agricultural practices. One future development would be to evaluate what could be the emerging risk at cultivating different crops, one of which may stimulate germination while the other offers opportunities for haustorium fixation. This would open the work on parasitism toward more community ecology and what can be considered the realistic nature of parasitism.

AUTHOR CONTRIBUTIONS

MF-A wrote the paper. XR and SG-L additional text, editing, and comments.

ACKNOWLEDGMENTS

This work was cofunded by the European Union and INRA, in the framework of the Marie-Curie FP7 COFUND People Program, through the award of an AgreenSkills' fellowship (under grant agreement n° PCOFUND-GA-2010-267196) to MF-A with additional support by the INRA Division "Santé des Plantes et Environnement."

[2]http://www.isb.vt.edu

REFERENCES

Abbasher, A. A., and Sauerborn, J. (1992). *Fusarium nygamai* a potential bioherbicide for *Striga hermonthica* control in sorghum. *Biol. Control* 2, 291–296. doi: 10.1016/1049-9644(92)90021-5

Abbes, Z., Kharrat, M., Delavault, P., Chaïbi, W., and Simier, P. (2009). Nitrogen and carbon relationships between the parasitic weed *Orobanche foetida* and susceptible and tolerant faba bean lines. *Plant Physiol. Biochem.* 47, 153–159. doi: 10.1016/j.plaphy.2008.10.004

Abbes, Z., Kharrat, M., Pouvreau, J. B., Delavault, P., Chaibi, W., and Simier, P. (2010). The dynamics of faba bean (*Vicia faba* L.) parasitism by *Orobanche foetida*. *Phytopathol. Mediterr.* 49, 239–248.

Aber, M., Fer, A., and Salle, G. (1983). Transfer of organic substances from the host plant *Vicia faba* to the parasite *Orobanche crenata* Forsk. *Z. Planzenphysiol.* 112, 297–308. doi: 10.1016/S0044-328X(83)80047-6

Abu-Irmaileh, B. E. (1994). Nitrogen reduces branched broomrape (*Orobanche ramosa*) seed germination. *Weed Sci.* 42, 57–60.

Abu-Irmaileh, B. E., and Labrada, R. (2009). *The Problem of Orobanche spp. in Africa and Near East*. Available at: www.fao.org/ag/AGP/AGPP/IPM/Weeds/Issues/orobanche.htm

Acharya, B. D., Khattri, B. G., Chettri, M. K., and Srivastava, X. (2002). Effect of *Brassica campestris* var. toria as a catch crop on *Orobanche aegyptiaca* seed bank. *Crop Prot.* 21, 533–537. doi: 10.1016/S0261-2194(01)00137-5

Ahonsi, M. O., Berner, D. K., Emechebe, A. M., Lagoke, S. T., and Sangina, N. (2003). Potential of ethylene-producing pseudomonads in combination with effective N2-fixing bradyrhizobial strains as supplements to legume rotation for *Striga hermonthica* control. *Biol. Control* 28, 1–10. doi: 10.1016/S1049-9644(03)00051-3

Akiyama, K., Matsuzaki, K. I., and Hayashi, H. (2005). Plant sesquiterpenes induce hyphal branching in arbuscular mycorrhizal fungi. *Nature* 435, 824–827. doi: 10.1038/nature03608

Albrecht, H., Yoder, J. I., and Phillips, D. A. (1999). Flavonoids promote haustoria formation in the root parasite *Triphysaria versicolor*. *Plant Physiol.* 119, 585–591. doi: 10.1104/pp.119.2.585

Aly, R. (2007). Conventional and biotechnological approaches for control of parasitic weeds. *In Vitro Cell. Dev. Biol. Plant* 43, 304–317. doi: 10.1007/s11627-007-9054-5

Aly, R., Plakhin, D., and Achdari, G. (2006). Expression of sarcotoxin IA gene via a root-specific tob promoter enhanced host resistance against parasitic weeds in tomato plants. *Plant Cell Rep.* 25, 297–303. doi: 10.1007/s00299-005-0052-y

Amsellem, Z., Zidack, N. K., Quimby, Jr P. C, and Gressel, J. (1999). Long term dry preservation of active mycelia of two mycoherbicidal organisms. *Crop Prot.* 18, 643–649. doi: 10.1016/S0261-2194(99)00070-8

Antonova, T. S., and Ter Borg, S. J. (1996). The role of peroxidase in the resistance of sunflower against *O. cumana* in Russia. *Weed Res.* 36, 113–121. doi: 10.1111/j.1365-3180.1996.tb01807.x

Atsatt, P. R. (1977). The insect herbivore as a predictive model in parasitic seed plant biology. *Am. Nat.* 111, 579–586. doi: 10.1086/283185

Auger, B., Pouvreau, J. B., Pouponneau, K., Yoneyama, K., Montiel, G., Le Bizec, B., et al. (2012). Germination stimulants of *Phelipanche ramosa* in the rhizosphere of *Brassica napus* are derived from the glucosinolate pathway. *Mol. Plant Microbe Interact.* 25, 993–1004. doi: 10.1094/MPMI-01-12-0006-R

Aviv, D., Amsellem, Z., and Gressel, J. (2002). Transformation of carrots with mutant acetolactate synthase for *Orobanche* (broomrape) control. *Pest Manag. Sci.* 58, 1187–1193. doi: 10.1002/ps.567

Aybeke, M., Şen, B., and Ökten, S. (2015). Pesta granule trials with *Aspergillus alliaceus* for the biocontrol of *Orobanche* spp. *Biocontrol Sci. Technol.* 25, 803–813. doi: 10.1080/09583157.2015.1018813

Babiker, A. G. T. (2008). *Orobanche crenata* in Sudan: history, distribution and management. *Haustorium* 53, 1–3.

Babiker, A. G. T., Ahmed, E. A., Dawoud, D. A., and Abdrella, N. K. (2007). *Orobanche* species in Sudan: history, distribution and management. *Sudan J. Agric. Res.* 10, 107–114.

Babiker, A. G. T., Hamdoun, A. M., Rudwan, A., Mansi, N. G., and Faki, H. H. (1987). Influence of soil moisture on activity and persistence of the strigol analogue GR 24. *Weed Res.* 27, 173–178. doi: 10.1111/j.1365-3180.1987.tb00751.x

Babiker, A. G. T., Ibrahim, N. E., and Edwards, W. G. (1988). Persistence of GR7 and *Striga* germination stimulant(s) from *Euphorbia aegyptiaca* Boiss. in soils and in solutions. *Weed Res.* 28, 1–6. doi: 10.1111/j.1365-3180.1988.tb00778.x

Bandaranayake, P. C. G., and Yoder, J. I. (2013). "Haustorium initiation and early development," in *Parasitic Orobanchaceae*, eds D. M. Joel, L. J. Musselman, and J. Gressel (Berlin: Springer), 61–74.

Barghouthi, S., and Salman, M. (2010). Bacterial inhibition of *Orobanche aegyptiaca* and *Orobanche cernua* radical elongation. *Biocontrol Sci. Technol.* 20, 423–435. doi: 10.1080/09583150903340544

Barker, E. R., Press, M. C., Scholes, J. D., and Quick, W. P. (1996). Interactions between the parasitic angiosperm *Orobanche aegyptiaca* and its tomato host: growth and biomass allocation. *New Phytol.* 133, 637–642. doi: 10.1111/j.1469-8137.1996.tb01932.x

Barkman, T. J., McNeal, J. R., Lim, S. H., Coat, G., Croom, H. B., Young, N. D., et al. (2007). Mitochondrial DNA suggests at least 11 origins of parasitism in angiosperms and reveals genomic chimerism in parasitic plants. *BMC Evol. Biol.* 7:248. doi: 10.1186/1471-2148-7-248

Bar-Nun, N., Ben-Hod, G., Lavi, E., and Mayer, A. M. (1996). Purification of pectin methylesterase from *Orobanche aegyptiaca*. *Phytochemistry* 41, 403–406. doi: 10.1016/0031-9422(95)00594-3

Bar-Nun, N., and Mayer, A. M. (1993). Preconditioning and germination of *Orobanche* seeds: respiration and protein synthesis. *Phytochemistry* 34, 39–45. doi: 10.1016/S0031-9422(00)90779-9

Bar-Nun, N., and Mayer, A. M. (2002). Composition of and changes in storage compounds in *Orobanche aegyptiaca* seeds during preconditioning. *Isr. J. Plant Sci.* 50, 277–279. doi: 10.1560/E2KB-FM11-X4U2-YC9J

Bar-Nun, N., Sachs, T., and Mayer, A. M. (2008). A role for IAA in the infection of *Arabidopsis thaliana* by *Orobanche aegyptiaca*. *Ann. Bot.* 101, 261–265. doi: 10.1093/aob/mcm032

Ben-Hod, G., Losner-Goshen, D., Joel, D. M., and Mayer, A. M. (1993). Pectin methylesterase in calli and germinating seeds of *Orobanche aegyptiaca*. *Phytochemistry* 32, 1399–1402. doi: 10.1016/0031-9422(93)85145-H

Bennett, J. R., and Mathews, S. (2006). Phylogeny of the parasitic plant family Orobanchaceae inferred from phytochrome A. *Am. J. Bot.* 93, 1039–1051. doi: 10.3732/ajb.93.7.1039

Berner, D. K., Schaad, N. W., and Volksch, B. (1999). Use of ethylene producing bacteria for stimulating of *Striga* spp. seed germination. *Biol. Control* 15, 274–282. doi: 10.1006/bcon.1999.0718

Bhattacharya, C., Bonfante, P., Deagostino, A., Kapulnik, Y., Larini, P., Occhiato, E. G., et al. (2009). A new class of conjugated strigolactone analogues with fluorescent properties: synthesis and biological activity. *Org. Biomol. Chem.* 7, 3413–3420. doi: 10.1039/b907026e

Boari, A., and Vurro, M. (2004). Evaluation of *Fusarium* spp. and other fungi as biological control agents of broomrape (*Orobanche ramosa*). *Biol. Control* 30, 212–219. doi: 10.1016/j.biocontrol.2003.12.003

Cala, A., Rial, C., Fernandez-Aparicio, M., Molinillo, J. M. G., Varela, R. M., Rubiales, D., et al. (2015). "Phthalimide-lactones stimulate germination of parasitic weeds," in *Proceedings of the XXXV Biennial Meeting of the Spanish Royal Society of Chemistry*, eds J. A. S. López, E. I. Martínez, T. R. Blas, M. C. López, and J. P. Sestelo (A Coruña: Dario Prada-Rodríguez of University of A Coruña), 688.

Cezard, R. (1973). "Quelques aspects particuliers de la biologie des *Orobanches*," in *Proceedings of the European Weed Research Council on Parasitic Weeds*, eds W. G. H. Edwards, L. Kasasian, C. Parker, A. R. Saghir, and W. van der Zweep (Malta: Royal University of Malta), 55–67.

Chae, S. H., Yoneyama, K., Takeuchi, Y., and Joel, D. M. (2004). Fluridone and norflurazon, carotenoid-biosynthesis inhibitors, promote seed conditioning and germination of the holoparasite *Orobanche minor*. *Physiol. Plant.* 120, 328–337. doi: 10.1111/j.0031-9317.2004.0243.x

Cimmino, A., Fernandez-Aparicio, M., Andolfi, A., Basso, S., Rubiales, D., and Evidente, A. (2014). Effect of fungal and plant metabolites on broomrapes (*Orobanche* and *Phelipanche* spp.) seed germination and radicle growth. *J. Agric. Food Chem.* 62, 10485–10492. doi: 10.1021/jf504609w

Cimmino, A., Fernandez-Aparicio, M., Avolio, F., Andolfi, A., Rubiales, D., Yoneyama, K., et al. (2015). Ryecyanatines A and B and ryecarbonitrilines A and B, substituted cyanatophenol, cyanato-benzo[1,3] diole, and benzo[1,3]dioxolecarbonitriles from rye (*Secale cereale* L.) root exudates: new metabolites with allelophatic activity on *Orobanche* seed germination and radicle growth. *Phytochemistry* 109, 57–65. doi: 10.1016/j.phytochem.2014.10.034

Conn, C. E., Bythell-Douglas, R., Neumann, D., Yoshida, S., Whittington, B., Westwood, J. H., et al. (2015). Convergent evolution of strigolactone perception enabled host detection in parasitic plants. *Science* 349, 540–543. doi: 10.1126/science.aab1140

Dadon, T., Nun, N. B., and Mayer, A. M. (2004). A factor from *Azospirillum brasilense* inhibits germination and radicle growth of *Orobanche aegyptiaca*. *Isr. J. Plant Sci.* 52, 83–86. doi: 10.1560/Q3BA-8BJW-W7GH-XHPX

Das, M., Fernández-Aparicio, M., Yang, Z. Z., Huang, K., Wickett, N. J., Alford, S., et al. (2015). Parasitic plants *Striga* and *Phelipanche* dependent upon exogenous strigolactones for germination have retained genes for strigolactone biosynthesis. *Am. J. Plant Sci.* 6, 1151–1166. doi: 10.4236/ajps.2015.68120

De Candolle, A. P. (1813). *Théorie Elémentaire de la Botanique*. Paris: Déterville.

Dor, E., and Hershenhorn, J. (2003). The use of several phytopathogenic fungi for broomrape control. *Phytoparasitica* 31, 422.

Dor, E., and Hershenhorn, J. (2009). Evaluation of the pathogenicity of microorganisms isolated from Egyptian broomrape (*Orobanche aegyptiaca*) in Israel. *Weed Biol. Manage.* 9, 200–208. doi: 10.1111/j.1445-6664.2009.00340.x

Dörr, I. (1997). How *Striga* parasitizes its host: a TEM and SEM study. *Ann. Bot.* 79, 463–472. doi: 10.1006/anbo.1996.0385

Dörr, I., and Kollmann, R. (1995). Symplasmic sieve element continuity between *Orobanche* and its host. *Bot. Acta* 108, 47–55. doi: 10.1111/j.1438-8677.1995.tb00830.x

Draie, R., Péron, T., Pouvreau, J.-B., Véronési, C., Jégou, S., Delavault, P., et al. (2011). Invertases involved in the development of the parasitic plant *Phelipanche ramosa*: characterization of the dominant soluble acid isoform, PrSAI1. *Mol. Plant Pathol.* 12, 638–652. doi: 10.1111/j.1364-3703.2010.00702.x

Ehleringer, J. R., and Marshall, J. D. (1995). "Water relations," in *Parasitic Plants*, eds M. C. Press and J. Graves (London: Chapman and Hall), 125–140.

Eizenberg, H., Aly, R., and Cohen, Y. (2012). Technologies for smart chemical control of broomrape (*Orobanche* spp. and *Phelipanche* spp.). *Weed Sci.* 60, 316–323. doi: 10.1614/WS-D-11-00120.1

Eizenberg, H., Colquhoun, J. B., and Mallory-Smith, C. A. (2006). Imazamox application timing for small broomrape (*Orobanche minor*) control in red clover. *Weed Sci.* 54, 923–927. doi: 10.1614/WS-05-151R.1

Eizenberg, H., Lande, T., Achdari, G., Roichman, A., and Hershenhorn, J. (2007). Effect of Egyptian broomrape (*Orobanche aegyptiaca*) burial depth on parasitism dynamics and chemical control in tomato. *Weed Sci.* 51, 152–156. doi: 10.1614/WS-06-135

Evidente, A., Cimmino, A., Fernández-Aparicio, M., Andolfi, A., Rubiales, D., and Motta, A. (2010). Polyphenols, including the new peapolyphenols A–C, from root exudates stimulate *Orobanche foetida* seed germination. *J. Agric. Food Chem.* 58, 2902–2907. doi: 10.1021/jf904247k

Evidente, A., Cimmino, A., Fernández-Aparicio, M., Rubiales, D., Andolfi, A., and Melck, D. (2011). Soyasapogenol B and trans-22-dehydrocamposterol from common vetch (*Vicia sativa* L.) root exudates stimulate broomrape seed germination. *Pest Manag. Sci.* 67, 1015–1022. doi: 10.1002/ps.2153

Evidente, A., Fernández-Aparicio, M., Cimmino, A., Rubiales, D., Andolfi, A., and Motta, A. (2009). Peagol and peagoldione, two new strigolactone like metabolites isolated from pea root exudates. *Tetrahedron Lett.* 50, 6955–6958. doi: 10.1016/j.tetlet.2009.09.142

Fernandez, J., and Ingber, D. (2013). Bioinspired chitinous material solutions for environmental sustainability and medicine. *Adv. Funct. Mater.* 23, 4454–4466. doi: 10.1002/adfm.201300053

Fernández-Aparicio, M., Andolfi, A., Evidente, A., Pérez-de-Luque, A., and Rubiales, D. (2008a). Fenugreek root exudates show species-specific stimulation of *Orobanche* seed germination. *Weed Res.* 48, 163–168. doi: 10.1111/j.1365-3180.2007.00609.x

Fernández-Aparicio, M., Emeran, A. A., and Rubiales, D. (2008b). Control of *Orobanche crenata* in legumes intercropped with fenugreek (*Trigonella foenum-graecum*). *Crop Prot.* 27, 653–659. doi: 10.1016/j.cropro.2007.09.009

Fernández-Aparicio, M., Pérez-de-Luque, A., Prats, E., and Rubiales, D. (2008c). Variability of interactions between barrel medic (*Medicago truncatula*) genotypes and *Orobanche* species. *Ann. Appl. Biol.* 153, 117–126. doi: 10.1111/j.1744-7348.2008.00241.x

Fernández-Aparicio, M., Emeran, A. A., and Rubiales, D. (2010a). Inter-cropping with berseem clover (*Trifolium alexandrinum*) reduces infection by *Orobanche crenata* in legumes. *Crop Prot.* 29, 867–871. doi: 10.1016/j.cropro.2010.03.004

Fernández-Aparicio, M., García-Garrido, J. M., Ocampo, J. A., and Rubiales, D. (2010b). Colonization of field pea roots by arbuscular mycorrhizal fungi reduces *Orobanche* and *Phelipanche* species seed germination. *Weed Res.* 50, 262–268. doi: 10.1111/j.1365-3180.2010.00771.x

Fernández-Aparicio, M., Flores, F., and Rubiales, D. (2009a). Field response of *Lathyrus cicera* germplasm to crenate broomrape (*Orobanche crenata*). *Field Crops Res.* 113, 321–327. doi: 10.1016/j.fcr.2009.06.009

Fernández-Aparicio, M., Flores, F., and Rubiales, D. (2009b). Recognition of root exudates by seeds of broomrape (*Orobanche* and *Phelipanche*) species. *Ann. Bot.* 103, 423–431. doi: 10.1093/aob/mcn236

Fernández-Aparicio, M., Soto, M. J., Rubiales, D., Ocampo, J. A., and García-Garrido, J. M. (2009c). The potential of *Rhizobium* mutants for biological control of *Orobanche crenata*. *Haustorium* 54, 3–4.

Fernández-Aparicio, M., Flores, F., and Rubiales, D. (2012a). Escape and true resistance to crenate broomrape (*Orobanche crenata* Forsk.) in grass pea (*Lathyrus sativus* L.) germplasm. *Field Crops Res.* 125, 92–97. doi: 10.1016/j.fcr.2011.09.003

Fernández-Aparicio, M., Moral, A., Kharrat, M., and Rubiales, D. (2012b). Resistance against broomrapes (*Orobanche* and *Phelipanche* spp.) in faba bean (*Vicia faba*) based in low induction of broomrape seed germination. *Euphytica* 186, 897–905. doi: 10.1021/jf5027235

Fernández-Aparicio, M., Kisugi, T., Xie, X., Rubiales, D., and Yoneyama, K. (2014). Low strigolactone root exudation: a novel mechanism of broomrape (*Orobanche* and *Phelipanche* spp.) resistance available for faba bean breeding. *J. Agric. Food Chem.* 62, 7063–7071. doi: 10.1021/jf5027235

Fernández-Aparicio, M., and Rubiales, D. (2012). Differential response of pea (*Pisum sativum*) to *Orobanche crenata*, *Orobanche foetida* and *Phelipanche aegyptiaca*. *Crop Prot.* 31, 27–30. doi: 10.1093/jxb/err246

Fernández-Aparicio, M., Sillero, J. C., and Rubiales, D. (2007). Intercropping with cereals reduces infection by *Orobanche crenata* in legumes. *Crop Prot.* 26, 1166–1172. doi: 10.1016/j.cropro.2006.10.012

Fernández-Aparicio, M., Yoneyama, K., and Rubiales, D. (2011). The role of strigolactones in host specificity of *Orobanche* and *Phelipanche* seed germination. *Seed Sci. Res.* 21, 55–61. doi: 10.1017/S0960258510000371

Fernández-Aparicio, M., Cimmino, A., Evidente, A., and Rubiales, D. (2013). Inhibition of *Orobanche crenata* seed germination and radicle growth by allelochemicals identified in cereals. *J. Agric. Food Chem.* 61, 9797–9803. doi: 10.1021/jf403738p

Finch-Savage, W. E., and Leubner-Metzger, G. (2006). Seed dormancy and the control of germination. *New Phytol.* 171, 501–523. doi: 10.1111/j.1469-8137.2006.01787.x

Foy, C. L., Jain, R., and Jacobsohn, R. (1989). Recent approaches for chemical control of broomrape (*Orobanche* spp.) – a review. *Rev. Weed Sci.* 4, 123–152.

Ghersa, C. M., and Martinez-Ghersa, M. A. (2000). Ecological of weed seed size and persistence in the soil under different tilling systems: implications for weed management. *Field Crops Res.* 67, 141–148. doi: 10.1016/S0378-4290(00)00089-7

Gibot-Leclerc, S., Abdennebi-Abdemessed, N., Reibel, C., and Colbach, N. (2013). Non-host facilitators, a new category that unexpectedly favours parasitic weeds. *Agron. Sustain. Dev.* 33, 787–793. doi: 10.1007/s13593-013-0153-x

Gibot-Leclerc, S., Corbineau, F., Sallé, G., and Côme, D. (2004). Responsiveness of *Orobanche ramosa* L. seeds to GR24 as related to temperature, oxygen availability and water potential during preconditioning and subsequent germination. *Plant Growth Regul.* 43, 63–71. doi: 10.1023/B:GROW.0000038242.77309.73

Goldwasser, Y., Kleifeld, Y., Golan, S., Bargutti, A., and Rubin, B. (1995). Dissipation of metham-sodium from soil and its effect on the control of *Orobanche aegyptiaca*. *Weed Res.* 35, 445–452. doi: 10.1111/j.1365-3180.1995.tb01641.x

Gomez-Roldan, V., Fermas, S., Brewer, P. B., Puech-Pages, V., Dun, E. A., Pillot, J. P., et al. (2008). Strigolactone inhibition of shoot branching. *Nature* 455, 189–194. doi: 10.1038/nature07271

Gonsior, G., Buschmann, H., Szinicz, G., Spring, O., and Sauerborn, J. (2004). Induced resistance - an innovative approach to manage branched broomrape (*Orobanche ramosa*) in hemp and tobacco. *Weed Sci.* 52, 1050–1053. doi: 10.1614/WS-04-088R1

Gozzo, F. (2003). Systemic acquired resistance in crop protection: from nature to chemical approach. *J. Agric. Food Chem.* 51, 4487–4503. doi: 10.1021/jf030025s

Grenz, J. H., Manschadi, A. M., Uygurc, F. N., and Sauerborn, J. (2005). Effects of environment and sowing date on the competition between faba bean (*Vicia faba*) and the parasitic weed *Orobanche crenata*. *Field Crops Res.* 93, 300–313. doi: 10.1016/j.fcr.2004.11.001

Grenz, J. H., and Sauerborn, J. (2007). Mechanisms limiting the geographical range of the parasitic weed *Orobanche crenata*. *Agric. Ecosyst. Environ.* 122, 275–281. doi: 10.1016/j.agee.2007.01.014

Gressel, J. (2009). Crops with target-site herbicide resistance for *Orobanche* and *Striga* control. *Pest Manag. Sci.* 65, 560–565. doi: 10.1002/ps.1738

Gworgwor, N. A., and Weber, H. C. (1991). Effect of N-application on sorghum growth, *Striga* infestation and the osmotic presure of the parasite in relation to the host. *J. Plant Physiol.* 139, 194–198.

Hamamouch, N., Westwood, J. H., Banner, I., Cramer, C. L., Gepstein, S., and Aly, R. (2005). A peptide from insects protects transgenic tobacco from a parasitic weed. *Transgenic Res.* 14, 227–236. doi: 10.1007/s11248-004-7546-1

Harb, A. M., Hameed, K. M., and Shibli, R. A. (2004). Effect of triiodobenzoic acid on broomrape (*Orobanche ramosa*) infection and development in tomato plants. *Plant Pathol. J.* 20, 81–84. doi: 10.5423/PPJ.2004.20.2.081

Hasabi, V., Askari, H., Alavi, S. M., and Zamanizadeh, H. (2014). Effect of amino acid application on induced resistance against citrus canker disease in lime plants. *J. Plant Prot. Res.* 54, 144–149. doi: 10.2478/jppr-2014-0023

Hearne, S. J. (2009). Control - the *Striga conundrum*. *Pest Manag. Sci.* 65, 603–614. doi: 10.1002/ps.1735

Hershenhorn, J., Eizenberg, H., Dor, E., Kapulnik, Y., and Goldwasser, Y. (2009). *Phelipanche aegyptiaca* management in tomato. *Weed Res.* 49(Suppl. 1), 34–37. doi: 10.1111/j.1365-3180.2009.00739.x

Hershenhorn, J., Goldwasser, Y., Plakhine, D., Lavan, Y., Blumenfeld, T., Bucsbaum, H., et al. (1998). *Orobanche aegyptiaca* control in tomato fields with sulfonylurea herbicides. *Weed Res.* 38, 343–349. doi: 10.1046/j.1365-3180.1998.00105.x

Hibberd, J. M., Quick, W. P., Press, M. C., and Scholes, J. D. (1998). Can source-sink relations explain responses of tobacco to infection by the root holoparasitic angiosperm *Orobanche cernua*? *Plant Cell Environ.* 21, 333–340. doi: 10.1046/j.1365-3040.1998.00272.x

Hibberd, J. M., Quick, W. P., Press, M. C., Scholes, J. D., and Jeschke, W. D. (1999). Solute fluxes from tobacco to the parasitic angiosperm *Orobanche cernua* and the influence of infection on host carbon and nitrogen relations. *Plant Cell Environ.* 22, 937–947. doi: 10.1046/j.1365-3040.1999.00462.x

Hiraoka, Y., Ueda, H., and Sugimoto, Y. (2009). Molecular responses of *Lotus japonicus* to parasitism by the compatible species *Orobanche aegyptiaca* and the incompatible species *Striga hermonthica*. *J. Exp. Bot.* 60, 641–650. doi: 10.1093/jxb/ern316

Jain, R., and Foy, C. L. (1992). Nutrient effects on parasitism and germination of Egyptian broomrape (*Orobanche aegyptiaca*). *Weed Technol.* 6, 269–275.

Joel, D. M. (2000). The long-term approach to parasitic weeds control: manipulation of specific developmental mechanisms of the parasite. *Crop Prot.* 19, 753–758. doi: 10.1016/S0261-2194(00)00100-9

Joel, D. M. (2009). The new nomenclature of *Orobanche* and *Phelipanche*. *Weed Res.* 49, 6–7. doi: 10.1111/j.1365-3180.2009.00748.x

Joel, D. M. (2013). "The haustorium and the life cycles of parasitic Orobanchaceae," in *Parasitic Orobanchaceae*, eds D. M. Joel, J. Gressel, and L. J. Musselman (Heidelberg: Springer Berlin), 21–23.

Joel, D. M., Back, A., Kleifeld, Y., and Gepstein, S. (1991). "Seed conditioning and its role in *Orobanche* seed germination: inhibition by paclobutrazol," in *Progress in Orobanche Research. Proceedings of the International Workshop on Orobanche Research*, eds K. Wegmann and L. J. Musselman (Obermarchtal, FRG: Eberhard Karls Universität), 147–156.

Joel, D. M., Bar, H., Mayer, A. M., Plakhine, D., Ziadne, H., Westwood, J. H., et al. (2012). Seed ultrastructure and water absorption pathway of the root-parasitic plant *Phelipanche aegyptiaca* (Orobanchaceae). *Ann. Bot.* 109, 181–195. doi: 10.1093/aob/mcr261

Joel, D. M., Chaudhuri, S. K., Plakhine, D., Ziadna, H., and Steffens, J. C. (2011). Dehydrocostus lactone is exuded from sunflower roots and stimulates germination of the root parasite *Orobanche cumana*. *Phytochemistry* 72, 624–634. doi: 10.1016/j.phytochem.2011.01.037

Joel, D. M., Hershenhorn, J., Eizenberg, H., Aly, R., Ejeta, G., Rich, P. J., et al. (2007). Biology and management of weedy root parasites. *Hortic. Rev.* 33, 267–349. doi: 10.1002/9780470168011.ch4

Joel, D. M., Kleifeld, Y., Losner-Goshen, D., Herzlinger, G., and Gressel, J. (1995). Transgenic crops against parasites. *Nature* 374, 220–221. doi: 10.1038/374220a0

Joel, D. M., and Losner-Goshen, D. (1994). The attachment organ of the parasitic angiosperms *Orobanche cumana* and *O. aegyptiaca* and its development. *Can. J. Bot.* 72, 564–574. doi: 10.1139/b94-075

Joel, D. M., and Portnoy, V. H. (1998). The angiospermous root parasite *Orobanche* L. (Orobanchaceae) induces expression of a pathogenesis related (PR) gene in susceptible tobacco roots. *Ann. Bot.* 81, 779–781. doi: 10.1006/anbo.1998.0629

Johnson, A. W., Rosebery, G., and Parker, C. (1976). A novel approach to *Striga* and *Orobanche* control using synthetic germination stimulants. *Weed Res.* 16, 223–227. doi: 10.1111/j.1365-3180.1976.tb00406.x

Katan, J. (1981). Solar heating (solarization) control of soilborne pests. *Ann. Rev. Phytopathol.* 19, 211–236. doi: 10.1146/annurev.py.19.090181.001235

Kebreab, E., and Murdoch, A. J. (1999). A quantitative model for loss of primary dormancy and induction of secondary dormancy in imbibed seeds of *Orobanche* spp. *J. Exp. Bot.* 50, 211–219. doi: 10.1093/jxb/50.331.211

Kebreab, E., and Murdoch, A. J. (2001). Simulation of integrated control strategies for *Orobanche* spp. based on a life cycle model. *Exp. Agric.* 37, 37–51. doi: 10.1017/S001447970100401X

Keyes, W. J., O'Malley, R. C., Kim, D., and Lynn, D. G. (2000). Signalling organogenesis in parasitic angiosperms: xenognosin generation, perception, and response. *J. Plant Growth Regul.* 19, 217–231.

Keyes, W. J., Palmer, A. G., Erbil, W. K., Taylor, J. V., Apkarian, R. P., Weeks, E. R., et al. (2007). Semagenesis and the parasitic angiosperm *Striga asiatica*. *Plant J.* 51, 707–716. doi: 10.1111/j.1365-313X.2007.03171.x

Klein, O., and Kroschel, J. (2002). Biological control of *Orobanche* spp. with *Phytomyza orobanchia*, a review. *Biocontrol* 47, 245–277. doi: 10.1023/A:1015654429456

Kroschel, J., Mueller-Stoever, D., Elzein, A., and Sauerborn, J. (2000). "The development of mycoherbicides for the management of parasitic weeds of the genus *Striga* and *Orobanche*—a review and recent results," in *Proceedings of the X International Symposium on Biological Control of Weeds*, ed. N. R. Spencer (Bozeman, MT: Montana State University), 139.

Kuijt, J. (1969). *The Biology of Parasitic Fowering Plants*. Berkeley, CA: University of California Press.

Kusumoto, D., Goldwasser, Y., Xie, X., Yoneyama, K., and Takeuchi, Y. (2007). Resistance of red clover (*Trifolium pratense*) to the root parasitic plant *Orobanche minor* is activated by salicylate but not by jasmonate. *Ann. Bot.* 100, 537–544. doi: 10.1093/annbot/mcm148

Labrousse, P., Arnaud, M. C., Griveau, Y., Fer, A., and Thalouarn, P. (2004). Analysis of resistance criteria of sunflower recombined inbred lines against *Orobanche cumana* Wallr. *Crop Prot.* 23, 407–413. doi: 10.1016/j.cropro.2003.09.013

Labrousse, P., Arnaud, M. C., Seryes, H., Berville, A., and Thalouarn, P. (2001). Several mechanisms are involved in resistance of *Helianthus* to *Orobanche cumana* Wallr. *Ann. Bot.* 88, 859–868. doi: 10.1006/anbo.2001.1520

Labrousse, P., Delmail, D., Arnaud, M. C., and Thalouarn, P. (2010). Mineral nutrient concentration influences sunflower infection by broomrape (*Orobanche cumana*). *Botany* 88, 839–849. doi: 10.1139/B10-057

Lechat, M. M., Brun, G., Montiel, G., Veronesi, C., Simier, P., Thoiron, S., et al. (2015). Seed response to strigolactone is controlled by abscisic acid-independent DNA methylation in the obligate root parasitic plant, *Phelipanche ramosa* L. Pomel. *J. Exp. Bot.* 6, 3129–3140. doi: 10.1093/jxb/erv119

Lechat, M. M., Pouvreau, J. B., Péron, T., Gauthier, M., Montiel, G., Veronesi, C., et al. (2012). PrCYP707A1, an ABA catabolic gene, is a key component of *Phelipanche ramosa* seed germination in response to the strigolactone analogue GR24. *J. Exp. Bot.* 63, 5311–5322. doi: 10.1093/jxb/ers189

Lee, J. A., and Stewart, G. R. (1978). Ecological aspects of nitrogen assimilation. *Adv. Bot. Res.* 6, 1–43. doi: 10.1016/S0065-2296(08)60328-6

Lieberman, M. (1979). Biosynthesis and action of ethylene. *Annu. Rev. Plant Physiol.* 30, 533–591. doi: 10.1146/annurev.pp.30.060179.002533

Linke, K. H., and Saxena, M. C. (1991). "Study on viability and longevity of *Orobanche* seed under laboratory conditions," in *Proceedings of the International Workshop on Orobanche Research: Progress in Orobanche Research*, eds K. Wegmann and L. J. Musselman (Obermarchtal: Eberhard-Karls Universitat), 110–114.

Lins, R. D., Colquhoun, J. B., and Mallory-Smith, C. A. (2007). Effect of small broomrape (*Orobanche minor*) on red clover growth and dry matter partitioning. *Weed Sci.* 55, 517–520. doi: 10.1614/WS-07-049.1

Liu, Q., Zhang, Y., Matusova, R., Charnikhova, T., Amini, M., Jamil, M., et al. (2014). *Striga hermonthica* MAX2 restores branching but not the very low fluence response in the *Arabidopsis thaliana* MAX2 mutant. *New Phytol.* 202, 531–541. doi: 10.1111/nph.12692

Logan, D., and Stewart, G. R. (1995). Thidiazuron stimulates germination and ethylene production in *Striga hermonthica* – comparison with the effects of GR24, ethylene and 1-aminocyclopropane-1-carboxylic acid. *Seed Sci. Res.* 5, 99–108. doi: 10.1017/S0960258500002671

López-Bellido, R. J., Benítez-Vega, J., and López-Bellido, L. (2009). No-tillage improves broomrape control with glyphosate in faba-bean. *Agron. J.* 101, 1394–1399. doi: 10.2134/agronj2009.0014

López-Granados, F., and García-Torres, L. (1996). Effects of environmental factors on dormancy and germination of crenate broomrape (*Orobanche crenata*). *Weed Sci.* 44, 284–289.

López-Granados, F., and García-Torres, L. (1999). Longevity of crenate broomrape (*Orobanche crenata*) seed under soil and laboratory conditions. *Weed Sci.* 47, 161–166.

López-Ráez, J. A., Charnikhova, T., Fernandez, I., Bouwmeester, H., and Pozo, M. J. (2011). Arbuscular mycorrhizal symbiosis decreases strigolactone production in tomato. *J. Plant Physiol.* 168, 294–297. doi: 10.1016/j.jplph.2010.08.011

Losner-Goshen, D., Portnoy, V. H., Mayer, A. M., and Joel, D. M. (1998). Pectolytic activity by the haustorium of the parasitic plant *Orobanche* L. (Orobanchaceae) in host roots. *Ann. Bot.* 81, 319–326. doi: 10.1006/anbo.1997.0563

Louarn, J., Carbonne, F., Delavault, P., Becard, G., and Rochange, S. (2012). Reduced germination of *Orobanche cumana* seeds in the presence of arbuscular mycorrhizal fungi or their exudates. *PLoS ONE* 7:e49273. doi: 10.1371/journal.pone.0049273

Mabrouk, Y., Mejri, S., Hemissi, I., Simier, P., Delavault, P., Saidi, M., et al. (2010). Bioprotection mechanisms of pea plant by *Rhizobium leguminosarum* against *Orobanche crenata*. *Afr. J. Microbiol. Res.* 4, 2570–2575.

Mabrouk, Y., Simier, P., Arfaoui, A., Sifi, B., Delavault, P., Zourgui, L., et al. (2007a). Induction of phenolic compounds in pea (*Pisum sativum* L.) inoculated by *Rhizobium leguminosarum* and infected with *Orobanche crenata*. *J. Phytopathol.* 155, 728–734. doi: 10.1111/j.1439-0434.2007.01307.x

Mabrouk, Y., Simier, P., Delavault, P., Delgrange, S., Sifi, B., Zourgui, L., et al. (2007b). Molecular and biochemical mechanisms of defence induced in pea by *Rhizobium leguminosarum* against *Orobanche crenata*. *Weed Res.* 47, 452–460. doi: 10.1111/j.1365-3180.2007.00583.x

Mabrouk, Y., Zourgui, L., Sifi, B., Delavault, P., Simier, P., and Belhadj, O. (2007c). Some compatible *Rhizobium leguminosarum* strains in peas decrease infections when parasitized by *Orobanche crenata*. *Weed Res.* 47, 44–53. doi: 10.1111/j.1365-3180.2007.00548.x

Manschadi, A. M., Kroschel, J., and Sauerborn, J. (1996). Dry matter production and partitioning in the host-parasite association *Vicia faba–Orobanche crenata*. *J. Appl. Bot.* 70, 224–229.

Matvienko, M., Wojtowicz, A., Wrobel, R., Jamison, D., Goldwasser, Y., and Yoder, J. I. (2001). Quinone oxidoreductase message levels are differentially regulated in parasitic and non-parasitic plants exposed to allelopathic quinones. *Plant J.* 25, 375–387. doi: 10.1046/j.1365-313x.2001.00971.x

Mauro, R. P., Lo Monaco, A., Lombardo, S., Restuccia, A., and Mauromicale, G. (2015). Eradication of *Orobanche/Phelipanche* spp. seedbank by soil solarization and organic supplementation. *Sci. Hortic.* 193, 62–68. doi: 10.1016/j.scienta.2015.06.038

Mauromicale, G., Lo Monaco, A., and Longo, M. G. A. (2008). Effect of branched broomrape (*Orobanche ramosa*) infection on the growth and photosynthesis of tomato. *Weed Sci.* 56, 574–581. doi: 10.1614/WS-07-147.1

Mauromicale, G., Restuccia, G., and Marchese, A. (2001). Soil solarization, a non-chemical technique for controlling *Orobanche crenata* and improving yield of faba bean. *Agronomie* 21, 757–765. doi: 10.1051/agro:2001167

Mayer, A. M., and Bar-Nun, N. (1994). "Metabolism during preconditioning and germination of *Orobanche aegyptiaca*," in *Proceedings of the 3rd International Workshop on Orobanche and related Striga Research: Biology and management of Orobanche*, eds A. H. Pieterse, J. A. C. Verkleij, and S. C. ter Borg (Amsterdam: Royal Tropical Institute), 146–156.

Mayer, A. M., and Bar-Nun, N. (1997). "Germination of *Orobanche* seeds: some aspects of metabolism during preconditioning," in *Basic and Applied Aspects of Seed Biology*, eds R. H. Ellis, M. Black, A. J. Murdoch, and T. D. S. Hing (Dordrecht: Kluwer Academic Publishers), 633–639.

McNally, S. F., Orebamjo, T. O., Hirel, B., and Stewart, G. R. (1983). Glutamine synthetase isozymes of *Striga hermonthica* and other angiosperm root parasites. *J. Exp. Bot.* 34, 610–619. doi: 10.1093/jxb/34.5.610

Metzger, J. (1983). The promotion of germination of dormant weed seeds by substituted phthalimides and gibberellic acid. *Weed Sci.* 31, 285–289.

Mohamed, K. I., Papes, M., Williams, R., Benz, B. W., and Peterson, A. T. (2006). Global invasive potential of 10 parasitic witchweeds and related Orobanchaceae. *Ambio* 35, 281–288. doi: 10.1579/05-R-051R.1

Müller-Stöver, D. (2001). Possibilities of biological control of *Orobanche crenata* and *O. cumana* with *Ulocladium botrytis* and *Fusarium oxysporum* f. sp. orthoceras. *Agroecology* 3, 174.

Müller-Stöver, D., Buschmann, H., and Sauerborn, J. (2005). Increasing control reliability of *Orobanche cumana* through integration of a biocontrol agent with a resistance-inducing chemical. *Eur. J. Plant Pathol.* 111, 193–202. doi: 10.1007/s10658-004-2814-8

Murdoch, A. J., and Kebreab, A. (2013). "Germination ecophysiology," in *Parasitic Orobanchaceae*, eds D. M. Joel, J. Gressel, and L. J. Musselman (Heidelberg: Springer Berlin), 195–219.

Musselman, L. J. (1980). The biology of *Striga*, *Orobanche* and other root parasitic weeds. *Annu. Rev. Phytopathol.* 18, 463–489. doi: 10.1146/annurev.py.18.090180.002335

Musselman, L. J., and Dickison, W. C. (1975). The structure and development of the haustorium in parasitic Scrophulariaceae. *Bot. J. Linn. Soc.* 70, 183–212. doi: 10.1111/j.1095-8339.1975.tb01645.x

Mwakaboko, A. S., and Zwanenburg, B. (2011). Strigolactone analogs derived from ketones using a working model for germination stimulants as a blueprint. *Plant Cell Physiol.* 52, 699–715. doi: 10.1093/pcp/pcr031

Nandula, V. K., Foster, J. G., and Foy, C. L. (2000). Impact of Egyptian broomrape (*Orobanche aegyptiaca* (Pers.) parasitism on amino acid composition of carrot (*Daucus carota* L.). *J. Agric. Food Chem.* 48, 3930–3934. doi: 10.1021/jf991145w

Panetta, F. D., and Lawes, R. (2005). Evaluation of weed eradication programs: the delimitation of extent. *Divers. Distrib.* 11, 435–442. doi: 10.1111/j.1366-9516.2005.00179.x

Parker, C. (2009). Observations on the current status of *Orobanche* and *Striga* problems worldwide. *Pest Manag. Scj.* 65, 453–459. doi: 10.1002/ps.1713

Parker, C. (2014). *Orobanche crenata* in UK- an update. *Haustorium* 65, 5–6.

Parker, C., and Riches, C. R. (1993). *Parasitic Weeds of the World: Biology and Control*. Wallingford: CAB International.

Pérez-de-Luque, A., Fondevilla, S., Pérez-Vich, B., Aly, R., Thoiron, S., Simier, P., et al. (2009). Understanding *Orobanche* and *Phelipanche*-host plant interactions and developing resistance. *Weed Res.* 49, 8–22. doi: 10.1111/j.1365-3180.2009.00738.x

Pérez-de-Luque, A., Jorrín, J., Cubero, J. I., and Rubiales, D. (2005). Resistance and avoidance against *Orobanche crenata* in pea (*Pisum* spp.) operate at different developmental stages of the parasite. *Weed Res.* 45, 379–387. doi: 10.1111/j.1365-3180.2005.00464.x

Pérez-de-Luque, A., Jorrín, J., and Rubiales, D. (2004). Broomrape (*Orobanche crenata* Forsks.) control in pea (*Pisum sativum* L.) by foliar applications of benzothiadiazole (BTH). *Phytoparasitica* 32, 21–29. doi: 10.1007/BF02980855

Pérez-de-Luque, A., Moreno, M. T., and Rubiales, D. (2008). Host plant resistance against broomrapes (*Orobanche* spp.): defence reactions and mechanisms of resistance. *Ann. Appl. Biol.* 152, 131–141. doi: 10.1111/j.1744-7348.2007.00212.x

Pérez-de-Luque, A., and Rubiales, D. (2009). Nanotechnology for parasitic plant control. *Pest Manang. Sci.* 65, 540–545. doi: 10.1002/ps.1732

Pérez-Vich, B., Velasco, L., Rich, P. J., and Ejeta, G. (2013). "Marker-assisted and physiology-based breeding for resistance to root parasitic Orobanchaceae," in *Parasitic Orobanchaceae*, eds D. M. Joel, J. Gressel, and L. J. Musselman (Heidelberg: Springer Berlin), 369–391.

Péron, T., Véronési, C., Mortreau, E., Pouvreau, J. B., Thoiron, S., Leduc, N., et al. (2012). Role of the sucrose synthase encoding PrSus1 gene in the development of the parasitic plant *Phelipanche ramosa* L. (Pomel). *Mol. Plant Microbe Interact.* 25, 402–411. doi: 10.1094/MPMI-10-11-0260

Plakhine, D., Eizenberg, H., Hershenhorn, J., Goldwasser, Y., and Kleifeld, Y. (2001). "Control of *Orobanche aegyptiaca* with sulfonylurea herbicides in tomato—polyethylene bag studies," in *International Parasitic Weed Symposium*, eds A. Fer, P. Thalouarn, D. M. Joel, C. Musselman, and J. A. C. Verkleij (Nantes: University of Nantes), 294–295.

Plakhine, D., Ziadna, H., and Joel, D. M. (2009). Is seed conditioning essential for *Orobanche* germination? *Pest Manag. Sci.* 65, 492–496. doi: 10.1002/ps.1716

Privat, G. (1960). Recherches sur les phanerogames parasites (etude d'*Orobanche* hederae Duby). *Ann. Sci. Nat. Bot.* 12, 722–865.

Quimby, P. C. Jr., Zidack, N. K., and Boyette, C. D. (1999). A simple method for stabilizing and granulating fungi. *Biocontrol Sci. Technol.* 9, 5–8. doi: 10.1080/09583159929857

Reda, F. (2006). *Orobanche crenata* in Ethiopia. *Haustorium* 49, 3.

Rich, P. J., Grenier, C., and Ejeta, G. (2004). *Striga* resistance in the wild relatives of sorghum. *Crop Sci.* 44, 2221–2229. doi: 10.2135/cropsci2004.2221

Riopel, J. L., and Timko, M. P. (1995). "Haustorial initiation and differentiation," in *Parasitic Plants*, eds M. C. Press and J. D. Graves (London: Chapman and Hall), 39–79.

Rubiales, D., Alcántara, C., Pérez-de-Luque, A., Gil, J., and Sillero, J. C. (2003a). Infection of chickpea (*Cicer arietinum*) by crenate broomrape (*Orobanche crenata*) as influenced by sowing date and weather conditions. *Agronomie* 23, 359–362. doi: 10.1051/agro:2003016

Rubiales, D., Pérez-de-Luque, A., Joel, D. M., Alcantara, C., and Sillero, J. C. (2003b). Characterization of resistance in chickpea to crenate broomrape (*Orobanche crenata*). *Weed Sci.* 51, 702–707. doi: 10.1614/P2002-151

Rubiales, D., Fernández-Aparicio, M., Pérez-de-Luque, A., Castillejo, M. A., Prats, E., Sillero, J., et al. (2009a). Breeding approaches for crenate broomrape (*Orobanche crenata* Forsk.) management in pea (*Pisum sativum* L.). *Pest Manag. Sci.* 65, 553–559. doi: 10.1002/ps.1740

Rubiales, D., Fernández-Aparicio, M., Wegmann, K., and Joel, D. (2009b). Revisiting strategies for reducing the seedbank of *Orobanche* and *Phelipanche* spp. *Weed Res.* 49, 23–33. doi: 10.1111/j.1365-3180.2009.00742.x

Rubiales, D., Fernandez-Aparicio, M., and Rodriguez, M. J. (2008). First report of crenate broomrape (*Orobanche crenata*) on lentil (*Lens culinaris*) and common vetch (*Vicia sativa*) in Salamanca Province, Spain. *Plant Dis.* 92, 1368. doi: 10.1094/PDIS-92-9-1368B

Saghir, A. R. (1986). "Dormancy and germination of *Orobanche* seeds in relation to control methods," in *Proceedings of a Workshop in Wageningen: Biology and Control of Orobanche*, ed. S. J. Ter Borg (Wageningen: LH/VPO), 25–34.

Sands, D. C., and Pilgeram, A. L. (2009). Methods for selecting hypervirulent biocontrol agents of weeds: why and how? *Pest Manag. Sci.* 65, 581–587. doi: 10.1002/ps.1739

Sarosh, B. R., Sivaramakrishnan, S., and Shetty, H. S. (2005). Elicitation of defense related enzymes and resistance by L-methionine in pearl millet against downy mildew disease caused by *Sclerospora graminicola*. *Plant Physiol. Biochem.* 43, 808–815. doi: 10.1016/j.plaphy.2005.06.009

Sauerborn, J. (1991). "The economic importance of the phytoparasites *Orobanche* and *Striga*," in *Proceedings of the Fifth Symposium on Parasitic Weeds, Nairobi*, eds J. K. Ransom, L. J. Musselman, A. D. Worsham, and C. Parker (Nairobi: CIMMYT), 137–143.

Sauerborn, J., Linke, K. H., Saxena, M. C., and Koch, W. (1989). Solarization, a physical control method for weeds and parasitic plants (*Orobanche* spp.) in Mediterranean agriculture. *Weed Res.* 29, 391–393. doi: 10.1111/j.1365-3180.1989.tb01310.x

Schneeweiss, G. M. (2007). Correlated evolution of life history and host range in the nonphotosynthetic parasitic flowering plants *Orobanche* and *Phelipanche* (Orobanchaceae). *J. Evol. Biol.* 20, 471–478. doi: 10.1111/j.1420-9101.2006.01273.x

Seo, M., Nambara, E., Choi, G., and Yamaguchi, S. (2009). Interaction of light and hormone signals in germinating seeds. *Plant Mol. Biol.* 69, 463–472. doi: 10.1007/s11103-008-9429-y

Sholmer-Ilan, A. (1993). Germinating seeds of the root parasite *Orobanche aegyptiaca* Pers. excrete enzymes with carbohydrase activity. *Symbiosis* 15, 61–70.

Sillero, J. C., Moreno, M. T., and Rubiales, D. (2005). Sources of resistance to crenate broomrape among species of *Vicia*. *Plant Dis.* 89, 23–27. doi: 10.1094/PD-89-0023

Singh, A., and Singh, M. (1993). Cell wall-degrading enzyme in *Orobanche aegyptiaca* and its host *Brassica campestris*. *Physiol. Plant.* 89, 177–181. doi: 10.1111/j.1399-3054.1993.tb01802.x

Slavov, S., Valkov, V., Batchvarova, R., Atanassova, S., Alexandrova, M., and Atanassov, A. (2005). Chlorsulfuron resistant transgenic tobacco as a tool for broomrape control. *Transgenic Res.* 14, 273–278. doi: 10.1007/s11248-004-8081-9

Song, W. J., Zhou, W. J., Jin, Z. L., Cao, D. D., Joel, D. M., Takeuchi, Y., et al. (2005). Germination response of *Orobanche* seeds subjected to conditioning temperature, water potential and growth regulator treatments. *Weed Res.* 45, 467–476. doi: 10.1111/j.1365-3180.2005.00477.x

Southwood, O. R. (1971). The effects of superphosphate application, 2,4-DB and grazing on broomrape (*Orobanche minor* Sm.) in a subterranean clover pasture. *Weed Res.* 11, 240–246. doi: 10.1111/j.1365-3180.1971.tb01006.x

Stewart, G. R., and Press, M. C. (1990). The physiology and biochemistry of parasitic angiosperms. *Annu. Rev. Plant Physiol. Plant Mol. Biol.* 41, 127–151. doi: 10.1146/annurev.pp.41.060190.001015

Suttle, J. C. (1983). N-substituted phthalimides as plant bioregulants. *News Bull. Br. Plant Growth Regul. Group* 6, 11–19.

Suttle, J. C., and Schreiner, D. R. (1982). The biological activity of AC-94, 377 [1-(3-chlorophthalimido)-cyclohexane-arboxamide]. *J. Plant Growth Regul.* 1, 139–146.

Takeuchi, Y., Omigawa, Y., Ogasawara, M., Yoneyama, K., Konnai, M., and Worsham, A. D. (1995). Effects of brassinosteroids on conditioning and germination of clover broomrape (*Orobanche minor*) seeds. *Plant Growth Regul.* 16, 153–160. doi: 10.1007/BF00029536

Tan, S., Evans, R. R., Dahmer, M. L., Sing, B. K., and Shaner, D. (2005). Imidazolinone-tolerant crops: history, current status and future. *Pest Manag. Sci.* 61, 246–257. doi: 10.1002/ps.993

Tank, D. C., Beardsley, P. M., Kelchner, S. A., and Olmstead, R. G. (2006). Review of the systematics of Scrophulariaceae s.l. and their current disposition. *Aust. Syst. Bot.* 19, 289–307. doi: 10.1071/SB05009

Thomas, H., Heller, A., Sauerborn, J., and Müller-Stöver, D. (1999). *Fusarium oxysporum* f. sp. orthoceras, a potential mycoherbicide, parasitizes seeds of *Orobanche cumana* (Sunflower broomrape): a cytological study. *Ann. Bot.* 83, 453–458. doi: 10.1006/anbo.1998.0847

Toh, S., Kamiya, Y., Kawakami, N., Nambara, E., McCourt, P., and Tsuchiya, Y. (2012). Thermoinhibition uncovers a role for strigolactones in *Arabidopsis* seed germination. *Plant Cell Physiol.* 53, 107–117. doi: 10.1093/pcp/pcr176

Tomilov, A., Tomilova, N., Shin, D. H., Jamison, D., Torres, M., Reagan, R., et al. (2006). "Chemical signalling between plants: mechanistic similarities between phytotoxic allelopathy and host recognition by parasitic plants," in *Chemical Ecology: From Gene to Ecosystem*, eds M. Dicke and W. Takken (Dordrecht: Springer), 55–69.

Umehara, M., Hanada, A., Yoshida, S., Akiyama, K., Arite, T., Takeda-Kamiya, N., et al. (2008). Inhibition of shoot branching by new terpenoid plant hormones. *Nature* 455, 195–200. doi: 10.1038/nature07272

USEPA (2004). *List of Inert Pesticide Ingredients List 4b. U. S. Environmental Protection Agency*. Available at: www.epa.gov/opprd001/inerts/inerts_list4Bname.pdf

Van Delft, G. J., Graves, J. D., Fitter, A. H., and Van Ast, A. (2000). *Striga* seed avoidance by deep planting and no-tillage in sorghum and maize. *Int. J. Pest Manag.* 46, 251–256. doi: 10.1080/09670870050206019

van Hezewijk, M. J., and Verkleij, J. A. C. (1996). The effect of nitrogenous compounds on in vitro germination of *Orobanche crenata* Forsk. *Weed Res.* 36, 395–404. doi: 10.1111/j.1365-3180.1996.tb01669.x

Vaucher, J. P. (1823). *Mémoire sur la Germination des Orobanches*, Vol. 10. Paris: Mémoires du Museum d'Histoire Naturelle, 261–273.

Veronesi, C., Bonnin, E., Benharrat, H., Fer, A., and Thalouarn, P. (2005). Are pectinolytic activities of *Orobanche cumana* seedlings related to virulence towards sunflower? *Isr. J. Plant Sci.* 53, 19–27. doi: 10.1560/ETEL-C34X-Y6MG-YT0M

Veronesi, C., Bonnin, E., Calvez, S., Thalouarn, P., and Simier, P. (2007). Activity of secreted cell wall-modifying enzymes and expression of peroxidase-encoding gene following germination of *Orobanche ramosa*. *Biol. Plant* 51, 391–394. doi: 10.1007/s10535-007-0084-y

Vurro, M., Boari, A., Evidente, A., Andolfi, A., and Zermane, N. (2009). Natural metabolites for parasitic weed management. *Pest Manag. Sci.* 65, 566–571. doi: 10.1002/ps.1742

Vurro, M., Boari, A., Pilgeram, A. L., and Sands, D. C. (2006). Exogenous amino acids inhibit seed germination and tubercle formation by *Orobanche ramosa* (broomrape): potential application for management of parasitic weeds. *Biol. Control* 36, 258–265. doi: 10.1016/j.biocontrol.2005.09.017

Westwood, J. H. (2013). "The physiology of the established parasite-host association," in *Parasitic Orobanchaceae*, eds D. M. Joel, J. Gressel, and L. J. Musselman (Berlin: Springer), 87–114.

Westwood, J. H., dePamphilis, C. W., Das, M., Fernández-Aparicio, M., Honaas, L. A., Timko, M. P., et al. (2012). The parasitic plant genome project: new tools for understanding the biology of *Orobanche* and *Striga*. *Weed Sci.* 60, 295–306. doi: 10.1614/WS-D-11-00113.1

Westwood, J. H., and Foy, C. L. (1999). Influence of nitrogen on germination and early development of broomrape (*Orobanche* spp.). *Weed Sci.* 47, 2–7.

Westwood, J. H., Yu, X., Foy, C. L., and Cramer, C. L. (1998). Expression of a defense-related 3-hydroxy-3-methylglutaryl CoA reductase gene in response to parasitism by *Orobanche* spp. *Mol. Plant Microbe Interact.* 11, 530–536. doi: 10.1094/MPMI.1998.11.6.530

Xie, X., Yoneyama, K., and Yoneyama, K. (2010). The strigolactone story. *Annu. Rev. Phytopathol.* 48, 93–117. doi: 10.1146/annurev-phyto-073009-114453

Yang, Z., Wafula, E. K., Honaas, L. A., Zhang, H., Das, M., Fernandez-Aparicio, M., et al. (2015). Comparative transcriptome analyses reveal core parasitism genes and suggest gene duplication and repurposing as sources of structural novelty. *Mol. Biol. Evol.* 32, 767–790. doi: 10.1093/molbev/msu343

Yoder, J. I., and Scholes, J. D. (2010). Host plant resistance to parasitic weeds; recent progress and bottlenecks. *Curr. Opin. Plant Biol.* 13, 478–484. doi: 10.1016/j.pbi.2010.04.011

Yoneyama, K., Xie, X., Kim, H. I., Kisugi, T., Nomura, T., Sekimoto, H., et al. (2012). How do nitrogen and phosphorus deficiencies affect strigolactone production and exudation? *Planta* 235, 1197–1207. doi: 10.1007/s00425-011-1568-8

Yoneyama, K., Xie, X., Kusumoto, D., Sekimoto, H., Sugimoto, Y., Takeuchi, Y., et al. (2007a). Nitrogen deficiency as well as phosphorus deficiency in sorghum promotes the production and exudation of 5-deoxystrigol, the host recognition signal for arbuscular mycorrhizal fungi and root parasites. *Planta* 227, 125–132. doi: 10.1007/s00425-007-0600-5

Yoneyama, K., Yoneyama, K., Takeuchi, Y., and Sekimoto, H. (2007b). Phosphorus deficiency in red clover promotes exudation of orobanchol, the signal for mycorrhizal symbionts and germination stimulant for root parasites. *Planta* 225, 1031–1038. doi: 10.1007/s00425-006-0410-1

Zehhar, N., Ingouff, M., Bouya, D., and Fer, A. (2002). Possible involvement of gibberellins and ethylene in *Orobanche ramosa* germination. *Weed Res.* 42, 464–469. doi: 10.1046/j.1365-3180.2002.00306.x

Zhang, Y., Luc, J. E., and Crow, W. T. (2010). Evaluation of amino acids as turfgrass nematicides. *J. Nematol.* 42, 292–297.

Zwanenburg, B., Mwakaboko, A. S., Reizelman, A., Anilkuma, G., and Sethumadhavan, D. (2009). Structure and function of natural and synthetic signaling molecules in parasitic weed germination. *Pest Manag. Sci.* 65, 478–491. doi: 10.1002/ps.1706

Conflict of Interest Statement: The authors declare that the research was conducted in the absence of any commercial or financial relationships that could be construed as a potential conflict of interest.

6

Identification of *Striga hermonthica*-Resistant Upland Rice Varieties in Sudan and their Resistance Phenotypes

Edited by:
Monica Fernandez-Aparicio,
Institut National de la Recherche Agronomique, France

Reviewed by:
Abu Hena Mostafa Kamal,
University of Texas at Arlington, USA
Leonardo Velasco,
Institute for Sustainable Agriculture – Consejo Superior de Investigaciones Científicas, Spain
Jonne Rodenburg,
Africa Rice Center, Tanzania
Chris Parker,
Formerly affiliated with Long Ashton Research Station, UK

***Correspondence:**
Hiroaki Samejima
samehiro@agr.nagoya-u.ac.jp

†Present address:
Abdel G. Babiker,
National Center for Research,
P.O. Box 2404, Khartoum, Sudan

Hiroaki Samejima[1,2], Abdel G. Babiker[3†], Ahmed Mustafa[4] and Yukihiro Sugimoto[1]*

[1] Graduate School of Agricultural Science, Kobe University, Kobe, Japan, [2] International Cooperation Center for Agricultural Education, Nagoya University, Nagoya, Japan, [3] College of Agricultural Studies, Sudan University of Science and Technology, Khartoum North, Sudan, [4] Gezira Research Station, Agricultural Research Corporation, Wad Madani, Sudan

Rice has become a major staple cereal in sub-Saharan Africa. Currently, upland rice cultivation is expanding particularly in rainfed areas where the root parasitic weed *Striga hermonthica*, a major constraint to cereal production, is endemic. Laboratory, pot, and semi-controlled open air experiments were performed to evaluate resistance of selected rice varieties in Sudan to a resident *S. hermonthica* population. In the laboratory, 27 varieties were screened for post-attachment resistance using the rhizotron technique. Varieties displaying high post-attachment resistance, Umgar, NERICA5, and NERICA13 together with NERICA4, NERICA18, and Nipponbare, a lowland rice variety, were further evaluated for performance and *Striga* resistance in pot and semi-controlled open air experiments and for germination inducing activity in a laboratory. In addition, comparative studies on reaction of Umgar, Kosti1 and Kosti2, released varieties for commercial production in Sudan, to the parasite were performed in two pot experiments. In the pot experiments Umgar and NERICA5, consistently, sustained the lowest *Striga* emergence (<2.2 *Striga* plants per pot), while NERICA13 and NERICA4 supported 1.8–5.7 and 8.7–16.4 *Striga* plants per pot, respectively. In an artificially *Striga*-infested field, number of emergent *Striga* plants per 10 rice hills, at harvest, was 2.0, 2.0, 4.8, 13.5, 13.3, and 18.3 on Umgar, NERICA5, NERICA13, NERICA4, NERICA18, and Nipponbare, respectively. *Striga* had no adverse effects on total above-ground parts and panicle dry weight in Umgar and NERICA5. Germination-inducing activity of root exudates, at 14 days after sowing onward, was markedly lower for Umgar than for NERICA5, NERICA13, NERICA4, and NERICA18. Based on these findings, Umgar has both pre and post-attachment resistance to a resident *Striga* population in Sudan. Kosti1 and Kosti2 did not exhibit *Striga*-resistance at the same level as Umgar. Further the resistance of NERICA5, a variety reported to be endowed with a broad spectrum resistance to *Striga* species and ecotypes, at least to one resident *Striga* population in Sudan was clearly indicated.

Keywords: NERICA, parasitic weed, pre-attachment, post-attachment, rhizotron

---

Abbreviations: ARC, Agricultural Research Corporation; DAI, days after inoculation; DAS, days after sowing; DW, dry weight; GDD, growing degree days.

INTRODUCTION

Rice is becoming an important staple crop in Africa. From 1984 to 2013 the area dedicated to rice cultivation increased by 124%, corresponding to a 215% increase in production (FAO, 2015). Upland rice varieties that are interspecific cultivars between *Oryza sativa* and *Oryza glaberrima* were developed by the Africa Rice Center under the name NEw RICe for Africa (NERICA; Jones et al., 1997a,b). These varieties promoted rice cultivation in the region and are partly responsible for the recent expansion of rice into rain-fed areas (Rodenburg et al., 2006; Balasubramanian et al., 2007; Wopereis et al., 2008). These areas are sometimes infested by the root-parasitic weed *Striga hermonthica* (Delile) Benth. Rice is a genuine host for *S. hermonthica* and heavy infestations by the parasite result in severe yield losses. *Striga* infestation and damage are expected to be exacerbated by the predominant low soil fertility, irregular rains and low-input (subsistence farming) conditions (Babiker, 2007).

Available control measures against *Striga*, including cultural, chemical, and biological methods, are either too expensive or too knowledge-intensive for subsistence farmers (Babiker, 2007). Under the prevailing farming conditions in sub-Saharan Africa, resistant crop varieties have been proposed as the most cost-effective and easy to adopt or deploy as an integral component of an integrated *Striga* management strategy (Ejeta, 2007). Various responses of rice varieties to *Striga* have been identified, although the number of resistant varieties is still limited (Rodenburg et al., 2010). The slow pace of development and deployment of *Striga*-resistant varieties is, majorly, due to a paucity of resistance sources, the complex genetics of resistance and limited knowledge regarding the specific mechanisms associated with resistance phenotypes (Amusan et al., 2008). Understanding the different types of resistance phenotypes facilitates development of selection methods and strategies to improve crop resistance to the parasite.

Research on rice has revealed several types of resistance phenotypes that are expressed in different stages of the *Striga* life cycle. As obligate hemiparasites, *Striga* seeds germinate only when exposed to host root-derived stimulants (López-Ráez et al., 2009). Host plants that produce lower amounts or less effective types of germination stimulants prevent parasite attachment, i.e., they have pre-attachment resistance. In a study using 18 NERICA varieties and their parents, NERICA1 and CG14 were shown to produce fewer germination stimulants and exhibit less frequent *Striga* infections compared with the other varieties (Jamil et al., 2011). Furthermore, host plants with post-attachment resistance prevent parasite development. Cissoko et al. (2011) reported that NERICA1 and NERICA10 exhibited the highest post-attachment resistance to *S. hermonthica* and *Striga asiatica* (L.) Kuntze. Recently Rodenburg et al. (2015) using a set of 25 rice varieties reported consistency between field and laboratory data on resistance to *S. hermonthica* and *S. asiatica*. A lowland rice cultivar, Nipponbare, has been reported to exhibit post-attachment resistance associated with an incompatibility in the cortex, endodermis, and root stele (Gurney et al., 2006; Swarbrick et al., 2008). Yoshida and Shirasu (2009) conducted detailed anatomical studies and proposed that mechanisms occurring subsequent to establishment of vascular connections contributed to the resistance of Nipponbare to *Striga*.

Although there are several *Striga*-resistant rice varieties, as noted above, crop resistance against one *S. hermonthica* population may not always be effective against other populations (Huang et al., 2012). *S. hermonthica*, an obligate outcrosser, is renowned for its genetic diversity and virulence variability (Yoshida et al., 2010). Furthermore, development of a virulent population over time from a subset of *Striga* individuals within a seed bank may lead to beakdown of resistance (Huang et al., 2012; Rodenburg et al., 2015). Cissoko et al. (2011) reported that NERICA1 showed high resistance to *S. hermonthica* collected from maize (*Zea mays* L.) in Kenya and sorghum [*Sorghum bicolor* (L.) Moench] in Sudan and to *S. asiatica* from maize in the USA and rice (cultivar Supa) in Kenya. However, it was susceptible to *S. hermonthica* collected from a field previously planted with NERICA1. In addition to differences in *Striga* populations, growth conditions such as *Striga* seed bank size, soil moisture, fertility, and temperature may influence apparent resistance in rice varieties (Kim, 1996)

Sudan is attempting to boost its domestic rice production and the area under the crop increased from 2,100 ha in 1984 to 7,562 ha in 2013 with a corresponding increase in production from 2,000 t to 25,000 t (FAO, 2015). Several NERICA varieties were introduced to Sudan and evaluated for growth and yield, but not for resistance to *Striga*, in the country (Somado and Guei, 2008). *Striga*, currently pre-dominantly on sorghum and millet, is expected to be a major production constraint to upland rice in Sudan as well. Expanding upland rice production without due attention to the *Striga* problem is fraught with an element of risk. Upland rice varieties which combine resistance to resident *Striga* populations and adaptability to Sudanese rice upland growing environments is therefore urgently needed.

The present investigation was therefore undertaken to (1) identify *Striga*-resistant upland rice varieties in a series of laboratory, pot and semi-controlled open air experiments using a resident *Striga* population and (2) determine, broadly, resistance phenotype(s) of the selected varieties. In Sudan, two distinct *S. hermonthica* strains, populations from sorghum fields and from millet fields, have been reported (Ali et al., 2009). We used a population from a sorghum field in this study, because millet cultivation area is often too dry for rice cultivation.

MATERIALS AND METHODS

Plant Materials

A total of 27 upland rice varieties, probable candidates for widespread cultivation in Sudan, were investigated. Three rice varieties, Umgar (YUNLU 30 originating from China), Kosti1 and Kosti2, released for commercial production in Sudan (Nyachae, 2011), were obtained from the ARC, Sudan. Eighteen upland NERICA varieties and their parents, CG14, WAB56-50, WAB56-104, and WAB181-18, were supplied by the Africa Rice Center, Benin. Two lowland varieties, Nipponbare (*Striga*-resistant) and Kasalath (*Striga*-susceptible or -tolerant; Gurney et al., 2006; Swarbrick et al., 2009; Yoshida and Shirasu, 2009;

Yoder and Scholes, 2010), were used as controls. Nipponbare and Kasalath were obtained from the GenBank of the National Institute of Agrobiological Science, Japan. A sorghum variety, Abu70, a high *Striga* germination stimulant producer and a positive control in evaluation of germination-inducing activity of root exudates, was provided by the ARC. *S. hermonthica* seeds were collected from a sorghum field near Wad Madani, Sudan, where upland rice cultivation is being conducted (Mustafa and Elsheikh, 2007).

Rhizotron Experiment

Two rhizotron experiments were carried out under laboratory conditions as described by Hiraoka and Sugimoto (2008). The first experiment was conducted to evaluate post-attachment resistance of 27 rice varieties. The second was conducted to evaluate the repeatability and reliability of rhizotron technique using a subset of the varieties comprising NERICA5, Umgar, Nipponbare, Kasalath, and NERICA4. Seeds of each variety, sown on a moist filter paper and placed in a Petri dish, were incubated in the dark at 30°C for 4 days. Each seedling was transferred to a 10-mL test tube and grown hydroponically in 40% Long Ashton solution for 6 days. The seedlings were maintained in a growth chamber at 30°C with a 12-h photoperiod using fluorescent lights (photosynthetic active radiation of 220 μmol m^{-2} s^{-1}). A 10-days-old seedling of each variety (with five replicates) was transferred to a rhizotron comprising a 15-cm Petri dish filled with rock wool overlaid by glass fiber filter paper watered with 40% Long Ashton solution. The seedlings were grown in the same growth chamber for additional 10 days. *Striga* seeds were surface sterilized by immersion in 0.75% (w/v) NaClO containing a few drops of Tween 20 and then subjected sonication for 3 min in an ultrasonic cleaner. After rinsing with distilled water and drying in a laminar hood, seeds were pre-treated (conditioned) for 12 days on 8-mm glass fiber filter paper disk (approximately 50 seeds each) placed on distilled water-saturated filter paper. Conditioned *Striga* seeds were treated, in separate Petri-dishes, with GR24 at 0.34 μM and incubated in the dark at 30°C for 1 day prior to inoculation. Roots of each 20-days-old rice seedling, in a rhizotron, were inoculated with 30 pre-germinated *Striga* seeds and subsequently incubated in the same growth chamber.

Growth of *Striga* seedlings was observed at 14 and 21 DAI using a stereomicroscope. Growth progression of *Striga* was classified into four developmental stages: stage I, failure to attach to the root due to lack of haustorium formation (**Figure 1a**); stage II, successful attachment but no shoot elongation (**Figures 1b–d**); stage III, shoot necrosis after reaching the four or more leaf pairs stage (**Figures 1e,f**); stage IV, successful parasitism with healthy shoots (**Figures 1g,h**).

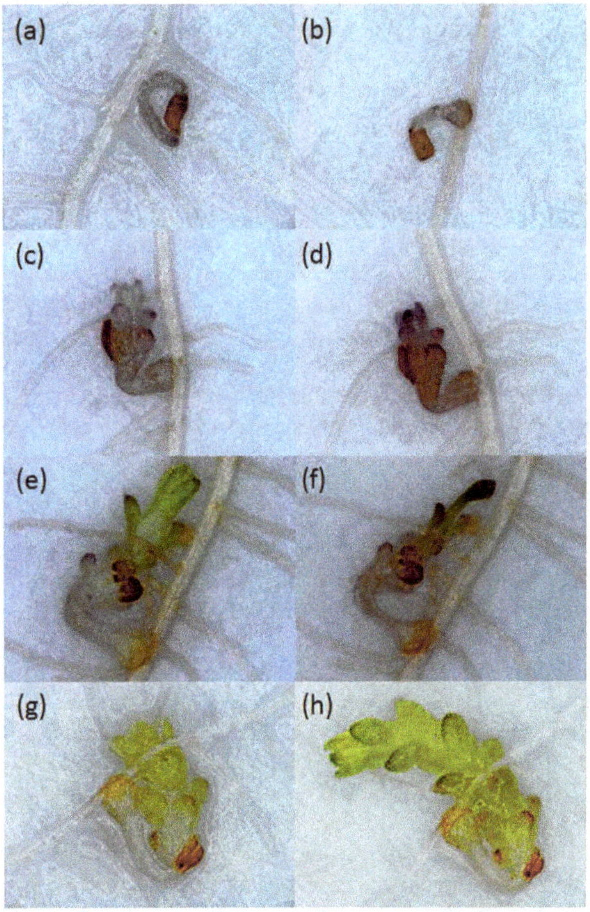

FIGURE 1 | Growth of *Striga hermonthica* seedlings inoculated onto rice roots in a rhizotron under 100× magnification. (a) A seedling that failed to attach to the host root at 14 DAI. **(b)** A seedling that attached to the root but showed no shoot emergence at 21 DAI, or no shoot elongation from **(c)** 14 DAI to **(d)** 21 DAI. A seedling that **(e)** reached the four or more leaf pairs stage before 14 DAI and then **(f)** died before 21 DAI. A seedling that parasitized successfully with steady healthy shoot growth from **(g)** 14 DAI to **(h)** 21 DAI.

Pot Evaluation

Three pot experiments were carried out in the open air at the Sudan University of Science and Technology, Khartoum North (15°39′N lat, 32°31′E long) to compare resistance and performance of selected rice varieties under the same *Striga* seed bank size. Three rice varieties Kosti1, Kosti2, and Umgar were used for the first experiment. For the second experiment, six rice varieties were used. NERICA5, NERICA13, and Umgar were selected as potential candidates showing post-attachment resistance, based on the results of the rhizotron evaluation. NERICA18 was selected as a susceptible variety and NERICA4 was selected as it is widely used in several sub-Saharan African countries (Somado et al., 2008; Wopereis et al., 2008). The *Striga*-resistant variety Nipponbare, albeit a lowland type, was included as a resistant control. For the third experiment, NERICA4, NERICA5, NERICA13, Umgar, Kosti1, and Kosti2 were used. Nipponbare was not included because of lack of adaptability to upland rice growing environment confirmed in the second pot experiment and the semi-controlled open air experiment mentioned below here.

Plastic pots, 30-cm internal diameter and 30-cm height, perforated at the bottom were, each, filled with 9 kg of soil (vertisol, pH 8.31, EC 0.14 mS cm^{-1}, total N 0.48 g kg^{-1}, Olsen-P 8.15 mg kg^{-1}, organic-C 8.9 g kg^{-1}). Soil was collected from a *Striga* free plot within the college experimental farm. *Striga* seeds (1, 2, 4, 8, and 16 mg for the first experiment, 16 mg for the second and 16, 32, and 48 mg for the third) were mixed into the top 5 cm of soil in each pot. *Striga*-infested and *Striga*-free controls were included, as relevant, for comparison. Treatments were organized in a randomized complete block design with three replicates in the first experiments and six replicates in the second and third experiments. One day prior to sowing, 100 g of compost was mixed with the soil in each pot. On the day of sowing, 0.92 g of N as urea and 0.17 g of P_2O_5 as single superphosphate were added to each pot. Five rice seeds were sown in each pot on 30 June 2010, 5 September 2012, and 1 July 2013 for the first, second and third experiments, respectively. The seedlings were thinned to two per pot 14 DAS. Weather data were not recorded in the first experiment. During the experimental period, daily mean air temperature ranged between 18.1 and 34.0°C for the second experiment and between 24.4 and 36.4 for the third experiment. Total precipitation was 33.4 mm for the second experiment and 111.2 mm for the third and the average daily solar radiation was 16.5 and 18.6 MJ m^{-2} d^{-1}, respectively. Rice was grown aerobically. Prior to heading, all pots were irrigated at 2-days intervals. After heading, plants were irrigated daily. Panicles were covered by paper bags 2 weeks after heading to prevent bird damage.

Emergent *Striga* plants per pot were counted weekly from 49 to 119 DAS for the first experiment, at 1140, 1453, 1907, 2112, 2305, 2493, 2689, 2875, 3023, and 3178 GDDs for the second experiment and at 1111, 1286, 1485, 1696, 1908, 2121, 2339, 2530, 3013, and 3203 GDD for the third. Rice was harvested by cutting the above-ground parts at ground level for the second and third experiments and no rice sampling was conducted in the first experiment. NERICA5 was harvested at 117 DAS for the second experiment and 98 DAS for the third and the other varieties were harvested at 125 DAS and 118 DAS, respectively. The rice above-ground parts, severed into panicles and vegetative parts, were oven dried at 80°C for 3 days and weighed. Panicle DW was used as an index for grain yield as described in studies of responses of sorghum varieties to *S. hermonthica* (Lendzemo and Kuyper, 2001; Van Ast and Bastiaans, 2006) and for rice in several studies (Kato et al., 2007; Ohe et al., 2010; Uga et al., 2011; Niones et al., 2012). Total above-ground parts DW was used as an index for rice growth, because it is an important factor for improving grain yield of rice (Fageria, 2007).

Semi-Controlled Open Air Experiment

An experiment was conducted at the college farm to validate the results of the pot experiments under conditions that were relatively similar to farmer's fields. During the experimental period (16 July 2012 to 14 November 2012) daily mean air temperature ranged between 26.1 and 34.8°C, relative humidity was 52.8% on average hourly measurements from sowing to 60 DAS. Low relative humidity, 37.5% on average hourly measurement, was predominant from 61 DAS onward. Total precipitation was 95.8 mm and the average daily solar radiation was 17.8 MJ m^{-2} d^{-1}.

The same six rice varieties used in the second pot experiment were grown in adjacent *Striga*-free and *Striga*-infested plots. The *Striga*-infested plot was established artificially in the *Striga*-free area in 2011 using the same parasite population used in the previous experiments. The plots were used for sorghum cultivation in 2011 and before. *Striga* infestation in the infested plots was further augmented with 1 mg of the parasite seeds placed in each rice-planting hole. Treatments were arranged in a split-plot design. The *Striga* treatment was the main plot. The main plot was subsequently divided into four blocks. Rice planting and husbandry practices were as recommended by ARC, Sudan for upland rice production under supplementary irrigation. Each subplot was surrounded by ridges and rice was grown aerobically using flush irrigation. The crop was irrigated at 3- to 4-days intervals. The rice varieties were randomly allocated to the subplots (0.75 m × 3.75 m) within each block. Rice seeds were sown in holes (five per hole) with a spacing of 25 cm × 25 cm. P_2O_5, as single superphosphate (110 kg P_2O_5 ha^{-1}) was applied at sowing. Nitrogen, as urea (165 kg N ha^{-1}), was applied as split equal doses at 14, 35, and 49 DAS. Weeds other than *Striga* were manually removed. The field was covered with a nylon net to evade bird damage.

Emergent *Striga* plants around 10 fixed rice hills in each subplot were counted at 926, 1357, 1827, 2436, 2846, and 3333 GDD. Rice above-ground parts from the same 10 hills were harvested by cutting at ground level. NERICA5, NERICA13, NERICA18, and Nipponbare were harvested at 107 DAS. NERICA4 and Umgar were harvested at 111 and 121 DAS, respectively. The rice above-ground parts were severed into panicles and vegetative parts, oven dried at 80°C for 3 days, and weighed.

Germination-Inducing Activity of Root Exudates

Seeds of the six rice varieties, used in the second pot and semi-controlled open air experiments, and those of a sorghum (cv Abu70), were surface sterilized by immersion in 1% (w/v) NaClO containing a few drops of Tween 20 and thoroughly rinsed with distilled water. The seeds sown on moist filter paper in Petri dishes were incubated in the dark at 30°C for 3–4 days for rice and 2 days for sorghum. The seedlings were transplanted each, to a 50-mL plastic tube wrapped in aluminum foil to exclude light. Plants were grown hydroponically in 40% Long Ashton solution in a growth chamber at 30°C with a 12-h photoperiod using fluorescent lights (photosynthetic active radiation of 220 μmol m^{-2} s^{-1}) throughout growth duration after the transplanting except during root exudates samplings. The root exudate was sampled at 7, 14, 21, 28, and 35 DAS. One day before each sampling date, the roots were thoroughly rinsed with tap water and the nutrient solution in the plastic tubes was replaced with tap water. After 1 day incubation and prior to extraction, water loss was replenished in each tube and 100-μL of the aqueous solution containing root exudates was sampled, hence forth referred to as original exudates. The remaining aqueous solution in each tube was extracted with ethyl acetate (3 × 10 mL). The respective ethyl acetate extracts, pooled, dried over anhydrous sodium sulfate, were evaporated *in vacuo* at 40°C to dryness. The residues, were dissolved, each, in 1 mL of ethyl

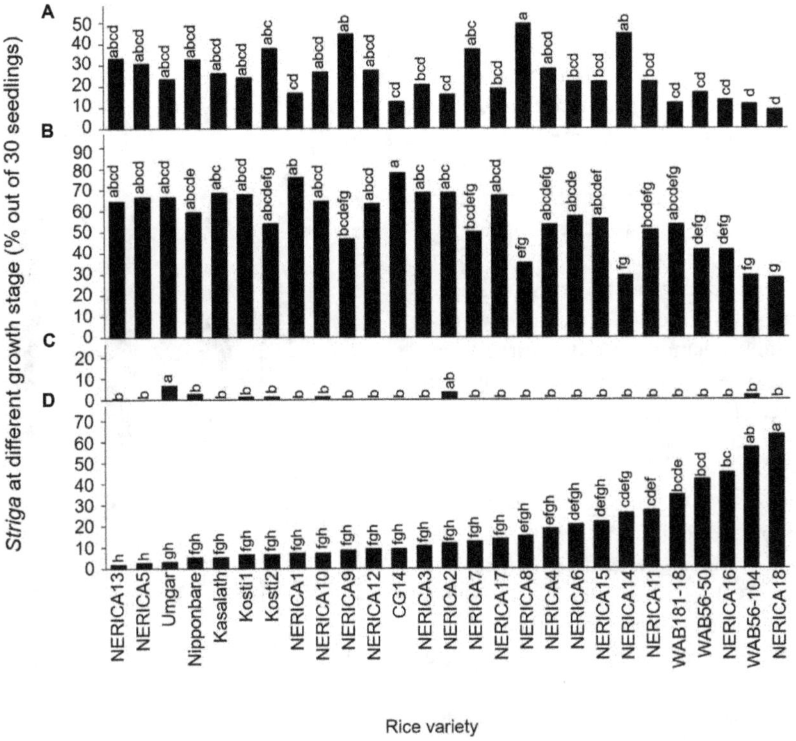

FIGURE 2 | Percentage out of 30 seedlings of *S. hermonthica* at stage (A) I, failure to attach to the root due to a lack of haustorium formation (Figure 1a); (B) II, successful attachment but no shoot elongation (Figures 1b–d); (C) III, shoot necrosis after reaching the four or more leaf pairs stage (Figures 1e,f); and (D) IV, successful parasitism with healthy shoots (Figures 1g,h) on the roots of 27 rice varieties in the first rhizotron experiment at 21 DAI. Means followed by a common letter are not significantly different by Tukey's HSD test in each growth stage ($p < 0.05$).

acetate. The original and concentrated root exudates were stored at −20°C, for at most 3 days, until completion of evaporating all samples at each sampling time. The germination bioassay followed the protocol described by Motonami et al. (2013) using *Striga* seeds that had been conditioned under sterilized conditions for 12 days on 8 mm diameter glass fiber disks (approximately 50 seeds per disk). Disks containing conditioned *Striga* seeds, dapped on a filter paper to remove excess water, were treated with an aliquot (20-μL each) of the original root exudates. Aliquots 20, 10, and 5-μL each, equivalent to 50, 25, and 10-time concentrated root exudates, respectively) of the concentrated extract were applied to a new 8-mm glass fiber disk. The disks were allowed to dry at room temperature for 2 h. Each treated disk was overlaid by a disk containing conditioned *Striga* seeds and moistened with distilled water (40 μL). The disks were incubated in the dark at 30°C for 1 day and examined for germination using a stereomicroscope. The synthetic germination stimulant GR24 was used at 0.34 μM as a positive control and tap water and distilled water were used as a negative controls. Three plants per variety were used and the root exudate from each plant was tested in five technical replicates (five disks).

Statistical Analysis

The statistical package R (Version 3.0.2, R Core Team, 2015) was used for comparison of treatment means. In the rhizotron evaluation, the number of *Striga* seedlings in each growth stage was examined by Tukey's HSD test with rice variety as a factor. In the pot and semi-controlled open air experiments, the number of emergent *Striga* plants and DW of rice varieties were analyzed by Student's *t*-test or Tukey's HSD test. The percentage values in germination tests were transformed to arcsine, compared by Tukey's HSD test and back-transformed (Kgosi et al., 2012).

RESULTS

Rhizotron Evaluation of Post-Attachment Resistance

Developmental arrest of the parasite at stage I, failure in attachment, was common among all varieties (**Figure 2A**). Of the seedling inoculums, failure to attach to the host roots was less than 20, 20–30, 30–40 and more than 40% in 9, 10, 5, and 3 varieties, respectively. The developmental arrest at this stage was highest in NERICA8 (49.3%) and lowest in NERICA18 (8.7%).

Developmental arrest of the parasite at stage II, successful attachment but no shoot elongation, was common among all varieties (**Figure 2B**). Proportions of *Striga* seedling showing arrested development at this stage were less than 30, 30–40, 40–50, 50–60, and more than 60% in 3, 1, 3, 8, and

12 varieties, respectively. Successful attachment but no shoot elongation of *Striga* was highest in NERICA1 and CG14 (76.0–78.0%) and lowest in NERICA18 (28%; **Figure 2B**). Other than NERICA1 and CG14, there were 10 varieties, including NERICA13, NERICA5, and Umgar, in which more than 60% of the inoculated parasites showed arrested development at stage II (successful attachment but no shoot elongation).

The developmental arrest of *Striga* at stage III, necrosis of the parasite shoot, was observed in seven varieties (**Figure 2C**) and was highest in Umgar (6.7%). In this variety, 90% of the inoculated parasites underwent developmental arrest at stage I (failure in attachment) or stage II (successful attachment but no shoot elongation). These findings demonstrate that three *Striga* seedlings (10% of 30 seedlings) developed elongated shoots, but two of them (6.7% of 30 seedlings) died on Umgar. In other words, two-third of *Striga* seedlings that developed elongated shoots died on the variety.

The *Striga* seedlings at stage IV, classified as successful parasitism with healthy shoots, varied widely among varieties and represented 2–63.3% of the original inoculums (**Figure 2D**). Of the inoculated seedlings less than 10, 10–20, 20–30, 40–50, and more than 50% were maintained up to successful parasitism by 12, 6, 5, 2, and 2 varieties, respectively. NERICA18 retained a considerable proportion (63.3%) of the inoculated seedlings. In contrast, Nipponbare supported only 5.3% of the inoculated seedlings. Three varieties, NERICA13, NERICA5, and Umgar retained less than 5% of the inoculated seedlings up to stage IV (successful parasitism).

The results in the second rhizotron evaluation with a subset of the varieties (**Figure 3**) were similar to those in the first rhizotron evaluation (**Figure 2**). Developmental arrest of the parasite at stage I, failure in attachment, was much lower than that at Stage II, successful attachment but no shoot elongation in both evaluation (**Figures 2A,B** and **3A,B**). Developmental arrest at stage III, necrosis of the parasite shoot, was observed repeatedly in Umgar and Nipponbare (**Figures 2C** and **3C**). *Striga* seedlings at stage IV, successful parasitism, on NERICA5, Umgar, Nipponbare, and Kasalath (2.7, 3.3, 5.3, and 5.3% in the first evaluation and 12.0, 5.4, 8.7, and 9.3% in the second evaluation, respectively) was much lower than on NERICA4 (18.7 and 38.7% in the first and second evaluation, respectively), albeit significant differences between the former four varieties and NERICA4 was observed only in the second evaluation (**Figures 2D** and **3D**).

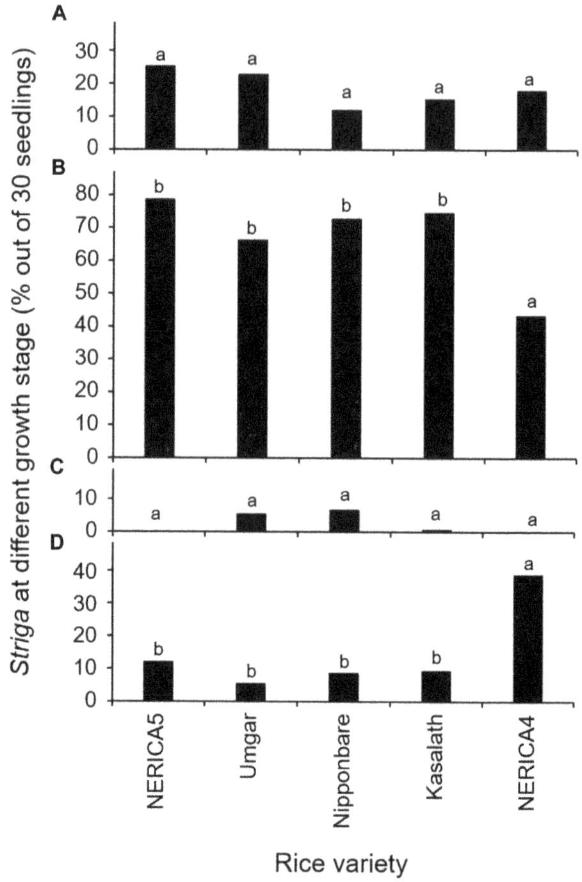

FIGURE 3 | Percentage out of 30 seedlings of *S. hermonthica* at stage (A) I, failure to attach to the root due to a lack of haustorium formation (Figure 1a); (B) II, successful attachment but no shoot elongation (Figures 1b–d); (C) III, shoot necrosis after reaching the four or more leaf pairs stage (Figures 1e,f); and (D) IV, successful parasitism with healthy shoots (Figures 1g,h) on the roots of five rice varieties in the second rhizotron experiment at 21 DAI. Means followed by a common letter are not significantly different by Tukey's HSD test in each growth stage ($p < 0.05$).

Pot and Semi-Controlled Open Air Evaluations of Selected Rice Varieties

In the first pot experiments, Kosti1 supported 0, 0, 0.3, 0.7, 1.0, and 3.7 emergent *Striga* plants per pot at seed bank size of 0, 1, 2, 4, 8, and 16 mg *Striga* (**Figure 4**). The corresponding numbers for Kosti2 were 0, 0, 0.3, 1.7, 1.0, and 1.7, respectively. Umgar supported no emergent *Striga* plants at seed bank size of 0–8 mg. It supported 1.3 emergent *Striga* plants per pot at a seed bank size of 16 mg. However, the emergent *Striga* plants in Umgar pots died over a short period of time.

In the second pot experiment, irrespective of rice variety, no *Striga* plants appeared in the *Striga*-free controls. Emergence of the parasite in the *Striga*-infested pots showed dependence on time and rice variety (**Figure 5**). Up to 1907 GDD, *Striga* emergence was lower than three plants per pot, irrespective of rice variety. However, the final parasite emergence reached 0, 0.5, 2.5, 4.7, 6.7, and 13.0 plants per pot for Umgar, NERICA5, NERICA13, Nipponbare, NERICA4, and NERICA18, respectively. There were significant differences ($p < 0.05$) between Umgar and NERICA18, between NERICA5 and NERICA18 and between NERICA13 and NERICA18. The difference between Nipponbare and NERICA18 was also significant ($p < 0.10$). *Striga* had no significant adverse effects ($p > 0.10$) on the total above-ground parts and panicle DW in Umgar, NERICA5, and NERICA13 (**Table 1**). However, significant reductions ($p < 0.01$ or 0.05) in total above-ground parts and panicle DW were observed in NERICA4, NERICA18, and Nipponbare (**Table 1**).

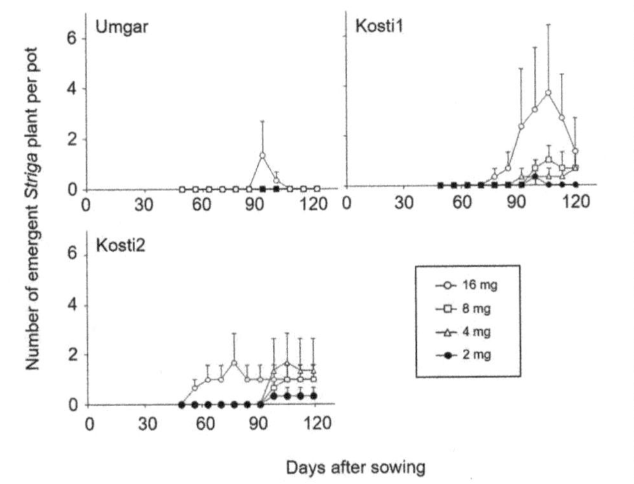

FIGURE 4 | Number of emergent *S. hermonthica* plants per pots planted with rice. Umgar, Kosti1, and Kosti2 were used in the first pot experiment. The *Striga* seeds (2, 4, 8, and 16 mg) were mixed into top 5 cm of soil in each pot. Bars indicate standard errors of each mean value of three replications.

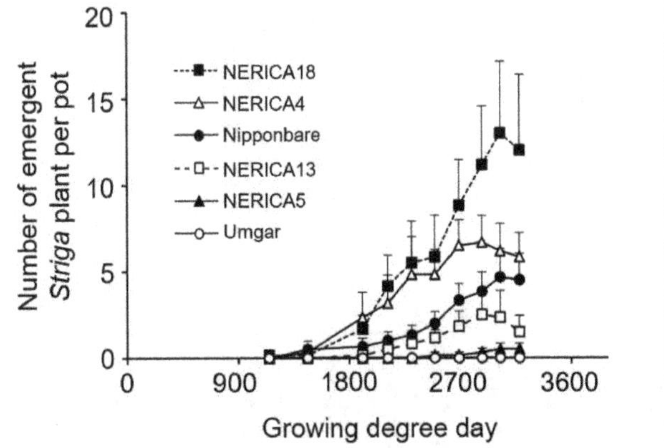

FIGURE 5 | Number of emergent *S. hermonthica* plants per pot planted with rice. Umgar, NERICA5, NERICA13, Nipponbare, NERICA4, and NERICA18 were used in the second pot experiment. The *Striga* seeds (16 mg) were mixed into top 5 cm of soil in each pot. Bars indicate standard errors of each mean value of six replications.

In the third pot experiment, irrespective of rice variety, no *Striga* plants appeared in the *Striga*-free controls. *Striga* emergence in the *Striga*-infested pots was lowest (2.2 plants per pot) on Umgar and NERICA5, irrespective of *Striga* seed bank size (**Figure 6**). NERICA13, at seed bank size of 16, 32, and 48 mg *Striga*, supported 1.8, 5.7, and 3.8 emergent *Striga* plants per pot, respectively. The corresponding numbers of emergent *Striga* plants were 11.7, 16.4, and 8.7 for NERICA4, 7.4, 5.7, and 3.8 for Kostil and 5.7, 15.3, and 10.0 for Kosti2, respectively. The differences between seed bank size of 16 and 32 mg was significant ($p < 0.05$) in NERICA13 and Kosti2. The deference between 32 and 48 mg was also significant for NERICA4 ($p < 0.10$). *Striga*, significantly ($p < 0.05$), reduced total above-ground parts DW of all varieties (**Table 2**) and the observed reductions were 17.2–36.8, 27.9–44.8, 74.5–84.9, 78.4–88.1, 84.6–86.1, and 71.8–75.9% for Umgar, NERICA5, NERICA13, NERICA4, Kostil, and Kosti2, respectively.

In the semi-controlled open air evaluation, irrespective of rice variety, no *Striga* plants appeared in the *Striga*-free controls. The number of emergent *Striga* plants per 10 rice hills in the *Striga*-infested field reached 2, 2, 4.8, 13.3, 13.5, and 18.3 for Umgar, NERICA5, NERICA13, NERICA18, NERICA4, and Nipponbare, respectively (**Figure 7**). There were significant differences ($p < 0.05$) between Umgar and Nipponbare, between NERICA5 and Nipponbare and between NERICA13 and Nipponbare. The differences between Umgar and NERICA18, between Umgar and NERICA4, between NERICA5 and NERICA18 and between NERICA5 and NERICA4 were also significant ($p < 0.10$). *Striga* had no significant adverse effects ($p > 0.10$) on total above-ground parts and panicle DW of Umgar, NERICA5, and NERICA13 (**Table 3**). However, the parasite reduced the total above-ground parts DW of NERICA4, NERICA18, and Nipponbare by 60.4% ($p < 0.05$), 43.1% ($p < 0.05$), and 45.3% ($p < 0.10$), respectively. Panicle DW was reduced by 69.5% ($p < 0.10$), 56.0% ($p < 0.05$), and 58.2% ($p < 0.10$) in NERICA4, NERICA18, and Nipponbare, respectively.

TABLE 1 | Effects of *Striga hermonthica* on rice total above-ground parts and panicle DW in the second open-air pot experiment in 2012.

Variety	Total above-ground parts DW per pot (g)		Panicle DW per pot (g)	
	***Striga*-infested**	***Striga*-free**	***Striga*-infested**	***Striga*-free**
Umgar	39.8 ± 7.1	45.4 ± 5.0 NS	10.5 ± 1.2	8.8 ± 1.2 NS
NERICA5	18.8 ± 3.6	23.8 ± 4.8 NS	3.9 ± 1.1	4.6 ± 1.5 NS
NERICA13	23.0 ± 4.7	32.9 ± 7.5 NS	3.7 ± 1.5	3.5 ± 1.5 NS
NERICA4	10.3 ± 3.2	39.6 ± 9.3*	1.6 ± 0.7	6.4 ± 1.0**
NERICA18	15.1 ± 4.3	51.1 ± 3.9**	3.3 ± 0.8	7.5 ± 1.4*
Nipponbare	17.8 ± 4.5	31.5 ± 2.7*	2.9 ± 1.0	6.5 ± 1.2*

*Data are mean ± standard error of six replications. * and ** indicate significant differences at p < 0.01 and 0.05, respectively, by Student t-test within a variety. NS indicates not significant (p > 0.10) within a variety.*

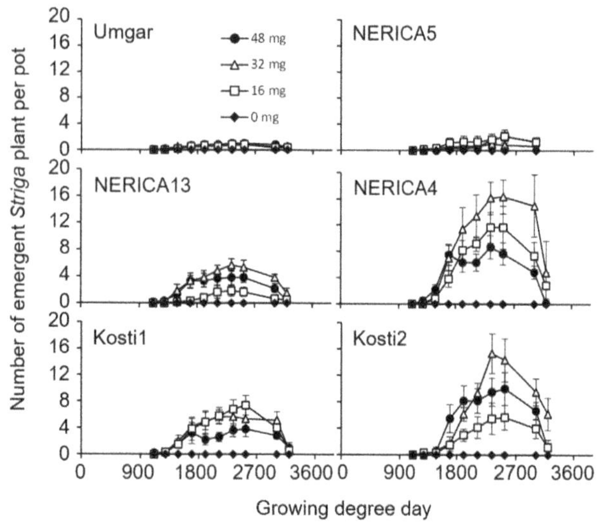

FIGURE 6 | Number of emergent *S. hermonthica* plants per pots planted with rice. Umgar, NERICA5, NERICA13, NERICA4, Kosti1, and Kosti2 were used in the third pot experiment. The *Striga* seeds (0, 16, 32, and 48 mg) were mixed into top 5 cm of soil in each pot. Bars indicate standard errors of each mean value of six replications.

Evaluation of Germination-Inducing Activity as Pre-Attachment Resistance

In all experiments tap water and distilled water induced negligible to little (0–5.4%) germination, while GR24 at 0.34 μM elicited over 90% germination (data not shown). Original root exudates of sorghum induced more than 80% germination. The 10, 25, and 50-time concentrated root exudates of sorghum induced 89.4, 96.4, and 95.1% germination, respectively. This confirms that the method used in our study could extract germination stimulants from hydroponic culture solutions.

The original root exudates from the rice varieties induced negligible to little (0.1–16.3%) germination (**Figure 8A**). Pooled across sampling dates, there was no significant ($p > 0.05$) difference among rice varieties. The 10 and 25-time concentrated root exudates from Umgar exhibited lower activity (1.9–29.6%) than those from NERICA4, NERICA13, and NERICA18 (14.7–65.0%) on all sampling dates (**Figures 8B,C**). Pooled across sampling dates, root exudates from Umgar exhibited significantly lower ($p < 0.05$) activities compared to those from NERICA4, NERICA13, and NERICA18 and there was no significant ($p > 0.05$) difference among Umgar, Nipponbare, and NERICA5. The 50-time concentrated root exudate from Umgar exhibited lower activity compared to those from the other rice varieties on all sampling dates, except at 7 DAS (**Figure 8D**). Pooled across sampling date, root exudates from Umgar and Nipponbare exhibited significantly lower ($p < 0.05$) activities compared to those from NERICA4, NERICA13, and NERICA18. Germination inducing activity from NERICA5 was intermediate among the six rice varieties.

DISCUSSION

Expanding upland rice cropping using irrigation water is one of options to boost domestic rice production in Sudan. Parasitic weeds thrive in production systems characterized by degraded soils with a lack of water control (Rodenburg et al., 2010). However, in the present study, obvious adverse effects from parasitism of a resident *Striga* population were observed on NERICA4 and NERICA18, probable candidates for widespread cultivation in Sudan, repeatedly in two pot experiments and a semi-controlled open air experiment with irrigation. These findings underpin the prediction that expanding upland rice production using supplementary irrigation in Sudan without due attention to the *Striga* problem is fraught with an element of risk. Identification of rice varieties with resistance to resident *Striga* populations and adaptability to the target environment is imperative to evade the risk.

The results of the rhizotron evaluation indicated that resistance to *S. hermonthica* is an outcome of several restrictions that block the parasite ingress and development. Failure of the parasite to attach to the host roots (stage I) due to lack of haustorium formation, is common to all varieties, however, inter-varietal differences were considerable. Inter-varietal differences

TABLE 2 | Effects of *S. hermonthica* seed density on rice total above-ground parts DW in the third open-air pot experiment in 2013.

Variety	Total above-ground parts DW per pot (g)			
	Amount or *Striga* seeds per pot (mg)			
	0	16	32	48
Umgar	48.2 ± 2.9a	39.9 ± 5.6ab	30.7 ± 3.9b	30.4 ± 3.8b
NERICA5	23.2 ± 2.1a	13.3 ± 3.1b	12.8 ± 1.4b	16.7 ± 1.9ab
NERICA13	32.3 ± 7.7a	8.2 ± 1.0b	4.8 ± 1.1b	4.9 ± 1.3b
NERICA4	29.1 ± 2.9a	3.7 ± 0.8b	6.3 ± 1.5b	3.5 ± 1.7b
Kosti1	22.3 ± 4.2a	3.1 ± 0.6b	3.4 ± 1.0b	3.2 ± 0.9b
Kosti2	22.5 ± 2.8a	5.9 ± 1.5b	6.4 ± 1.4b	5.4 ± 1.4b

Data are mean ± standard error of six replications. Means within a variety followed by a common letter are not significantly different by Tukey's HSD test ($p < 0.05$).

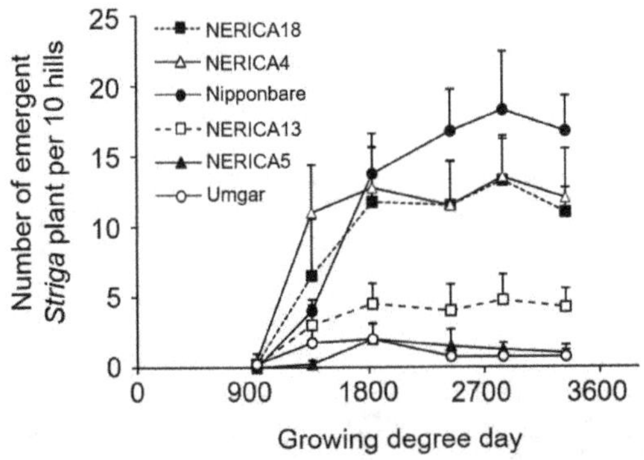

FIGURE 7 | Number of emergent *S. hermonthica* plants per 10 rice hills in an artificially infested field. Umgar, NERICA5, NERICA13, Nipponbare, NERICA4, and NERICA18 were used in a semi-controlled open air experiment. Bars indicate standard errors of each mean value of four replications.

in haustorium initiation is at variance with earlier views that haustorium initiation is not host discriminate and that host specificity is not associated with haustorium initiation, but rather with the ability of haustoria to functionally establish after invading the host (Riopel et al., 1990). However, recent research associated differential pre-attachment resistance to *S. asiatica* in wild sorghum with variability in haustorium-inducing activity (Rich et al., 2004; Bandaranayake and Yoder, 2013).

Developmental arrest of the parasite at stage II (successful attachment but no shoot elongation) contributed to the high post-attachment resistance in NERICA13, NERICA5, and Umgar. A similar resistance phenotype, namely successful attachment but no shoot elongation, has been reported in other rice varieties including Nipponbare, Koshihikari, CG14, NERICA1, and NERICA10 (Gurney et al., 2006; Swarbrick et al., 2008; Yoshida and Shirasu, 2009; Cissoko et al., 2011). However, the reasons for the developmental arrest are unclear and need to be ascertained. Although deposition of dense staining material(s) at the interface between NERICA10 and *Striga* has been observed (Cissoko et al., 2011), the mechanism was not always applicable to other cases. In addition, root endodermis structure did not explain the difference between a resistant variety (Nipponbare) and a susceptible variety (Kasalath; Gurney et al., 2006). Furthermore, hypersensitivity, which has been observed in sorghum (Mohamed et al., 2003), was not observed in this study.

The high post-attachment resistance of Umgar was partly attributable to the developmental arrest of *Striga* at stage III, namely shoot necrosis after reaching the four or more leaf pairs stage. The developmental arrest at this stage suggests inhibition following successful establishment of a vascular connection between the host and the parasite. Resistance following successful connection with host vascular system has been observed with several *Striga* and *Orobanche* species and is attributed to blockage of the host vessels and disruption of the flux of water, nutrients and carbohydrates from host to parasite (Arnaud et al., 1999; Amusan et al., 2008; Timko and Scholes, 2013). This resistance phenotype was not observed in NERICA13 and NERICA5, demonstrating that Umgar has additional resistance phenotype(s) which come(s) into play after establishment of connection with the host vascular system.

The results from the pot and semi-controlled open air experiments corroborated the strong *Striga*-resistance of NERICA5 as identified in the rhizotron experiment. NERICA5, irrespective of *Striga* seed bank size, sustained the lowest *Striga* emergence. The sharp decline in *Striga* emergence at the highest seed bank size (48 mg per pot) notable with the susceptible variety NERICA4 and Kosti2 could be due to intra-specific competition among the attached parasite seedlings and to severe inhibition of host growth. Interestingly, NERICA5 was reported as a variety with broad spectrum resistance to *Striga* species and ecotypes (Cissoko et al., 2011; Jamil et al., 2011; Rodenburg et al., 2015). The results of the current study, invariably, indicate the resistance of NERICA5 to another population of *S. hermonthica*.

The strong *Striga*-resistance of Umgar to a resident *Striga* population was also confirmed repeatedly in the rhizotron, the pot and the semi-controlled open air experiments undertaken in this study. Umgar is an upland variety released in Sudan for commercial production (Nyachae, 2011). Two other released rice varieties in Sudan, Kostil and Kosti2, did not exhibit *Striga*-resistance at the same level as Umgar. It must be taken into account that *Striga* seeds were mixed into the top 5 cm of soil in the pot experiments and the infested plot was further augmented with the parasite seeds placed in each rice-planting hole in the semi-controlled open air experiment. The possibility of having

TABLE 3 | Effects of *S. hermonthica* on rice total above-ground parts and panicle DW in a semi-controlled open air experiment in 2012.

Variety	Total above-ground parts DW (g m^{-2})		Panicle DW (g m^{-2})	
	Striga-infested	*Striga*-free	*Striga*-infested	*Striga*-free
Umgar	1010 ± 115	1000 ± 79 NS	293 ± 91	308 ± 41 NS
NERICA5	420 ± 65	378 ± 44 NS	128 ± 34	116 ± 25 NS
NERICA13	552 ± 82	543 ± 71 NS	143 ± 26	208 ± 55 NS
NERICA4	354 ± 140	894 ± 140*	92 ± 55	302 ± 80+
NERICA18	366 ± 70	643 ± 51*	84 ± 15	191 ± 32*
Nipponbare	196 ± 46	358 ± 62+	41 ± 11	98 ± 24+

*Data are mean ± standard error of four replications. * and + indicate significant differences at $p < 0.05$ and 0.10, respectively, by Student t-test within a variety. NS indicates not significant ($p > 0.10$) within a variety.*

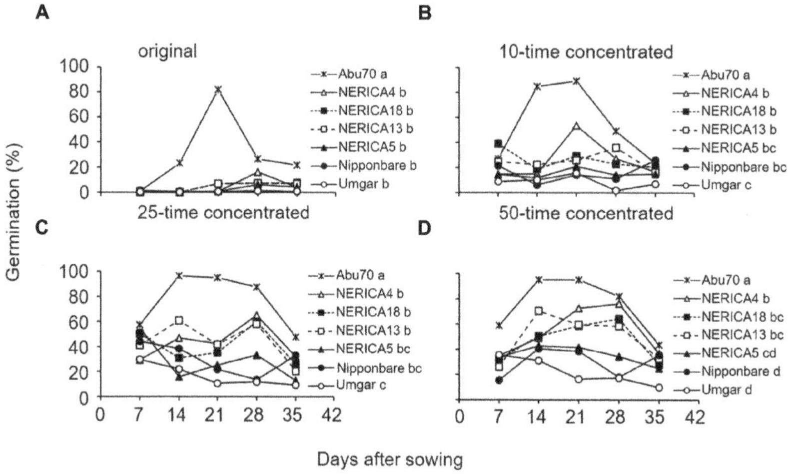

FIGURE 8 | Germination inducing activity of root exudates from Umgar, NERICA5, NERICA13, Nipponbare, NERICA4, and NERICA18 grown in a growth chamber. Original **(A)**, 10-time concetrated **(B)**, 25-time concentrated **(C)** and 50-time concetrated **(D)** root exudates were used in the germintion test. The percentage values were arcsine transformed, compared by multiple mean-separation tests among rice and sorghum varieties and back-transformed. Different letters next to variety name indicate significant difference ($p < 0.05$) by Tukey's HSD test pooled across sampling date.

differential response of the rice varieties to *Striga* due to variations in root size and architecture and/or vertical distribution of the parasite seeds cannot be ruled out.

The lower germination inducing activity of root exudates from Umgar showed pre-attachment resistance. Although the germination inducing activity from NERICA5 did not show any significant difference ($p > 0.05$) to that from Umgar, the former was always higher than the latter, except with the test using original root exudates which would be too diluted to induce *Striga* germination. This finding suggest that Umgar has higher pre-attachment resistance than NERICA5. The mechanism of low induction of *Striga* seed germination by Umgar could be further investigated. The germination test using serial concentration ruled out the possibility of inhibition of *Striga* seed germination by high concentration of the stimulant. However, involvement of germination inhibitors cannot be ruled out as the solvent extraction may exclude hydrophilic inhibitors (Yoneyama et al., 2010).

NERICA13, as revealed by the rhizotron experiment, showed post-attachment resistance comparable to Umgar and NERICA5. However, it sustained higher *Striga* emergence in the pot and semi-controlled open air experiments. The differential performance of these varieties may be attributed to the relatively high germination inducing activity of root exudates of NERICA13 compared to the other varieties.

The results from rhizotron, pot and semi-controlled open air evaluations and germination test were at variance for Nipponbare. The rhizotron evaluation and germination test confirmed previous reports of a high *Striga* resistance (Gurney et al., 2006; Swarbrick et al., 2008; Yoshida and Shirasu, 2009). However, the results from pot and semi-controlled open air evaluations demonstrated a high *Striga* emergence. Notably, Nipponbare displayed poor growth and reached 50% heading very early at 44 DAS in the semi-controlled open air experiment (data not shown). The notable high *Striga* emergence could thus be due to the attendant stress and/or differences in ecotypes of *Striga* used by the previous workers (Gurney et al., 2006; Swarbrick et al., 2008; Yoshida and Shirasu, 2009) and the one used in this experiment. The poor growth and early flowering could be attributed, among other factors, to photoperiodic sensitivity (Matsubara et al., 2008). Nipponbare is a lowland variety and not adapted to aerobic conditions. The results also varied for Kasalath. Although, it was evaluated as *Striga*-susceptible or -tolerant in previous reports (Gurney et al., 2006; Swarbrick et al., 2009; Yoshida and Shirasu, 2009; Yoder and Scholes, 2010), it exhibited resistance comparable to Nipponbare repeatedly in the rhizotron experiments. The variance could be attributable to differences in *Striga* populations used in the previous studies and the present one. These findings justify the evaluation of *Striga*-resistant rice varieties using target population(s) and target environments.

CONCLUSION

The relatively low germination inducing activity of root exudate together with the notable developmental arrest of the inoculated seedlings during stage II (successful attachment but no shoot elongation) and stage III (shoot necrosis) as revealed by the rhizotron experiments showed that Umgar possesses both pre- and post-attachment resistance. The high *Striga* resistance of Umgar was confirmed in pot and semi-controlled open air experiments conducted in Sudan. Its multiple resistance phenotypes to *Striga* would confer a wider ecological amplitude. NERICA5, which was reported as a variety with broad spectrum of resistance to *Striga* species and ecotypes, exhibited resistance against a population of *S. hermonthica* resident in Sudan.

However, further studies including grain yield data under farmers' conditions where *Striga* is pandemic is essential to evaluate environmental adaptability and usefulness of Umgar and NERICA5 in Sudan and to compare the results to other studies. Other rice varieties not included in pot and semi-controlled open air experiments in this study could express high *Striga* resistance, and therefore screening other rice varieties is highly recommended. Furthermore, spatial and temporal stability and durability of *Striga* resistance have to be further ascertained taking into account variability in environmental conditions and virulence between *S. hermonthica* ecotypes.

AUTHOR CONTRIBUTIONS

YS and HS conceived and designed the experiments with help of AB. HS performed the experiments with guidance of AM. HS, YS, and AB wrote the manuscript; all authors contributed to the discussion and approved the final manuscript.

FUNDING

This study was supported by a JST/JICA SATREPS (Japan Science and Technology Agency, Tokyo Japan and International Cooperation Agency, Tokyo, Japan) and JSPS KAKENHI Grant Numbers 23790015 and 26450019 (Japan Society for the Promotion of Science, Tokyo, Japan).

ACKNOWLEDGMENT

The authors are indebted to Mr. Ismail Ibrahim Elmunsor and Ms. Chizu Yoshimoto for daily maintenance of plant materials used in the experiments.

REFERENCES

Ali, R. A., El-Hussein, A. A., Mohamed, K. I., and Babiker, A. G. T. (2009). Specificity and genetic relatedness among *Striga hermonthica* strains in Sudan. *Life Sci. Int. J.* 3, 1159–1166.

Amusan, I. O., Rich, P. J., Menkir, A., Housley, T., and Ejeta, G. (2008). Resistance to *Striga hermonthica* in a maize inbred line derived from *Zea diploperennis*. *New Phytol.* 178, 157–166. doi: 10.1111/j.1469-8137.2007.02355.x

Arnaud, M. C., Véronési, C., and Thalouarn, P. (1999). Physiology and histology of resistance to *Striga hermonthica* in *Sorghum bicolor* var. Framida. *Aust. J. Plant Physiol.* 26, 63–70. doi: 10.1071/PP98070

Babiker, A. G. T. (2007). *Striga*: the spreading scourge in Africa. *Regul. Plant Growth Dev.* 42, 74–87.

Balasubramanian, V., Sie, M., Hijmans, R. J., and Otsuka, K. (2007). Increasing rice production in sub-Saharan Africa: challenges and opportunities. *Adv. Agron.* 94, 55–133. doi: 10.1016/S0065-2113(06)94002-4

Bandaranayake, P. C. G., and Yoder, J. I. (2013). "Haustorium initiation and early development," in *Parasitic Orobanchaceae. Parasitic Mechanisms and Control Strategies*, eds D. M. Joel, J. Gressel, and L. J. Musselman (Heidelberg: Springer), 61–74. doi: 10.1007/978-3-642-38146-1_4

Cissoko, M., Boisnard, A., Rodenburg, J., Press, M. C., and Scholes, J. D. (2011). New Rice for Africa (NERICA) cultivars exhibit different levels of post-attachment resistance against the parasitic weeds *Striga hermonthica* and *Striga asiatica*. *New Phytol.* 192, 952–963. doi: 10.1111/j.1469-8137.2011.03846.x

Ejeta, G. (2007). "The *Striga* scourge in Africa: a growing pandemic," in *Integrating New Technologies for Striga Control: Towards Ending the Witch-Hunt*, eds G. Ejeta and J. Gressel (Singapore: World Scientific Publishing), 3–16. doi: 10.1142/9789812771506_0001

Fageria, N. K. (2007). Yield physiology of rice. *J. Plant Nutr.* 30, 843–879. doi: 10.1080/15226510701374831

FAO (2015). *FAOSTAT*. Available at: http://faostat3.fao.org/download/Q/QC/E [accessed March 9, 2015].

Gurney, A. L., Slate, J., Press, M. C., and Scholes, J. D. (2006). A novel form of resistance in rice to the angiosperm parasite *Striga hermonthica*. *New Phytol.* 169, 199–208. doi: 10.1111/j.1469-8137.2005.01560.x

Hiraoka, Y., and Sugimoto, Y. (2008). Molecular responses of *Sorghum* to purple witchweed (*Striga hermonthica*) parasitism. *Weed Sci.* 56, 356–363. doi: 10.1614/WS-07-136.1

Huang, K., Whitlock, R., Press, M. C., and Scholes, J. C. (2012). Variation for host range within and among populations of the parasitic plant *Striga hermonthica*. *Heredity (Edinb.)* 108, 96–104. doi: 10.1038/hdy.2011.52

Jamil, M., Rodenburg, J., Charnikhova, T., and Bouwmeester, H. J. (2011). Pre-attachment *Striga hermonthica* resistance of new rice for Africa (NERICA) cultivars based on low strigolactone production. *New Phytol.* 192, 964–975. doi: 10.1111/j.1469-8137.2011.03850.x

Jones, M. P., Dingkuhn, M., Aluko, G. K., and Semon, M. (1997a). Interspecific *Oryza sativa* L. X *O. glaberrima* Steud. progenies in upland rice improvement. *Euphytica* 94, 237–246. doi: 10.1023/A:1002969932224

Jones, M. P., Mande, S., and Aluko, K. (1997b). Diversity and potential of *Oryza glaberrima* Steud in upland rice breeding. *Breed. Sci.* 47, 395–398. doi: 10.1270/jsbbs1951.47.395

Kato, Y., Kamoshita, A., and Yamagishi, J. (2007). Evaluation the resistance of six rice cultivars to drought: restriction of deep rooting and the use of raised beds. *Plant Soil* 300, 149–161. doi: 10.1007/s11104-007-9397-z

Kgosi, R. L., Zwanenburg, B., Mwakaboko, A. S., and Murdoch, A. J. (2012). Strigolactone analogues induce suicidal seed germination of *Striga* spp. in soil. *Weed Res.* 52, 197–203. doi: 10.1111/j.1365-3180.2012.00912.x

Kim, S. K. (1996). Horizontal resistance: core to a research breakthrough to combat *Striga* in Africa. *Integr. Pest Manage. Rev.* 1, 229–249. doi: 10.1007/BF00139766

Lendzemo, V. W., and Kuyper, T. W. (2001). Effects of arbuscular mycorrhizal fungi on damage by *Striga hermonthica* on two contrasting cultivars of *Sorghum*, *Sorghum bicolor*. *Agric. Ecosyst. Environ.* 87, 29–35. doi: 10.1016/S0167-8809(00)00293-0

López-Ráez, J. A., Matusova, R., Cardoso, C., Jamil, M., Charnikhova, T., Kohlen, W., et al. (2009). Strigolactones: ecological significance and use as a target for parasitic plant control. *Pest Manag. Sci.* 65, 471–477. doi: 10.1002/ps.1692

Matsubara, K., Kono, I., Hori, K., Nonoue, Y., Ono, N., Shomura, A., et al. (2008). Novel QTLs for photoperiodic flowering revealed by using reciprocal backcross inbred lines from crosses between japonica rice cultivars. *Theor. Appl. Genet.* 117, 935–945. doi: 10.1007/s00122-008-0833-0

Mohamed, A., Ellicott, A., Housley, T. L., and Ejeta, G. (2003). Hypersensitive response to *Striga* infection in *Sorghum*. *Crop Sci.* 43, 1320–1324. doi: 10.2135/cropsci2003.1320

Motonami, N., Ueno, K., Nakashima, H., Nomura, S., Mizutani, M., Takikawa, H., et al. (2013). The bioconversion of 5-deoxystrigol to sorgomol by the *Sorghum*, *Sorghum bicolor* (L.) Moench. *Phytochemistry* 93, 41–48. doi: 10.1016/j.phytochem.2013.02.017

Mustafa, M. A., and Elsheikh, M. A. Y. (2007). Variability, correlation and path co-efficient analysis for yield and its components in rice. *Afr. Crop Sci. J.* 15, 183–189.

Niones, J. M., Suralta, R. R., Inukai, Y., and Yamauchi, A. (2012). Field evaluation on functional roles of root plastic responses on dry matter production and grain yield of rice under cycles of transient soil moisture stresses using chromosome segment substitution lines. *Plant Soil* 359, 107–120. doi: 10.1007/s11104-012-1178-7

Nyachae, C. (2011). *Sudanese Seed Sector. A Baseline Study/Survey*. Available at: http://afsta.org/media/baseline-studies/ [accessed February 29, 2016].

Ohe, M., Okita, N., and Daimon, H. (2010). Effects of deep-flooding irrigation on growth, canopy structure and panicle weight yield under different planting patterns in rice. *Plant Prod. Sci.* 13, 193–198. doi: 10.1626/pps.13.193

R Core Team (2015). *R: A Language and Environment for Statistical Computing.* Vienna: R Foundation for Statistical Computing. Available at: http://www.r-project.org/ [accessed April 30, 2015].

Rich, P. J., Grenier, C., and Ejeta, G. (2004). *Striga* resistance in the wild relatives of *Sorghum. Crop Sci.* 44, 2221–2229. doi: 10.2135/cropsci2004.2221

Riopel, J. L., Baird, W. V., Chang, M., and Lynn, D. G. (1990). "Haustorial development in *Striga asiatica*," in *Witchweed Research and Control in the United States*, eds P. F. Sand, R. E. Eplee, and R. G. Westbrroks (Champaign IL: Weed Science Society of America), 27–36.

Rodenburg, J., Cissoko, M., Kayeke, J., Dieng, I., Khan, Z. R., Midega, C. A. O., et al. (2015). Do NERICA rice cultivars express resistance to *Striga hermonthica* (Del.) Benth. and *Striga asiatica* (L.) Kuntze under field conditions? *Field Crop Res.* 170, 83–94. doi: 10.1016/j.fcr.2014.10.010

Rodenburg, J., Diagne, A., Oikeh, S., Futakuchi, K., Kormawa, P. M., Semon, M., et al. (2006). Achievements and impact of NERICA on sustainable rice production in sub-Saharan Africa. *Int. Rice Comm. Newsl.* 55, 45–58.

Rodenburg, J., Riches, C. R., and Kayeke, J. M. (2010). Addressing current and future problems of parasitic weeds in rice. *Crop Prot.* 29, 210–221. doi: 10.1016/j.cropro.2009.10.015

Somado, E. A., and Guei, R. G. (2008). "The role of INGER-Africa," in *NERICA: The New Rice for Africa—A Compendium*, eds E. A. Somado, R. G. Guei, and S. O. Keya (Cotonou: Africa Rice Center (WARDA)), 35–40.

Somado, E. A., Guei, R. G., and Nguyen, N. (2008). "NERICA: origins, nomenclature and identification characteristics," in *NERICA: The New Rice for Africa—A Compendium*, eds E. A. Somado, R. G. Guei, and S. O. Keya (Cotonou: Africa Rice Center (WARDA)), 10–30.

Swarbrick, P. J., Huang, K., Liu, G., Slate, J., Press, M. C., and Scholes, J. D. (2008). Global patterns of gene expression in rice cultivars undergoing a susceptible or resistant interaction with the parasitic plant *Striga hermonthica. New Phytol.* 179, 515–529. doi: 10.1111/j.1469-8137.2008.02484.x

Swarbrick, P. J., Scholes, J. D., Press, M. C., and Slate, J. (2009). A major QTL for resistance of rice to the parasitic plant *Striga hermonthica* is not dependent on genetic background. *Pest Manag. Sci.* 65, 528–532. doi: 10.1002/ps.1719

Timko, M. P., and Scholes, J. D. (2013). "Host reaction to attack by root parasitic plants," in *Parasitic Orobanchaceae. Parasitic Mechanisms and Control Strategies*, eds D. M. Joel, J. Gressel, and L. J. Musselman (Heidelberg: Springer), 115–141. doi: 10.1007/978-3-642-38146-1_7

Uga, Y., Okuno, K., and Yano, M. (2011). Drol, a major QTL involved in deep rooting of rice under upland field conditions. *J. Exp. Bot.* 62, 2485–2495. doi: 10.1093/jxb/erq429

Van Ast, A., and Bastiaans, L. (2006). The role of infection time in the differential response of *Sorghum* cultivars to *Striga hermonthica* infection. *Weed Res.* 46, 264–274. doi: 10.1111/j.1365-3180.2006.00507.x

Wopereis, M. C. S., Diagne, A., Rodenburg, J., Sié, M., and Somado, E. A. (2008). Why NERICA is a successful innovation for African farmers. A response to Orr et al from the Africa rice center. *Outlook Agric.* 37, 169–176. doi: 10.5367/000000008785915502

Yoder, J. I., and Scholes, J. D. (2010). Host plant resistance to parasitic weeds; recent progress and bottlenecks. *Curr. Opin. Plant Biol.* 13, 478–484. doi: 10.1016/j.pbi.2010.04.011

Yoneyama, K., Awad, A. A., Xie, X., Yoneyama, K., and Takeuchi, Y. (2010). Strigolactones as germination stimulants for root parasitic plants. *Plant Cell Physiol.* 51, 1095–1103. doi: 10.1093/pcp/pcq055

Yoshida, S., Ishida, J. K., Kamel, N. M., Ali, A. M., Namba, S., and Shirasu, K. (2010). A full-length enriched cDNA library and expressed sequence tag analysis of the parasitic weed, *Striga hermonthica. BMC Plant Biol.* 10:55. doi: 10.1186/1471-2229-10-55

Yoshida, S., and Shirasu, K. (2009). Multiple layers of incompatibility to the parasitic witchweed, *Striga hermonthica. New Phytol.* 183, 180–189. doi: 10.1111/j.1469-8137.2009.02840.x

Conflict of Interest Statement: The authors declare that the research was conducted in the absence of any commercial or financial relationships that could be construed as a potential conflict of interest.

7

Differences in Crenate Broomrape Parasitism Dynamics on Three Legume Crops using a Thermal Time Model

Edited by:
Paul Christiaan Struik,
Wageningen University and Research Centre, Netherlands

Reviewed by:
Grama Nanjappa Dhanapal,
University of Agricultural Sciences, Bangalore, India
Hanan Eizenberg,
Agricultural Research Organization, Israel

***Correspondence:**
Alejandro Pérez-de-Luque
alejandro.perez.luque@juntadeandalucia.es

Alejandro Pérez-de-Luque[1], Fernando Flores[2] and Diego Rubiales[3]*

[1] Área de Mejora y Biotecnología, IFAPA, Centro Alameda del Obispo, Córdoba, Spain, [2] Departamento de Ciencias Agroforestales, ETSI La Rábida, Universidad de Huelva, Huelva, Spain, [3] Institute for Sustainable Agriculture, Consejo Superior de Investigaciones Científicas (CSIC), Córdoba, Spain

Root parasitic weeds are a major limiting production factor in a number of crops, and control is difficult. Genetic resistance and chemical control lead the fight, but without unequivocal success. Models that help to describe and even predict the evolution of parasitism underground are a valuable tool for herbicide applications, and even could help in breeding programs. Legumes are heavily affected by *Orobanche crenata* (crenate broomrape) in the Mediterranean basin. This work presents a descriptive model based on thermal time and correlating growing degree days (GDD) with the different developmental stages of the parasite. The model was developed in three different legume crops (faba bean, grass pea and lentil) attacked by crenate broomrape. The developmental stages of the parasite strongly correlated with the GDD and differences were found depending on the host crop.

Keywords: *Orobanche crenata*, parasitic weeds, GDD, faba bean, lentil, grass pea, legumes

INTRODUCTION

Broomrapes (*Orobanche* and *Phelipanche* spp.) are holoparasitic plants which attach to the host roots and grow at the expense of the host plant's resources. They parasitize a large number of crops, legumes being some of the most severely affected (Joel et al., 2007; Parker, 2013). Because the pathogenesis and most of the growing process take place underground long before diagnosis of the infection, it hampers the development of effective control strategies. In addition, the large amount of seeds released by a single individual (more than 100,000) provides the parasite with a great genetic adaptability to environmental changes (Parker and Riches, 1993; Press and Graves, 1995; Joel et al., 2007). *Orobanche crenata* Forsk. (crenate broomrape) is widely spread through the Mediterranean basin, and remains to date as the most important threat against legumes in the area (Parker, 2009, 2013; Rubiales and Heide-Jørgensen, 2011). Other broomrape species such *Orobanche foetida* Poir. or *Phelipanche aegyptiaca* (Pers.) Pomel can also infect legumes, but are of local importance only.

Although *O. crenata* has a broad host range infecting most legume crops, levels of infection vary between the different species and between accessions within species (Rubiales et al., 2003a,b; Pérez-de-Luque et al., 2004; Román et al., 2007). No complete resistance to *O. crenata* is available in any legume so far, but existing levels of incomplete resistance in some accessions (Rubiales et al., 2006, 2014; Pérez-de-Luque et al., 2007; Fernández-Aparicio et al., 2008, 2012) can delay the growth of already established broomrapes, because the presence of defense mechanisms interferes with the normal development of the parasite (Pérez-de-Luque et al., 2004, 2005).

In addition to the host, the environment can greatly influence the development of the parasitic weed on a given host. Although water availability may be an important factor for broomrape development, particularly in Mediterranean areas (Rubiales et al., 2003a,b; Pérez-de-Luque et al., 2004), it seems that temperature is the main factor affecting broomrape development (Mesa-García and García-Torres, 1986; Castejón-Muñoz et al., 1993; Eizenberg et al., 2004; Ephrath and Eizenberg, 2010). Increased temperature is often associated with an increase in broomrape parasitism, whereas low temperature correlates with lower infections. Cool winters reduce broomrape infection in several crops such as sunflower (*Helianthus annuus* L.), chickpea (*Cicer arietinum* L.), lentil (*Lens culinaris* Med..), pea (*Pisum sativum* L.), vetch (*Vicia sativa* L.) or faba bean (*Vicia faba* L.) (Arjona-Berral et al., 1987; Castejón-Muñoz et al., 1993; Rubiales et al., 2003a,b).

In recent years, several experiments accomplished by Eizenberg and coworkers have correlated the growth of broomrape (*Phelipanche aegyptiaca*, *O. minor* and *O. cumana*) with the thermal time, measured as growing day degrees (GDD) (Eizenberg et al., 2004, 2005; Ephrath and Eizenberg, 2010; Ephrath et al., 2012). Even more, the thermal time can be used to predict broomrape growth on crop roots under the soil, and mathematical models can be developed based on GDD for optimization of herbicide timing application (Eizenberg et al., 2006). This can be a powerful tool for the chemical control of broomrapes, because one of the main problems to date has been establishing the right time for herbicide applications, as to date no consistent relationships between the growth stages of the different hosts and the parasite development have been found (Arjona-Berral et al., 1987).

However, there are no works to date using a thermal time relation as a tool for showing differences in parasite development on different hosts. If temperature affects both, host and parasite development, a thermal time model could be used as a descriptive method for differences in infection levels integrating the two main factors affecting broomrape development: the host and the temperature. For that reason, the relationship between soil temperature and *O. crenata* growth on three different legume crops has been studied and a putative descriptive model based on work on *O. minor* / *Trifolium* by Eizenberg et al. (2005) is proposed. The hypotheses are that a) there are differences in parasite virulence between different host species, b) a sigmoidal relation between thermal time and parasite development exists and c) such relation will show the differences in parasite development between hosts.

MATERIALS AND METHODS

Plant Material

Lentils, faba beans and grass peas (*Lathyrus sativus* L.) were included in the study. The three tested species are natural hosts for *O. crenata*, and the cultivars selected ("Rubia de la Armuña" for lentil, "Prothabon" for faba bean, breeding line "ICARDA 2" for grass pea) were susceptible to the parasitic weed attack.

Field Assays

Field trials were conducted over two cropping seasons (2006–2007 and 2007–2008) at Alameda del Obispo experimental station, in Córdoba (southern Spain). The fields located there present a deep loam soil (Typic Xerofluvent). The selected crops were sown by the end of November in a field highly infested with *O. crenata*, in 2 m length rows, with 0.7 m between rows and 20 plants per row. Four plots were set for each crop and every year (12 plots per year), surrounded and delimited by faba bean rows. The experiment was performed as a randomized, complete block design with four replications.

Temperature Data Recording

Soil temperature was recorded hourly starting January 1st and using data loggers (Gemini Tinytag Plus Data Logger) buried at 5 cm depth and converted to GDD according to McMaster and Wilhelm (1997) using the following equation:

$$GDD = \sum \left[\frac{T_{max} + T_{min}}{2} - T_{base} \right] \quad (1)$$

where T_{max} and T_{min} are the maximum and minimum daily temperatures, respectively, and T_{base} is the base temperature, all of them measured in degree Celsius (°C). T_{base} was fixed as 0°C for faba bean and grass pea (Stützel, 1995; Rao and Northup, 2008), and 1.5°C for lentil (Ellis and Barrett, 1994).

Plant Sampling and Broomrape Development

Beginning the first week of February, five plants were extracted from each crop plot every week. The number of attached broomrapes was recorded and their developmental phase classified according to seven stages (Pérez-de-Luque et al., 2004): Stage 1 (T1), tubercles smaller than 2 mm; Stage 2 (T2), tubercles greater than 2 mm, without root development; Stage 3 (T3), tubercles with crown roots, without shoot formation; Stage 4 (T4), shoot formation, remaining underground; Stage 5 (T5), shoot emergence; Stage 6 (T6), flowering; Stage 7 (T7), setting of seeds.

Statistical Analysis

Broomrape tubercle number was log transformed (log[number of broomrapes per plant]) and the transformed data presented because of nonhomogeneity in variance. In order to model each stage (T1 to T7) of broomrape development, log-transformed broomrape number was nonlinearly regressed to GDD using a three-parameter logistic function (Brown and Mayer, 1988):

$$Y = \frac{a}{1 + \left(\frac{x}{x_o}\right)^b} \quad (2)$$

where Y represents log-transformed broomrape tubercle number, a represents the upper asymptote (maximum broomrape number), x represents GDD, x_o represents the GDD when the Y is 50% of maximum, and b represents the slope at x_o. A t-test was used for comparison of the regression parameter estimates in each parasitism stage between crops

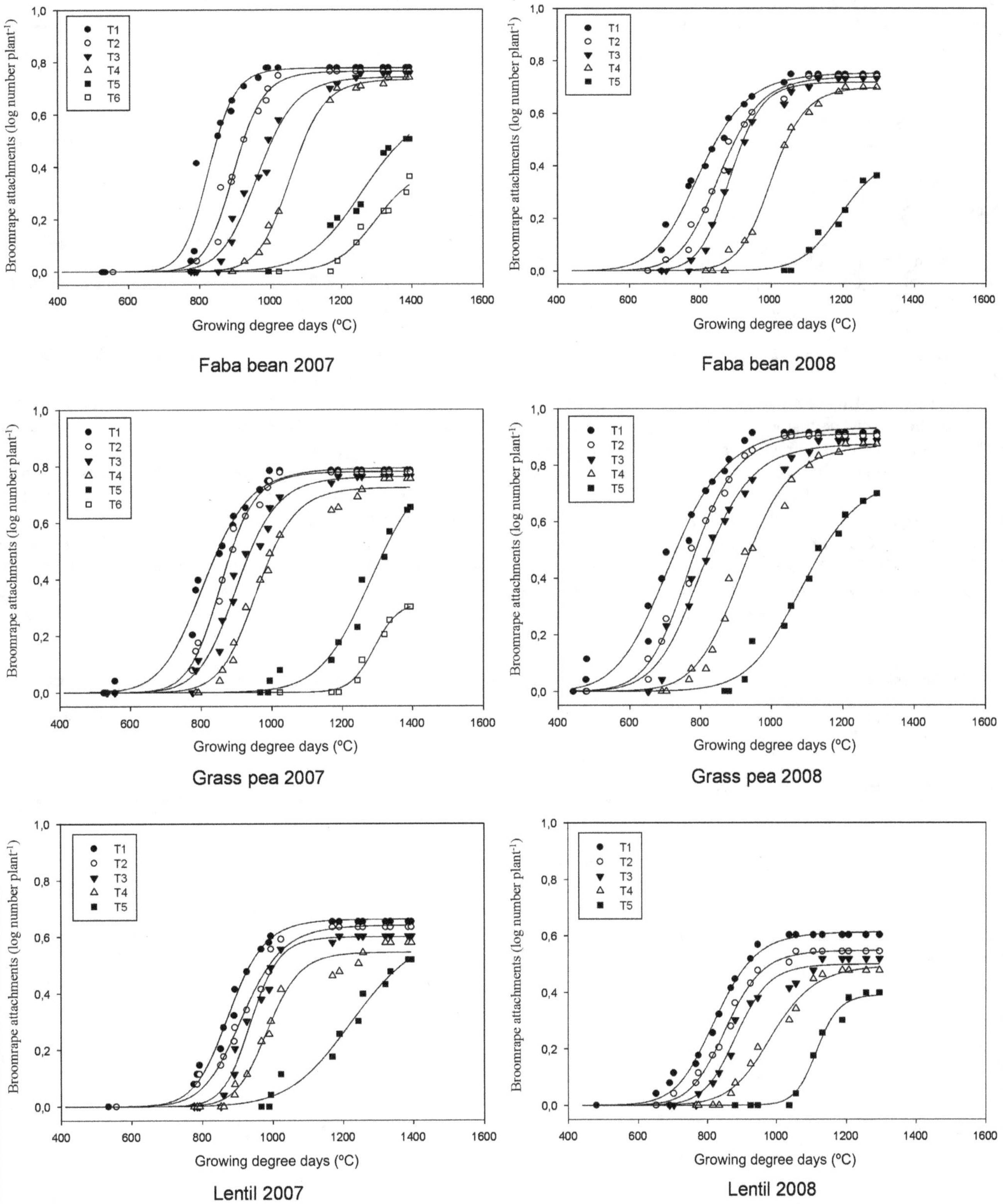

FIGURE 1 | Three-parameter sigmoid nonlinear regressions between log-transformed broomrape attachment number and growing degree days (GDD) in three crops (faba bean, grass pea and lentil) grouped by broomrape growth stage in two years (2007 and 2008).

(Zar, 1999). A t-test compares a difference with the standard error of that difference. We used the standard error reported by nonlinear regression. The numerator is the difference between fit values. The denominator is an estimate of the standard error of that difference, computed as the square root of the sum of the squares of the two standard error values. This is a reasonable estimate if the number of data points in the two curves is equal, or nearly so.

RESULTS

The combined analysis of variance showed that the interaction between species and year was significant for all the developmental stages of broomrape. A strong relation was found for the broomrape stage and GDD (**Figure 1**; **Tables 1, 2**). The model can be divided into three phases for each parasite stage: lag, log and maximum. During the lag phase, the parasite population grows very slowly. After that, the log phase covers an exponential growth of the parasite population. Finally, a maximum is reached and the parasite population remains stable. The lag phase varied slightly with the crop, the parasite stage and the season. For example, the lag phase in faba bean for T3 stage lasted until 850 GDD in 2007 and 750 GDD in 2008; in grass pea until 750 GDD in 2007 and 650 GDD in 2008; and in lentil until 850 GDD in 2007 and 750 GDD in 2008. In general terms, a delay of about 100 GDD was observed in the 2007 season compared to the 2008 season, and the attachments started to develop first in grass pea and later in faba bean and lentil. For that same stage (T3), de log phase ended at about 1150 GDD in 2007 and 1050 GDD in 2008 for faba bean (lasted for about 300 GDD); at 1050 GDD in 2007 and 950 GDD in 2008 for grass pea (lasted for 300 GDD); and at 1050 GDD in 2007 and 950 GDD in 2008 for lentil (lasted for 200 GDD).

The median (x_o) in crenate broomrape parasitism (the GDD in the log phase where the infection was 50% of the maximum) varied also with the crop, the parasitism stage and the season (**Tables 1, 2**). This index gives an indication about the speed of parasite growth on the host: a low value means that less GDD are needed to reach 50% of the maximum number of attachments, so the parasite grows faster. In general, grass pea presented the lower values, followed by lentil and then faba bean, with some exceptions mainly in season 2007. Broomrape grew faster on grass pea than on the other crops and needed less GDD for developing the shoot (T4, 920–958) than those on lentil (985–987) or faba bean (1001–1058). During season 2007, broomrapes needed some more GDD to reach the median (x_o) of all the developmental stages (except T4 in lentil) than during season 2008. Such differences were more accentuated in grass pea (near 100 GDD for most of the stages) than in faba bean or lentil (near 50–60 GDD for most of the stages).

The maximum parasite number was reached in stages T1 to T4 at about 1000 GDD for T1 and T2, 1100–1200 GDD for T3, and 1200 GDD for T4. It was similar for the three crops and the two seasons. However, the number of broomrapes per plant

TABLE 1 | Coefficients of a three-parameter sigmoid nonlinear regression between log-transformed broomrape number and GDD for season 2006-07.

Crop/Parasitism stage	Coefficients						Regression	
	a	SE	*b*	SE	x_0	SE	RSME	*P*
FABA BEAN								
T1	0.7769	0.0170	−22.51	3.05	826.41	6.07	0.0561	<0.0001
T2	0.7635	0.0119	−21.75	1.88	897.35	3.88	0.0344	<0.0001
T3	0.7420	0.0167	−18.35	2.06	965.15	6.32	0.0435	<0.0001
T4	0.7330	0.0077	−22.48	1.38	1058.24	4.61	0.0159	<0.0001
T5	0.6425	0.1132	−15.43	3.79	1267.77	36.10	0.0353	<0.0001
T6	0.4020	0.0914	−22.03	6.86	1300.33	35.29	0.0294	<0.0001
GRASS PEA								
T1	0.7941	0.0109	−12.89	1.07	811.79	4.73	0.0304	<0.0001
T2	0.7807	0.0074	−18.37	0.96	857.33	2.64	0.0221	<0.0001
T3	0.7631	0.0125	−16.27	1.27	903.04	4.72	0.0339	<0.0001
T4	0.7236	0.0134	−18.02	1.62	958.74	5.21	0.0350	<0.0001
T5	0.8359	0.1606	−16.84	4.07	1288.31	36.30	0.0422	<0.0001
T6	0.3194	0.0233	−37.42	6.65	1291.46	9.31	0.0182	<0.0001
LENTIL								
T1	0.6618	0.0098	−17.33	1.31	876.24	4.06	0.0275	<0.0001
T2	0.6406	0.0102	−16.07	1.19	914.13	4.61	0.0271	<0.0001
T3	0.5996	0.0106	−23.50	2.06	937.46	4.39	0.0298	<0.0001
T4	0.5446	0.0125	−21.62	2.98	985.00	6.06	0.0334	<0.0001
T5	0.6304	0.0820	−13.05	2.46	1232.95	31.55	0.0342	<0.0001

T1, tubercles smaller than 2 mm; T2, tubercles greater than 2 mm, without root development; T3, tubercles with crown roots, without shoot formation; T4, shoot formation, remaining underground; T5, shoot emergence; T6, flowering; a, upper asymptote (maximum broomrape number); x_0, GDD when the Y is 50% of maximum; b, slope at x_0; RSME, root mean square error.

TABLE 2 | Coefficients of a three-parameter sigmoid nonlinear regression between log-transformed broomrape number and GDD for season 2007–08.

Crop/Parasitism stage	Coefficients						Regression	
	a	SE	*b*	SE	x_0	SE	RSME	*P*
FABA BEAN								
T1	0.7539	0.0126	−12.07	0.89	800.03	5.32	0.0306	<0.0001
T2	0.7387	0.0106	−15.08	1.03	855.39	4.09	0.0254	<0.0001
T3	0.7180	0.0102	−20.14	1.56	883.76	3.71	0.0262	<0.0001
T4	0.6997	0.0108	−21.06	1.26	1001.48	4.18	0.0193	<0.0001
T5	0.4383	0.0855	−21.09	5.64	1196.03	28.9	0.0294	<0.0001
GRASS PEA								
T1	0.9343	0.0176	−9.19	0.82	718.41	6.77	0.0419	<0.0001
T2	0.9126	0.0099	−12.24	0.61	774.06	3.54	0.0256	<0.0001
T3	0.8764	0.0187	−11.78	1.08	806.34	6.79	0.0440	<0.0001
T4	0.8725	0.0226	−13.67	1.31	920.03	7.96	0.0400	<0.0001
T5	0.7680	0.0591	−13.71	1.95	1092.56	17.58	0.0376	<0.0001
LENTIL								
T1	0.6132	0.0084	−14.11	0.92	823.53	4.05	0.0210	<0.0001
T2	0.5477	0.0061	−16.36	0.92	853.67	3.10	0.0154	<0.0001
T3	0.4993	0.0129	−18.56	2.50	883.62	6.94	0.0319	<0.0001
T4	0.4948	0.0156	−15.68	1.53	987.14	9.05	0.0233	<0.0001
T5	0.3911	0.0163	−35.02	6.22	1116.44	6.24	0.0226	<0.0003

T1, tubercles smaller than 2 mm; T2, tubercles greater than 2 mm, without root development; T3, tubercles with crown roots, without shoot formation; T4, shoot formation, remaining underground; T5, shoot emergence; a, upper asymptote (maximum broomrape number); x_0, GDD when the Y is 50% of maximum; b, slope at x_0; RSME, root mean square error.

was different depending on the crop: grass pea and faba bean supported a higher number of broomrapes than lentil (**Figure 1**). In addition, the upper asymptote (or maximum broomrape number, *a*) reached by stages T1 to T4 for each crop was similar between years: 0.7–0.8 (5.0–6.3 broomrapes per plant) for faba bean, 0.7–0.9 (5.0–7.9 broomrapes per plant) for grass pea, 0.5–0.6 (3.2–4.0 broomrapes per plant) for lentil (**Tables 1, 2**). With respect to the emergence of broomrape (>T5), although the upper asymptote was not reached completely at the end of the experiment for all the crops, the tendency indicates a maximum below the threshold reached by the other stages: 04–0.6 (2.5–4.0 broomrapes per plant) for faba bean, 0.7–0.8 (5.0–6.3 broomrapes per plant) for grass pea, 0.4–0.6 (2.5–4.0 broomrapes per plant) for lentil. Finally, only in faba bean and grass pea during the 2007 season, broomrapes at the T6 stage were detected, with a lag phase at about 1100–1200 GDD, a median (x_o) of 1300 and 1292 (respectively), and a maximum of 0.4 (2.5 broomrapes per plant) and 0.3 (2.0 broomrapes per plant) (respectively).

When the model was compared between crops, significant differences were found for the three parameters (*a*, *b*, and x_o) regarding broomrape development (**Table 3**). The maximum broomrape number (*a*) was significantly different for the underground stages (T1–T4) in lentils respect to the other two crops in 2007 and for the three crops in 2008. For aboveground stage (T5) only grass pea was significantly different from faba bean and lentil in 2008. The median (x_o) was significantly different for the three crops considering the underground stages in 2007, except for stage T1 in faba bean and grass pea. In 2008, grass pea values were different from those of faba bean and lentil, except for T1 stage, which values were significant for all the crops. The values for aboveground stages (T5) were significant only in 2008 for faba bean respect to the other two crops. The slope at x_o (*b*) showed the most variable results, with significant differences varying between the season and the parasitism stage considered.

DISCUSSION

The present study confirms that the growth of crenate broomrape on the three tested legume crops is highly temperature-related. The thermal time measured as GDD appears as a valuable tool for describing the parasite growth and establishing the developmental stage of the infection, as previously shown for other crops and broomrape species (Eizenberg et al., 2004, 2005; Ephrath and Eizenberg, 2010; Ephrath et al., 2012).

Although we found differences between years, such differences do not seem to be a problem for developing a descriptive model because, in general terms, they were about 100 GDD or less. Broomrapes growing on grass pea appear as the most affected by year to year variation, with a faster rate and a higher number of infections in the second season (2008) than in the previous one (2007). There are previous reports about differences in broomrape infection depending on the crop and the environmental conditions (Pérez-de-Luque et al., 2004). Such differences could be explained by the uneven rainfall during both seasons. The distribution of the rainfall during both seasons showed some differences (Supplementary Figure 1), with a remarkable higher precipitation during January and April in 2008 compared to 2007, and a low rainfall during

TABLE 3 | Comparison of the coefficients for the three-parameter sigmoid nonlinear regression between the different crops.

Parasitism stage	Diff. between crops	Parameter differences					
		Season 2006–2007			Season 2007–2008		
		a – *P*-value	b – *P*-value	X_0 – *P*-value	a – *P*-value	b – *P*-value	X_0 – *P*-value
T1	Fb-Lt	0.3989 ns	0.0046***	0.0637 ns	0.0000***	0.0219*	0.0000***
	Fb-Ln	0.0000***	0.1446 ns	0.0000***	0.0000***	0.1196 ns	0.0011***
	Lt-Ln	0.0000***	0.0112**	0.0000***	0.0000***	0.0002***	0.0000***
T2	Fb-Lt	0.2178 ns	0.1076 ns	0.0000***	0.0000***	0.0243*	0.0000***
	Fb-Ln	0.0000***	0.0157*	0.0080**	0.0000***	0.3602 ns	0.7395 ns
	Lt-Ln	0.0000***	0.1360 ns	0.0000***	0.0000***	0.0005***	0.0000***
T3	Fb-Lt	0.3123 ns	0.3962 ns	0.0000***	0.0000***	0.0001***	0.0000***
	Fb-Ln	0.0000***	0.0857 ns	0.0010***	0.0000***	0.5959 ns	0.9859 ns
	Lt-Ln	0.0000***	0.0055**	0.0000***	0.0000***	0.0203*	0.0000***
T4	Fb-Lt	0.5749 ns	0.0509*	0.0000***	0.0000***	0.0004***	0.0000***
	Fb-Ln	0.0000***	0.7955 ns	0.0000***	0.0000***	0.0125**	0.1650 ns
	Lt-Ln	0.0000***	0.2977 ns	0.0023***	0.0000***	0.3227 ns	0.0000***
T5	Fb-Lt	0.3553 ns	0.8053 ns	0.6952 ns	0.0041***	0.2484 ns	0.0044***
	Fb-Ln	0.9304 ns	0.5926 ns	0.4743 ns	0.6036 ns	0.1301 ns	0.0285*
	Lt-Ln	0.2667 ns	0.4338 ns	0.2621 ns	0.0000***	0.0067**	0.2217 ns
T6	Fb-Lt	0.3933 ns	0.1268 ns	0.8134 ns	–	–	–
	Fb-Ln	–	–	–	–	–	–
	Lt-Ln	–	–	–	–	–	–

*Fb, faba bean; Lt, grass pea; Ln, lentil; T1, tubercles smaller than 2 mm; T2, tubercles greater than 2 mm, without root development; T3, tubercles with crown roots, without shoot formation; T4, shoot formation, remaining underground; T5, shoot emergence; T6, flowering; a, upper asymptote (maximum broomrape number); x_0, GDD when the Y is 50% of maximum; b, slope at x_0; */**/***: level of significance at 5, 1, and 0.1% respectively; ns, non-significant.*

March 2008 respect to 2007. The higher rainfall during January 2008 could positively have affected germination of broomrape and the beginning of the infection (allowing a better spreading of germination stimulants), because such conditions have been previously reported as a factor increasing parasitism (Pérez-de-Luque et al., 2004). It is possible that including soil moisture measurements, in addition to thermal time, might improve this kind of models, mainly in dryland fields with no irrigation (highly dependent on rainfall).

Regarding crop species, differences were found in the development of the parasite stages. Most of the parasites evolved from one stage into the next one (T1 to T2, etc.). However, the upper asymptote (*a*) (the maximum) was lower as the developmental stage advances. In the T5 stage emergence differences between crop species become more marked: a lower proportion of broomrapes emerged in faba bean compared with the other two crops. It seems that broomrapes attached to grass pea or lentil can evolve more easily into more advanced developmental stages compared to faba bean, which could be explained by a better availability of resources from such host. It is possible that severity of the attack and the development of broomrapes could be related to physiological traits of the host plant determining the allocation of nutrients into the parasite, as has been shown in *Striga* spp. (Arnaud et al., 1999). In addition, the size of emerged broomrapes on faba bean was bigger than those on grass pea and lentil. As big individuals are more competitive for resources than small ones, intra-specific competition might play a role here.

The comparison of the descriptive thermal model between the three crops confirms significant differences regarding the evolution of the developmental stages of broomrape. The most consistent differences are found when the parameters (mainly *a* and x_0) of the underground stages (T1-T4) are considered. The thermal models are usually associated to a specific host (Eizenberg et al., 2004, 2005; Ephrath and Eizenberg, 2010; Ephrath et al., 2012) and they link phenological events with temperature (GDD). This means that the models include physiological traits of the host (for example, taking into account the crop's base temperature). Considering this, differences in the parameters of the model for different crops are pointing out differences in the virulence of the parasite against each crop. In the same way, such differences could be used for differentiating susceptible from resistant genotypes: resistant accessions have been shown to delay broomrape development in the field (Rubiales et al., 2003a; Pérez-de-Luque et al., 2004). Additionally, resistance to broomrape has been reported as temperature dependent in some cases (Eizenberg et al., 2003), so a model considering the environmental temperature would be desirable for evaluating sources of resistance in a breeding programme. Future work should involve validation of this kind of descriptive model in order to develop a predictive one, using different locations and cultivars with different degrees of resistance against crenate broomrape, in order to use it as a tool for plant breeding.

AUTHOR CONTRIBUTIONS

AP and DR: Conceived and designed the experiments; AP: Developed the experiment; FF performed the statistical analysis; AP, FF, and DR: Discussed and interpreted the results; AP: Wrote a first draft of the manuscript and DR and FF helped with the writing and further edition and correction process; DR: Contributed materials, equipment, and analysis tools.

ACKNOWLEDGMENTS

The authors acknowledge the financial support from the EU project FP6-2002-FOOD-1-5062232004-2008 and Spanish Ministerio de Economia and Competitividad project ALG2014-52871-R confinanced by FEDER.

REFERENCES

Arjona-Berral, A., Mesa-García, J., and García-Torres, L. (1987). Phenology and growth of *Orobanche crenata* Forsk. *(broomrape)* in four *legume crops.* Weed Res. 27, 349–360. doi: 10.1111/j.1365-3180.1987.tb00772.x

Arnaud, M. C., Véronési, C., and Thalouarn, P. (1999). Physiology and histology of resistance to *Striga hermonthica* in *Sorghum bicolor* var. *Framida. Aust. J. Plant Physiol.* 26, 63–70. doi: 10.1071/PP98070

Brown, R. F., and Mayer, D. G. (1988). Representing cumulative germination. The use of the Weibull function and other empirically derived curves. *Ann. Bot.* 61, 127–138.

Castejón-Muñoz, M., Romero-Muñoz, F., and García-Torres, L. (1993). Effect of planting date on broomrape (*Orobanche cernua* Loefl.) infection on sunflower (*Helianthus annuus* L.). *Weed Res.* 33, 171–176. doi: 10.1111/j.1365-3180.1993.tb01930.x

Eizenberg, H., Colquhoun, J. B., and Mallory-Smith, C. A. (2004). The relationship between growing degree days and small broomrape (*Orobanche minor*) parasitism in red clover (*Trifolium pratense*). *Weed Sci.* 52, 735–741. doi: 10.1614/WS-03-157R

Eizenberg, H., Colquhoun, J. B., and Mallory-Smith, C. A. (2005). A predictive degree-days model for small broomrape (*Orobanche minor*) parasitism in red clover in Oregon. *Weed Sci.* 53, 37–40. doi: 10.1614/WS-04-018R1

Eizenberg, H., Colquhoun, J. B., and Mallory-Smith, C. A. (2006). Imazamox application timing for small broomrape (*Orobanche minor*) control in red clover. *Weed Sci.* 54, 923–927. doi: 10.1614/WS-05-151R.1

Eizenberg, H., Plakhine, D., Hershenhorn, J., Kleifeld, Y., and Rubin, B. (2003). Resistance to broomrape (*Orobanche* spp.) in sunflower (*Helianthus annuus* L.) is temperature dependent. *J. Exp. Bot.* 54, 1305–1311. doi: 10.1093/jxb/erg129

Ellis, R. H., and Barrett, S. (1994). Alternating temperatures and rate of seed germination in lentil. *Ann. Bot.* 74, 519–524. doi: 10.1006/anbo.1994.1149

Ephrath, J. E., and Eizenberg, H. (2010). Quantification of the dynamics of *Orobanche cumana* and *Phelipanche aegyptiaca* parasitism in confectionery sunflower. *Weed Res.* 50, 140–152. doi: 10.1111/j.1365-3180.2010.00768.x

Ephrath, J. E., Hershenhorn, J., Achdari, G., Bringer, S., and Eizenberg, H. (2012). Use of logistic equation for detection of the initial parasitism phase of Egyptian broomrape (*Phelipanche aegyptiaca*) in tomato. *Weed Sci.* 60, 57–63. doi: 10.1614/WS-D-11-00070.1

Fernández-Aparicio, M., Flores, F., and Rubiales, D. (2012). Escape and true resistance to crenate broomrape (*Orobanche crenata* Forsk.) in grass pea (*Lathyrus sativus* L.) germplasm. *Field Crops Res.* 125, 92–97. doi: 10.1016/j.fcr.2011.09.003

Fernández-Aparicio, M., Sillero, J. C., Pérez-de-Luque, A., and Rubiales, D. (2008). Identification of sources of resistance to crenate broomrape (*Orobanche crenata*) in Spanish lentil (*Lens culinaris*) germplasm. *Weed Res.* 48, 85–94. doi: 10.1111/j.1365-3180.2008.00604.x

Joel, D. M., Hershenhorn, J., Eizenberg, H., Aly, R., Ejeta, G., Rich, P. J., et al. (2007). Biology and management of weedy root parasites. *Hortic. Rev.* 33, 267–349. doi: 10.1002/9780470168011.ch4

McMaster, G. S., and Wilhelm, W. W. (1997). Growing degree-days: one equation, two interpretations. *Agric. Forest Meteorol.* 87, 291–300. doi: 10.1016/S0168-1923(97)00027-0

Mesa-García, J., and García-Torres, L. (1986). Effect of planting date on parasitism of broadbean (*Vicia faba*) by crenate broomrape (*Orobanche crenata*). *Weed Sci.* 34, 544–550.

Parker, C. (2009). Observations on the current status of *Orobanche* and *Striga* problems worldwide. *Pest Manag. Sci.* 65, 453–459. doi: 10.1002/ps.1713

Parker, C. (2013). "The parasitic weeds of the Orobanchaceae," in *Parasitic Orobanchaceae*, eds D. M. Joel, J. Gressel, and L. Musselman (Berlin: Springer-Verlag), 313–344.

Parker, C., and Riches, C. R. (1993). *Parasitic Weeds of the World: Biology and Control.* Wallingford, CT: CAB International.

Pérez-de-Luque, A., Jorrín, J., Cubero, J. I., and Rubiales, D. (2005). *Orobanche crenata* resistance and avoidance in pea (*Pisum* spp.) operate at different developmental stages of the parasite. *Weed Res.* 45, 379–387. doi: 10.1111/j.1365-3180.2005.00464.x

Pérez-de-Luque, A., Lozano, M. D., Moreno, M. T., Testillano, P. S., and Rubiales, D. (2007). Resistance to broomrape (*Orobanche crenata*) in faba bean (*Vicia faba*): cell wall changes associated with pre-haustorial defensive mechanisms. *Ann. Appl. Biol.* 151, 89–98. doi: 10.1111/j.1744-7348.2007.00164.x

Pérez-de-Luque, A., Sillero, J. C., Moral, A., Cubero, J. I., and Rubiales, D. (2004). Effect of sowing date and host resistance on the establishment of *Orobanche crenata* in faba bean and common vetch. *Weed Res.* 44, 282–288. doi: 10.1111/j.1365-3180.2004.00401.x

Press, M. C., and Graves, J. D. (1995). *Parasitic Plants.* London: Chapman and Hall.

Rao, S., and Northup, B. (2008). Planting date affects production and quality of grass pea forage. *Crop Sci.* 48, 1629–1635. doi: 10.2135/cropsci2007.11.0622

Román, B., Satovic, Z., Alfaro, C., Moreno, M. T., Kharrat, M., Pérez-de-Luque, A., et al. (2007). Host differentiation in *Orobanche foetida* Poir. *Flora* 202, 201–208. doi: 10.1016/j.flora.2006.07.003

Rubiales, D., Alcántara, C., Pérez-de-Luque, A., Gil, J., and Sillero, J. C. (2003a). Infection of chickpea (*Cicer arietinum*) by crenate broomrape (*Orobanche crenata*) as influenced by sowing date and weather conditions. *Agronomie* 23, 359–362. doi: 10.1051/agro:2003016

Rubiales, D., Flores, F., Emeran, A. S., Kharrat, M., Amri, M., Rojas-Molina, M. M., et al. (2014). Identification and multi-environment validation of resistance against broomrapes (*Orobanche crenata* and *O. foetida)* in faba bean *(Vicia faba). Field Crops Res.* 166, 58–65. doi: 10.1016/j.fcr.2014.06.010

Rubiales, D., and Heide-Jørgensen, H. S. (2011). "Parasitic plants," in: *eLS* (Chichester: John Wiley & Sons Ltd). Available online at: http://www.els.net; http://onlinelibrary.wiley.com/doi/10.1002/9780470015902.a0021271/full

Rubiales, D., Pérez-de-Luque, A., Cubero, J. I., and Sillero, J. C. (2003b). Crenate broomrape (*Orobanche crenata*) infection in field pea cultivars. *Crop Prot.* 22, 865–872. doi: 10.1016/S0261-2194(03)00070-X

Rubiales, D., Pérez-de-Luque, A., Sillero, J. C., Román, B., Kharrat, M., Khalil, S., et al. (2006). Screening techniques and sources of resistance against parasitic weeds in grain legumes. *Euphytica* 147, 187–199. doi: 10.1007/s10681-006-7399-1

Stützel, H. (1995). A simple model for simulation of growth and development in faba beans (*Vicia faba* L.), 1. Model description. *Eur. J. Agron.* 4, 175–185. doi: 10.1016/S1161-0301(14)80044-0

Zar, H. J. (1999). *Biostatistical Analysis*. New Jersey, NJ: Prentice-Hall Inc.

Conflict of Interest Statement: The authors declare that the research was conducted in the absence of any commercial or financial relationships that could be construed as a potential conflict of interest.

8

Thermal Time Model for Egyptian Broomrape (*Phelipanche aegyptiaca*) Parasitism Dynamics in Carrot (*Daucus carota* L.): Field Validation

Edited by:
Marcello Mastrorilli,
Consiglio per la Ricerca in Agricoltura e l'analisi dell'Economia Agraria, Italy

Reviewed by:
Grama Nanjappa Dhanapal,
University of Agricultural Sciences, India
Paola Leonetti,
National Research Council (CNR), Italy

***Correspondence:**
Hanan Eizenberg
eizenber@agri.gov.il

***Amnon Cochavi*[1,2], *Baruch Rubin*[2], *Guy Achdari*[1] and *Hanan Eizenberg*[1*]**

[1] *Department of Phytopathology and Weed Research, Newe Ya'ar Research Center, Agricultural Research Organization, Ramat Yishay, Israel,* [2] *R. H. Smith Institute of Plant Sciences and Genetics in Agriculture, Faculty of Agricultural, Food and Environmental Sciences, The Hebrew University of Jerusalem, Rehovot, Israel*

Carrot, a highly profitable crop in Israel, is severely damaged by *Phelipanche aegyptiaca* parasitism. Herbicides can effectively control the parasite and prevent damage, but for optimal results, knowledge about the soil–subsurface phenological stage of the parasite is essential. Parasitism dynamics models have been successfully developed for the parasites *P. aegyptiaca*, *Orobanche cumana,* and *Orobanche minor* in the summer crops, tomato, sunflower, and red clover, respectively. However, these models, which are based on a linear relationship between thermal time and the parasitism dynamics, may not necessarily be directly applicable to the *P. aegyptiaca*–carrot system. The objective of the current study was to develop a thermal time model to predict the effect of *P. aegyptiaca* parasitism dynamics on carrot growth. For development and validation of the models, data was collected from a temperature-controlled growth experiment and from 13 plots naturally infested with *P. aegyptiaca* in commercial carrot fields. Our results revealed that *P. aegyptiaca* development is related to soil temperature. Moreover, unlike *P. aegyptiaca* parasitism in sunflower and tomato, which could be predicted both a linear model, *P. aegyptiaca* parasitism dynamics on carrot roots required a nonlinear model, due to the wider range of growth temperatures of both the carrot and the parasite. Hence, two different nonlinear models were developed for optimizing the prediction of *P. aegyptiaca* parasitism dynamics. Both models, a beta function model and combined model composed of a beta function and a sigmoid curve, were able to predict first *P. aegyptiaca* attachment. However, overall *P. aegyptiaca* dynamics was described more accurately by the combined model (RMSE = 14.58 and 10.79, respectively). The results of this study will complement previous studies on *P. aegyptiaca* management by herbicides to facilitate optimal carrot growth and handling in fields infested with *P. aegyptiaca*.

Keywords: broomrape, growing degree days model, beta function, sigmoid curve, cross validation

INTRODUCTION

Carrot (*Daucus carota*, Apiaceae) is widely grown throughout Europe and the Mediterranean area, including Israel, where it has become a high-income cash crop. In the Mediterranean area, where carrot is grown all the year round, two species of broomrape, *Orobanche crenata* and *Phelipanche aegyptiaca,* which are chlorophyll-deficient root holoparasites, parasitize the carrot taproot, and roots (Parker, 2012). In highly infested fields, broomrape can cause a total yield loss (Bernhard et al., 1998). Although both species parasitize carrot, there are some differences between them, such as host range and base temperature for germination (Kebreab and Murdoch, 1999), with the germination temperature being crucial for parasite development and affecting the parasite's ability to attach to the crop species. *O. crenata* also parasitizes plants of Fabaceae family, while *P. aegyptiaca* also parasitizes plants of the Solanaceae family (Heide-Jørgensen, 2013). Since many Solanaceae crops are grown in the Mediterranean area, *P. aegyptiaca* is widely spread in this area.

Chemical control has been found to be effective for managing broomrape damage to crops (Eizenberg et al., 2012a). The various application methods that are in widespread practice differ mainly in terms of the broomrape developmental stage at which they are applied. In one of the most commonly used methods, the herbicide is applied during the broomrape seed germination stage with the aim to prevent elongation of the parasite haustorium; for example, for tomato, sulfosulfuron is applied to prevent haustorium elongation in *P. aegyptiaca* seeds (Eizenberg et al., 2012a). In this method, the herbicide is applied to the crop foliage but is immediately washed to the soil by sprinkler irrigation. In another, more complicated method, the underlying idea is that the sprayed herbicide must move through the host plant vessels to reach the parasite tubercle after the parasite has established a connection with the host xylem and phloem system. Since broomrape is a strong sink, absorbing water and nutrients from the host, the herbicide will indeed move rapidly through the plant vascular system to the parasite (Eizenberg et al., 2012b). In view of these application methods, there is a need to establish the optimal conditions for effective and efficient broomrape control without causing damage to the host. A previous study has shown that control of *P. aegyptiaca* growing on carrots can be achieved by application of low doses of glyphosate to the carrot foliage (Jacobsohn and Kelman, 1980). The findings of that study were confirmed by recent studies in Israel that demonstrated effective glyphosate control of *P. aegyptiaca* parasitism at different infestation levels under different climate regimes (Cochavi et al., 2015, 2016).

As discussed above, both methods of control require knowledge of the broomrape developmental stage, with correct timing of herbicide application being crucial for effective broomrape control. For example, if a systemic herbicide is applied before the broomrape seedlings have attached to the host roots, the parasite will not be controlled, and if the herbicide is applied after *P. aegyptiaca* biomass has accumulated, herbicide efficiency will be reduced (Eizenberg et al., 2006). The importance of timing was illustrated by Castejon-Muñoz et al. (1993), who found that bringing forward the sowing dates of sunflower (*Helianthus annuus*) seeds reduced *Orobanche cernua* parasitism and yield loss (Castejon-Muñoz et al., 1993). Another factor having a significant influence on the broomrape developmental stage—and hence on the herbicide application regime—is temperature. Kebreab and Murdoch (1999), for example, found that four different broomrape species germinated in different temperature ranges. They showed that for *O. crenata* the highest germination rate of 70% was obtained at 15–20°C, whereas for *P. aegyptiaca* seeds almost 100% germination was obtained at temperatures ranging between 15 and 35°C (Kebreab and Murdoch, 1999). The base temperature for development—a critical factor affecting parasitism—is particularly important in models for herbicide application, as we will elaborate in detail below. In such models, the base temperature is taken as the lower threshold below which there is no development of the parasite. For instance, Ephrath et al. (2012) defined 10°C as the minimal temperature for development of *P. aegyptiaca* on tomato roots. Murdoch and Kebreab (2013) demonstrated that germination of *Orobanche* seeds cannot occur below a temperature of 4.9°C. Mesa-Garcia and Garcia-Torres found that *O. crenata* develops faster on broad bean roots under low winter temperatures (Mesa-García and García-Torres, 1986). However, unlike *O. crenata, P. aegyptiaca* parasitizes crops the year over at all temperatures.

In the Northern hemisphere, the carrot growing season extends for 120–180 days after sowing. The season can start in July (summer) and end in January (winter) or start in December (winter) and end in May (summer). Therefore, carrots are grown under hugely disparate temperature regimes, with temperatures ranging from winter lows falling to 0°C to extremely high summer temperatures of above 35°C. Finch-Savage et al. (1998) found that the minimal temperature for carrot seed germination and seedling emergence is 2°C. Carrots sown between April and July (spring/summer) are not parasitized by *O. crenata* and *P. aegyptiaca*, but the main sowing season is during fall and winter, which gives maximal yields: it was found that even when plants were infested, the yield was higher than that for non-infested plants grown through the summer (Eizenberg et al., 2001).

To determine the optimal protocol for precise and effective herbicide application for broomrape control that takes into account factors such as those described above, a number of studies have focused on developing prediction models based on thermal time as measured in growing degree days (GDDs). Some studies have described a positive correlation between soil temperature and broomrape parasitism on different species. Ephrath et al. (2012), for example, found a positive linear correlation between increased temperatures and *P. aegyptiaca* development on tomato roots. A similar correlation was described between *Orobanche minor* and *Orobanche cumana* parasitism on red clover and sunflower roots, respectively (Eizenberg et al., 2005a, 2012b). According to these models, decision support systems for effective parasite management were developed for each crop separately. There is, however, no prediction model for correlation between soil temperature and *P. aegyptiaca* parasitism in carrot. The objective of the current study was thus to develop a robust thermal time model for the quantification and prediction of the parasitism dynamics of *P. aegyptiaca* in carrot.

MATERIALS AND METHODS

Plant Material

Carrot cv. "Nairobi" (Bejo Seeds, Oceano, CA, USA) was used in all experiments. *Phelipanche aegyptiaca* inflorescences were collected in 2008 from a broomrape-infested tomato field (Mevo Hama, Israel). Seeds were sieved through a 300-mesh sieve and stored in the dark at 4°C until use. To determine the germination potential of broomrape seeds at the beginning of the experiments, a germination test was performed under standard conditions at 25°C with the standard synthetic stimulant GR24 (Yoneyama et al., 2013). GR24 was applied at a concentration of $10\,g{\cdot}kg^{-1}$ soil (10 ppm) after 12 days of pre-conditioning, resulting in a germination rate of 84%. Seeds of the same lot were used for soil infestation in controlled environment experiments and for planting above transparent tubes in the field experiments for monitoring *in-situ* parasitism using a minirhizotron, according to Eizenberg et al. (2005b).

Controlled-Environment Experiment

Carrot seeds were sown in 2-L pots (two seeds per pot) in infested ($15\,mg{\cdot}g^{-1}$ soil) and non-infested Newe Ya'ar soil [Chromic Haploxerert (a fine-clayey, montmorillonitic, thermic soil; 55% clay, 25% silt, and 20% sand, 2% organic matter, pH 7.2)]. Plants were grown under different temperatures regimes (16/10, 22/16, 28/22, and 34/28°C day/night) and different day lengths (16 h/8 h and 8 h/16 h day/night); each treatment included five replicates. At the end of the experiment, after 150 days, emerged broomrape inflorescences were counted, and biomass was measured. Thereafter, carrot taproots were removed from the pots and washed free of soil for biomass measurement. In addition, parasite total biomass, including weight of the inflorescences, was measured.

Field Observations

The field trials were performed over the years 2009–2012. With the aim to collect wide time and space temperature range patterns, 13 field trials were conducted in commercial carrot fields in Israel (**Table 1**). At each location, four transparent minirhizotron tubes were placed in the soil according to Eizenberg et al. (2005b). Ten carrot seeds were sown in the soil above each tube and germinated together with the seeds in the entire field. Once a week from carrot germination until broomrape emergence, broomrape parasitism in the soil sub-surface for each tube was monitored using a minirhizotron camera (Bartz Technology, Carpinteria, CA, USA). Hourly soil temperature at a depth of 10 cm was recorded with temperature data logger (Onset®, Hobo data loggers, Cape Cod, MA, USA).

To optimize temperature measurements, soil temperature values were summed in two methodologies: (a) daily average temperature calculated as the average of the daily minimum and maximum temperatures minus the base temperature; (b) hourly measurements summed by their relative parts, i.e., each measurement was divided by 24.

Broomrape Appearance and Development Analysis

Minirhizothron images were analyzed to determine the number of broomrape tubercles formed on each tube. First tubercle attachment was defined as the first image in which a small tubercle could be seen. To prevent false identification, the determination was confirmed by identifying the same tubercle in the next session a week later. The number of broomrape tubercles was considered as maximal when no additional attachments were observed (**Figure 1**). To obtain the partial development value of the parasite, the number of tubercles recorded in each session was divided by the maximal number of observed tubercles on the specific tube (Eizenberg et al., 2005b).

Models for Predicting *P. aegyptiaca* Development Dynamics

Data for *P. aegyptiaca* development and soil temperatures at each location were collected using the minirhizotron system. Several GDD models were tested with the aim to find the correlation between soil temperature and broomrape appearance, as described below.

Linear Model

A linear model was developed according to Eizenberg et al. (2005a):

$$T_{daily} = \frac{T_{max} + T_{min}}{2} - T_{base} \tag{1}$$

where T_{daily} is the calculated temperature, T_{max} and T_{min} are the daily maximum and minimum temperatures, respectively, and T_{base} is the minimum temperature for development (in carrot 2°C; Finch-Savage et al., 1998). Daily GDDs were summed to the limit value:

$$GDD_{limit} = \sum\nolimits_{i=1}^{n} T_n \tag{2}$$

The model predicts that higher temperatures accelerate the development rate.

Beta-Function Model

A four-parameter beta-function equation was used to calculate the effect of temperature on the parasitism dynamics:

$$r = \left(\left(\frac{T - T_b}{T_0 - T_b} \right) * \left(\frac{T_m - T}{T_m - T_0} \right)^{\frac{T_m - T_0}{T_0 - T_b}} \right)^a \tag{3}$$

where r is the calculated partial development rate, T denotes the hourly measured temperature, T_b denotes the minimal temperature for development, T_0 denotes the optimal temperature for development (i.e., the highest development rate), T_m refers to the maximal temperature for development, and a denotes the shape of the slope (**Figure 2**); all parameters values determent during model optimization to data (Yin and Kropff, 1996).

TABLE 1 | Details of field experiments conducted between 2009 and 2012 in 13 locations in Israel.

Location	Sowing date	Days to appearance	Average temperature (°C)	Max. temperature (°C)	Min. temperature (°C)
1	17/11/2009	41	15.39	28.26	11.63
2	10/12/2010	61	13.35	26.49	5.14
3	13/10/2010	36	21.53	32.39	14.23
4	10/11/2010	42	16.9	27.37	8.18
5	21/12/2010	63	13.48	21.47	0.00
6	20/09/2011	31	21.9	28.56	16.43
7	10/08/2011	39	28.84	46	23.39
8	06/10/2011	38	19.22	26.88	12.01
9	26/10/2011	55	14.23	28.85	4.73
10	14/11/2011	63	12.75	23.1	0.01
11	30/11/2011	69	11.76	21	4.10
12	02/09/2010	40	27.70	36.4	20.23
13	25/07/2012	33	27.91	36.08	22.33

FIGURE 1 | Detection of *Phelipanche aegyptiaca* tubercles with a minirhizotron camera.

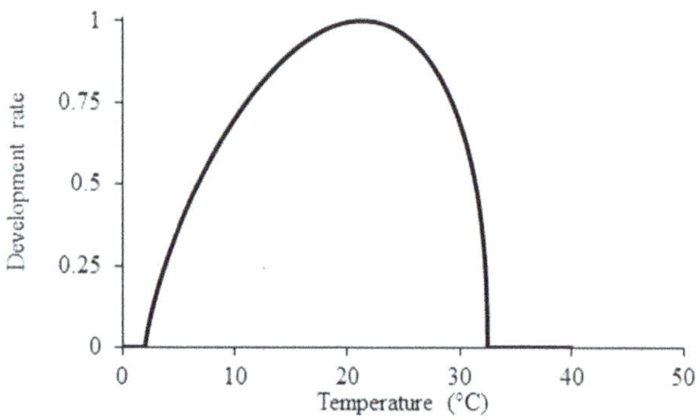

FIGURE 2 | Mathematical model for the description of *P. aegyptiaca* development rate as temperatures rise according to beta function model.

The r -value was multiplied by the measured soil temperature and summed to estimate the effect of temperature on parasitism dynamics, where R is the accumulated GDDs:

$$R = \sum_{n=1}^{i} ri * Ti \tag{4}$$

According to this model, the parasitism dynamics rate increases as the temperature rises from a minimal value to the optimal value. Beyond the optimal temperature, the parasitism dynamics rate decreases until the maximal temperature is reached. Temperatures below the minimal and beyond the maximal values did not contribute to the model and were therefore computed as zero.

Beta Function Model Combined with a Sigmoid Curve (BTSG)

A third model was developed for prediction of *P. aegyptiaca* parasitism dynamics by combining the beta function Equation (3) with a sigmoid curve at the optimal temperature:

$$r = -\frac{1}{1 + e^{\frac{-T+X_0}{b}}} + 1 \tag{5}$$

where X_0 is the inflection point, b is the curve width, and a is the upper asymptote (in this case it is 1, the maximal rate). The sigmoid equation was combined with Equation (3) to develop Equation (6):

$$\begin{cases} T < To \;\; r = [3] \\ T > To \;\; r = [5] \end{cases} \tag{6}$$

According to Equation (6), the parasitism dynamics rate at high temperatures (above optimal temperature) will be calculated according to the sigmoid curve and not be assumed to be zero (**Figure 3**).

P. aegyptiaca Parasitism Dynamic Model

Partial development values were fitted to the different GDD models. For each model (linear, beta function and BTSG), a dynamic curve was fitted by four parameters (Weibull lag-phase equation):

$$f(x) = \left(1 - \exp\left(-\left(\frac{GDD - lag}{b}\right)^c\right)\right) \quad (7)$$

where $f(x)$ denotes the partial number of attachments out of the total number, a is the upper asymptote (in this case $a = 100$), c is defined as the slope of the curve, b is a scale parameter regardless of the shape value, lag is defined as the minimal GDD number for the appearance of attachments, and GDD is the calculated value on the X-axis (Ephrath and Eizenberg, 2010).

Curve fitting was performed according to root mean square error values (RMSE), where a RMSE-value represents the error between the observed and expected value, computed by:

$$RMSE = \sqrt{\frac{1}{n\sum_{i=1}^{n}(x_i - y_i)^2}} \quad (8)$$

where x_i is the observed partial number of attachments number, y_i is the predicted number of attachments, and n is the number of observations. Lower RMSE-values indicate a better-fitted model (Lati et al., 2013).

A value that may be used for comparison of models is the Akaike information criterion correction (AICc), which is applied to take into consideration the model complexity and modeling accuracy according to the number of parameters in the model. In other words, AICc provides an estimate of the expected discrepancy between the model generating the data and a fitted candidate model (Cavanaugh, 1997). The AICc-value, with a normal distribution, is obtained by:

$$AICc = 2m + n\ln\left(\frac{RSS}{n}\right) + \frac{2 * m(m+1)}{n - m - 1} \quad (9)$$

where m is the number of model coefficients, n is the number of observations, and RSS is the residual sum of squares. A lower $AICc$-value indicates a better-fitted GDD prediction model (Lati et al., 2011).

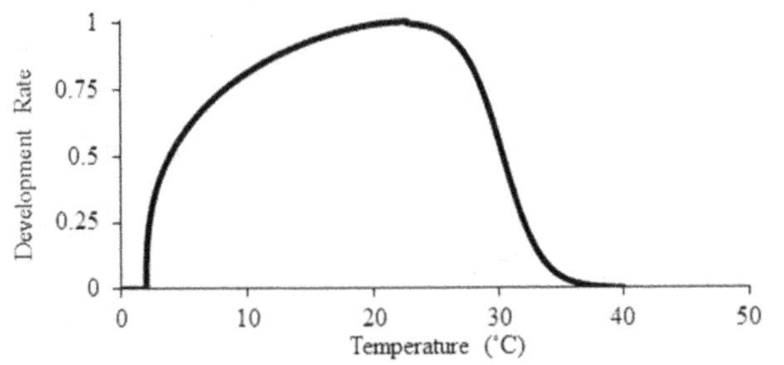

FIGURE 3 | Mathematical models for the description of *P. aegyptiaca* development rate as temperatures rise according to the beta function model combined with a sigmoid curve.

Validation of the Models

To validate the different models, a leave-one-out cross validation procedure was conducted (Hawkins et al., 2003). By using cross validation, it is possible to use data as a learning group and a testing group simultaneously. For each specific computed model (beta function and BTSG), prediction was computed according to 12 out of 13 locations. The estimation of the one-out location was computed, following the learning group fitted parameters. The mean ± standard error for 12 locations was defined as the prediction area of the calculated model. A prediction was defined as true when the leave-one-out value lay in the prediction area (1), otherwise it was defined as false (0). The test was repeated for all 13 locations, using each location both as part of the learning group and as part of the validation group. The difference in days between the prediction and the observed result was found by matching the expected and the observed results to the calculated GDD from the data logger.

Statistical Analysis

The controlled environment experiments were arranged in a two-factorial design (temperature, day length) and repeated twice. ANOVA and comparison of the means were conducted by Tukey–Kramer HSD test ($\alpha < 0.05$) using JMP software (vers. 7, SAS). Optimization of the parameters of models (3) and (6) (i.e., finding the parameters that give the lowest variance between locations) for the first of appearance broomrape was achieved by using a simulated annealing method (Aarts et al., 1997), since the large number of parameters in each equation [four in Equation (3) and seven in Equation (6)] and the large number of locations (13) did not enable a pure solution for optimization. This method is designed to avoid local optimum solution fixation and search forward until the optimal highest solution is found. Computing was performed with MATLAB (version 2009b, MathWorks, Natick, MA, USA). Lag-phase curve fitting was performed with Sigmaplot software (vers. 11, SPSS, Chicago, IL, USA). Leave-one-out cross validation was performed with MATLAB.

RESULTS

Controlled-Environment Experiment

The effect of day length and temperature on carrot taproot and *P. aegyptiaca* development were examined under controlled-environment conditions. No mutual effect was found between temperature and day length. Temperature, however, did affect carrot biomass of both *P. aegyptiaca* infested and non-infested taproots. The maximal biomass of non-infested carrot taproots was obtained for a 28/22°C (day/night) temperature regime. However, maximal biomass of infested taproots was obtained for the 22/16°C (day/night) temperature regime. Maximal *P. aegyptiaca* biomass was obtained at 28/22°C (day/night). The lowest biomasses of carrot taproots (infested and non-infested) and of *P. aegyptiaca* were obtained at 34/28°C (day/night) (**Figure 4**).

P. aegyptiaca Field Observations

The development of *P. aegyptiaca* was followed in 13 commercial carrot fields by using a minirhizotron camera. No correlation was found between the time after carrot seed planting and the appearance of the first *P. aegyptiaca* attachment on carrot roots (**Table 1**). Average temperatures at the different locations varied between 11 and 28°C, emphasis the temperature range (i.e. from maximum of 46°C to minimum of 0°C) in carrot growth as well as the temperature range that the model have to be valid.

Models

Temperature data were converted to thermal time (GDD) using three different models: (a) linear model, (b) beta function model, and (c) BTSG model. The examined models were fitted to the same data set to identify the best approach for predicting *P. aegyptiaca* parasitism dynamics. The fitting process was done in three steps: (a) optimization of model parameters for first *P. aegyptiaca* appearance for minimal variance between locations, (b) dynamic curve fitting to the overall *P. aegyptiaca* development process, and (c) comparison of models and validation using a cross-validation procedure.

For prediction of the first *P. aegyptiaca* attachment, the linear model was fitted to the soil temperature data. The mean value (±*SD*) for the first *P. aegyptiaca* attachment was 759 ± 146 (*SD*) GDD (**Table 2**). The values for locations 7 and 12 were more than 2 *SD* units from the average value (**Figure 5A**). A Weibull lag-phase curve was fitted to data for demonstrating the parasitism dynamics, except for the higher asymptote (*a*), all other coefficients were not significant (**Figure 5B**, **Table 3**). Although, the model was found to be significant, with RMSE and AICc-values of 22.84 and 348.33, respectively, the model cannot predict *P. aegyptiaca* development dynamics due to the non-significance of its parameters.

Fitting of the beta-function curve to the data set showed that the development rate was inhibited after the optimum point had been reached and stopped after the maximal temperature (Equation 3). According to the model, *P. aegyptiaca* first attachment was observed at 656 ± 55 (*SD*) GDD (**Table 2**, **Figure 6A**). A four-parameter Weibull equation with lag phase adoption curve was fitted to the beta function adjusted values

TABLE 2 | Calculation of the appearance *P. aegyptiaca* first attachment as seen with the minirhizotron system, according to recorded soil temperature in 13 different locations.

	Mean	Variance	Standard deviation
Linear	759.85	21,409.30	146.31
Beta	656.19	3121.17	55.86
Beta-sigmoid	700.73	2890.03	53.75

Model values were optimized to minimal variance by using Matlab.

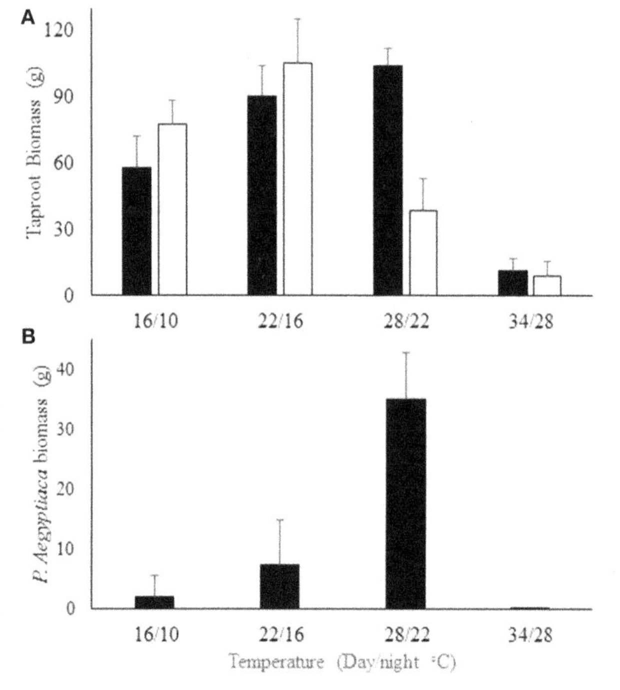

FIGURE 4 | (A) Development of infested (□) and non-infested (■) carrot taproot biomass. **(B)** *P. aegyptiaca* biomass on carrot roots grown under different temperature regimes. Plants were grown in 2-L pots, two plants per pot. Bars indicate standard error, $n = 5$.

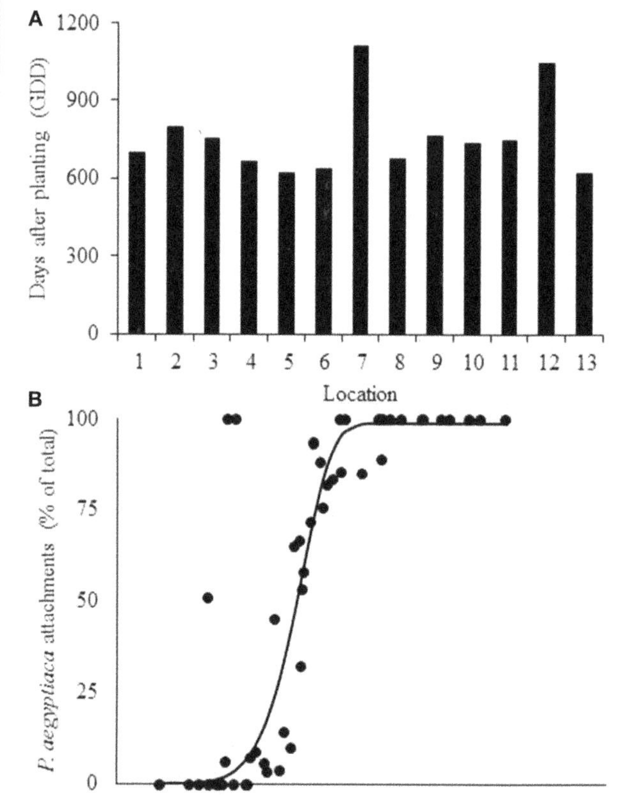

FIGURE 5 | (A) Estimation of *Phelipanche aegyptiaca* parasitism with a linear model. **(A)** Detection of first *P. aegyptiaca* attachment according to the linear model using growing degree days (GDD). **(B)** *P. aegyptiaca* development dynamics according to the linear model. Curve parameters are presented in **Table 3**.

TABLE 3 | Non-linear four-parameter lag equation for describing the dynamics of *P. aegyptiaca* attachment on carrot roots according to the different models.

Method	a*	SE (a)	P (a)	b**	SE (b)	P (b)	c***	SE (c)	P (c)	LAG+	SE (LAG)	P (LAG)	P	R^2	RMSE++	AICc+++
Linear	98.66	5.61	<0.0001	653.99	1621.11	0.68	7.06	19.13	0.71	$1.23*10^{-13}$	1617.29	1	<0.0001	0.73	22.84	348.33
Beta	96.75	4.72	<0.0001	85.61	20.96	0.0002	0.84	0.29	0.006	607.82	13.28	0.006	<0.0001	0.89	14.58	304.82
Beta-sigmoid	100	4.09	<0.0001	101.69	18.7	<0.0001	0.92	0.26	0.0009	583.83	13.64	<0.0001	<0.0001	0.94	10.79	278.58

Curves were fitted with Sigmaplot software.
**Represents the maximal asymptote.*
***The scale parameter regardless of the shape value.*
****Represents the shape parameter that determines the skewness and kurtosis of the equation.*
+Represents the lag phase until P. aegyptiaca attachment was initiated.
++Root mean square error.
+++Akaike Information Criterion (Corrected).

(**Figure 6B**, **Table 3**). The first *P. aegyptiaca* attachment was calculated to occur after 607 GDD; the RMSE-value was 14.58, while the AICc-value was 304.82, i.e., the values were lower than those for the linear model.

The BTSG model (Equation 6) was also fitted to the data (**Figure 7B**, **Table 3**). This model estimated that development does not cease after the maximal temperature, but decreases until it becomes asymptotically close to zero at higher temperatures. Fitting the model of the first *P. aegyptiaca* attachment for minimal variance gave a mean value ($\pm$ *SD*) of 700 $\pm$ 53 GDD (**Table 2**, **Figure 7A**). In the second step, as was done for the other models, a dynamic curve was fitted to the *P. aegyptiaca* development observations. The first *P. aegyptiaca* attachments appeared after 583 GDD. The RMSE-value was 10.79, while the AICc-value was 278.58, both being lower than the values obtained with the other two models. For both beta function and BTSG models, data was calculated in two ways: hourly measurements and daily average. While hourly measurements resulted in good estimation with low variance, daily average measurements demonstrated poor estimation with high variance between locations. Therefore, further analyses of the models were done using hourly measurements.

Comparison and Field Validation of the Models' Prediction Power

The two best models were compared; the beta function and the BTSG models. These two models were validated for prediction of *P. aegyptiaca* first attachment by the leave-one-out cross validation method. Both models were able to predict *P. aegyptiaca* parasitism in all 13 locations (**Tables 4, 5**). The average variances of the beta function and the BTSG were 6228 and 5801, respectively. The average differences between the expected first *P. aegyptiaca* attachment and the observed attachment were 2.16 and 2.35 days for the beta function and the combined models, respectively. The maximal time gap between the observed *P. aegyptiaca* first attachment and the prediction made by the model were 5.13 and 5 days for the beta function and the BTSG models, respectively.

DISCUSSION

Most weed species develop above the soil surface and can therefore be observed and quantified (e.g., weed height or biomass). However, in the case of root parasitic weed, broomrape, development, and biomass accumulation take place in the soil sub-surface and are therefore extremely difficult to quantify. Several methods have been developed for broomrape detection. One of the most promising relies on thermal time or GDD models. These models assume a positive linear correlation between soil temperatures and broomrape development (Eizenberg et al., 2012a). However, the current study found that *P. aegyptiaca* parasitism on carrot roots responded differently from that reported in previous studies, in which low temperatures (below 18°C) were found to be essential for *P. aegyptiaca* and *O. crenata* development on carrot roots (Eizenberg et al., 2001). In the current work, results from the

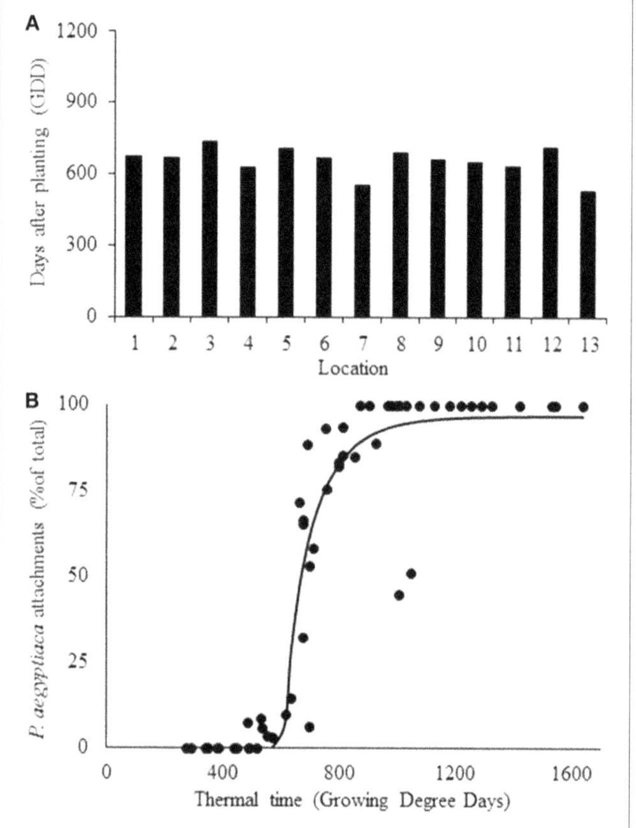

FIGURE 6 | Estimation of *Phelipanche aegyptiaca* parasitism according to the beta function model. (A) Calculation of first *P. aegyptiaca* attachment according to the beta model. **(B)** *P. aegyptiaca* development dynamics according to the beta model. Observation were made in 13 locations. Curve parameters are presented in **Table 3**.

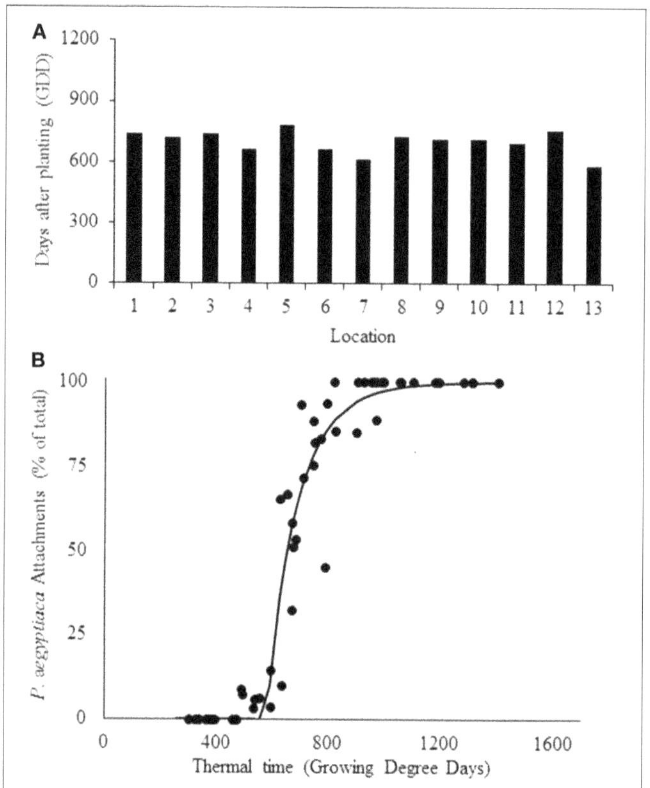

FIGURE 7 | Estimation of *Phelipanche aegyptiaca* parasitism with the combined beta function and sigmoid model. (A) Calculation of first *P. aegyptiaca* attachment calculated according to the combined model. **(B)** *P. aegyptiaca* development dynamic according to the combined model, Observations in 13 locations. Curve parameters are presented in **Table 3**.

controlled-environment experiments demonstrated that the *P. aegyptiaca* development rate on carrot roots accelerated until an optimal temperature (28°C/22°C day/night) was reached; at higher temperatures *P. aegyptiaca* development was inhibited.

The controlled-environment experiments on *P. aegyptiaca*–infested carrot plants demonstrated that the optimal temperatures for parasite and carrot development were similar, being 28°C/22°C and 22°C/16°C, respectively, for *P. aegyptiaca* infested carrot taproots. At high temperatures development of both *P. aegyptiaca* and carrot taproot was inhibited. In previous research, a linear model was found to adequately describe the relations between thermal time and parasitism dynamics for *O. minor* on red clover, *O. cumana* on sunflower, and *P. aegyptiaca* on tomato (Eizenberg et al., 2005a, 2012b; Ephrath et al., 2012). However, model fitting to the current data using the linear model resulted in poor prediction. Assessment of first *P. aegyptiaca* attachment by using the linear model resulted in high variation, i.e., 2 out of the 13 locations gave higher values than the average predicted by the linear model. Moreover, the equation describing the parasitism dynamics failed to describe the first appearance of *P. aegyptiaca* attachment according to the linear model [the parameter lag phase on the Weibull Equation (7)]. The controlled-environment experiment confirmed that carrot optimal growth rate rise until 28°C/22°C, and thereafter carrot growth decreases above the optimal temperature. It has been reported that *P. aegyptiaca* seed germination is also inhibited at high temperatures (Kebreab and Murdoch, 1999). According to these results, a linear model cannot describe the parasitism dynamics. Thus, these findings emphasize the need for a different model for prediction at temperatures beyond the optimal values.

The beta function model addresses the above problem by adding the maximal temperature for development parameter (T_m). According to the model, development begins above the minimal temperature (T_b), rises until the optimal temperature (T_o), decreases between the optimal and the maximal temperatures, and ceases completely beyond Tm (Yin and Kropff, 1996). Using the beta function model, we were able to reduce the variance in the first appearance of *P. aegyptiaca* between the different locations, compared to the linear model. The minimal temperature for development (T_{base}), according to the fitted model, was found to be 2°C, as was previously reported for the development of carrot seeds and seedlings (Finch-Savage et al., 1998). This T_{base} was appropriate for carrots but for not *P. aegyptiaca* germination, which was found to be 4–5°C (Murdoch and Kebreab, 2013). This disparity can be explained by the understanding that the model refers to overall parasitism dynamics, which include *P. aegyptiaca* that is parasitized carrot.

TABLE 4 | Leave-one-out cross validation for prediction of first *P. aegyptiaca* attachment on carrot roots according to beta-function-calculated subsoil temperatures.

Site out number	Mean	One-out value	Variance	*SD*	Yes/No	Difference	Days
1	689.75	708.48	6145.69	78.39	1	18.73	1.00
2	694.95	683.09	6183.63	78.63	1	11.86	1.25
3	679.12	749.84	6276.35	79.22	1	70.72	3.17
4	700.41	652.74	6027.26	77.63	1	47.67	3.75
5	690.13	727.74	6370.49	79.81	1	37.61	1.33
6	711.01	678.6	6178.69	78.6	1	32.41	1.79
7	694.16	692.82	6115.01	78.19	1	1.34	0.83
8	690.49	717.12	6314.86	79.46	1	26.63	1.33
9	692.86	677.78	6269.45	79.17	1	15.08	1.25
10	695.99	663.18	6332.43	79.57	1	32.81	5.13
11	694.25	642.06	5950.98	77.14	1	52.19	5.08
12	700.94	682.79	6266.83	79.16	1	18.15	1.21
13	689.56	711.78	6535.71	80.48	1	22.22	1.00
Average	694.12	691.38	6228.26	78.88	1	29.80	2.16

Means and SD were computed with 12 sites, while the site out value was computed by the estimated parameters. Difference refers to calculated GDDs gap between the mean and the one-out value. Days represent the difference converted into days.

TABLE 5 | Leave-one-out cross validation for prediction of first *P. aegyptiaca* attachment on carrot roots according to beta function combined with sigmoid equation calculated sub soil temperatures.

Site-out number	Mean	One-out value	Variance	*SD*	Yes/No	Difference	Days
1	702.74	729.37	5852.42	76.50	1	26.63	1.83
2	709.71	741.68	5855.55	76.71	1	31.97	2.29
3	691.15	751.16	5721.73	75.64	1	60.01	2.92
4	709.07	666.9	5754.78	75.86	1	42.17	2.96
5	698.29	767.48	5567.25	74.61	1	69.19	5.00
6	726.48	679.24	5807.28	76.20	1	47.24	2.50
7	703.96	669.74	5834.79	76.38	1	34.22	4.22
8	698.9	719.77	5859.39	76.54	1	20.87	1.21
9	704.18	716.76	5866.98	76.59	1	12.58	1.00
10	707.05	723.31	5868.64	76.60	1	16.26	2.04
11	707.32	709.62	5879.59	76.67	1	2.3	0.21
12	704.31	630.88	5697.78	75.48	1	73.43	4.21
13	703.88	701.71	5855.47	76.52	1	2.17	0.17
Average	705.15	708.27	5801.66	76.17	1	33.77	2.35

Means and SD were computed with 12 sites, while the site out value was computed by the estimated parameters. Difference refers to calculated gap in GDDs between the mean and the one-out value. Days represent the difference converted into days.

Therefore, we can assume that seed germination, a short part of the attachment process, plays a negligible role in the parasitism dynamics. Moreover, the effect of temperature on host development is more important than its effect on the parasite, because the germination of *P. aegyptiaca* seeds is autonomic, while its development (i.e., biomass accumulation) is dependent on the host conditions; therefore, the effect of soil temperature on host development is more significant for *P. aegyptiaca* biomass accumulation than for its germination.

Previous research has shown different *P. aegyptiaca* base temperatures for development on different crops (Ephrath and Eizenberg, 2010; Ephrath et al., 2012). Unlike the linear model, the dynamics curve according to the beta function can predict the entire parasitism progression. However, it cannot be assumed that there is complete inhibition of parasite development beyond the optimal temperature. Moreover, although previous research has described optimal parasite germination at temperatures up to 30°C, parasite germination is still possible above this temperature (Murdoch and Kebreab, 2013). In addition, parasite genetic diversity should be taken into consideration, and therefore we cannot assume unequivocally that no parasite development has occurred above the maximal temperature. Accordingly, an adjusted model was developed for use at temperatures above the optimal temperature, namely, a beta function model combined with a sigmoid curve, the BSTG model. According to the BTSG model, beyond the optimal temperature there is a moderate decline of the development rate, instead of a rapid decline to zero. This model describes non-symmetric development under rising temperatures. As demonstrated in the controlled environment experiment and in the field minirhizotron observations, development at low temperatures differs from that at high temperatures. Therefore, the BTSG model—with a particular development rate up to the optimal temperature and a different rate beyond it—can provide a good estimation of parasitism dynamics.

Cross validation analysis of both models confirmed the utility of the BTSG model. The BTSG model and the beta function model made similar predictions for the first *P. aegyptiaca* attachment, but prediction of overall *P. aegyptiaca* parasitism dynamics was found to be superior with the combined model. RMSE- and AICc-values confirm the robustness of using the combined model, even though it contains more parameters.

Although chemical control is a practical tool for broomrape management (Eizenberg et al., 2012a), effective use of herbicides requires accurate application at a specific parasite development stage; for example, for red clover infected by *O. minor*, imazamox application to the foliage according to thermal time model reduced the damage caused to crop by the parasite. Moreover, herbicide application at an early parasite developmental stage has the dual advantage of requiring a lower dose of the chemical than that at a later stage and remaining effective for a longer period (Eizenberg et al., 2006). Therefore, accurate prediction of initiation of *P. aegyptiaca* attachment is a prerequisite for optimal herbicide application (Cochavi et al., 2016). However, field experiments have shown that the best results are obtained using sequential herbicide application (Cochavi et al., 2015), and it is therefore essential to have in hand a parasitism dynamics model for accurately predicting the timing of herbicide application. Previously developed models are not suitable, as they pertain to short-term crops grown over the summer or greenhouse crops grown mostly under moderately high temperature regimes.

Carrot was originally a moderate temperature crop, but nowadays it is grown the year round. Therefore, the crop is exposed to a wide range of temperatures, including supra optimal temperatures, reducing its development rate during the growth period. It is well-known from studies on non-parasitic weeds that the development rate at high temperatures is not linear and tends to decline (White et al., 2015). Hence, the prediction model should deal differently with the crop–parasite relation at different temperatures. The beta-function model, unlike the linear model, gives an accurate prediction of the first establishment of *P. aegyptiaca* on carrot roots, due to its ability to deal with the parasite development at supra-optimal temperatures. However, it is known from previous research that beneficial *P. aegyptiaca* management in the field requires at least three sequential applications of glyphosate, not only at the first attachment but also at later stages (Cochavi et al., 2015). Therefore, although the BTSG model contains more parameters than the beta function model and can predict the *P. aegyptiaca* first attachment at the same level, its ability to predict more accurately the total parasitism process is superior, as is needed for optimal herbicide application in the field.

CONCLUSIONS

A number of models have been developed to describe the parasitism dynamics of broomrape species in their hosts, and in all a linear equation for computing GDD was fit to the parasitism dynamics. The point to be stressed is that all these models are seasonal and quantify the parasitism dynamics in the spring when temperatures begin to increase. However, carrot are cultivated the year round, with sowing in the summer and growth into the winter or be sowing in the winter and growth into the summer. Therefore, for this host, it was necessary to develop a robust temperature-sensitive model for predicting the parasitism dynamics. Hence, the BTSG model was found to be effective for prediction of *P. aegyptiaca* development in all temperature regimes, including high temperatures. The model found to be effective for predicting the entire parasitism dynamics and not only for predicting the first attachment therefore can be used for precise temporal chemical management of in carrot.

Further research should test herbicide efficiency according to the suggested model, meaning the best timing of applications and the optimal number of applications for the development of a sustainable method for *P. aegyptiaca* management in carrot.

AUTHOR CONTRIBUTIONS

AC: MSc student, performed field studies, data analysis, and writing the manuscript. GA: Research Assistant, technical support in field studies. BR: Supervisor, support in data analysis and writing. HE: Principle investigator, head of the lab of parasitic weeds.

ACKNOWLEDGMENTS

The authors wish to thank the Chief Scientist of the Israel Ministry of Agriculture and The Plant Production and Marketing Board for funding this project (Grant No. 132-1499-11). We would like to thank Adir Katz from The William Davidson Faculty of Industrial Engineering and Management, Technion-Israel Institute of Technology for modeling and statistical support.

REFERENCES

Aarts, E. H. L., Korts, J. H. M., and Van Laarhoven, P. J. M. (1997). "Simulated annealing," in *Local Search in Combinatorial Optimization*, eds E. H. L. Aarts and J. K. Lenstra (New York, NY: John Wiley & Sons Inc.), 91–120.

Bernhard, R. H., Jensen, J. E., and Andreasen, C. (1998). Prediction of yield loss caused by *Orobanche* spp. in carrot and pea crops based on the soil seedbank. *Weed Res.* 38, 191–197. doi: 10.1046/j.1365-3180.1998.00089.x

Castejon-Muñoz, M., Romero-Muñoz, F., and Garcia-Torres, L. (1993). Effect of planting date on broomrape (*Orobanche cernua* Loefl.) infections in sunflower

(*Helianthus annuus* L.). *Weed Res.* 33, 171–176. doi: 10.1111/j.1365-3180.1993.tb01930.x

Cavanaugh, J. E. (1997). Unifying the derivations for the akaike and corrected akaike information criteria. *Stat. Probab. Lett.* 33, 201–208. doi: 10.1016/S0167-7152(96)00128-9

Cochavi, A., Achdari, G., Smirnov, Y., Rubin, B., and Eizenberg, H. (2015). Egyptian Broomrape (*Phelipanche aegyptiaca*) management in carrot under field conditions. *Weed Technol.* 29, 519–528. doi: 10.1614/WT-D-14-00140.1

Cochavi, A., Rubin, B., Smirnov, E., Achdari, G., and Eizenberg, H. (2016). Factors affecting *Phelipanche aegyptiaca* (Egyptian Broomrape) control in carrot. *Weed Sci.* 64, 321–330. doi: 10.1614/WS-D-15-00123.1

Eizenberg, H., Aly, R., and Cohen, Y. (2012a). Technologies for smart chemical control of broomrape (*Orobanche* spp. and *Phelipanche* spp.). *Weed Sci.* 60, 316–323. doi: 10.1614/WS-D-11-00120.1

Eizenberg, H., Colquhoun, J. B., and Mallory-Smith, C. A. (2006). Imazamox application timing for small broomrape (*Orobanche minor*) control in red clover. *Weed Sci.* 54, 923–927. doi: 10.1614/WS-05-151R.1

Eizenberg, H., Colquhoun, J., and Mallory-Smith, C. (2005a). A predictive degree-days model for small broomrape (*Orobanche minor*) parasitism in red clover in Oregon. *Weed Sci.* 53, 37–40. doi: 10.1614/WS-04-018R1

Eizenberg, H., Hershenhorn, J., Achdari, G., and Ephrath, J. E. (2012b). A thermal time model for predicting parasitism of *Orobanche cumana* in irrigated sunflower - field validation. *F. Crop Res.* 137, 49–55. doi: 10.1016/j.fcr.2012.07.020

Eizenberg, H., Shtienberg, D., Silberbush, M., and Ephrath, J. E. (2005b). A new method for *in-situ* monitoring of the underground development of *Orobanche cumana* in sunflower (*Helianthus annuus*) with a mini-rhizotron. *Ann. Bot.* 96, 1137–1140. doi: 10.1093/aob/mci252

Eizenberg, H., Tanaami, Z., Jacobsohn, R., and Rubin, B. (2001). Effect of temperature on the relationship between *Orobanche* spp. and carrot (*Daucus carota* L.). *Crop Prot.* 20, 415–420. doi: 10.1016/S0261-2194(00)00165-4

Ephrath, J. E., and Eizenberg, H. (2010). Quantification of the dynamics of *Orobanche cumana* and *Phelipanche aegyptiaca* parasitism in confectionery sunflower. *Weed Res.* 50, 140–152. doi: 10.1111/j.1365-3180.2010.00768.x

Ephrath, J. E., Hershenhorn, J., Achdari, G., Bringer, S., and Eizenberg, H. (2012). Use of logistic equation for detection of the initial parasitism phase of Egyptian Broomrape (*Phelipanche aegyptiaca*) in tomato. *Weed Sci.* 60, 57–63. doi: 10.1614/WS-D-11-00070.1

Finch-Savage, W., Steckel, J. R., and Phelps, K. (1998). Germination and post-germination growth to carrot seedling emergence: predictive threshold models and sources of variation between sowing occasions. *New Phytol.* 139, 505–516. doi: 10.1046/j.1469-8137.1998.00208.x

Hawkins, D. M., Basak, S. C., and Mills, D. (2003). Assessing model fit by cross-validation. *J. Chem. Inf. Comput. Sci.* 43, 579–586. doi: 10.1021/ci025626i

Heide-Jørgensen, H. S. (2013). "Introduction: The parasitic syndrome in higher plants," in *Parasitic Orobanchaceae*, eds D. M. Joel, J. Gressel, and L. J. Musselman (Berlin, Heidelberg: Springer), 1–14.

Jacobsohn, R., and Kelman, Y. (1980). Effectiveness of glyphosate in broomrape (*Orobanche* spp.) control in four crops. *Weed Sci.* 28, 692–698.

Kebreab, E., and Murdoch, A. J. (1999). A model of the effects of a wide range of constant and alternating temperatures on seed germination of four *Orobanche* species. *Ann. Bot.* 84, 549–557. doi: 10.1006/anbo.1998.0948

Lati, R. N., Filin, S., and Eizenberg, H. (2011). Temperature- and radiation-based models for predicting spatial growth of purple nutsedge (*Cyperus rotundus*). *Weed Sci.* 59, 476–482. doi: 10.1614/WS-D-11-00007.1

Lati, R. N., Filin, S., and Eizenberg, H. (2013). Estimating plant growth parameters using an energy minimization-based stereovision model. *Comput. Electron. Agric.* 98, 260–271. doi: 10.1016/j.compag.2013.07.012

Mesa-García, J., and García-Torres, L. (1986). Effect of planting date on parasitism of broadbean (*Vicia faba*) by crenate broomrape (*Orobanche crenata*). *Weed Sci.* 34, 544–550.

Murdoch, A. J., and Kebreab, E. (2013). "Germination ecophysiology," in *Parasitic Orobanchaceae: Parasitic Mechanism and Control Strategy*, eds D. M. Joel, L. J. Musselman, and J. Gressel (Heidelberg: Springer), 195–220. doi: 10.1007/978-3-642-38146-1_11

Parker, C. (2012). Parasitic weeds: a world challenge. *Weed Sci.* 60, 269–276. doi: 10.1614/WS-D-11-00068.1

White, S. N., Boyd, N. S., and Van Acker, R. C. (2015). Temperature thresholds and growing-degree-day models for red sorrel (*Rumex acetosella*) ramet sprouting, emergence, and flowering in wild blueberry. *Weed Sci.* 63, 254–263. doi: 10.1614/WS-D-14-00048.1

Yin, X., and Kropff, M. J. (1996). Use of the beta function to quantify effects of photoperiod on flowering and leaf number in rice. *Agric. For. Meteorol.* 81, 217–228. doi: 10.1016/0168-1923(95)02324-0

Yoneyama, K., Ruyter-spira, C., and Bouwmeester, H. J. (2013). "Induction of germination," in *Parasitic Orobanchaceae*, eds D. M. Joel, J. Gressel, and L. Musselmam (Heidelberg; New York, NY: Springer), 167–194. doi: 10.1007/978-3-642-38146-1_10

Conflict of Interest Statement: The authors declare that the research was conducted in the absence of any commercial or financial relationships that could be construed as a potential conflict of interest.

9

Investigation of Amino Acids as Herbicides for Control of *Orobanche minor* Parasitism in Red Clover

Edited by:
Antonio Elia,
University of Foggia, Italy

Reviewed by:
Giovanni Mauromicale,
University of Catania, Italy
Cesare Accinelli,
Università di Bologna, Italy

***Correspondence:**
Mónica Fernández-Aparicio
monica.fernandez@ias.csic.es

Mónica Fernández-Aparicio[1,2]*, Alexandre Bernard[3], Laurent Falchetto[3], Pascal Marget[1,3], Bruno Chauvel[1], Christian Steinberg[1], Cindy E. Morris[4], Stephanie Gibot-Leclerc[1], Angela Boari[5], Maurizio Vurro[5], David A. Bohan[1], David C. Sands[6] and Xavier Reboud[1]

[1] Agroécologie, AgroSup Dijon, INRA, Université Bourgogne Franche-Comté, Dijon, France, [2] CSIC, Institute for Sustainable Agriculture, Córdoba, Spain, [3] INRA, UE0115 Domaine Expérimental d'Epoisses, Bretenière, France, [4] INRA, UR0407 Pathologie Végétale, Montfavet, France, [5] CNR, Institute of Sciences of Food Production, Bari, Italy, [6] Department of Plant Sciences & Plant Pathology, Montana State University, Bozeman, MT, United States

Certain amino acids induce inhibitory effects in plant growth due to feedback inhibition of metabolic pathways. The inhibition patterns depend on plant species and the plant developmental stage. Those amino acids with inhibitory action on specific weeds could be utilized as herbicides, however, their use for weed control has not been put into practice. *Orobanche minor* is a weed that parasitizes red clover. *O. minor* germination is stimulated by clover root exudates. The subsequent seedling is an obligated parasite that must attach quickly to the clover root to withdraw its nutrients. Early development of *O. minor* is vulnerable to amino acid inhibition and therefore, a series of *in vitro*, rhizotron, and field experiments were conducted to investigate the potential of amino acids to inhibit *O. minor* parasitism. In *in vitro* experiments it was found that among a collection of 20 protein amino acids, lysine, methionine and tryptophan strongly interfere with *O. minor* early development. Field research confirmed their inhibitory effect but revealed that methionine was more effective than lysine and tryptophan, and that two successive methionine applications at 308 and 543 growing degree days inhibited *O. minor* emergence in red clover up to 67%. We investigated additional effects with potential to influence the practical use of amino acids against broomrape weeds, whether the herbicidal effect may be reversible by other amino acids exuded by host plants or may be amplified by inducing host resistance barriers against *O. minor* penetration. This paper suggests that amino acids may have the potential to be integrated into biorational programs of broomrape management.

Keywords: parasitic weeds, methionine, alternatives for crop protection, animal feed supplements, germination, host attachment, herbicidal activity of amino acids, sustainable agriculture

INTRODUCTION

Broomrape weeds (*Orobanche* and *Phelipanche* spp.) are angiosperms that lack photosynthetic competence and a root system. They thrive in agricultural ecosystems parasitizing a wide range of host crops in the Apiaceae, Asteraceae, Brassicaceae, Fabaceae, and Solanaceae (Parker and Riches, 1993). Among them, *Orobanche minor* Sm., is widely distributed inducing major economic damage in clover and lucerne crops (Lins et al., 2007; Parker, 2013). In eastern France, the main host crop is red clover (*Trifolium pratense* L.) in which, besides a strong reduction in clover harvest,

O. minor seeds contaminate clover forage and seed yield resulting in a significant problem in crops for commercial seed production.

Broomrape seed germination is stimulated by host root exudates. Once germination is initiated, an infective radicle arises from the seed coat that grows to make contact with the host plant. Upon host root contact, a haustorium is induced at the tip of the broomrape radicle, with functions of host root invasion, vascular connection and nutrient withdrawal (Vaucher, 1823; Kuijt, 1969; Joel, 2013). Once the resource sink is established by the young parasite, the parasite initiates storage of host-derived nutrients in an underground parasitic organ called tubercle. This underground stage of parasitic development produces a strong injury in the crop (Eizenberg et al., 2005). Near the end of parasitic life cycle, underground shoots are developed in the tubercles that quickly emerge through the soil surface for flowering and seed set. Each flowering stalk can set up more than half million dust-like seeds easily dispersed by wind, farm machinery and crop seeds and therefore the build-up of parasitic seed bank is very efficient.

The control of broomrape weeds remains an elusive task in agriculture (Rubiales et al., 2009). Early diagnosis of broomrape infection is difficult due to the mostly underground growing habit of this weed. In addition, the close physical and chemical associations established by broomrape with the parasitized crop difficult the selective control of the weed without injuring the crop. In absence of host crop stimulation to germinate, broomrape seeds can last viable in agricultural soils for several decades being their seed bank extremely persistent. Once germinated the broomrape seedling does not survive longer than few days without host-derived nutritive supply. Control strategies interfering with broomrape germination and radicle growth inhibit host infection and broomrape reproduction producing a decline in the parasitic seed bank while meeting short-term farming expectations of crop productivity (Fernández-Aparicio et al., 2016). Once the parasitic seedling has made contact with the host root surface, host resistance to parasite invasion either natural, genetically engineered or elicited by biotic- or abiotic-inducing agents leads to death of the invading parasite before it causes strong injury in the crop (Gonsior et al., 2004; Aly, 2007; Kusumoto et al., 2007). Once the parasite has established resource sink and the tubercle started to develop, the control of broomrape weeds can be achieved by using glyphosate and imidazolinone herbicides (Hershenhorn et al., 2009). Those herbicides inhibit enzymes that catalyze key steps in the biosynthesis of aromatic and branched-chain amino acid, respectively. For efficacious action on broomrape weeds, the herbicide is applied on the foliage of herbicide-tolerant crop cultivars and then translocated systemically in active form towards the underground parasite via the haustorium (Lins et al., 2005; Eizenberg et al., 2006). Despite broomrape parasitism can be inhibited very efficiently by cultivation of resistant crops to broomrape infection or application of systemic herbicides, parasite resistant- or herbicide tolerant-cultivars are not always available for many crop species.

In addition, due to increasing concern about overdependence on synthetic herbicides there is a great demand for novel approaches in weed management programs. As a green alternative for weed control, the enzymes targeted in weeds by glyphosate and imidazolinone herbicides, can also be inhibited by the exogenous addition of certain amino acids (Sands and Pilgeram, 2009). The rationale is based in the mode of enzymatic regulation in amino acid biosynthetic pathways. Amino acid biosynthesis is an energy expensive process, and it is regulated by feedback inhibition of key enzymes by some of the amino acid products of the pathways they regulate. This feedback inhibition leads in plant growth inhibition (Bright et al., 1978), and it can be counteracted by exogenous addition of certain related amino acids in the same biosynthetic pathway. In addition, amino acids from unrelated pathways can rescue the plant growth by giving general growth stimulus or reactivating the targeted enzyme in the case of end product of competitive pathways. Active amino acids for plant growth inhibition and plant growth rescue are dependent on the plant species, the plant developmental stage and concentration of application (Henke et al., 1974) and therefore, the development of this strategy for weed management systems requires the determination of which amino acids and concentrations are efficacious against the targeted weed species and whether the inhibition is reversible by other amino acids potentially present in the soil (Sands and Pilgeram, 2009). Properly implemented this strategy assumes low toxicity for plants other than the targeted weed and low persistence in the soil through microbial metabolization. In addition, because the targeted enzymes are absent in animals it is also assumed low toxicity in them. Although this approach has been reported effective for control of Canada thistle, red bromegrass, kudzu and cannabis, the direct application of amino acids as herbicides has not been put into field practice yet (Sands and Pilgeram, 2009).

Germination and radicle growth of the parasitic weeds *Phelipanche ramosa* and *Orobanche crenata* have been found to be inhibited by certain amino acids (Vurro et al., 2006; Fernández-Aparicio et al., 2013). Therefore, amino acids could have the potential to be implemented into broomrape management systems as herbicides. Given the obligated parasitic growth of these weeds, the implementation of this approach requires to consider additional factors which could influence the herbicidal treatment. The growth habit of broomrape before host attachment is restricted to the crop rhizosphere because its germination is stimulated by host root exudates. In addition to germination factors, root exudates of host crop species are a significant source of amino acids. Glycine, serine, and alanine are abundant in root exudates, but also concentrations of glutamic acid, arginine, tyrosine, glutamine, or valine have been described to be significant in crop root exudates (Lesuffleur et al., 2007). Therefore, for development of this approach, it should be determined which of those amino acids could have the potential to counteract the herbicidal effect of the broomrape-inhibitory amino acids in case they were deposited in the soil by clover roots.

Another amino acid effect against parasitic weeds could be through the potential elicitation of broomrape resistance in the crop. It has been found that certain amino acids such as methionine induce resistance in other pathosystems (Sarosh et al., 2005). Induction of resistance barriers in the crop roots to parasitic weed penetration could extend the amino acid protective effect against the parasitic weed to parasitic stages beyond its attachment. Broomrape infection is susceptible to be inhibited by other abiotic elicitors of resistance in the host crop (Gonsior et al., 2004; Kusumoto et al., 2007) but the potential of methionine to elicit crop resistance to broomrape parasitism has never been investigated before.

Given the predicted low persistence of amino acid treatment in the soil (Sands and Pilgeram, 2009), the herbicidal effect of amino acids would be maximized if the timing of application is based on the occurrence of susceptible parasitic life stages. Broomrape development depends on temperature (Eizenberg et al., 2005) and therefore, the relation between temperature and underground phenology should be recorded locally under field conditions in order to ease the targeting of vulnerable parasitic life stages by amino acid treatment.

This research conducted laboratory and field experiments to evaluate the effect of direct application of amino acid as herbicides for field control of *O. minor* parasitism in red clover. This paper reports (i) the influence of single amino acids and combination of amino acids on *O. minor* growth previous to clover attachment. To achieve this, amino acids were applied *in vitro* to seeds and germination rate and radicle length were measured; (ii) the indirect effects of methionine on growth of *O. minor* after its attachment to clover roots. This was achieved by treating clover with methionine and measuring its resistance to *O. minor* haustorium penetration; (iii) to study under open field conditions the inhibitory effect of amino acids on *O. minor* parasitism in clover. To achieve this the amino acid was applied by irrigation at clover post-emergence during the occurrence of amino acid-sensitive *O. minor* stages. This paper proposes that the amino acid herbicidal approach could have the potential to be integrated into sustainable programs of broomrape management.

MATERIALS AND METHODS

In Vitro Experiments

In Vitro Identification of Amino Acids with Inhibitory Activity on *O. minor* Parasitism at Host Pre-attached Stages

Sensitivity of *O. minor* germination and radicle growth to alanine, arginine, aspartate, cysteine, glutamic acid, glutamine, glycine, histidine, homoserine, isoleucine, leucine, lysine, methionine, phenylalanine, proline, serine, threonine, tryptophan, tyrosine, and valine was assessed *in vitro*. The *O. minor* seeds used were obtained in 2014 from *O. minor* plants infesting red clover field at INRA Experimental Unit of Epoisses, Bretenières (France). Seeds were collected from mature, dry *O. minor* inflorescences using a 0.5 mm mesh-size sieve and stored dry in the dark at room temperature until use.

In vitro bioassays were conducted to study the amino acid inhibitory activity on *O. minor* germination and radicle growth. One-year old *O. minor* seeds were surface sterilized by immersion in 1.5% (w/v) NaOCl, and 0.02% (v/v) Tween 20, and sonication for 2 min, rinsed thoroughly with sterile distilled water and dried in a laminar air flow cabinet. Inhibitory activity was tested as reported previously (Fernández-Aparicio et al., 2013). For germination, *Orobanche* seeds require a warm stratification period called conditioning, followed by chemical stimulation with germination-inducing factors (Lechat et al., 2012). Approximately 100 seeds of *O. minor* were placed separately on each of 183 glass fiber filter paper (GFFP) disks of 9 mm-diameter moistened with 50 μL of sterile distilled water. *O. minor* seeds were conditioned in the dark at 22°C for 7 days. The germination stimulant GR24 (Johnson et al., 1976) was used to promote germination of conditioned seeds. Test solutions of each amino acid were prepared in GR24 solution at three decreasing concentrations of the amino acid (5, 2.5, and 1.25 mM) but keeping constant the GR24 concentration (10^{-6} M) in order to allow comparisons between treatments. GR24 solution (10^{-6} M) without amino acid was used as control. GFFP disks containing the conditioned *O. minor* seeds were transferred inside laminar flow cabinet to sterile sheet of filter paper to remove excess of moisture and transferred to 10 cm-diameter Petri dishes. Aliquots of 50 μl of each test solution were applied to triplicated GFFP disks containing the *O. minor* conditioned seeds. Petri dishes were sealed with Parafilm and stored in the dark at 22°C to promote germination and radicle growth. Seven days later, *O. minor* seeds were observed under stereoscopic microscope at which time germination percentage was assessed by scoring the number of of seeds with and emerged radicle from 100 seeds in each of three replicate GFFP disks and radicle length was determined in 15 randomly selected germinated seeds in each of the three replicate GFFP disks. Amino acid-mediated reductions of radicle growth were expressed as a percent of the untreated control (GR24).

Inhibitory Action of Amino Acids on *O. minor* Parasitism at Host Post-attached Stages

We tested the hypothesis that amino acids could elicit a mechanism of clover resistance to *O. minor* penetration. We make this hypothesis based on the facts that certain amino acids such as methionine induce resistance in other pathosystems (Sarosh et al., 2005) and that abiotic elicitors of induced resistance have been shown effective to inhibit broomrape parasitism (Gonsior et al., 2004; Kusumoto et al., 2007). The tetraploid red clover cultivar Trevvio (RAGT semences) was chosen due to its high susceptibility to *O. minor* parasitism. Seeds of Trevvio were treated with 20 mM concentration of L-methionine (Sigma-Aldrich, St. Louis, MO, United States) at 22°C for 9 h. Then, imbibed clover seeds were placed in pots with perlite to allow clover germination and seedling growth during 10 days prior being transplanted to the minirhizotron system described by Fernández-Aparicio et al. (2008). One clover seedling developed from L-methionine imbibed seeds were transplanted per square Petri dishes filled with sterile perlite and covered in the upper surface by a sheet of GFFP. Clover seeds imbibed with distilled water (control treatment) were cultivated under the same conditions for comparisons. *O. minor* seeds collected, sterilized and conditioned as described in Section "*In Vitro* Identification of Amino Acids with Inhibitory Activity on *O. minor* Parasitism at Host Pre-attached Stages", were treated with the germination stimulant GR24 and distributed at a proximate density of 50 seeds per cm^2 on the GFFP on which the clover plants were developing their roots. After 27 days of *O. minor* inoculation, parasitic seedlings that made contact with clover roots were inspected under a stereoscopic microscope and classified as either (i) *O. minor* radicles attached to clover root surface which died as a consequence of unsuccessful penetration and vascular connection showing hypersensitive-like response at the attachment point, (ii) *O. minor* radicle that penetrated the host root and formed a healthy tubercle as consequence of successful nutrient transfer, (iii) *O. minor* radicles that penetrated the host root and formed tubercle that quickly turned necrotic and later died as a consequence of host-parasite late incompatible interaction, and (iv) "spider" stage, i.e., tubercles with rapid and healthy development as seen by initiation of crown roots. The effects of the methionine imbibition treatment on *O. minor* infection was determined by calculation of the following four parameters: infection success (percent of contacted *O. minor* radicles that successfully invaded the host root), hypersensitive-like response (percent of contacted *O. minor* radicles that died before vascular connection showing browning area at the host attachment point), necrosis development in tubercle (percent of total formed tubercles that died at early stages of development becoming necrotic) and 'spider' development in tubercle (percent of formed tubercles that initiated crown roots as consequence of successful parasitism).

In Vitro Validation of Inhibitory Activity of Amino Acids Formulated as Supplements in Animal Feed

Among the group of *O. minor*-inhibitory amino acids identified in Section "*In vitro* Identification of Amino Acids with Inhibitory Activity on *O. minor* Parasitism at Host Pre-attached Stages", tryptophan and the aspartate-derived amino acids: lysine, methionine and threonine were selected as candidates for field control. This choice was on the basis of the preliminary results and took into account the ready availability of these amino acids at large commercial scale as animal feed supplements. Prior to field research, the inhibitory activity of these amino acids was validated in the form of lysine-, threonine- and tryptophan-based animal feed formulations (Ajinomoto Heartland, Inc, Japan). In addition, methionine hydroxyl analog with 100% equivalency to DL-methionine (Novus International, St. Charles, IL, United States) was tested in two different formulations: formulation 1: Alimet®; and formulation 2: non-commercialized experimental formulation. Activity of each formulation was tested at three decreasing concentrations of the amino acid (10, 5, and 2.5 mM) on *O. minor* seed germination and radicle growth as described in Section "*In Vitro* Identification of Amino Acids with Inhibitory Activity on *O. minor* Parasitism at Host Pre-attached Stages".

Determination Whether Inhibitory Action of Candidate Amino Acids for Field Control is Reversed by Amino Acids Commonly Exuded by Crop Roots

O. minor seeds and seedlings placed in the soil within the potentially host infective distance are exposed to crop root exudates which may contain significant amounts of amino acids (Lesuffleur et al., 2007). To assess if herbicidal effects of lysine, methionine and tryptophan could be reversible by other amino acids, their respective inhibitory activity was studied mixed in pairs with the amino acid collection tested in Section "*In Vitro* Identification of Amino Acids with Inhibitory Activity on *O. minor* Parasitism at Host Pre-attached Stages". Rescue of 2.5 mM lysine-, methionine-, and tryptophan-toxicity was assessed by adding in pairs the other 19 amino acids at 2.5 mM concentration and the joint activity of each mixture on *O. minor* germination and radicle growth compared with the inhibitory action performed by their respective control: single solutions of lysine, methionine, or tryptophan at 2.5 mM. Bioassay and data collection were performed as described in Section "*In Vitro* Identification of Amino Acids with Inhibitory Activity on *O. minor* Parasitism at Host Pre-attached Stages".

Field Validation of Amino Acid Inhibitory Activity on *O. minor* Parasitism

Field experimentation was performed at INRA Experimental Unit of Epoisses, Bretenières, France (5°05'56"E; 47°14'20"N) with long history of approximately uniform seed bank infestation level of *O. minor* specialized in red clover. The soil at the experimental site (35 first cm layer) has a silt clay texture (sand 6.4%, silt 50.6%, clay 43%), with pH 7.97, 2.5% organic matter, C/N ratio of 10.5 and a Cation Exchange Capacity of 29.6 cmol(+) kg^{-1} soil. The red clover cultivar Trevvio was chosen due to its high susceptibility to *O. minor* parasitism. On April 7, 2015, 84 plots (3 m× 1.5 m) were mechanically sown with Trevvio seeds at a density of 37.5 kg/ha. In addition, in order to ensure dense and uniform *O. minor* seed bank, the sowing machine (Plotseed, Wintersteiger equipped with "Accord CX" disk sowing equipment) distributed 0.5 ml (approximately 63000) *O. minor* seeds per plot.

The timing of amino acid treatment was determined by inspecting the *O. minor* underground development in control plots. Starting at red clover sowing, soil and air temperature was recorded every hour with temperature data loggers, respectively, placed at 5 cm below the soil surface and 1 m above soil surface in control plots. Approximately ten clover plants were randomly sampled every 3–4 days and brought to lab. Roots were submerged in water allowing the excess of soil gently separate from the roots but maintaining the rhizosheath intact as much as possible. Then, clover roots were inspected under stereoscopic microscope. Dates were recorded for each of the following *O. minor* phenological stages: germination, radicle adhesion to host root surface, radicle penetration, vascular connection, tubercle development, development of "spider" stage (**Figure 1**). For each date, temperatures were converted to growing degree day (GDD) accumulated from clover sowing date considering base temperature as 0°C (McMaster and Wilhelm, 1997; Eizenberg et al., 2005). The timing of amino acid treatment was chosen according with observations of *O. minor* underground phenology. Two amino acid treatments were applied at 22 and 36 days after clover sowing. They, respectively, corresponded to clover-induced *O. minor* germination (282 GDD air/ 308 GDD soil) and initiation of *O. minor* radicle penetration into clover root (494 GDD air/543 GDD soil). For each application date, 10 treatments were included as follows. (A) Four animal feed formulations described in Section "*In Vitro* Validation of Inhibitory Activity of Amino Acids Formulated as Supplements in Animal Feed": methionine hydroxyl analog formulations 1 and

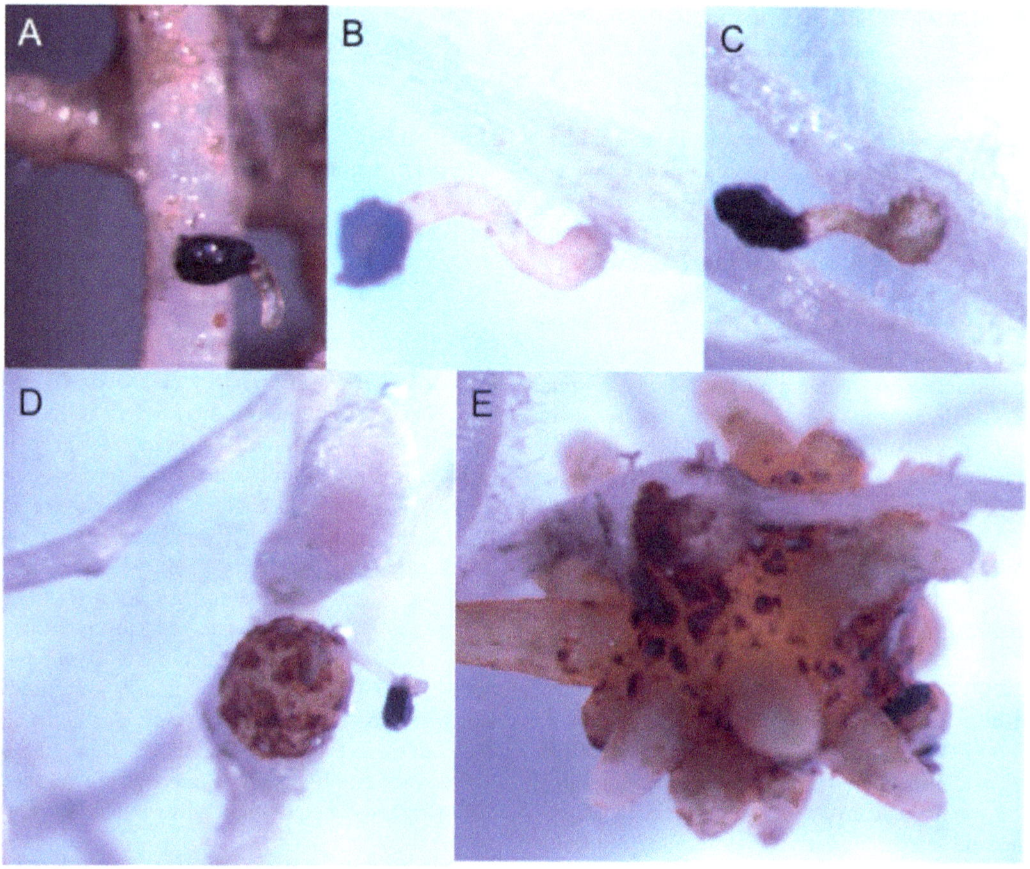

FIGURE 1 | *Orobanche minor* seedlings at successive stages of the infection of clover roots observed during a field trial in Burgundy, France. (A) *O. minor* germination. **(B)** *O. minor* adhesion to host root. **(C)** Host penetration. **(D)** Vascular connection. **(E)** Tubercle development with initiation of 'spider' stage.

2 (Novus International, St. Charles, IL, United States), lysine and tryptophan (Ajinomoto Heartland, Inc, Japan) treatments at 10 and 20 mM concentrations. (B) L-methionine (Sigma) treatment at 20 mM concentration, and (C) water (control treatment). For each treatment, the corresponding amino acids formulation were mixed with water at the described concentrations and used to irrigate the plots at a rate of 6 L m^{-2} of amino acid solution. For each treatment, the test solution was prepared based on the amino acid content in each formulation: L-methionine (100%), methionine hydroxyl analog formulation 1 (88%), methionine hydroxyl analogue formulation 2 (95%), lysine (50%), tryptophan (98%). The experimental design was a randomized block with four replications. Each treated plot including the control-plot was surrounded by two untreated plots at their immediate right and left side. Plots were managed under traditional culture. Weeds other than *O. minor* were carefully removed by hand.

The level of *O. minor* parasitism on each plot was assessed counting the number of emerged *O. minor* plants per plot. In order to avoid the error due to unavoidable heterogeneous distribution of seeds across the trial, this parameter was standardized as '*O. minor* emergence referred to the control' calculated as percentage to the mean value of emerged broomrapes per respective control plots located at right and left of each test plots.

For statistical analysis, percentages were transformed to arcsine square roots (transformed value = 180/Π x arcsine [$\sqrt{(\%/100)}$]) to normalize data and stabilize variance throughout the data range and subjected to analysis of variance using SAS 9.4 (SAS Institute Inc., Cary, NC, United States). The significance of mean differences between each treatment against the control was evaluated by the two-sided Dunnett test. The null hypothesis was rejected at the level of 0.05.

RESULTS

Sensitivity of *O. minor* Germination and Radicle Growth to Exogenous Addition of Amino Acids

The first experiment was conducted to identify which amino acids inhibit *O. minor* host-preattached stages. Twenty amino acids in pure form were individually applied to seeds of *O. minor* in combination with the germination stimulant GR24 and levels of germination and radicle growth inhibition rated in

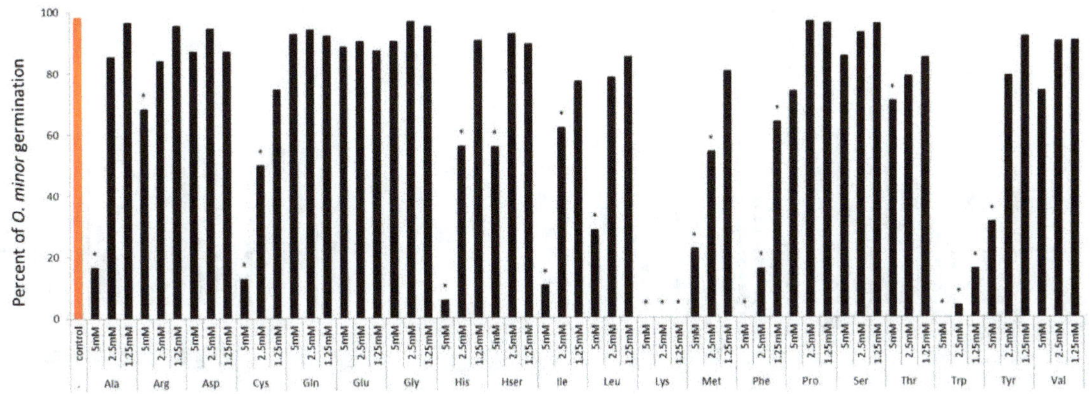

FIGURE 2 | Influence of amino acid treatments on *O. minor in vitro* germination. Treatment solutions of alanine, arginine, aspartate, cysteine, glutamic acid, glutamine, glycine, histidine, homoserine, isoleucine, leucine, lysine, methionine, phenylalanine, proline, serine, threonine, tryptophan, tyrosine, and valine along with a germination stimulant GR24 were applied to *O. minor* seeds at a range of 5–1.25 mM and their inhibitory activity in germination rated 7 days later as percentage of germination in each of three replicated disks. Analysis of variance was applied to transformed replicate data. *Indicates differences of each treatment compared with the control GR24 assessed by Dunnett's test at the 0.05 level.

comparison with the control (GR24 with no amino acid). The sensitivity of *O. minor* germination to amino acids is shown in **Figure 2**. Arginine, aspartic acid, glutamic acid, glutamine, glycine, homoserine, proline, serine, threonine, and valine caused low or non-significant reduction in percent seed germination in comparison with that of the control. The magnitude of reduction in germination was greater in alanine, cysteine, histidine, isoleucine, leucine, methionine and tyrosine which significantly inhibited *O. minor* germination at concentration of 5 mM although lower concentrations of these amino acids induced low or negligible effect. Lastly, the greatest inhibitory activity was identified in lysine, phenylalanine, and tryptophan that completely inhibited *O. minor* germination at all concentrations tested.

The sensitivity of *O. minor* radicle growth to exogenous amino acids is shown in **Figure 3**. Except for glutamine, all amino acids inhibited radicle growth at 5 mM and therefore this concentration did not allow discrimination between amino acids for their inhibitory action. At 1.25 mM, the lowest concentration tested, it was possible to identify strong inhibition activity in histidine, homoserine, isoleucine, leucine, lysine, methionine, phenylalanine, serine, tryptophan, tyrosine, and valine. Although little effect was induced at 1.25 mM by alanine, arginine, proline, and threonine, they caused more than 50% radicle growth inhibition at concentration of 2.5 mM and higher. A group of low or non-inhibitory amino acids of radicle growth was formed by aspartic acid, cysteine, glutamic acid, glutamine, and glycine, which induced less than 50% radicle growth inhibition at concentration range of 2.5–1.25 mM.

In Vitro Efficacy of Amino Acid Supplements for Animal Feed as Readily Deployable Formulations for Field Application

In vitro tests were used to validate the herbicidal activity of methionine, lysine, threonine and tryptophan in the form of animal feed-based formulation. **Figure 4** shows that lysine-based formulation induced the highest inhibition of *O. minor* germination and radicle growth at all concentration tested being this strong herbicidal activity in agreement with activity found by lysine in the pure form (**Figures 2, 3**). **Figure 4** also shows that the activity of methionine- and tryptophan-based feed supplements was slightly lower than that observed when tested in pure form. *O. minor* development was inhibited by their commercial formulations at 10 and 5 mM but not at the 2.5 mM concentration which was inhibitory when those amino acids were tested in pure form. Threonine-based animal feed supplements displayed the lowest inhibitory activity being only active at the highest concentration tested and therefore it was not included for further tests.

Potential Reversible Effects of Amino Acid Herbicidal Activity by Other Amino Acids

An *in vitro* experiment was conducted to verify that the herbicidal activity identified in lysine, methionine, and tryptophan could not be reversed by other amino acids potentially present in the crop rhizosphere. **Figures 5, 6** show that methionine inhibition of *O. minor* germination and radicle growth was not reversed by any other amino acid tested resulting the amino acid mixtures in equal or higher *O. minor* inhibition levels when compared to their respective methionine control solutions. The lysine or tryptophan control solutions induced complete inhibition of *O. minor* germination and radicle growth and this herbicidal activity was not counteracted by addition of any other amino acid (data not shown).

Timing of Amino Acid Application

Amino acid inhibition of weed parasitism targets early parasitism stages, i.e., germination and host attachment. The kinetics of clover infection by *O. minor* in relation with temperature under field conditions in Burgundy (France) was determined by regular

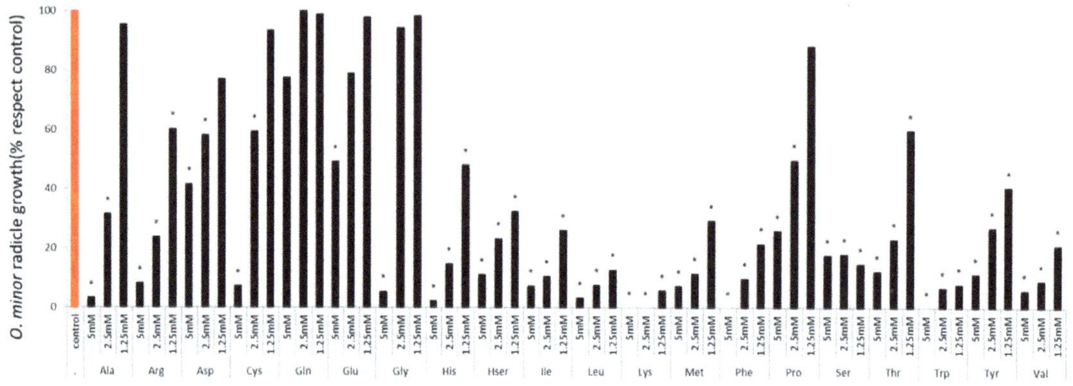

FIGURE 3 | Influence of amino acid treatments on *O. minor* radicle growth *in vitro*. Treatment solutions of alanine, arginine, aspartate, cysteine, glutamic acid, glutamine, glycine, histidine, homoserine, isoleucine, leucine, lysine, methionine, phenylalanine, proline, serine, threonine, tryptophan, tyrosine, and valine along with a germination stimulant GR24 were applied to *O. minor* seeds at a range of 5–1.25 mM and their activity in growth reduction rated 7 days later in each of three replicated disks and expressed as a percent of the untreated control activity. Analysis of variance was applied to transformed replicate data. *Indicates differences of each treatment compared with the control assessed by Dunnett's test at the 0.05 level.

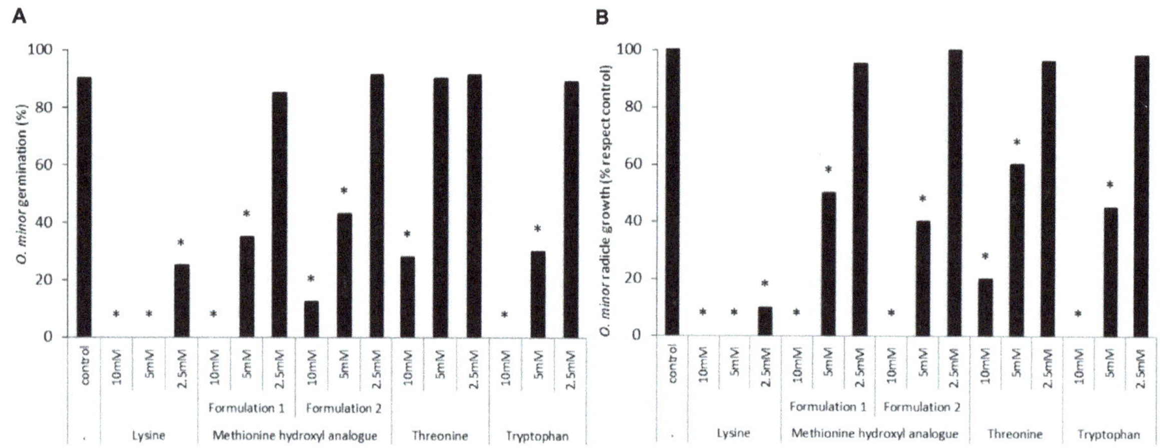

FIGURE 4 | Effects of lysine-, methionine hydroxyl analog-, threonine-, and tryptophan-based feed supplements in *O. minor in vitro* development. (A) Inhibition of *O. minor* germination. **(B)** Inhibition of *O. minor* radicle growth. *Indicates differences of each treatment compared with the control (2.5 mM methionine treatment) assessed by Dunnett's test at the 0.05 level.

inspections of clover root system from clover sowing. **Table 1** shows dates transformed to soil and air GDD for *O. minor* germination, radicle adhesion to clover root surface, host penetration, vascular connection, tubercle formation, full spider stage development (**Figure 1**). In addition to characterization of temperature-related underground development of *O. minor* in real farming conditions in Burgundy, these measurements allowed us to determine an adequate timing for amino acid treatments at 282 GDD air/308 GDD soil (corresponding to clover-induced *O. minor* germination) and 494 GDD air/543 GDD soil (corresponding to initiation of *O. minor* radicle penetration into clover root).

Effect of Amino Acid Field Treatments in the Parasitism of *O. minor* in Red Clover

A field experiment in Burgundy was conducted to study the effects of amino acid treatments (lysine, methionine, methionine hydroxyl analog, and tryptophan) at concentration (20 and 10 mM) on *O. minor* field emergence (**Figure 7**). In control-plots, an average of 21 *O. minor* plants emerged per square meter of soil surface. The stronger herbicidal activity observed *in vitro* by lysine and tryptophan relative to that induced by methionine was not validated in the field. Lysine- and tryptophan-based treatments at 10 mM did not reduce *O. minor* emergence in field. Increasing the concentration of lysine and tryptophan up to 20 mM lead, respectively, to 37 and 39% reductions of *O. minor* emergence with respect to the *O. minor* emergence observed in the control. By contrast, methionine-based formulations were more effective in reducing *O. minor* field parasitism. Methionine at 20 mM either as L-methionine pure form or as methionine hydroxyl analog (formulations 1 and 2) respectively, reduced by 67, 65, and 60% the emergence of *O. minor* in comparison with the control. The inhibitory activity was maintained over 40% in methionine hydroxyl analog formulations when applied

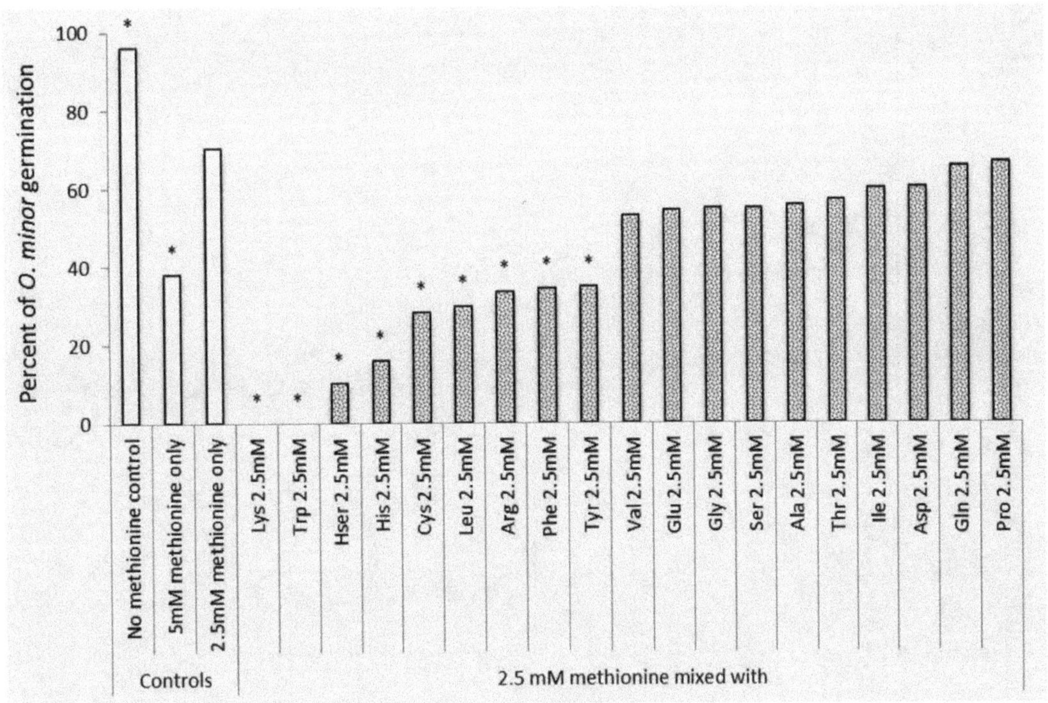

FIGURE 5 | Rescue of methionine inhibition of *O. minor* germination induced by application of 2.5mM of alanine, arginine, aspartate, cysteine, glutamic acid, glutamine, glycine, histidine, homoserine, isoleucine, leucine, lysine, methionine, phenylalanine, proline, serine, threonine, tryptophan, tyrosine, and valine. *Indicates differences of each treatment compared with the control (germination observed in 2.5 mM methionine treatment) assessed by Dunnett's test at the 0.05 level.

at 10 mM. None of the tested amino acid formulations were inhibitory to clover plants.

Induction of Clover Resistance: A Potential Enhancement of Herbicidal Efficacy

We studied the possibility that methionine could inhibit the invasion of *O. minor* haustorium into clover roots by inducing a resistant reaction in clover. **Table 2** shows the results of microscope observation of *O. minor* infection on roots of methionine-treated clover plants. Despite methionine-treated clover plants did not show significant symptoms of hypersensitive-like response at the host-parasite interface, the *O. minor* infection success m as the percentage of successful host-parasite vascular connections out of total parasitic radicles that made contact with host roots was slightly but significantly reduced in methionine-treated clover plants in comparison with control clover plants. The parasitic development beyond vascular connection measured as the percentage of vascular connections that developed into spider stage was not affected by methionine treatment.

DISCUSSION

Herbicidal activity of amino acids has been studied previously being the amino acid delivered either by root exudation of allelopathic plants (Bertin et al., 2007) or by over-excreting pathogens (Nzioki et al., 2016). Under laboratory conditions, exogenous supply of amino acids to tomato rhizosphere inhibited *Phelipanche* parasitism (Vurro et al., 2006). The present work is consistent with those reports certifying the value of amino acids for weed control by studying for the first time the direct field delivery of amino acids as a management strategy against *O. minor* parasitism in red clover.

We have investigated some of the main mechanisms potentially involved in the weed inhibition activity. Previously, the germination sensitivity of the parasitic plant *P. ramosa* to amino acid inhibition was described *in vitro* (Vurro et al., 2006). We tested the hypothesis that amino acid inhibition and effective doses vary depending on the broomrape species targeted. In *in vitro* assays, *O. minor* germination was strongly inhibited by lysine, phenylalanine and tryptophan but not affected by any other amino acid at the concentration range tested except when assayed at the highest concentration at which *O. minor* seeds showed sensitivity to alanine, cysteine, histidine, isoleucine, leucine, methionine and tyrosine. Although the effective doses differ, our results are in agreement with those reported for *P. ramosa* by Vurro et al. (2006) in that lysine, methionine, alanine, and histidine have the potential to inhibit broomrape germination. In contrast, *O. minor* sensitivity differs in that several amino acids such as proline and glycine, reported to be strong inhibitors of *P. ramosa* (Vurro et al., 2006), do not inhibit *O. minor* germination certifying that patterns of germination

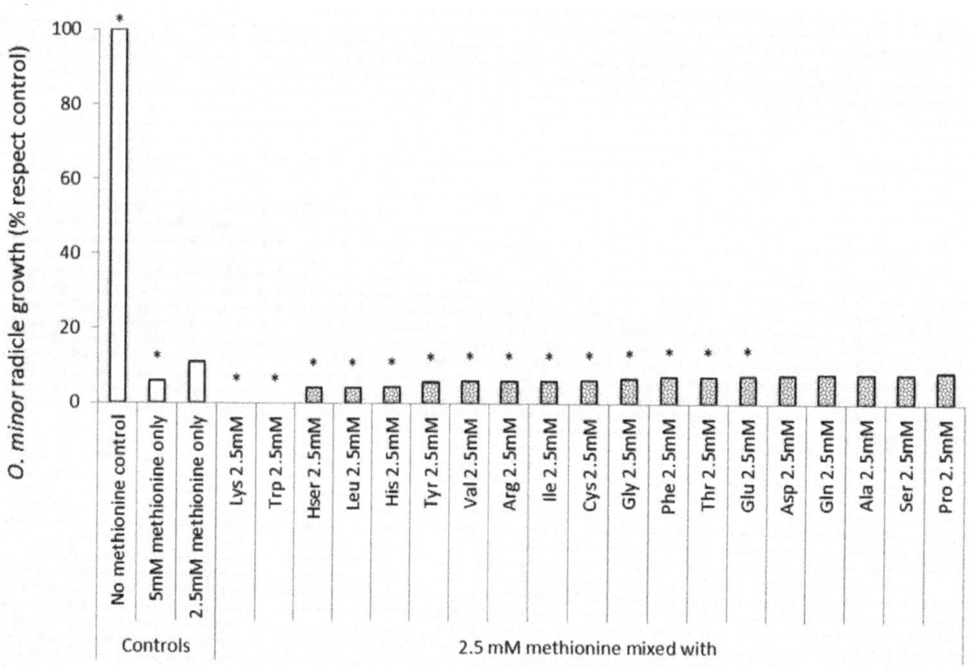

FIGURE 6 | Rescue of methionine inhibition of *O. minor* radicle growth induced by application of 2.5mM alanine, arginine, aspartate, cysteine, glutamic acid, glutamine, glycine, histidine, homoserine, isoleucine, leucine, lysine, methionine, phenylalanine, proline, serine, threonine, tryptophan, tyrosine, and valine in *in vitro* assays. * Indicates differences of each treatment compared with the control (radicle growth observed in 2.5 mM methionine treatment) assessed by Dunnett's test at the 0.05 level.

TABLE 1 | Growing degree-days (GDD) accumulated since clover sowing at each *O. minor* parasitism stage observed based on measurements of soil temperature (5 cm below the soil surface) and air temperature (1 m above soil surface).

		***O. minor* parasitism stages**						
Growing degree days (GDD) (C)	***Clover emergence***	**Germination**	**Radicle adhesion**	**Host penetration**	**Vascular connection established**	**Tubercle development**	**Spider stage**	**Shoot meristem**
	April 15, 2015	**April 29, 2015**	**May 5, 2015**	**May 13, 2015**	**May 19, 2015**	**May 28, 2015**	**June 2, 2015**	**June 9, 2015**
GDD Soil	110.2	307.5	412.9	542.6	654.1	797.3	886.9	1035.6
GDD Air	100.1	281.9	385.2	493.5	581.7	703.7	788.0	938.3

inhibition triggered by exogenous amino acids differ between broomrape species.

Although radicle growth inhibition by amino acids have been reported before (Vurro et al., 2006; Fernández-Aparicio et al., 2013), sensitivity of broomrape radicle growth to exogenous amino acids have not been systematically studied before the present work. Our *in vitro* screening demonstrated that the process of radicle growth is more sensitive than germination to amino acid inhibitory activity. In fact, every amino acid tested except for glutamine, glutamic acid and aspartic acid, promoted general levels of radicle growth inhibition at the highest concentration tested. Applying each amino acid at reduced concentrations allowed us to identify strong inhibitors, many of them also being active as inhibitors of germination, i.e., lysine, phenylalanine, and tryptophan but also alanine, isoleucine, histidine, leucine, methionine, and tyrosine. This indicates that the inhibition phenomenon could be affecting common metabolic pathways required for germination and radicle growth processes. Amino acids that differentially targeted either the processes of germination or radicle growth were identified less frequently. Among them serine, threonine and valine inhibited *O. minor* radicle growth but did not inhibit its germination at the concentration tested while cysteine inhibited germination but its activity in radicle growth was only observed at high concentrations in which general inhibition was observed.

Although the direct delivery of phytoinhibitory amino acids as alternative to synthetic herbicides is a promising weed control strategy, it lacks field optimization (Sands and Pilgeram, 2009). Among the *in vitro*-identified top inhibitory amino acids, the commercial availability at a large scale of lysine-, methionine-, and tryptophan-based supplements for animal feed enabled us to select those amino acids as candidates for field research. Lysine

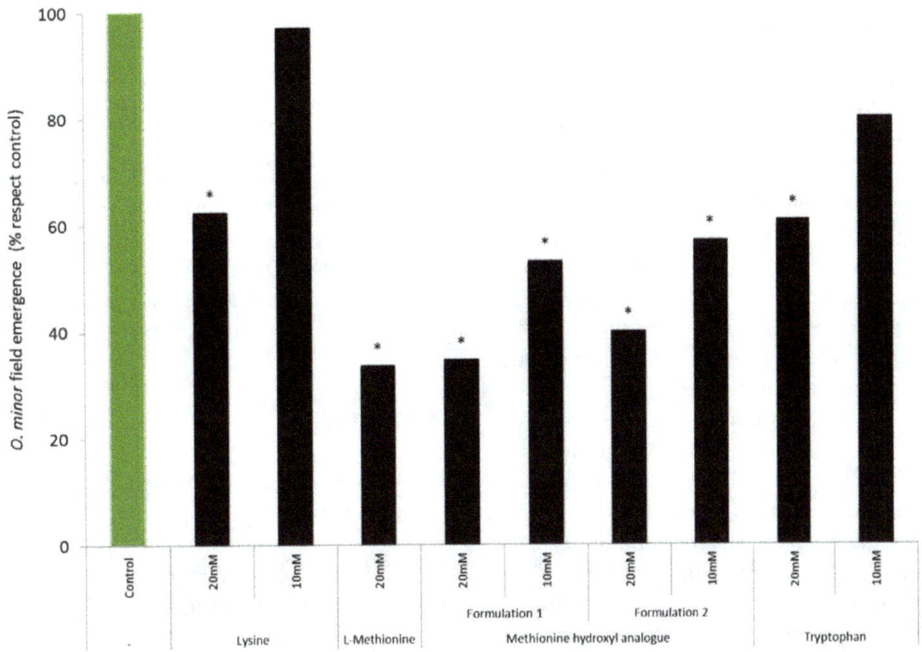

FIGURE 7 | Influence of lysine, L-methionine, methionine hydroxyl analog, and tryptophan field treatments in the emergence of *O. minor* in red clover during a field trial in Burgundy, France in 2014. *Indicates differences of each treatment compared with the control assessed by Dunnett's test at the 0.05 level.

TABLE 2 | Protective effect of methionine imbibition of red clover seeds against *O. minor* parasitism.

Clover seed imbibition treatment	Total number of *O. minor* tubercles per clover plant	Infection success (%)	Hypersensitive-like response (%)	Necrosis of tubercle (%)	'Spider' development (%)
Control	28.0	44.0	0.4	0.0	28.6
20mM L-methionine	17.4*	26.3*	1.5^{ns}	0.0	29.3^{ns}

and methionine are commonly called aspartate-derived amino acids produced from aspartate in plants via a branched pathway (Jander and Joshi, 2010). Tryptophan is an aromatic amino acid synthesized via the shikimate pathway followed by the branched aromatic amino acid metabolic pathway (Tzin and Galili, 2010).

For broomrape control, the inhibitory amino acid needs to be delivered in the crop rhizosphere, the narrow region of soil in which broomrape seed germination is triggered by crop root exudates and the infectious radicle invades the crop root. Crop root exudates are also a source of amino acids (Lesuffleur et al., 2007) which could reverse the herbicidal effect due to complex regulation events (Bonner and Jensen, 1997; Curien et al., 2005; Jander and Joshi, 2010). Therefore, the broomrape vulnerability observed *in vitro* to amino acid inhibition might have little consequences in the natural environment if the herbicidal amino acid co-exists in the rhizosphere with another amino acid that abolishes the growth inhibition. Our *in vitro* results suggest that the *O. minor* inhibition caused by lysine, methionine, and tryptophan is not reversed by amino acids frequently exuded by roots of clover and other crops.

The *in vitro* screening identified lysine and tryptophan as strong inhibitors of *O. minor* germination and radicle growth. Those biological processes are crucial in the life cycle of obligated parasitic weeds enabling host root recognition and invasion. The germination inhibitory activity of methionine was low, however, it strongly reduces the maximum length of the infective radicle which also may lead to inhibition of broomrape parasitism by inhibiting the capacity of the parasite to reach and therefore invade the host (Vurro et al., 2006; Fernández-Aparicio et al., 2013). Because amino acid treatments target those highly coordinated but ephemeral parasitic stages the timing of application is instrumental for the success of this technique especially in light of the fact that the persistence of amino acids in agricultural soils is predicted to be low due to microbial metabolization (Sands and Pilgeram, 2009). Development of broomrape life cycle is highly synchronized with host phenology and for each broomrape-crop association, temperature is the main factor affecting such synchronization (Eizenberg et al., 2005). GDD-based models for *O. minor* field development from tubercle formation have been efficient in assisting the decision of application timing of systemic herbicides (Eizenberg et al., 2005; Lins et al., 2005). However, our amino acid treatment targets earlier stages in *Orobanche* parasitism and therefore in order to standardize adequate timing of amino acid field applications, we monitored in the field from clover sowing the early underground stages of *O. minor* parasitism. From the field screening we determined that two applications of methionine hydroxyl analog at the parasitic stages of germination and host invasion suppressed *O. minor* parasitism. The suppressive effect of methionine hydroxyl analog in field was better than lysine and tryptophan formulations despite the fact that lysine and tryptophan were better inhibitors of *O. minor in vitro* development. Although faster microbial metabolization could influence the lower effect of lysine and tryptophan relative to methionine, we considered the possibility that besides the direct effect of methionine in *O. minor* preattached seedling, an additional mechanism could be protecting the crop through a methionine-elicited mechanism of induced resistance against *O. minor* penetration. In fact, methionine has been described to elicit defense related enzymes and resistance in other crop-pathogen associations (Sarosh et al., 2005). In order to dissect the effect of methionine as potential elicitor of clover resistance against *O. minor* from the methionine direct inhibitory effect in *O. minor* germination and radicle growth we imbibed clover seeds with methionine and cultivated the subsequent seedlings in rhizotrons in absence of methionine. A hypersensitive response was not observed at the *O. minor*-clover interface. However, despite no difference being observed in the development of *O. minor* pre-attached seedlings relative to the control, the number of established parasites relative to the number of parasitic seedlings that made contact with the host root was significantly reduced in methionine-protected clover plants. This study provides justification for molecular studies to unveil the precise cause of the infection inhibition by methionine and offers the possibility to refine the application conditions of amino acids such as methionine-coated clover seeds.

AUTHOR CONTRIBUTIONS

MF-A designed, implemented the studies, collected and analyzed the data and wrote the manuscript. ABe, LF, PM, BC, CS, CM, XR contributed in field design and implementation. ABe, LF, PM, SG-L, ABo, MV, DB, DS, XR provided materials and laboratory equipment. BC, CS, CM, MV, DS, XR reviewed and discussed the final version of the manuscript.

FUNDING

This work was funded by INRA 'Division Santé des Plantes et Environnement. Appel à projets scientifiques 2015. Catégorie Biocontrôle'. MF-A was supported by the Marie-Curie FP7 COFUND People Programme, through the award of an AgreenSkills fellowship (under grant agreement n° PCOFUND-GA-2010-267196).

ACKNOWLEDGMENTS

We thank Graciela B. Arhancet and Matt Mahoney from Novus International Inc, Jacob van Milgen from INRA UMR1348 and Evonik Degussa for providing amino acid formulations as animal feed supplements.

REFERENCES

Aly, R. (2007). Conventional and biotechnological approaches for control of parasitic weeds. *Vitro Cell. Dev. Biol. Plant* 43, 304–317. doi: 10.3109/07388551.2012.743502

Bertin, C., Weston, L. A., Huang, T., Jander, G., Owens, T., Meinwald, J., et al. (2007). Grass roots chemistry: meta-Tyrosine, an herbicidal nonprotein amino acid. *Proc. Natl. Acad. Sci. U.S.A.* 104, 16964–16969. doi: 10.1073/pnas.0707198104

Bonner, C. A., and Jensen, R. A. (1997). Recognition of specific patterns of amino acid inhibition of growth in higher plants, uncomplicated by glutamine-reversible 'general amino acid inhibition'. *Plant Sci.* 130, 133–143. doi: 10.1016/S0168-9452(97)00213-6

Bright, S., Wood, E. A., and Miflin, B. J. (1978). The effect of aspartate derived amino acids (lysine, threonine, methionine) on the growth of excised embryos of wheat and barley. *Planta* 139, 113–117. doi: 10.1007/BF00387135

Curien, G., Ravanel, S., Robert, M., and Dumas, R. (2005). Identification of six novel allosteric effectors of *Arabidopsis thaliana* aspartate kinase-homoserine dehydrogenase isoforms: physiological context sets the specificity. *J. Biol. Chem.* 280, 41178–41183. doi: 10.1074/jbc.M509324200

Eizenberg, H., Colquhoun, J. B., and Mallory-Smith, C. A. (2005). A predictive degree-days model for small broomrape (*Orobanche minor*) parasitism in red clover in Oregon. *Weed Sci.* 53, 37–40. doi: 10.1614/WS-04-018R1

Eizenberg, H., Colquhoun, J. B., and Mallory-Smith, C. A. (2006). Imazamox application timing for small broomrape (*Orobanche minor*) control in red clover. *Weed Sci.* 54, 923–927. doi: 10.1614/WS-05-151R.1

Fernández-Aparicio, M., Pérez-de-Luque, A., Prats, E., and Rubiales, D. (2008). Variability of interactions between barrel medic (*Medicago truncatula*) genotypes and *Orobanche* species. *Ann. Appl. Biol.* 153, 117–126. doi: 10.1111/j.1744-7348.2008.00241.x

Fernández-Aparicio, M., Reboud, X., and Gibot-Leclerc, S. (2016). Broomrape Weeds. Underground mechanisms of parasitism and associated strategies for their control: a review. *Front. Plant Sci.* 7:135. doi: 10.3389/fpls.2016.00135

Fernández-Aparicio, M., Cimmino, A., Evidente, A., and Rubiales, D. (2013). Inhibition of *Orobanche crenata* seed germination and radicle growth by allelochemicals identified in cereals. *J. Agric. Food. Chem.* 61, 9797–9803. doi: 10.1021/jf403738p

Gonsior, G., Buschmann, H., Szinicz, G., Spring, O., and Sauerborn, J. (2004). Induced resistance–an innovative approach to manage branched broomrape (*Orobanche ramosa*) in hemp and tobacco. *Weed Sci.* 52, 1050–1053. doi: 10.1614/WS-04-088R1

Henke, R. R., Wilson, K. G., McClure, J. W., and Treick, R. W. (1974). Lysine methionine-threonine interactions in growth and development of *Mimulus cardinalis* seedlings. *Planta* 116, 333–345. doi: 10.1007/BF00390857

Hershenhorn, J., Eizenberg, H., Dor, E., Kapulnik, Y., and Goldwasser, Y. (2009). *Phelipanche aegyptiaca* management in tomato. *Weed Res.* 49(Suppl. 1), 34–37. doi: 10.1111/j.1365-3180.2009.00739.x

Jander, G., and Joshi, V. (2010). Recent progress in deciphering the biosynthesis of aspartate-derived amino acids in plants. *Mol. Plant* 3, 54–65. doi: 10.1093/mp/ssp104

Joel, D. M. (2013). "The haustorium and the life cycles of parasitic Orobanchaceae," in *Parasitic Orobanchaceae*, eds D. M. Joel, J. Gressel, and L. J. Musselman (New York, NY: Springer), 21–23.

Johnson, A. W., Rosebery, G., and Parker, C. (1976). A novel approach to *Striga* and *Orobanche* control using synthetic germination stimulants. *Weed Res.* 16, 223–227. doi: 10.1111/j.1365-3180.1976.tb00406.x

Kuijt, J. (1969). *The Biology of Parasitic Fowering Plants.* Berkeley, CA: University of California Press.

Kusumoto, D., Goldwasser, Y., Xie, X., Yoneyama, K., and Takeuchi, Y. (2007). Resistance of red clover (*Trifolium pratense*) to the root parasitic plant

Orobanche minor is activated by salicylate but not by jasmonate. *Ann. Bot.* 100, 537–544. doi: 10.1093/aob/mcm148

Lechat, M. M., Pouvreau, J. B., Péron, T., Gauthier, M., Montiel, G., Veronesi, C., et al. (2012). PrCYP707A1, an ABA catabolic gene, is a key component of *Phelipanche ramosa* seed germination in response to the strigolactone analogue GR24. *J. Exp. Bot.* 63, 5311–5322. doi: 10.1093/jxb/ers189

Lesuffleur, F., Paynel, F., Bataillé, M., Le Deunff, E., and Cliquet, J. B. (2007). Root amino acid exudation: measurement of high efflux rates of glycine and serine from six different plant species. *Plant Soil* 294, 235. doi: 10.1007/s11104-007-9249-x

Lins, R., Colquhoun, J. B., Cole, C. M., and Mallory-Smith, C. A. (2005). Postemergence small broomrape (*Orobanche minor*) in red clover. *Weed Technol.* 19, 411–415. doi: 10.1614/WT-04-175R1

Lins, R. D., Colquhoun, J. B., and Mallory-Smith, C. A. (2007). Effect of small broomrape (*Orobanche minor*) on red clover growth and dry matter partitioning. *Weed Sci.* 55, 517–520. doi: 10.1614/WS-07-049.1

McMaster, G. S., and Wilhelm, W. W. (1997). Growing degree-days: one equation, two interpretations. *Agric. For. Meteorol.* 87, 291–300. doi: 10.1016/S0168-1923(97)00027-0

Nzioki, H. S., Oyosi, F., Morris, C. E., Kaya, E., Pilgeram, A., Baker, C. S., et al. (2016). *Striga* biocontrol on a toothpick: a readily deployable and inexpensive method for smallholder farmers. *Front. Plant Sci.* 7:1121. doi: 10.3389/fpls.2016.01121

Parker, C. (2013). "The parasitic weeds of the Orobanchaceae," in *Parasitic Orobanchaceae*, eds D. M. Joel, J. Gressel, and L. J. Musselman (New York, NY: Springer), 333–343.

Parker, C., and Riches, C. R. (1993). *Parasitic Weeds of the World: Biology and Control.* Wallingford, CT: CAB International.

Rubiales, D., Fernández-Aparicio, M., Wegmann, K., and Joel, D. (2009). Revisiting strategies for reducing the seedbank of *Orobanche* and *Phelipanche* spp. *Weed Res.* 49, 23–33. doi: 10.1111/j.1365-3180.2009.00742.x

Sands, D. C., and Pilgeram, A. L. (2009). Methods for selecting hypervirulent biocontrol agents of weeds: why and how? *Pest Manag. Sci.* 65, 581–587. doi: 10.1002/ps.1739

Sarosh, B. R., Sivaramakrishnan, S., and Shetty, H. S. (2005). Elicitation of defense related enzymes and resistance by L-methionine in pearl millet against downy mildew disease caused by *Sclerospora graminicola*. *Plant Physiol. Biochem.* 43, 808–815. doi: 10.1016/j.plaphy.2005.06.009

Tzin, V., and Galili, G. (2010). The biosynthetic pathways for shikimate and aromatic amino acids in *Arabidopsis thaliana*. *Arabidopsis Book* 8:e0132. doi: 10.1199/tab.0132

Vaucher, J. P. (1823). Mémoire sur la germination des orobanches. *Mém. Mus. Hist. Nat. Paris* 10, 261–273.

Vurro, M., Boari, A., Pilgeram, A. L., and Sands, D. C. (2006). Exogenous amino acids inhibit seed germination and tubercle formation by *Orobanche ramosa* (broomrape): potential application for management of parasitic weeds. *Biol. Control* 36, 258–265. doi: 10.1016/j.biocontrol.2005.09.017

Conflict of Interest Statement: The authors declare that the research was conducted in the absence of any commercial or financial relationships that could be construed as a potential conflict of interest.

10

Host-Parasite-Bacteria Triangle: The Microbiome of the Parasitic Weed *Phelipanche aegyptiaca* and Tomato-*Solanum lycopersicum* (Mill.) as a Host

Edited by:
Maurizio Vurro,
National Research Council, Italy

Reviewed by:
Grama Nanjappa Dhanapal,
University of Agricultural Sciences,
Bengaluru, India
Katarzyna Turnau,
Jagiellonian University, Poland

***Correspondence:**
Radi Aly
radi@volcani.agri.gov.il

*Lilach Iasur Kruh[1,2], Tamar Lahav[3], Jacline Abu-Nassar[1], Guy Achdari[1], Raghda Salami[2], Shiri Freilich[3] and Radi Aly[1]**

[1] Department of Weed Science, Newe Ya'ar Research Center, Agricultural Research Organization, Ramat Yishay, Israel, [2] Department of Biotechnology Engineering, ORT Braude College, Karmiel, Israel, [3] The Institute of Plant Sciences, Newe Ya'ar Research Center, Agricultural Research Organization, Ramat Yishay, Israel

Broomrapes (*Phelipanche/Orobanche* spp.) are holoparasitic plants that subsist on the roots of a variety of agricultural crops, establishing direct connections with the host vascular system. This connection allows for the exchange of various substances and a possible exchange of endophytic microorganisms that inhabit the internal tissues of both plants. To shed some light on bacterial interactions occurring between the parasitic *Phelipanche aegyptiaca* and its host tomato, we characterized the endophytic composition in the parasite during the parasitization process and ascertained if these changes were accompanied by changes to endophytes in the host root. Endophyte communities of the parasitic weed were significantly different from that of the non-parasitized tomato root but no significant differences were observed between the parasite and its host after parasitization, suggesting the occurrence of bacterial exchange between these two plants. Moreover, the *P. aegyptiaca* endophytic community composition showed a clear shift from gram negative to gram-positive bacteria at different developmental stages of the parasite life cycle. To examine possible functions of the endophytic bacteria in both the host and the parasite plants, a number of unique bacterial candidates were isolated and characterized. Results showed that a *Pseudomonas* strain *PhelS10*, originating from the tomato roots, suppressed approximately 80% of *P. aegyptiaca* seed germination and significantly reduced *P. aegyptiaca* parasitism. The information gleaned in the present study regarding the endophytic microbial communities in this unique ecological system of two plants connected by their vascular system, highlights the potential of exploiting alternative environmentally friendly approaches for parasitic weed control.

Keywords: broomrapes, parasitic weed, endophytic bacteria, tomato (*Solanum lycopersicum*), biocontrol

INTRODUCTION

Phelipanche aegyptiaca (broomrapes) is an obligate holoparasite that attacks the roots of almost all economically important crops in the Solanaceae, Fabaceae, Compositae, Brassicaceae, and Umbelliferae plant families (Parker and Riches, 1993; Westwood et al., 2010). This parasitic weed invades host plants by using a highly specialized detection system for strigolactones (hormones secreted by the host roots), whose presence is essential for the germination of parasite seeds (Yoder, 1999; Bouwmeester et al., 2003; Joel et al., 2006; Cardoso et al., 2011). Lacking functional roots and a photosynthetic system, the parasite develops special intrusive organs (haustorium) that directly connect to the vascular system of the host plants (Westwood et al., 2010; Joel et al., 2013). By developing a metabolic sink stronger than that of the host, the parasite channels the flow of water and nutrients from the host, thereby damaging crop (its host plant) development. Following successful attachment to the host root, the adjacent broomrape tissue grows into a bulbous structure called a tubercle (spider stage) from which short root-like organs arise that can form secondary attachments to neighboring host roots. After approximately 4 weeks of growth, a floral meristem is produced (shoot stage), which emerges aboveground to flower and disseminate seeds.

This parasite is the main limiting factor in processing-tomato production in numerous Middle East countries (Dor et al., 2014). A wide variety of parasitic weed control methods have been applied in attempts to control broomrape (Joel et al., 2006; Aly, 2007; Aly et al., 2009; Cochavi et al., 2016), most of which are based on chemical sprays that can be windborne and toxic to non-target plants. Therefore there is a need to find alternative solutions to reduce plant–plant parasitization.

One such alternative is to harness endophytic bacteria that naturally inhabit the internal tissues of most plants (Hallmann, 2001). These bacteria often play important beneficial roles in numerous aspects of their host plant's biology, including enhanced host growth rate, acceleration of seed germination, increased stress tolerance, and the provision of critical nutrients to the host. Endophytic bacteria may also contribute resistance to their hosts by suppressing pathogens and enhancing the plant's immune system (Azevedo et al., 2000; Rosenblueth and Martínez-Romero, 2006; Ryan et al., 2008; East, 2013). Furthermore, as recently reviewed by Joel et al. (2013), plant endophytes may influence the interaction of their hosts with parasitic weeds, For example the bacterium *Azospirillum brasilense* inhibits seed germination and radical elongation in the broomrape *P. aegyptiaca*; *Pseudomonas fluorescens* reduces both the quantity and biomass of the broomrape *Orobanche foetida* (Zermane et al., 2007); and the bacterium *Rhizobium* spp. reduces not only seed germination of *O. foetida* but also the number of tubercles on its host's (chickpea) roots (Hemissi et al., 2013). Despite these reports, to date, there is no published information regarding endophytic bacteria inhabiting the parasitic weed during the establishment of parasitism. Furthermore, the parasite and host connect through their vascular system forming a unique ecological system that potentially enables movement of bacteria from one plant to another, resulting in a host-parasite-endophyte triangle. Therefore, the first step in clarifying the role of these bacteria in the parasitism is to investigate the endophytic communities during the phases of parasitization and compared them with parasitized and non- parasitized host tissues.

To do so, we used tomato (*Solanum lycopersicum*) as a host and *P. aegyptiaca* as the parasitic weed.

MATERIALS AND METHODS

Plant Materials

The parasitic weed and its host were grown as was previously described (Eizenberg et al., 2004): *Solanum lycopersicum* L. 'MP-1' plants were served as hosts for *P. aegyptiaca* parasitization. The parasitic seeds were collected from an infested tomato field near Qiryat Shemona (northern Israel), dried and kept at 8°C until use. The host plant was planted into 4 L pots filled with soil (light-medium clay with 63% sand, 12% silt, and 22% clay) and grown in a greenhouse under natural lighting with an average 14 h of daylight and a temperature of 20 ± 6°C. These plants were watered and fertilized as needed. Four developmental stages of *P. aegyptiaca* were sampled: seeds, pre-haustorium stage, tubercle (spider stage) and shoot. To ensure that only endophytic bacteria from the plant tissue were being examined, samples were surface-sterilized twice by 2-min incubation in 70% ethanol and 10 min in 6% sodium hypochlorite, followed by a double rinsing in sterile double-distilled water (DDW) (10 min each). To deprive contamination by external bacteria, 100 μl of DDW from the second wash was plated on nutrient agar by Drigalski spatula. In addition, we examined the plant surface by general bacterial probe using FISH analysis (Supplementary Figure S1). No external contamination was detected.

Characterization of Community Composition of Associated Endophytic Bacteria in *P. aegyptiaca* and its Host Root

DNA was extracted from plant tissue using the cetyltrimethylammonium bromide (CTAB) method (Chen and Ronald, 1999).

DNA from sterilized tissues of non-parasitized tomato roots, parasitized tomato roots (collected a few cm from parasite-attachment point) and from the tubercle of the parasite (spider stage). PCR was performed using general 16S rRNA bacterial primers, which reduce plastid amplification (63F+1401R) (Supplementary Table S1). The PCR products were sent for sequencing to the DNA Services Facility (Chicago, IL, USA) and 10 μl of each reaction was kept at −20°C as a reference. Sequencing was performed by high-throughput amplicon sequencing using the Illumina MiSeq platform at Research and Testing Laboratory (Lubbock, TX, USA) with the primers 515F+806R (Supplementary Table S1). We used five replicates for each treatment.

Characterization of Community Composition of Associated Microbes of *P. aegyptiaca* (the Parasitic Weed) Across its Developmental Stages

To identify the bacterial community associated with *P. aegyptiaca*, DNA was extracted at different stages of the parasite development (pre-haustorium, tubercles and shoots) and PCR analysis was performed using general 16S rRNA bacterial primers, which reduce plastid amplification (27F+783R). PCR products were isolated with AgencountAMPure® XP beads (Beckman Coulter, Nyon, Switzerland) and next generation sequencing was performed by Ion Torrent^{TM} sequencer (Life Technologies, Grand Island, NY, USA) using primers 27F+338R (Supplementary Table S1). We used three replicates for each treatment.

Two mass-sequencing methods and primer sets were applied to the parasitic weeds at the spider stage and the resultant community composition data were similar, indicating that both can be used interchangeably. Still, we treated these two dataset independently (**Figures 1, 2**).

Microbial Community Analysis

Retrieved sequences were analyzed using MOTHUR software (Schloss et al., 2009). Sequences shorter than 150 bp, as well as those of low quality (multiple N, chimeras, etc.), were omitted. Bacterial sequences were aligned using the Silva-compatible alignment database and a distance matrix was calculated. Sequences were grouped into OTUs at a 97% sequence similarity threshold (i.e., sequences that differed by 3% were clustered in the same OTU). A "sub.sample" command was performed and all samples were adjusted, by random selection, to the sample with the lowest number of sequences (13000 and 1100 sequences per sample for Illumina and ion torrent data, respectively). "Indicator" command was performed in order to identify the specific OTUs (indicator species) of each treatment. Diversity indices, principal component analysis (PCA) and ANOVA statistics (Bray–Curtis distance measure) were calculated using PAST software (Hammer et al., 2001). Diversity indices (Supplementary Tables S2, S3) showing observed OTUs, Chao-1 index of richness - representing the predicted number of OTUs in each sample, Dominance - representing relative abundance distribution among species within each sample and Shannon index of diversity.

Endophyte Isolation and Identification

In order to make sure that the bacteria in the current study are endophytes, we sterilized the examined plant tissue twice as specified above. Surface-sterilized tissue (100 mg) was homogenized with 1 ml of sterile saline solution (9% NaCl) by sterilized pestle. Lysate (100 μL) was spread on NA plates and kept at 28°C in the dark. Three days after incubation, bacterial isolates were counted and classified according to different colony morphologies. Five repeats were carried out for each treatment.

To determine the phylogenetic affiliation of the bacterial isolates, single colonies were collected and subjected to PCR analysis using 27F+1513R primers targeting the 16S rRNA gene by direct colony PCR method as was previously described (Iasur-Kruh et al., 2011) with primers 27F and 1513R (Supplementary Table S1). The PCR procedure was as follows: DNA was denatured at 95°C for 5 min, followed by 30 cycles at 95°C for 30 s each, 57°C for 30 s and 72°C for 1 min, followed by 5 min at 72°C. The PCR product was sequenced by 3130xl Genetic Analyzer (Applied Biosystems). Phylogenetic affiliation of the isolates was determined by comparing with sequences obtained from the NCBI GenBank database. These strains are kept in glycerol at −20°C in our laboratory collection.

In vitro Seed Germination

Parasite seeds were surface-sterilized, dispersed on Whatman GF/A glass-fiber filters (0.7 cm diameter), covered with the same filter and placed on a Petri dish. After 1 week GR24 (1 mg L^{-1}) was added to the disks. To examine the effect of different isolates on the germination, each isolate was grown in LB (Luria broth-Life Technologies, Israel) overnight at 28°C, centrifuged, the pellet was washed with DDW and the culture was adapted to an optical density at 600 nm (OD_{600}) of 0.3. The culture was added to the disks with the parasite seeds. After incubation in the dark at 26°C for 7 days, the seeds were rated for seed germination in comparison to seed germination control (without bacteria). The experiment was conducted in five repeats for each isolate.

In planta Test

We used an *in planta* system in which tomato plants were transplanted into polyethylene bags and kept moist with half-strength nutrient solution (Hoagland and Arnon, 1950). Plant growth conditions were 25°C, with 14 h light at 100 μE/s m^2. Surface-sterilized *P. aegyptiaca* seeds were applied on and around the host roots with care taken to achieve even inoculation of all plant roots. After allowing 7 days for parasite seed preconditioning, 10 mL of 2 mg L^{-1} GR24, a germination stimulant (Mangnus et al., 1992), was added to each bag to synchronize the germination of the *P. aegyptiaca* seeds. To examine the effect of chosen isolates on parasite development, it was grown in LB overnight at 28°C, centrifuged, the pellet was washed with DDW and the culture was adapted to an OD_{600} of 0.4 (equivalent to 1.00E+08 cfu\ml). The culture was added to the polyethylene bags. After 2–3 weeks, parasitism was evaluated by counting the number of live and dead tubercles on each plant using a binocular microscope. The experiment was conducted in five repeats for each treatment (control and endophytic isolate).

Sequence Accessions

Sequences obtained from mass sequencing were deposited in the European Nucleotide Archive (ENA) (study accession number: PRJEB7137). Sequences from isolated endophytes were deposited in GenBank (NCBI) (KP219403-KP219415).

RESULTS

To assess whether bacterial community composition changes following attachment of the parasite (*P. aegyptiaca*) to its host (tomato), we compared the endophytic communities in parasitized and non-parasitized host roots with those found in the parasite during ongoing parasitization (i.e., occurring during the spider stage, the stage of attachment to the host). The results showed that endophytic communities of the parasite and the non-parasitized host differed significantly (t-test; $p = 0.0048$). Furthermore, the overall community of the parasitized host root did not differ from that of the parasite on the one hand or from that of the non-parasitized host root (t-test; $p > 0.05$) (**Figure 1A**) on the other. A greater number of sphingobacterial sequences were found in the parasitized tomato roots and in the parasite than in the non-parasitized tomato roots (t-test; $p = 0.03$)

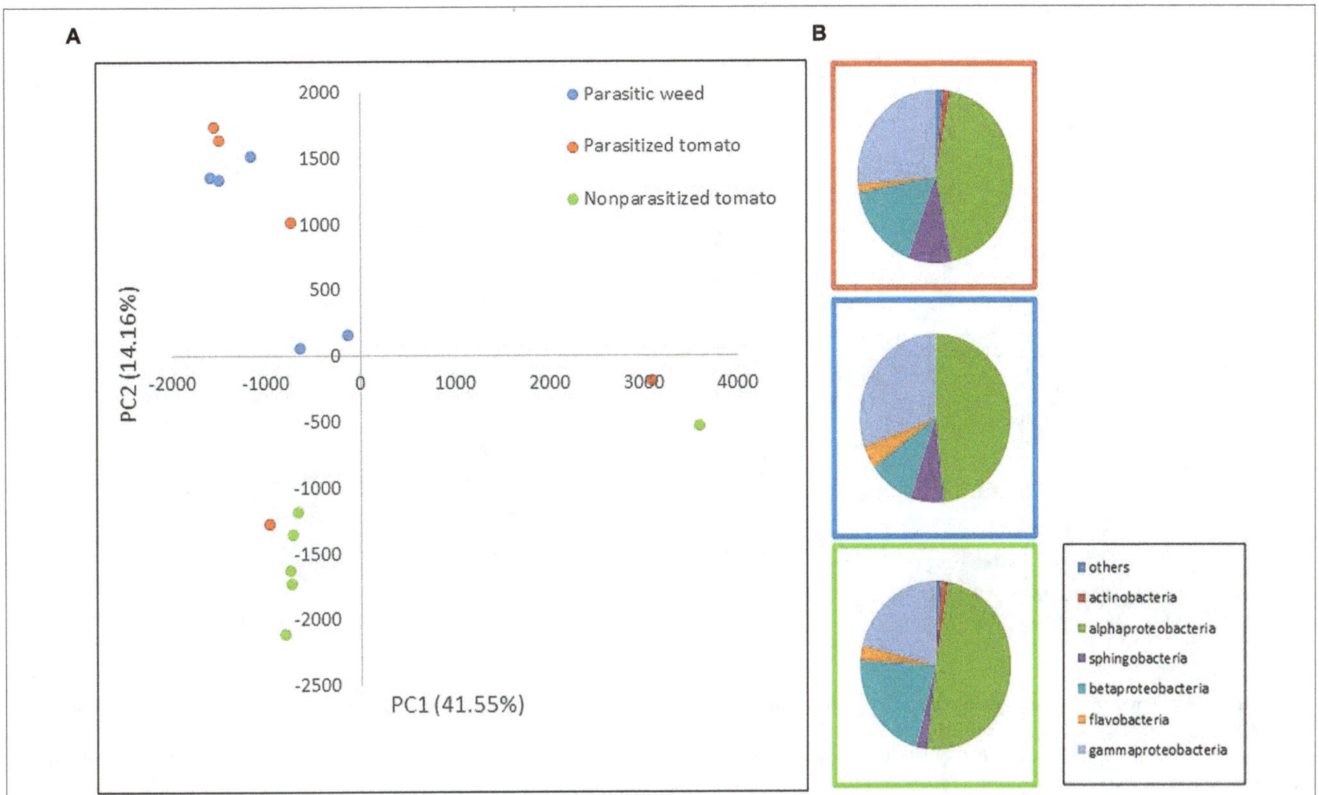

FIGURE 1 | Comparison of endophytic communities based on mass sequencing of ribosomal 16S rRNA gene. The bacterial community composition was determined for parasitized tomato root (red), parasite (*Phelipanche aegyptiaca* – spider stage) (blue) and non-parasitized tomato root (green) in five repeats each. Sequences were classified into operational taxonomic units using a 97% similarity threshold. **(A)** The first and second dimensions of PCA analysis. **(B)** Relative abundance of bacterial classes among treatments (an average of all five repeats).

(**Figure 1B**). We observed higher numbers of beta-Proteobacteria and fewer gamma-Proteobacteria in the tomato roots (both non-parasitized and parasitized tomato) in comparison to endophytic community structure of the parasitic weed. Furthermore, there was an introduction of additional taxonomic groups such as actinomycetes in the parasitic weed.

The most abundant OTUs, belonging to the proteobacteria genera; *Rhizobium, Pseudomonas, Comamonadaceae* sp., *Sphingomonas* and *Burkholderia*, are similarly represented in all three "treatments" (parasitized and non-parasitized tomato roots and parasitic weed) with no significant differences between them. However, examination of specific indicator genera showed that *Novosphingobium* and *Methylophilus* (representing 1 and 0.8% of the total sequence dataset, respectively) were specific for the parasitic weed and *Devosia* (representing 0.5% of the total sequence dataset) was significantly associated with the parasitized tomato root.

Next, we characterized the dynamics of the endophyte communities in the parasitic weed *across* its life cycle, that is, before and after attachment to the host. FISH analysis clearly showed that endophytic bacteria inhabit the inner tissue of all developmental stages of the parasitic weed (Supplementary Figure S1). The endophyte communities of the two stages of the parasite at the post-attachment stages (spider and shoot) were more similar to each other than to those detected in the parasite prior to attachment to the host (pre-haustorium stage) (**Figure 2A**). In the pre-attachment stage the dominant classes found in the parasite were alpha- and gamma-Proteobacteria (mostly represented by *Sphingomonas* and *Acinetobacter* species), comprising 55 and 35% of the bacterial community composition, respectively (**Figure 2B**). Just after attachment, at the spider stage, there was a decrease in the proportions of these classes and an increase in the fraction of Flavobacteria and beta-Proteobacteria (mostly representative of the *Flavobacterium* and *Methylophilus* genera). An increased community of Bacilli and Actinobacteria (40 and 25%, respectively), mostly representative of *Jeotgalibacillus* and *Propionibacterium*, were found at the shoot stage (**Figure 2B**).

In addition, we isolated bacteria from surface-sterilized host tomato roots and from different surface-sterilized developmental stages of *P. aegyptiaca*. Similarly, to the mass sequencing results the endophytic bacteria isolated from pre-haustorium stage belonged to alpha- and gamma-Proteobacteria genera; *Acinetobacter* and *Sphingobium* as well as *Roseomonas* and *Pseudomonas*. Gram positive bacteria were isolated from both the parasite tubercle (spider stage) and its host belonging to *Agrococcus* and *Bacillus* as well as *Roseomonas* and *Pseudomonas* which are gram negative. Unique isolates were identified for

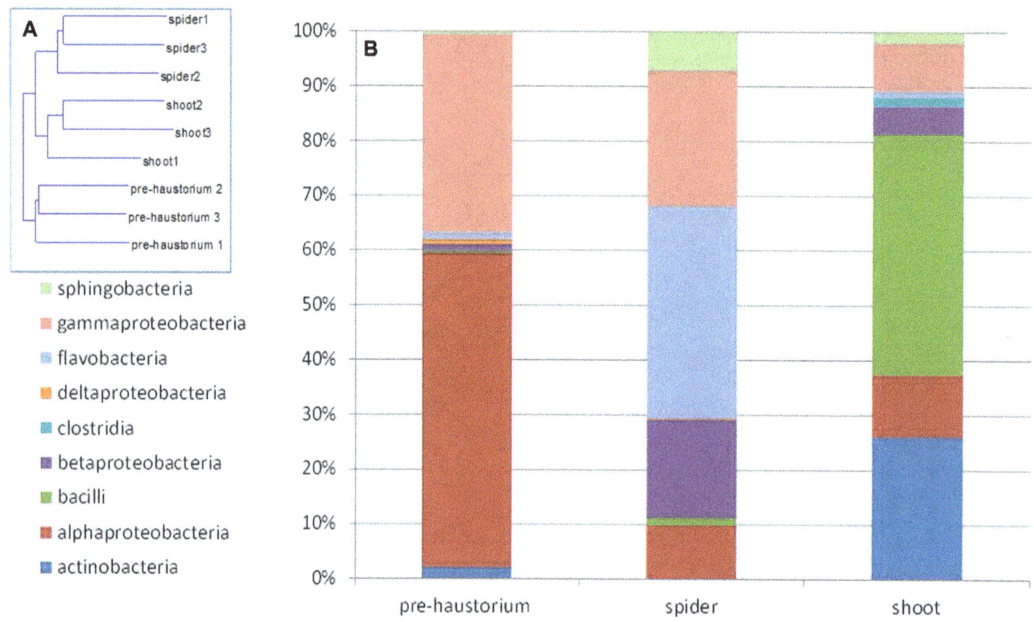

FIGURE 2 | Endophytic community composition at different stages of the parasite development as obtained from 16S rRNA gene mass sequencing. The bacterial community composition was determined for *P. aegyptiaca* germinated seeds before attachment to the host root (pre-haustorium), *P. aegyptiaca* tubercles (spider stage) and shoot, after attachment to the host root. Sequences were classified into operational taxonomic units using a 97% similarity threshold. **(A)** Neighbor-joining tree calculated by Rho. **(B)** Relative abundance of bacterial class at different parasite developmental stages.

the parasite: *Bacillus* sp. and *Rhizobium* sp. and for its host: *Pseudomonas* sp. The diversity of the isolated endophytes was reduced, containing only one *Bacillus* species in the shoot stage and in its host (**Table 1**).

To examine the effect of different endophytic isolates on the interactions between the parasitic weed and its host, we then examined the effect of the isolates on development of the parasitic weed. We screened for the ability of parasite's seeds to germinate *in vitro* in the presence of each isolate (**Table 1**) and found that two isolates (from the host) reduced the germination of the seeds. *Pseudomonas PhelS10* reduced germination by 80% and *Bacillus* sp. *PhelSh11* reduced it by 70%. These bacteria were then chosen for examination *in planta* using the polyethylene bag test (**Figure 3**). This test showed that the *Bacillus* strain dramatically reduced the viability of tomato roots coincident with the reduction of the parasite tubercles, while the *Pseudomonas*

TABLE 1 | Phylogenetic identification of bacteria isolated from internal tissues of different developmental stages (pre-haustorium, spider and shoot) of *Phelipanche aegyptica* and from host (tomato roots).

Isolates origin	Isolate name	Isolate closest match in NCBI (accession No. – % seq Identity)	Biological test – seed germination as effected by isolated endophytes
Pre-haustorium	*Acinetobacter* sp. *PhelS2*	*Acinetobacter johnsonii* (KP236314 – 99%)	0%
Pre-haustorium	*Sphingobium* sp. *PhelPH4*	*Sphingobium yanoikuyae* (KC355325 – 99%)	0%
Pre-haustorium	*Roseomonas* sp. *PhelPH5*	*Roseomonas musae* (NR_113233 – 98%)	0%
Pre-haustorium	*Pseudomonas* sp. *PhelS6*	*Pseudomonas stutzeri* (KU749990 – 99%)	0%
Spider and host	*Pseudomonas* sp. *PhelPH1*	*Pseudomonas rhizosphaerae* (KT825699 – 99%)	0%
Spider and host	*Agrococcus* sp. *PhelS3*	*Agrococcus jenensis* (EF672044 – 99%)	0%
Spider and host	*Chryseobacterium PhelS7*	*Chryseobacterium profundimaris* (NR_136427 – 99%)	22%
Spider and host	*Bacillus* sp. *PhelS6*	*Bacillus megaterium* (KU145691 – 99%)	42%
Spider	*Bacillus* sp. *PhelSs*	*Bacillus subtilis* (KX268132 – 99%)	20%
Spider	*Rhizobium* sp. *PhelS9*	*Rhizobium rhizoryzae* (NR_133844 – 99%)	0%
Host	*Pseudomonas* sp. *PhelS10*	*Pseudomonas aeruginosa* (KY203649 – 98%)	80%
Shoot and host	*Bacillus* sp. *PhelSh11*	*Bacillus oceanisediminis* (KX767124 – 99%)	70%

The effect of these isolates on P. aegyptica seeds in the presence of germination stimulant (GR24) was examined in comparison to control seed (without inoculation of bacteria): 0%- no effect- the same germination as control, 100%- no seed germination observed.

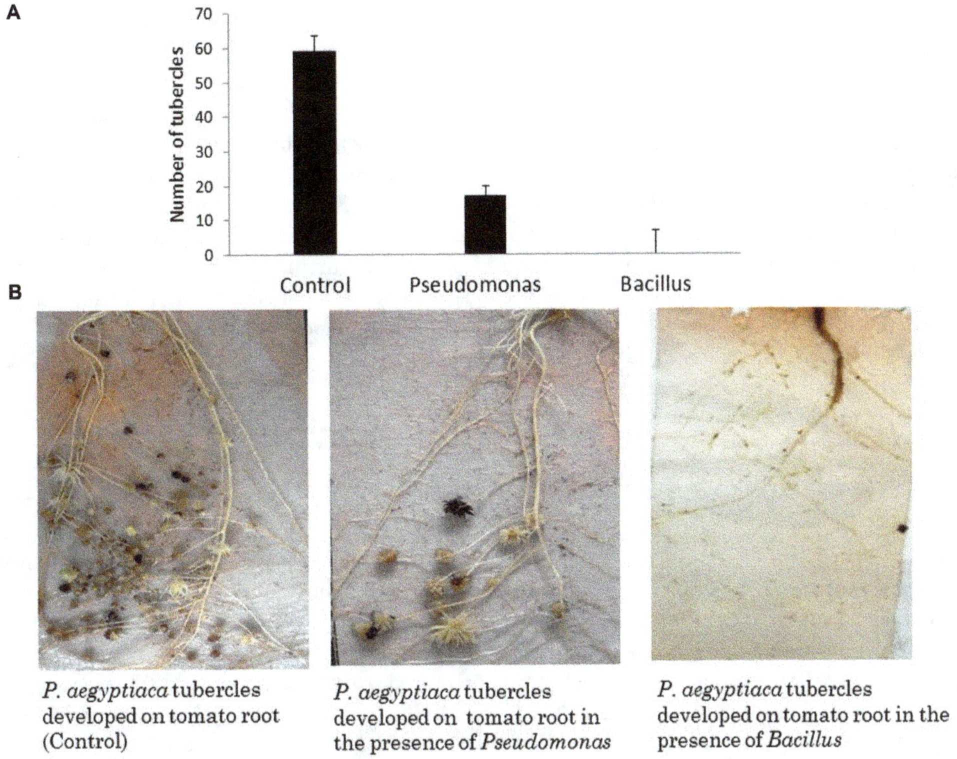

FIGURE 3 | The effect of *in planta* test of two isolated endophytes (approximately 1.00E+08 cfu\ml each) on parasitic weed development. (A) The number of *P. aegyptica* tubercles developing in the presence of selected isolates. **(B)** The effects of the bacterial isolates on *P. aegyptica* tubercles and on tomato roots in PE system. While *Pseudomonas PhelS10* reduced the number of tubercles without harming the host root, the *Bacillus* sp. *PhelSh11* destroyed the host root making the parasitic weed unable to develop. The results are an average of five repeats.

strain reduced the number of tubercles without harming the tomato roots (**Figure 3**).

DISCUSSION

Endophytes are microorganisms (bacteria, fungi, etc.) inhabiting the inner tissue of plants without causing apparent disease. Tomato is hosting both endophytic fungi and endophytic bacteria, some of them showing the ability to promote seed germination and protect tomato plants from pathogens (Larran et al., 2001; Pozo et al., 2002; Shittu et al., 2009; Xu et al., 2014). Furthermore, since endophytic mycorrhizal fungi contain a large amount of bacteria in the ectomycorrhizal root tips (Sbrana et al., 2002), it might spread endophytic bacteria while colonizing the tomato. de Vega et al. (2010) demonstrated tripartite association among a holoparasitic weed (*Cytinus*), its host Cistaceae species, and mycorrhizal fungi. However, studies on endophytic bacteria in parasitic weed have not been reported yet. In the current work we show that attachment of a parasitic weed to its host has an effect on the bacterial endophytic community composition of both plants. Parasitic weeds are considered ecosystem engineers, due to their ability to induce changes in the abundance and diversity of plants. Such changes may lead to up-stream effects on herbivores, pollinators and seed vectors (Press and Phoenix, 2005). Therefore changes in parasite endophytic communities may affect parasitism and as a consequence plant community makeup. In the current study, we showed for the first time, that bacterial endophytic communities changed with different stages of parasitism, and that they also affect the host's endophytic composition. The communities inhabiting the root of the host were examined in parallel to the parasite tubercle because this is the stage that the parasite is well-established on the host and the endophytic exchange may occur.

Endophytic bacteria are located *in planta*, where they are protected from the outside environment on the one hand, but are easily affected by changes occurring in the plant tissue on the other (Ryan et al., 2008). The fact that both the parasitic weed and its host endophytic communities changed following parasitization interaction and become more similar to each other can be explained by the following: the parasite affects the environment of the inner tissue of the host plant. Such changes may be exemplified by the reported changes in metabolic profile of the tomato host that occur following parasitization (Hacham et al., 2016). Thus, the changes in endophytic community occurring in the host plant may be affected by the parasite in an indirect manner.

Alternatively, the endophytic community composition is affected in direct manner by an exchange of endophytic bacteria between the *P. aegyptiaca* and its host. Indeed, in the present study the endophytic community of the parasitic weed did not differ from that of the parasitized tomato host but was significantly different from the non-parasitized host. However, the fact that bacterial community composition in the parasitized tomato root is not significantly different from either the non-parasitized tomato root or the parasitic weed itself, makes it difficult to determine the direction(s) in which the bacteria are being transferred. In general endophytic bacterial movement can be either apolplasitc via intercellular space (Maheshwari et al., 2013) or may be facilitated through xylem tubes (Rosenblueth and Martínez-Romero, 2006; Compant et al., 2010). Since the *P. aegyptiaca* tubercle is composed mostly of parenchyma cells which are traversed by both xylem and phloem (Joel et al., 2013), both movement types can serve as potential transmission routes between the parasitic weed and its host through the haustorium bridge.

To date, most of the data in the literature on endophytes and their biological impact on their host plants were established by classical microbiological methods resulting in the loss of much valuable information regarding uncultured endophytes (Hallmann, 2001; Zermane et al., 2007; Hemissi et al., 2013). By using molecular tools we were able to show that even though the dominant endophyte taxonomy in the parasite is similar to that in other known plant endophytes (Lodewyckx et al., 2002), the microbial ecology of these endophytic communities in *P. aegyptiaca* is greatly affected by its connection to the host.

The greater similarity between the endophyte communities of the parasitic weed in the different post-attachment stages to each other, than to those detected prior to attachment to the host (pre-haustorium stage) indicate that connection to the host is affecting the endophytic community composition of the parasite as well as the host. Isolation of and further investigation of indicator bacterial species from both the host and parasite may therefore be of interest. Indeed investigating indicator bacteria such as *Novosphingobium* and *Methylophilus*, from the host *P. aegyptiaca* will enable the understanding of their unique interaction with the parasitic weed.

Endophytes may be equipped with a rich arsenal of metabolites involved in defense, as well as in interaction with the plant (Bouwmeester et al., 2003; Brader et al., 2014), supporting their host plant in different aspects. Therefore, we further examined specific endophytic bacteria isolated from the tomato-parasitic weed system and assessed their effect on *P. aegyptiaca* development. In agreement to previous studies (reviewed by Joel et al., 2013), our results showed that two isolates, associated with tomato-plant, reduced the germination of the parasite's seeds. Similar mechanisms were shown in a number of studies (Azevedo et al., 2000; Eljounaidi et al., 2016), where bacteria obtained from different crops were shown to protect their hosts against pests and diseases. Furthermore, *Pseudomonas* sp. *PhelS10*, which originated from the inner tissue of the host tomato root, was able to suppress both seed germination (*in vitro*) and parasitism (*in vivo*) of *P. aegyptiaca*, suggesting it may be used by the host for protection against parasitic weed attack. This protection may be through secreting different substances that suppress the pest and\or enhancing the plant's immune system (ISR), enabling higher resistance against the parasite as was demonstrated by other beneficial endophytes against other pathogens (Berg et al., 2005; Compant et al., 2005; Innerebner et al., 2011). Therefore it may be possible to harness these naturally occurring microbial partners as possible bio-controls agents against parasitization, reducing the need for harmful chemical pesticide agents.

CONCLUSION

In this study, we showed that attachment to the host had a major effect on all tested bacterial parameters of the parasitic weed, suggesting an exchange of endophytes between the host and the parasite. Considering the impact of parasitic weeds on agriculture and the difficulty in establishing efficient control methods further research is required to characterize additional endophytic isolates and to fully optimize this mechanism of resistance. Moreover, a deeper understanding of the relationships in the host-parasite-endophyte triangle may lead to new weed control methods and help alleviate weed-related ecological, agricultural, and economic issues.

AUTHOR CONTRIBUTIONS

RA conceived, planned and supervised the work. LI designed and performed the experiments and analyzed the data. SF analyzed the data. TL, JA-N, GA, and RS contributed in data production.

FUNDING

Agricultural Research Organization, The Volcani Center, Bet-Dagan, Israel, number 566/17.

ACKNOWLEDGMENT

Thanks to Dr. Stefan Green, Dr. Asaf Sade, and Prof. Harry Paris for their suggestions and scientific support.

REFERENCES

Aly, R. (2007). Conventional and biotechnological approaches for control of parasitic weeds. *In Vitro Cell. Dev. Biol. Plant* 43, 304–317. doi: 10.3109/07388551.2012.743502

Aly, R., Cholakh, H., Joel, D. M., Leibman, D., Steinitz, B., Zelcer, A., et al. (2009). Gene silencing of mannose 6-phosphate reductase in the parasitic weed *Orobanche aegyptiaca* through the production of homologous dsRNA sequences in the host plant. *Plant Biotechnol.* 7, 487–498. doi: 10.1111/j.1467-7652.2009.00418.x

Azevedo, J. L., Maccheroni, W. Jr., Pereira, J. O., and De Araújo, W. L. (2000). Endophytic microorganisms: a review on insect control and recent advances on tropical plants. *Electron. J. Biotechnol.* 3, 15–16. doi: 10.2225/vol3-issue1-fulltext-4

Berg, G., Krechel, A., Ditz, M., Sikora, R. A., Ulrich, A., and Hallmann, J. (2005). Endophytic and ectophytic potato-associated bacterial communities differ in structure and antagonistic function against plant pathogenic fungi. *FEMS Microbiol. Ecol.* 51, 215–229. doi: 10.1016/j.femsec.2004.08.006

Bouwmeester, H. J., Matusova, R., Zhongkui, S., and Beale, M. H. (2003). Secondary metabolite signalling in host-parasitic plant interactions. *Curr. Opin. Plant Biol.* 6, 358–364. doi: 10.1016/S1369-5266(03)00065-7

Brader, G., Compant, S., Mitter, B., Trognitz, F., and Sessitsch, A. (2014). Metabolic potential of endophytic bacteria. *Curr. Opin. Biotechnol.* 27, 30–37. doi: 10.1016/j.copbio.2013.09.012

Cardoso, C., Ruyter-Spira, C., and Bouwmeester, H. J. (2011). Strigolactones and root infestation by plant-parasitic *Striga*, *Orobanche* and *Phelipanche* spp. *Plant Sci.* 180, 414–420. doi: 10.1016/j.plantsci.2010.11.007

Chen, D., and Ronald, P. (1999). A rapid DNA minipreparation method suitable for AFLP and other PCR applications. *Plant Mol. Biol. Rep.* 17, 53–57. doi: 10.1023/A:1007585532036

Cochavi, A., Rubin, B., Achdari, G., and Eizenberg, H. (2016). Thermal time model for Egyptian broomrape (*Phelipanche aegyptiaca*) parasitism dynamics in carrot (*Daucus carota* L.): field validation. *Front. Plant Sci.* 7:1807. doi: 10.3389/fpls.2016.01807

Compant, S., Clément, C., and Sessitsch, A. (2010). Plant growth-promoting bacteria in the rhizo- and endosphere of plants: their role, colonization, mechanisms involved and prospects for utilization. *Soil Biol. Biochem.* 42, 669–678. doi: 10.1016/j.soilbio.2009.11.024

Compant, S., Duffy, B., Nowak, J., and Cle, C. (2005). Use of plant growth-promoting bacteria for biocontrol of plant diseases: principles, mechanisms of action, and future prospects. *Appl. Environ. Microbiol.* 71, 4951–4959. doi: 10.1128/AEM.71.9.4951-4959.2005

Daims, H., Ramsing, N. B., Schleifer, K. H., and Wagner, M. (2001). Cultivation-independent, semiautomatic determination of absolute bacterial cell numbers in environmental samples by fluorescent in situ hybridization. *Appl. Environ. Microbiol.* 67, 5810–5818. doi: 10.1128/AEM.67.12.5810-5818.2001

de Vega, C., Arista, M., Ortiz, P. L., and Talavera, S. (2010). Anatomical relations among endophytic holoparasitic angiosperms, autotrophic host plants and mycorrhizal fungi: a novel tripartite interaction. *Am. J. Bot.* 97, 730–737. doi: 10.3732/ajb.0900147

Dor, E., Eizenberg, H., Joel, D. M., Smirnov, E., Achdari, G., and Hershenhorn, J. (2014). *Orobanche* palaestina: a potential threat to agricultural crops in Israel. *Phytoparasitica* 42, 285–291. doi: 10.1007/s12600-013-0368-z

Dowd, S. E., Callaway, T. R., Wolcott, R. D., Sun, Y., McKeehan, T., Hagevoort, R. G., et al. (2008). Evaluation of the bacterial diversity in the feces of cattle using 16S rDNA bacterial tag-encoded FLX amplicon pyrosequencing (bTEFAP). *BMC Microbiol.* 8:125. doi: 10.1186/1471-2180-8-125

East, R. (2013). Soil science comes to life. *Nature* 501, S18–S19. doi: 10.1038/501S18a

Eizenberg, H., Goldwasser, Y., Golan, S., Plakhine, D., and Hershenhorn, J. (2004). Egyptian broomrape (*Orobanche aegyptiaca*) control in tomato with sulfonylurea herbicides-greenhouse studies 1. *Weed Technol.* 18, 490–496. doi: 10.1614/WT-03-023R3

Eljounaidi, K., Lee, S. K., and Bae, H. (2016). Bacterial endophytes as potential biocontrol agents of vascular wilt diseases–review and future prospects. *Biol. Control* 103, 62–68. doi: 10.1016/j.biocontrol.2016.07.013

Hacham, Y., Hershenhorn, J., Dor, E., and Amir, R. (2016). Primary metabolic profiling of Egyptian broomrape (*Phelipanche aegyptiaca*) compared to its host tomato roots. *J. Plant Physiol.* 205, 11–19. doi: 10.1016/j.jplph.2016.08.005

Hallmann, J. (2001). *Plant Interactions with Endophytic Bacteria.* New York, NY: CABI Publishing, 87–119. doi: 10.1079/9780851995120.0087

Hammer, H. Q., Harper, D. A. T., and Ryan, P. D. (2001). PAST: saleontological statistics software package for education and data analysis. *Palaeontol. Electron.* 4, 9.

Harasawa, R., Pitcher, D. G., Ramírez, A. S., and Bradbury, J. M. (2004). A putative transposase gene in the 16S-23S rRNA intergenic spacer region of *Mycoplasma imitans. Microbiology* 150, 1023–1029. doi: 10.1099/mic.0.26629-0

Hemissi, I., Mabrouk, Y., Abdi, N., Bouraoui, M., Saidi, M., and Sifi, B. (2013). Growth promotion and protection against *Orobanche foetida* of chickpea (*Cicer* aerietinum) by two *Rhizobium* strains under greenhouse conditions. *Afr. J. Biotechnol.* 12, 1371–1377.

Hoagland, D. R., and Arnon, D. I. (1950). *The Water-Culture Method for Growing Plants without Soil.* Berkeley, CA: College of Agriculture, University of California.

Iasur-Kruh, L., Hadar, Y., and Minz, D. (2011). Isolation and bioaugmentation of an estradiol-degrading bacterium and its integration into a mature biofilm. *Appl. Environ. Microbiol.* 77, 3734–3740. doi: 10.1128/AEM.00691-11

Innerebner, G., Knief, C., and Vorholt, J. A. (2011). Protection of *Arabidopsis thaliana* against leaf-pathogenic *Pseudomonas syringae* by *Sphingomonas* strains in a controlled model system. *Appl. Environ. Microbiol.* 77, 3202–3210. doi: 10.1128/AEM.00133-11

Joel, D. M., Gressel, J., and Musselman, L. J. (2013). *Parasitic Orobanchaceae: Parasitic Mechanisms and Control Strategies.* Berlin: Springer-Verlag. doi: 10.1007/978-3-642-38146-1

Joel, D. M., Hershenhorn, Y., Eizenberg, H., Aly, R., Ejeta, G., Rich, P. J., et al. (2006). "Biology and management of weedy root parasites," in *Horticultural Reviews*, ed. J. Janick (Hoboken, NJ: John Wiley & Sons), 267–349.

Larran, S., Monaco, C., and Alippi, H. E. (2001). Endophytic fungi in leaves of *Lycopersicon esculentum* Mill. *World J. Microbiol. Biotechnol.* 17, 181–184. doi: 10.1023/A:1016670000288

Lodewyckx, C., Vangronsveld, J., Porteous, F., Moore, E., Taghavi, S., Mezgeay, M., et al. (2002). Endophytic bacteria and their potential applications. *CRC Crit. Rev. Plant Sci.* 21, 583–606. doi: 10.1080/0735-260291044377

Maheshwari, D. K., Meenu, S., and Abhinav, A. (2013). *Bacteria in Agrobiology: Crop Productivity.* Berlin: Springer Science & Business Media.

Mangnus, E. M., Van Vliet, L. A., Vandenput, D. A., and Zwanenburg, B. (1992). Structural modifications of strigol analogs. Influence of the B and C rings on the bioactivity of the germination stimulant GR24. *J. Agric. Food Chem.* 40, 1222–1229. doi: 10.1021/jf00019a030

Parker, C., and Riches, C. R. (1993). *Parasitic Weeds of the World: Biology and Control.* Wallingford, CT: CAB Int.

Pozo, M. J., Cordier, C., Dumas-Gaudot, E., Gianinazzi, S., Barea, J. M., and Azcón-Aguilar, C. (2002). Localized versus systemic effect of arbuscular mycorrhizal fungi on defence responses to *Phytophthora* infection in tomato plants. *J. Exp. Bot.* 53, 525–534. doi: 10.1093/jexbot/53.368.525

Press, M. C., and Phoenix, G. K. (2005). Impacts of parasitic plants on natural communities. *New phytol.* 166, 737–751. doi: 10.1111/j.1469-8137.2005.01358.x

Rosenblueth, M., and Martínez-Romero, E. (2006). Bacterial endophytes and their interactions with hosts. *Mol. Plant Microbe Interact.* 19, 827–837. doi: 10.1094/MPMI-19-0827

Ryan, R. P., Germaine, K., Franks, A., Ryan, D. J., and Dowling, D. N. (2008). Bacterial endophytes: recent developments and applications. *FEMS Microbiol. Lett.* 278, 1–9. doi: 10.1111/j.1574-6968.2007.00918.x

Sakai, M., Matsuka, A., Komura, T., and Kanazawa, S. (2004). Application of a new PCR primer for terminal restriction fragment length polymorphism analysis of the bacterial communities in plant roots. *J. Microbiol. Methods* 59, 81–89. doi: 10.1016/j.mimet.2004.06.005

Sbrana, C., Agnolucci, M., Bedini, S., Lepera, A., Toffanin, A., Giovannetti, M., et al. (2002). Diversity of culturable bacterial populations associated to *Tuber borchii* ectomycorrhizas and their activity on *T. borchii* mycelial growth. *FEMS Microbiol. Lett.* 211, 195–201. doi: 10.1111/j.1574-6968.2002.tb11224.x

Schloss, P. D., Westcott, S. L., Ryabin, T., Hall, J. R., Hartmann, M., Hollister, E. B., et al. (2009). Introducing mothur: open-source, platform-independent, community-supported software for describing and comparing microbial communities. *Appl. Environ. Microbiol.* 75, 7537–7541. doi: 10.1128/AEM.01541-09

Shittu, H. O., Castroverde, D. C. M., Nazar, R. N., and Robb, J. (2009). Plant-endophyte interplay protects tomato against a virulent *Verticillium*. *Planta* 229, 415–426. doi: 10.1007/s00425-008-0840-z

Weisburg, W. G., Barns, S. M., Pelletier, D. A., and Lane, D. J. (1991). 16S ribosomal DNA amplification for phylogenetic study. *J. Bacteriol.* 173, 697–703. doi: 10.1128/jb.173.2.697-703.1991

Westwood, J. H., Yoder, J. I., and Timko, M. P. (2010). The evolution of parasitism in plants. *Trends Plant Sci.* 15, 227–235. doi: 10.1016/j.tplants.2010.01.004

Xu, M., Sheng, J., Chen, L., Men, Y., Gan, L., Guo, S., et al. (2014). Bacterial community compositions of tomato (*Lycopersicum esculentum* Mill.) seeds and plant growth promoting activity of ACC deaminase producing *Bacillus subtilis* (HYT-12-1) on tomato seedlings. *World J. Microbiol. Biotechnol.* 30, 835–845. doi: 10.1007/s11274-013-1486-y

Yoder, J. I. (1999). A species-specific recognition system directs haustorium development in the parasitic plant *Triphysaria* (Scrophulariaceae). *Planta* 202, 407–413. doi: 10.1007/s004250050144

Zermane, N., Souissi, T., Kroschel, J., and Sikora, R. (2007). Biocontrol of broomrape (*Orobanche crenata* Forsk. and *Orobanche foetida* Poir.) by *Pseudomonas fluorescens* isolate Bf7-9 from the faba bean rhizosphere. *Biocontrol. Sci. Technol.* 17, 483–497. doi: 10.1080/09583150701309535

Conflict of Interest Statement: The authors declare that the research was conducted in the absence of any commercial or financial relationships that could be construed as a potential conflict of interest.

11

Molecular Identification of Broomrape Species from a Single Seed by High Resolution Melting Analysis

Edited by:
Monica Fernandez-Aparicio,
Institut National de la Recherche Agronomique (INRA), France

Reviewed by:
Paula Martins-Lopes,
University of Trás-os-Montes and Alto Douro, Portugal
Leonardo Velasco,
Institute for Sustainable Agriculture – CSIC, Spain
Manuel Miller,
Helmholtz Zentrum München, Germany

***Correspondence:**
Mathieu Rolland
mathieu.rolland@anses.fr

Mathieu Rolland[1], Aurélie Dupuy[1], Aude Pelleray[2] and Philippe Delavault[2]*

[1] GEVES, Beaucouzé, France, [2] Laboratoire de Biologie et Pathologie Végétales, Université de Nantes, Nantes, France

Broomrapes are holoparasitic plants spreading through seeds. Each plant produces hundreds of thousands of seeds which remain viable in the soils for decades. To limit their spread, drastic measures are being taken and the contamination of a commercial seed lot by a single broomrape seed can lead to its rejection. Considering that broomrapes species identification from a single seed is extremely difficult even for trained botanists and that among all the described species, only a few are really noxious for the crops, numerous seed lots are rejected because of the contamination by seeds of non-noxious broomrape species. The aim of this study was to develop and evaluate a High Resolution Melting assay identifying the eight most noxious and common broomrape species (*Phelipanche aegyptiaca*, *Orobanche cernua*, *O. crenata*, *O. cumana*, *O. foetida*, *O. hederae*, *O. minor*, and *P. ramosa*) from a single seed. Based on *trn*L and *rbc*L plastidial genes amplification, the designed assay successfully identifies *O. cumana*, *O. cernua*, *O. crenata*, *O. minor*, *O. hederae*, and *O. foetida*; *P. ramosa*, and *P. aegyptiaca* can be differentiated from other species but not from each other. Tested on 50 seed lots, obtained results perfectly matched identifications performed by sequencing. Through the analysis of common seed lots by different analysts, the reproducibility of the assay was evaluated at 90%. Despite an original sample preparation process it was not possible to extract enough DNA from some seeds (10% of the samples). The described assay fulfills its objectives and allows an accurate identification of the targeted broomrape species. It can be used to identify contaminants in commercial seed lots or for any other purpose. The assay might be extended to vegetative material.

Keywords: *Orobanche*, *Phelipanche*, parasitic weed, molecular diagnosis, HRM, *trn*L, *rbc*L

INTRODUCTION

Broomrapes (*Orobanche* and *Phelipanche* spp.) are angiosperms in the Orobanchaceae which have evolved into obligate root holoparasitic plants (Joel, 2009). Devoid of leaves, of chlorophyll as well as of functional roots, they entirely depend on their host for nutritional requirements (Westwood, 2013). One single broomrape plant can produce hundreds of thousands of extremely small seeds, between 200 and 300 μm, each weighing around 5 μg and composed of only 200 to 300 cells (Joel, 1987). They are easily dispersed mainly by wind and water, and remain viable in the soils for many years until their germination is triggered chemically by exudates released in the soil by the roots of potential host plants (Lopez-Granados and Garcia-Torres, 1996, 1999). At the vicinity of host roots, germinated seeds develop a haustorium that penetrates host tissues and establish a connection with its vascular tissues. This connection will constitute the source of the parasite for water and nutrients. Some species in the *Orobanche* and *Phelipanche* genera, the weedy broomrapes, are able to infect a large range of plant species, including many important crops. In severe cases, infection of the host plant can lead in reduction of crop yields up 100%. This makes broomrapes one of the most devastating parasitic weeds in the Mediterranean and western Asian regions but also in many other parts of the world (Parker, 2009). Thus, among these noxious parasitic species are the closely related *Phelipanche ramosa* L. and *P. aegyptiaca* Pers. (synonym *Orobanche ramosa* and *O. aegyptiaca*) (Joel, 2009), *O. cumana* Wallr., *O. cernua* Loefl., *O. crenata* Forsk., *O. foetida* Poir., and *O. minor* Sm., while *O. hederae* Vaucher ex Duby has no agronomical impact but is extremely common.

Strategies to control parasitic weeds can be classified in chemical, cultural, physical, and biological control methods (Fernández-Aparicio et al., 2016). Among them, breeding for crop resistance seems to be the best approach to manage this issue. However, sources of resistance to most parasitic plants are either scarce or of complex nature (Perez-De-Luque et al., 2009). Despite these difficulties, significant success has been made on some crops. All these approaches allow a control of the parasitic population or permit resistant crops to grow and yield on infested soils, however, the eradication of parasitic weeds remains extremely difficult. Considering that one major mean of field contamination is through contaminated crop seed lots, preventive measures have to be taken to avoid spreading parasite seeds, especially through global scale seed exchanges. This requires to detect efficiently the possible contaminations of crop seeds lots by broomrape seed (Dongo et al., 2012). Visual detection of broomrape seeds in crop seed lots is conducted by sieving and observation of the obtained residues. Characterization of broomrape seeds at the species level in contaminated crop seed lots is important giving the differential host ranges among broomrape weed species and the capacity of some broomrape weeds to thrive in non-parasitic weed species. However, due to their nuanced microscopic morphological features, this identification is extremely difficult and can only can be performed by high qualified specialists (Abu Sbaih and Jury, 1994; Plaza et al., 2004). Molecular tools have been developed to detect and identify broomrape species from soil and crop seed batches. Random amplified polymorphic DNA technique (RAPD) allowed the differentiation between species such as *P. aegyptiaca*, *P. ramosa*, *O. cernua*, *O. cumana*, and *O. crenata* (Katzir et al., 1996; Paran et al., 1997). This technique was even used on single seeds (Portnoy et al., 1997), however, the main drawback of RAPDs is their low reproducibility (Harris, 1999). Intersimple sequence repeats (ISSR) were latter used to discriminate closely related species such as *O. cumana* and *O. cernua* (Benharrat et al., 2002). A TaqMan assay was developed on internal transcribed spacers (ITS) with the aim of detecting and quantifying *P. ramosa* and *O. cumana* seeds in oilseed rape and sunflower seed lots, respectively (Dongo et al., 2012). Microsatellites were also developed to investigate intraspecific variations in *O. cumana* (Pineda-Martos et al., 2014). Due to its monoparental inheritance, plastid genome has a low intraspecific variability and seems to be an adequate target for species identification. In the case of *Orobanche* genus, a particular attention was paid to the pseudogene *rbc*L which showed important sequence divergences among species due to an evolution under purifying selection (Wolfe and dePamphilis, 1997; Benharrat et al., 2000; Manen et al., 2004). Recently, full broomrape plastid genome sequence was made available (Wicke et al., 2013; Cusimano and Wicke, 2015) providing new molecular markers for species identification.

High resolution melting (HRM) is a technique based on the real-time measure of double stranded DNA denaturation at a high resolution. It is suitable for gene scanning and genotyping (Gori et al., 2012) and allows the detection of genetic variations such as single nucleotide polymorphisms (SNP), mutations (Toi and Dwyer, 2008), or methylation (Wojdacz and Dobrovic, 2007). Used on PCR products during a post-PCR denaturation, it requires no tube opening, purification, or product separation. With a minimum manipulation, HRM minimizes the contamination risk, it is cost efficient, suitable for high-throughput, and can be performed in-house by laboratories with no sequencing facility (Reed et al., 2007). This technique has been extensively used on human tissues (Krypuy et al., 2007; Takano et al., 2008), for clinical or phytopathological diagnostic and food analysis (Druml and Cichna-Markl, 2014). It is increasingly used on plant tissues for species and cultivar differentiation (Mackay et al., 2008; Jaakola et al., 2010) or genotyping (Lochlainn et al., 2011).

The objective of this study is to combine the knowledge recently obtained on plastid genome and the HRM technique to develop a new application allowing the differentiation of the seven most noxious and common broomrape weed species (*P. aegyptiaca, O. cernua, O. crenata, O. cumana, O. foetida, O. minor*, and *P. ramosa*) and the widely distributed *O. hederae* species from a single seed. This new application should provide to laboratories, involved in seed certification, a decision-making tool to evaluate crop seed lots potentially contaminated by noxious broomrape species.

MATERIALS AND METHODS

Plant Material

Broomrape seeds (*P. aegyptiaca, O. cernua, O. crenata, O. cumana, O. foetida, O. hederae, O. minor*, and *P. ramosa*) were either obtained from international collections or collected during field sampling by GEVES, Syngenta, Terres Inovia, or University of Nantes. Available data concerning the tested seed lots are summarized in **Table 1**.

Single Seed Grinding Procedure and DNA Extraction

One of the technical challenges associated with the development of an assay aiming to characterize broomrape single seeds

TABLE 1 | Origin and identifications of the 50 seed lots tested during the study.

N°	Visual identification	Country (region)	Crop	Collector	Date	High resolution melting (HRM)			Sequencing
						trnL	rbcL	Identification	Identification
1	*P. ramosa*	Charente, France	Rape	LBPV	Unkn	trnL-C	NA	sub. ramosae	*P. ramosa*
2	*P. ramosa*	Aube, France	Hemp	LBPV	2011	trnL-C	NA	sub. ramosae	*P. ramosa*
3	*P. ramosa*	Bas-Rhin, France	Tobacco	LBPV	Unkn	trnL-C	NA	sub. ramosae	*P. ramosa*
4	*P. ramosa*	Vendée, France	Rape	LBPV	2009	trnL-C	NA	sub. ramosae	*P. ramosa*
5	*P. ramosa*	Charente-Maritime, France	Tobacco	LBPV	2012	trnL-B	rbcL-B	*O. minor*	*O. minor*
6	*P. ramosa*	Vendée, France	Rape	This study	2013	trnL- C	NA	sub. ramosae	*P. ramosa*
7	*P. ramosa*	Aube, France	Celery	This study	2013	NA	NA	NA	NA
8	*P. ramosa*	Charente-Maritime, France	Rape	This study	2014	trnL- C	NA	sub. ramosae	*P. ramosa*
9	*P. ramosa*	Vendée, France	Rape	LBPV	2011	trnL- C	NA	sub. ramosae	*P. ramosa*
10	*P. ramosa*	Aube, France	Hemp	LBPV	2007	trnL- C	NA	sub. ramosae	*P. ramosa*
11	*P. ramosa*	Unkn	Hemp	LBPV	2013	trnL- C	NA	sub. ramosae	*P. ramosa*
12	*P. aegyptiaca*	Israël	Unkn	LBPV	2011	NA	NA	NA	NA
13	*O. cumana*	Iazu, Romania	Sunflower	LBPV	Unkn	trnL- A	NA	*O. cumana*	*O. cumana*
14	*O. cumana*	Carmona, Spain	Sunflower	Authors	2014	trnL- A	NA	*O. cumana*	*O. cumana*
15	*O. cumana*	Haute-Garonne, France	Sunflower	Authors	2014	NA	NA	NA	NA
16	*O. cumana*	Vendée, France	Unkn	LBPV	2009	trnL- A	NA	*O. cumana*	*O. cumana*
17	*O. cumana*	Turcia, Turkey	Unkn	LBPV	Unkn	trnL- A	NA	*O. cumana*	*O. cumana*
18	*O. cernua*	Israël	Unkn	LBPV	Unkn	trnL- A bis	NA	*O. cernua*	*O. cernua*
19	*O. cernua*	Lucainena, Spain	Unkn	LBPV	2006	trnL- A bis	NA	*O. cernua*	*O. cernua*
20	*O. foetida*	Unkn	Chickpea	LBPV	Unkn	trnL- D	NA	*O. foetida*	*O. foetida*
21	*O. foetida*	Tunisia	Chickpea	LBPV	2014	trnL- D	NA	*O. foetida*	*O. foetida*
22	*O. minor*	Unkn	Unkn	Authors	1987	trnL- B	rbcL-B	*O. minor*	*O. minor*
23	*O. minor*	Vendée, France	Clover	Authors	2013	trnL- B	rbcL-B	*O. minor*	*O. minor*
24	*O. minor*	Utsunomiya, Japan	Unkn	LBPV	2013	trnL- B	rbcL-B	*O. minor*	*O. minor*
25	*O. minor*	Maine-et-Loire, France	Clover	Authors	2014	trnL- B	rbcL-B	*O. minor*	*O. minor*
26	*O. minor*	Côte-d'Or, France	Clover	Authors	2014	trnL- B	rbcL-B	*O. minor*	*O. minor*
27	*O. minor*	Nienberge, Germany	Unkn	Botanical garden of Munster	1995	trnL- B	rbcL-B	*O. minor*	*O. minor*
28	*O. minor* subsp. *maritima*	Unkn	Unkn	Authors	1987	trnL- B	rbcL-B	*O. minor*	*O. minor*
29	*O. crenata*	Meknes, Morocco	Peas	LBPV	2001	trnL- B	rbcL-A	*O. crenata*	*O. crenata*
30	*O. crenata*	Ariana, Tunisia	Unkn	LBPV	2007	trnL- B	rbcL-A	*O. crenata*	*O. crenata*
31	*O. crenata*	Unkn	Unkn	Botanical garden of Braunschweig	1994	NA	NA	NA	NA
32	*O. crenata*	Unkn	Unkn	Botanical garden of Zurich	1997	trnL- B	rbcL-A	*O. crenata*	*O. crenata*
33	*O. hederae*	Unkn	Unkn	Authors	1987	NA	NA	NA	NA
34	*O. hederae*	Maine-et-Loire, France	Climber	Authors	2012	trnL- B	rbcL-C	*O. hederae*	*O. hederae*
35	*O. hederae*	Maine-et-Loire, France	Climber	Authors	2012	trnL- B	rbcL-C	*O. hederae*	*O. hederae*
36	*O. hederae*	Maine-et-Loire, France	Climber	Authors	2014	trnL- B	rbcL-C	*O. hederae*	*O. hederae*

(Continued)

TABLE 1 | Continued

N°	Visual identification	Country (region)	Crop	Collector	Date	High resolution melting (HRM)			Sequencing
						trnL	rbcL	Identification	Identification
37	*O. hederae*	Finistère, France	Unkn	LBPV	1993	trnL- B	rbcL-C	*O. hederae*	*O. hederae*
38	*P. purpurea*	La palma, Spain	Unkn	Botanical garden of Zurich	Unkn	trnL- B	rbcL-C	*O. hederae*	*O. hederae*
39	*O. flava*	Unkn	*Aconitum variegatum*	Botanical garden of Zurich	Unkn	trnL- B	rbcL-C	*O. hederae*	*O. hederae*
40	*O. picridis*	Yvelines, France	Unkn	Authors	2014	trnL- B	NI	NI	minores spp
41	*O. artemisia campestris*	Maine-et-Loire, France	Unkn	Authors	2014	trnL- B	NI	NI	minores spp
42	*O. alsatica*	Binn, Switzerland	*Peucedanum cervaria*	Botanical garden of Zurich	Unkn	NI	NA	NI	NI
43	*O. alsatica*	Unkn	Unkn	Botanical garden of Frankfurt	2012	trnL-B	NI	NI	*O. bartlingii*
44	*P. arenaria*	Indre-et-Loire, France	Unkn	Botanical garden of Nantes	Unkn	NI	NA	NI	*P. arenaria*
45	*O. laevis*	Lax, Switzerland	*Artemisia campestris*	Botanical garden of Zurich	Unkn	NI	NA	NI	*P. arenaria*
46	*O. lucorum*	Unkn	*Berberis vulgaris*	Botanical garden of Zurich	Unkn	NI	NA	NI	*O. lucorum*
47	*O. gracilis*	Haute-Savoie, France	Unkn	Botanical garden of Alpin	Unkn	NI	NA	NI	*O. gracilis*
48	*O. caryophyllacea*	Binn, Switzerland	*Galium spec.*	Botanical garden of Zurich	Unkn	NI	NA	NI	NI
49	*P. mutelii*	Australia	Unkn	LBPV	Unkn	trnL-C	NA	sub. ramosae	*P. mutelii*
50	*P. mutelii*	Mannum south, Australia	Unkn	LBPV	2002	trnL- C	NA	sub. ramosae	*P. mutelii*

NA, not amplified; NI, not identified; Unkn, unknown.

is the ability to obtain enough DNA from seeds weighing an average of 5 μg in a reproducible manner. To this end, each seed was crushed between two microscopy glass slides in presence of 2 μl of ultrapure water and seed tissues were then collected in a microtube. In order to maximize the amount of collected DNA, the slides were rinsed with 400 μl of PL1 extraction buffer (Macherey-Nagel) and the rinse collected in the microtube. Total DNA extraction was then performed using the NucleoSpin® Plant II commercial kit (Macherey-Nagel) following the manufacturer's instructions (filtration columns were not used). A control of the quantity and quality of the extracted DNA was performed using a NanoVue^{TM} Spectrophotometer (GE Healthcare).

Sequencing

Thanks to previous studies on plastid genome sequences in broomrapes (Wicke et al., 2013; unpublished results), sequences corresponding to eight plastid genes (*rbc*L, *rps7*, *rps11*, *rpl36*, *rpl16*, *trn*Q, *trn*L, and *rrn23*) and one nuclear region (ITS) were obtained for the eight studied species. Sequences were aligned using the default alignment algorithm of Geneious v5.6.4. Two markers showing significant sequence divergence among the eight species were selected for subsequent HRM experiments: trnL and rbcL.

To design HRM primers and to control the identification of the species, pseudogenes *trn*L and *rbc*L were amplified and sequenced, respectively, using the primers (i) trnL C (F) and trnL HRM R, (ii) 1F and 1352R (**Table 2**). Amplification was performed on 5 μl of single seed total DNA extract, by 1 U of AmpliTaq Gold® (Life Technologies), in a total volume of 40 μl at the final concentration of 1X of the appropriate Buffer II, 0.3 μM of each primer, 1.5 mM of $MgCl_2$, and 0.2 mM of dNTP. PCR conditions were adjusted as follow, an initial denaturation of 10 min at 95°C, 40 cycles of 30 s at 95°C, 15 s at 58°C, and 1 min at 72°C, and a final extension of 10 min at 72°C. After migration in a 1.5% agarose gel at 180 V for 45 min and ethidium bromide staining, PCR products were visualized under UV light. Purification and sequencing of the PCR products was provided by Genoscreen.

Primer Design

Obtained sequences were aligned using the default alignment algorithm of Geneious v5.6.4 (some alignments are provided as **Supplementary Images 1** and **2**). Conserved regions and potential markers were identified visually. To achieve HRM identification of the species, primers surrounding the selected markers were designed using primer 3[1] with an estimated melting temperature of 60°C (**Table 2**). According to the tested species, the designed primers surround fragments of 315–463 bp for *trn*L and 345–389 bp for *rbc*L.

High Resolution Melting Analysis

HRM reactions were performed on 5 μl of single seed DNA extract, in a total volume of 20 μl, using the MeltDoctor Master mix (Life technologies) on a StepOnePlus instrument (Applied

[1]http://primer3.ut.ee/

TABLE 2 | Primers used for amplification, sequencing, and HRM.

Target	Name	Sequence (5′–3′)	Source	Purpose
trnL	trnL C (F)	CGAAATCGGTAGACGCTACG	Taberlet et al., 1991	PCR and sequencing
	trnL HRM R	GGGGATAGAGGGACTTGAACC		
rbcL	1F	ATGTCACCACAAACAGAAAC	Manen et al., 2004	PCR and sequencing
	1352R	CAGCAACTAGTTCAGGRCTCC		
trnL	trnL-Z1-F	CGGTAGACGCTACGGACTTA	This study	HRM
	trnL-Lg-2R	ATGGGACTCTATCTTTATTCTC		
rbcL	rbcL-lg-1-F	AACCTGAAGTTCCGCCTGAA	This study	HRM
	rbcL-Z2-R	AGTACATCCCAACAGGGGAC		

Biosystems) following the manufacturers recommendations. *TrnL* pseudogene was amplified using the primers trnL-Z1-F and trnL-lg-2R at the final concentration of 0.2 μM, *rbcL* by the primers rbcL-lg-1-F and rbcL-Z2-R at the final concentration of 0.15 μM (**Table 2**). PCR conditions were adjusted as follow, an initial denaturation of 10 min at 95°C, 45 cycles of 15 s at 95°C, and 1 min at 60°C, a complete denaturation of 10 s at 95°C, 1 min at 60°C, and a continuous melt rising from 60 to 90°C with 0.3% temperature increment every 15 s.

Each extract was run in duplicate, in the presence of the usual positive, negative and process controls and in the presence of reference materials used for HRM profiles analysis. One reference material is required for each HRM profile. These reference materials were previously prepared by extraction of identified seeds using the described protocol and control of the species by sequencing.

Considering the real-time amplification results, only the samples providing cycle threshold (C_t) values below 35 were considered for HRM results analysis. Analysis of the melting profiles was performed using High Resolution Melt Software v3.0 (Applied Biosystems).

RESULTS

DNA Extraction from Single Seeds

A simple methodology was developed to crush individual seeds between two microscopy glass slides and extract total DNA from this crushed material. For the 50 seed lots tested (**Table 1**), extractions and amplifications were performed separately from two single seeds. DNA concentration of the obtained extracts was too low to be measured using a Nanovolume spectrophotometer. For seed lots number 22, 29, 30, and 37, only one of the extracts allowed a proper amplification. For lots number 7, 12, 15, 31, and 33, respectively, harvested in 2013, 2011, 2014, 1994, and 1987, it was not possible to obtain any amplification. For these five last seed lots, single seed extraction was performed on two more seeds with the similar results. The failure of these seed batches was not species-specific associated. The viability of the different seed lots was not assessed.

Species Identification

For the selected markers *trnL* and *rbcL*, respectively, 9 and 11 primer pairs were designed and evaluated for their ability to provide a suitable assay. Results obtained with the best primers are reported. HRM primers were first selected according to their ability to differentiate the eight target species by providing distinct HRM profiles. Discrimination between species was possible because of sequence divergences (SNP and indel) between the amplicons. Targeted species show five different profiles when considering the high resolution melt curves of the *trnL* PCR product (**Figure 1A**). *O. cumana*, *O. cernua*, and *O. foetida* are easily identified using this HRM marker since each of these species is the only one associated with a profile (respectively, red, orange, and yellow). *O. crenata*, *O. minor*, and *O. hederae* are associated with the blue profile. They can be then differentiated from other species but not from each other. The same goes for species *P. ramosa* and *P. aegyptiaca* associated with the green profile. Among the eight species considered, the rbcL primers amplify only *O. crenata*, *O. minor*, and *O. hederae*. PCR products obtained from these three species show distinct and identifiable HRM profiles (respectively, red, blue, and green; **Figure 1A**). Considering the obtained results, an identification key is proposed to facilitate the analysis of the results (**Table 3**).

The second primers selection criterion was the consistency of the profiles between lots belonging to the same species. Amplification and HRM were performed on single seed DNA extracts obtained from the 37 (out of 50) available seed lots belonging to the eight targeted species (**Table 1**). **Figure 1B** presents the aspect of the obtained melting curve for the *trnL* and *rbcL* PCR products. Considering raw (not shown) or derivated melt curves, profiles obtained from samples of identical species show some variability. However, for identical profiles, when considering the aligned melt curves of both *trnL* and *rbcL*, the highest relative standard deviation of measured melting temperatures is 3.3%. The different profiles presented above are consistently reproduced between samples of identical species.

Specificity of the Assay

Besides the ability of HRM primers to discriminate among weedy broomrape species, both trnL and rbcL primer pairs amplified single seed DNA extracts obtained from additional 13 seed lots initially identified based on morphological characteristics of adult plant as belonging to eight wild *Orobanche* species and three wild *Phelipanche* species (**Table 1**; **Figure 1C**). Among the 11 wild species, nine showed original non-identified profiles (NI) which could be easily distinguished from the profiles of the weedy species. Samples 38 and 39, respectively, declared in

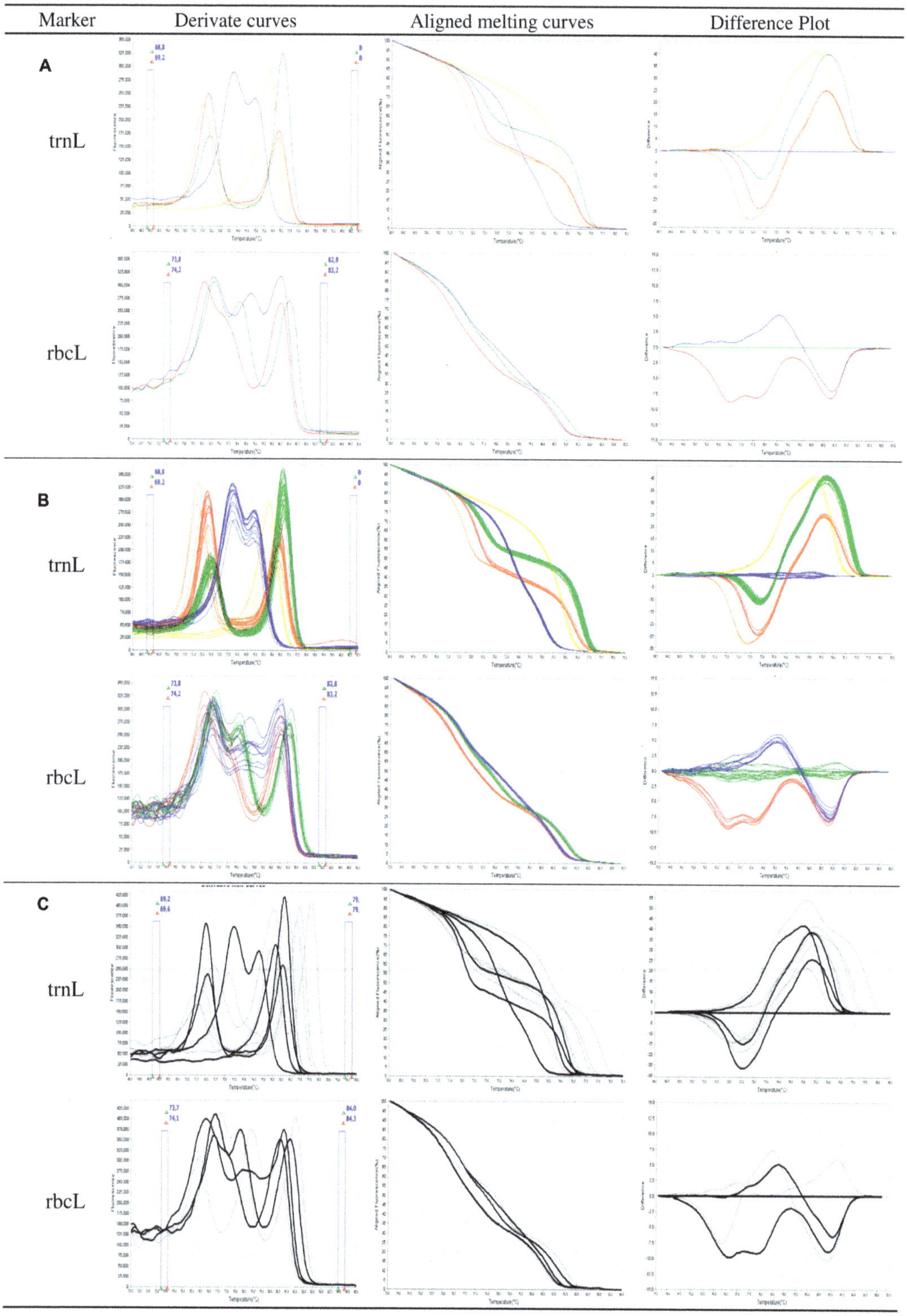

FIGURE 1 | High resolution melting (HRM) analysis of broomrape single seeds for the markers trnL and rbcL. (A) Ability of the assay to differentiate the targeted species as different profiles (i) trnL marker: red = *Orobanche cumana*; orange = *O. cernua*; blue = *O. crenata*, *O. minor*, or *O. hederae*; green = ramosae clade; yellow = *O. foetida*; (ii) rbcL marker: red = *O. crenata*; blue = *O. minor*; green = *O. hederae*. **(B)** Consistency of the profiles between samples of a same species. **(C)** Specificity of the assay, curves in black correspond to the targeted species, while curves in gray correspond to other tested species. Derivate, aligned melting curves, and difference plots correspond to three representations of the same data: aligned melting curves have been normalized by eliminating fluorescence variance out of the melt regions; difference plots are achieved by subtracting the normalized fluorescence data of a user-defined genotype from that of each of the other samples in the HRM analysis.

collection records as *P. purpurea* and *O. flava* were identified as *O. hederae* by HRM analysis. Samples 49 and 50 initially declared in collection records as *P. mutelii* show the same trnL profile than *P. ramosa* and *P. aegyptiaca*.

To control the identification made by HRM, sequencing of the pseudogenes *trn*L and *rbc*L was performed on all the analyzed DNA extracts. Sequences were submitted to GenBank and are available with accession numbers KX539159–KX539172. Comparison of the HRM and sequencing interpretations is presented in **Table 1**. Concerning the targeted species, results show a 100% match between interpretations obtained using both techniques. Furthermore, all the samples identified as non-target by sequencing are designated as non-identifiable by HRM. It is interesting to note that identifications performed by sequencing are not always consistent with primary identification based on visual criteria. For samples 38 and 39 visually identified as *O. purpurea* and *O. flava*, HRM identifications as *O. hederae* are consistent with sequencing results, suggesting that the HRM identification is correct and that the initial morphological identification failed. Sequencing confirmed the visual identity of the samples 49 and 50 as *P. mutelii*. HRM profile common to *P. ramosa* and *P. aegyptiaca* is therefore not specific to these two species but also includes close species.

Reproducibility of the Assay

The HRM profiles obtained with the trnL and rbcL primers can be considered as complex. Furthermore, reading a melting profile is performed by an analyst and is somehow subjective. To question the transferability of the technique, 20 seed lots have been analyzed by three analysts in two different laboratories (GEVES and Terres Inovia). Each analyst performed the experiment on one single seed of each seed lot. Obtained results are shown in **Table 4**. For 7 out of the 20 seed lots tested, at least one extraction did not allow the amplification. When DNA was properly extracted and amplified, obtained results were in accordance except in the cases of lots 13 and 16. In one case out of three, profiles associated with these seed lots of ramosae had the correct aspect but a different melting temperature and were noted as non-identified. These differences of melting temperature were consistently reproduced, however, sequencing of trnL PCR products showed no difference of sequence between these extracts and others. Excluding the failure of proper DNA extraction on some seeds, the reproducibility of the assay (defined as the percentage of agreements between two identifications performed on a same seed lot) is of 90.9%.

TABLE 3 | Correspondence between HRM profiles obtained using the trnL and rbcL markers and broomrape species.

trnL	rbcL	Species
Red	NA	*O. cumana*
Orange	NA	*O. cernua*
Blue	Red	*O. crenata*
Blue	Blue	*O. minor*
Blue	Green	*O. hederae*
Green	NA	ramosae clade
Yellow	NA	*O. foetida*
NI	NA or NI	Unkn
NA	NA or NI	Unkn

NA, not amplified; NI, not identified; Unkn, unknown.

DISCUSSION

Seed producers may face contaminations of their crop seed lots by seeds of noxious broomrapes. In case of trading, they require international seed lot certificates provided by official seed testing stations. This is mainly carried out through analysis of specific purity of seed lots. However, if this analysis can identify seed lots containing broomrape seeds, it cannot allow a clear identification of the parasite species. Indeed, broomrape species identification can be achieved thanks to seed coat morphological features observed under microscopy (Joel, 1987; Abu Sbaih and Jury, 1994), but this approach is extremely difficult even for trained botanists and requires an extensive expertise usually not available in most laboratories. Molecular markers such as ITS, ISSR, plastid genes, or RAPD were developed for identification of broomrape species, but all required large amount of seeds incompatible with an isolation of few seeds from a specific purity analysis. Thus, protocols allowing DNA extraction from broomrape single seed

TABLE 4 | Data of reproducibility generated by three analysts on 20 seed lots, each analyst analyzing one single seed of each lot.

Seed lot number	Analyst			Rep.
	1	2	3	
1	ramosae clade	ramosae clade	ramosae clade	3/3
2	ramosae clade	ramosae clade	ramosae clade	3/3
3	ramosae clade	NI	ramosae clade	1/3
4	ramosae clade	ramosae clade	ramosae clade	3/3
6	NI	ramosae clade	ramosae clade	1/3
11	ramosae clade	ramosae clade	ramosae clade	3/3
13	*O. cumana*	*NA*	*O. cumana*	1/1
15	*NA*	*NA*	*O. cumana*	0/0
18	*O. cernua*	*NA*	*O. cernua*	1/1
20	*O. foetida*	*O. foetida*	*O. foetida*	3/3
21	*O. foetida*	*O. foetida*	*O. foetida*	3/3
24	*O. minor*	*O. minor*	*O. minor*	3/3
29	*O. crenata*	*O. crenata*	*O. crenata*	3/3
32	*O. crenata*	*O. crenata*	*NA*	1/1
35	*O. hederae*	*O. hederae*	*O. hederae*	3/3
37	*O. hederae*	*NA*	*O. hederae*	1/1
38	*O. hederae*	*O. hederae*	*O. hederae*	3/3
44	NI	NI	NI	3/3
46	*NA*	NI	*NA*	0/0
48	NI	NI	*NA*	1/1
				90.9%

NA, not amplified; NI, not identified; Rep, reproducibility defined as the percentage of agreements between two identifications performed on a same seed lot (absence of amplifications have been excluded of the calculation).

were developed (Portnoy et al., 1997; Osterbauer and Rehms, 2002) and used with RAPD markers (Katzir et al., 1996). Seeds of five different species could be identified using these methods: *P. aegyptiaca*, *P. ramosa*, *O. cernua*, *O. cumana*, and *O. crenata*.

The protocol developed in this study is the first work describing the application of HRM curve analysis for differentiation of broomrape species. Compared to previous technologies, the proposed protocol and markers allow to extend the identification spectrum since it was able to differentiate between eight species, the five above mentioned plus *O. foetida*, *O. hederae*, and *O. minor*. The sequences of the root parasitic plants used in this study present variation generating divergences in the HRM patterns. Deletions, insertions, and several SNPs are responsible for the differences in the observed melting curves between the different species amplicons. The two plastid genes, *rbc*L, and *trn*L, have been already used as HRM markers for identification of plants (Madesis et al., 2012; Osathanunkul et al., 2015).

By using the HRM technology and by targeting plastid sequences, it was then possible to develop a simple, reliable, and cost effective assay to identify the seven main weedy species of broomrape potentially found in crop seed lots. In addition, it allows discrimination between these weedy species and 12 species lacking agronomic interest. The high level of divergence between species in the targeted sequences provided more complex profiles than for HRM assays targeting SNP (Toi and Dwyer, 2008) or microsatellites (Mackay et al., 2008). However, in most cases, DNA extracted from single seeds allowed a proper amplification and profiles could be identified by the analyst by comparison with the reference materials introduced in each experiment. On 45 amplified samples, the assay provided results perfectly matching with sequencing. The technique was used by several analysts in two laboratories using different HRM-capable real-time PCR machine and visual analysis of the HRM profiles. In these conditions, the technique shows a reproducibility of 90%. This rate of reproducibility is higher than the one received with RAPD markers, known to be weakly reproducible when employed in different laboratories with different PCR apparatus (Jones et al., 1997). The described assay will make then reliable identification much easier for any diagnostic or research purpose. It is also a fast close tube method not requiring post-PCR manipulation such as DNA gel electrophoresis like in RAPD analysis.

However, the success of the developed assay depends on the concentration and/or quality of the extracted DNA. Indeed for some seed lots it was not possible to reach the minimal concentration and/or quality from some tested seeds or even from any of the tested seeds. HRM technique also requires homogenous DNA extract compositions among samples to compare. Composition may indeed impact the melting temperature of the amplification products. By extracting DNA from single seeds, extracts are relatively homogeneous. However, during the evaluation of the reproducibility, some profiles showed the expected melting profile but with a different melting temperature. In a seed lot, heterogeneity of the seeds may therefore occasionally be an issue for profiles comparison.

Using the identified plastid targets, it was not possible to differentiate the species of the taxonomically difficult ramosa aggregate (*P. mutelii*, *P. ramosa*, and *P. aegyptiaca*) referred to as *ramosae clade*. Further development of the assay by adding a third marker could provide the ability to differentiate species in the ramosa aggregate. If a species identification is necessary after a *ramosae clade* or a NI result, the product obtained after the HRM amplification and denaturation can be used for sequencing as any regular PCR product.

The development of an assay able to identify broomrape species from single seeds allows testing of seeds found in commercial seed lots but also identification of mature plants from the field. Broomrape seeds are indeed a material easy to collect and transport, it can be stored at room temperature for many years. For the identification of plants at early stages (before the presence of seeds), the assay can be extended to vegetative material.

AUTHOR CONTRIBUTIONS

MR planned and designed the research; PD and AP performed the plastid genome sequences analysis; AD performed the experiments. MR wrote the paper with the help of PD.

FUNDING

Funding for this work was provided by the CASDAR "*Orobanche*" project (N° C-2012-07) financed by the French Ministère de l'Agriculture, de l'Agroalimentaire et de la Forêt (MAAF).

ACKNOWLEDGMENT

The authors want to thank Julien Carpezat (Terres Inovia) for his participation to the evaluation of the reproducibility of the assay.

IMAGE 1 | Alignment of partial sequences of the *trn*L gene obtained by amplification using the primers trnL C (F) and trnL HRM R of total DNA extracted from individual seeds belonging to the 8 species of interest.

IMAGE 2 | Alignment of partial sequences of the *rbc*L gene obtained by amplification of total DNA extracted from individual seeds belonging to 7 of the 8 species of interest.

REFERENCES

Abu Sbaih, H. A., and Jury, S. L. (1994). Seed micromorphology and taxonomy in *Orobanche* (Orobanchaceae). *Flora Mediterr.* 4, 41–48.

Benharrat, H., Delavault, P., Theodet, C., Figureau, C., and Thalouarn, P. (2000). rbcL plastid pseudogene as a tool for *Orobanche* (subsection Minores) identification. *Plant Biol.* 2, 34–39. doi: 10.1055/s-2000-9457

Benharrat, H., Veronesi, C., Theodet, C., and Thalouarn, P. (2002). *Orobanche* species and population discrimination using intersimple sequence repeat (ISSR). *Weed Res.* 42, 470–475. doi: 10.1046/j.1365-3180.2002.00305.x

Cusimano, N., and Wicke, S. (2015). Massive intracellular gene transfer during plastid genome reduction in nongreen Orobanchaceae. *New Phytol.* 210, 680–693. doi: 10.1111/nph.13784

Dongo, A., Leflon, M., Simier, P., and Delavault, P. (2012). Development of a high-throughput real-time quantitative PCR method to detect and quantify contaminating seeds of *Phelipanche ramosa* and *Orobanche cumana* in crop seed lots. *Weed Res.* 52, 34–41. doi: 10.1111/j.1365-3180.2011.00891.x

Druml, B., and Cichna-Markl, M. (2014). High resolution melting (HRM) analysis of DNA – Its role and potential in food analysis. *Food Chem.* 158, 245–254. doi: 10.1016/j.foodchem.2014.02.111

Fernández-Aparicio, M., Reboud, X., and Gibot-Leclerc, S. (2016). Broomrape weeds. Underground mechanisms of parasitism and associated strategies for their control: a review. *Front. Plant Sci.* 7:135. doi: 10.3389/fpls.2016.00135

Gori, A., Cerboneschi, M., and Tegli, S. (2012). High-Resolution melting analysis as a powerful tool to discriminate and genotype *Pseudomonas savastanoi* Pathovars and Strains. *PLoS ONE* 7:e30199. doi: 10.1371/journal.pone.0030199

Harris, S. (1999). RAPDs in systematics: a useful methodology. *Mol. Syst. Plant Evol.* 57, 211–228. doi: 10.1201/9781439833278.ch11

Jaakola, L., Suokas, M., and Haggman, H. (2010). Novel approaches based on DNA barcoding and high-resolution melting of amplicons for authenticity analyses of berry species. *Food Chem.* 123, 494–500. doi: 10.1016/j.foodchem.2010.04.069

Joel, D. M. (1987). Detection and identification of *Orobanche* seeds using fluorescence microscopy. *Seed Sci. Technol.* 15, 119–124.

Joel, D. M. (2009). The new nomenclature of *Orobanche* and *Phelipanche*. *Weed Res.* 49, 6–7. doi: 10.1111/j.1365-3180.2009.00748.x

Jones, C. J., Edwards, K. J., Castaglione, S., Winfield, M. O., Sala, F., van de Wiel, C., et al. (1997). Reproducibility testing of RAPD, AFLP and SSR markers in plants by a network of European laboratories. *Mol. Breed.* 3, 381–390. doi: 10.1023/a:1009612517139

Katzir, N., Portnoy, V., Tzuri, G., CastejonMunoz, M., and Joel, D. M. (1996). Use of random amplified polymorphic DNA (RAPD) markers in the study of the parasitic weed *Orobanche*. *Theor. Appl. Genet.* 93, 367–372. doi: 10.1007/s001220050290

Krypuy, M., Ahmed, A. A., Etemadmoghadam, D., Hyland, S. J., deFazio, A., Fox, S. B., et al. (2007). High resolution melting for mutation scanning of TP53 exons 5–8. *BMC Cancer* 7:168. doi: 10.1186/1471-2407-7-168

Lochlainn, S. O., Amoah, S., Graham, N. S., Alamer, K., Rios, J. J., Kurup, S., et al. (2011). High Resolution Melt (HRM) analysis is an efficient tool to genotype EMS mutants in complex crop genomes. *Plant Methods* 7, 43. doi: 10.1186/1746-4811-7-43

Lopez-Granados, F., and Garcia-Torres, L. (1996). Effects of environmental factors on dormancy and germination of crenate broomrape (*Orobanche crenata*). *Weed Sci.* 44, 284–289. doi: 10.2307/4045680

Lopez-Granados, F., and Garcia-Torres, L. (1999). Longevity of crenate broomrape (*Orobanche crenata*) seed under soil and laboratory conditions. *Weed Sci.* 47, 161–166. doi: 10.2307/4046192

Mackay, J. F., Wright, C. D., and Bonfiglioli, R. G. (2008). A new approach to varietal identification in plants by microsatellite high resolution melting analysis: application to the verification of grapevine and olive cultivars. *Plant Methods* 4, 8. doi: 10.1186/1746-4811-4-8

Madesis, P., Ganopoulos, I., Anagnostis, A., and Tsaftaris, A. (2012). The application of Bar-HRM (Barcode DNA-High Resolution Melting) analysis for authenticity testing and quantitative detection of bean crops (Leguminosae) without prior DNA purification. *Food Control* 25, 576–582. doi: 10.1016/j.foodcont.2011.11.034

Manen, J. F., Habashi, C., Jeanmonod, D., Park, J. M., and Schneeweiss, G. M. (2004). Phylogeny and intraspecific variability of holoparasitic *Orobanche* (Orobanchaceae) inferred from plastid rbcL sequences. *Mol. Phylogenet. Evol.* 33, 482–500. doi: 10.1016/j.ympev.2004.06.010

Osathanunkul, M., Madesis, P., and de Boer, H. (2015). Bar-HRM for authentication of plant-based medicines: evaluation of three medicinal products derived from Acanthaceae species. *PLoS ONE* 10:e0128476. doi: 10.1371/journal.pone.0128476

Osterbauer, N. K., and Rehms, L. (2002). Detecting single seeds of small broomrape (*Orobanche* minor) with a polymerase chain reaction. *Plant Health Progress* doi: 10.1094/PHP-2002-1111-01-RS

Paran, I., Gidoni, D., and Jacobsohn, R. (1997). Variation between and within broomrape (*Orobanche*) species revealed by RAPD markers. *Heredity* 78, 68–74. doi: 10.1038/hdy.1997.8

Parker, C. (2009). Observations on the current status of *Orobanche* and *Striga* problems worldwide. *Pest. Manag. Sci.* 65, 453–459. doi: 10.1002/ps.1713

Perez-De-Luque, A., Fondevilla, S., Perez-Vich, B., Aly, R., Thoiron, S., Simier, P., et al. (2009). Understanding *Orobanche* and *Phelipanche*-host plant interactions and developing resistance. *Weed Res.* 49, 8–22. doi: 10.1111/j.1365-3180.2009.00738.x

Pineda-Martos, R., Velasco, L., and Perez-Vich, B. (2014). Identification, characterisation and discriminatory power of microsatellite markers in the parasitic weed *Orobanche cumana*. *Weed Res.* 54, 120–132. doi: 10.1111/wre.12062

Plaza, L., Fernandez, I., Juan, R., Pastor, J., and Pujadas, A. (2004). Micromorphological studies on seeds of *Orobanche* species from the Iberian Peninsula and the Balearic Islands, and their systematic significance. *Ann. Bot.* 94, 167–178. doi: 10.1093/aob/mch124

Portnoy, V. H., Katzir, N., and Joel, D. M. (1997). Species identification of soil-borne *Orobanche* seeds by DNA fingerprinting. *Pestic. Biochem. Physiol.* 58, 49–54. doi: 10.1006/pest.1997.2281

Reed, G. H., Kent, J. O., and Wittwer, C. T. (2007). High-resolution DNA melting analysis for simple and efficient molecular diagnostics. *Pharmacogenomics* 8, 597–608. doi: 10.2217/14622416.8.6.597

Taberlet, P., Gielly, L., Pautou, G., and Bouvet, J. (1991). Universal primers for amplification of three non-coding regions of chloroplast DNA. *Plant Mol. Biol.* 17, 1105–1109. doi: 10.1007/bf00037152

Takano, E. A., Mitchell, G., Fox, S. B., and Dobrovic, A. (2008). Rapid detection of carriers with BRCA1 and BRCA2 mutations using high resolution melting analysis. *BMC Cancer* 8:59. doi: 10.1186/1471-2407-8-59

Toi, C. S., and Dwyer, D. E. (2008). Differentiation between vaccine and wild-type varicella-zoster virus genotypes by high-resolution melt analysis of single nucleotide polymorphisms. *J. Clin. Virol.* 43, 18–24. doi: 10.1016/j.jcv.2008.03.027

Westwood, J. H. (2013). "The physiology of the established parasite–host association," in *Parasitic Orobanchaceae: Parasitic Mechanisms and Control Strategies*, eds M. D. Joel, J. Gressel, and J. L. Musselman (Berlin: Springer Berlin Heidelberg), 87–114.

Wicke, S., Muller, K. F., Pamphilis, C. W. D., Quandt, D., Wickett, N. J., Zhang, Y., et al. (2013). Mechanisms of functional and physical genome reduction in photosynthetic and nonphotosynthetic parasitic plants of the broomrape family. *Plant Cell* 25, 3711–3725. doi: 10.1105/tpc.113.113373

Wojdacz, T. K., and Dobrovic, A. (2007). Methylation-sensitive high resolution melting (MS-HRM): a new approach for sensitive and high-throughput assessment of methylation. *Nucleic Acids Res.* 35, e41–e41. doi: 10.1093/nar/gkm013

Wolfe, A. D., and dePamphilis, C. W. (1997). Alternate paths of evolution for the photosynthetic gene rbcL in four nonphotosynthetic species of *Orobanche*. *Plant Mol. Biol.* 33, 965–977. doi: 10.1023/a:1005739223993

Conflict of Interest Statement: The authors declare that the research was conducted in the absence of any commercial or financial relationships that could be construed as a potential conflict of interest.

12

Non-Chemical Control of Root Parasitic Weeds with Biochar

Edited by:
Joshua L. Heazlewood,
University of Melbourne, Australia

Reviewed by:
Khawar Jabran,
Duzce University, Turkey
Grama Nanjappa Dhanapal,
University of Agricultural Sciences,
Bangalore, India
Kevin Gibson,
Purdue University, United States

***Correspondence:**
Ellen R. Graber
ergraber@agri.gov.il

*Hanan Eizenberg[1], Dina Plakhine[1], Hammam Ziadne[1], Ludmila Tsechansky[2] and Ellen R. Graber[2]**

[1] Department of Plant Pathology and Weed Research, Newe Ya'ar Research Center, Agricultural Research Organization, Ramat Yishai, Israel, [2] Institute of Soil, Water and Environmental Sciences, Agricultural Research Organization, The Volcani Center, Beit Dagan, Israel

This study tested whether soil-applied biochar can impact the seed germination and attachment of root parasitic weeds. Three hypotheses were evaluated: (i) biochar adsorbs host-exuded signaling molecules; (ii) biochar activates plants' innate system-wide defenses against invasion by the parasite; and (iii) biochar has a systemic influence on the amount of seed germination stimulant produced or released by the host plant. Three types of experiments were performed: (I) pot trials with tomato (*Solanum lycopersicum*) infested with *Phelipanche aegyptiaca PERS*. (Egyptian broomrape) and three different types of biochar at concentrations ranging from 0 to 1.5% weight, wherein tomato plant biomass, *P. aegyptiaca* biomass, and number of *P. aegyptiaca*-tomato root attachments were quantified; (II) split-root biochar/no-biochar experiments under hydroponic growing conditions performed in polyethylene bags with tomato plant rootings, wherein *P. aegyptiaca* seed germination percentage and radicle attachment numbers were quantified; and (III) germination trials, wherein the effect of biochar adsorption of GR-24 (artificial germination stimulant) on *P. aegyptiaca* seed germination was quantified. Addition of biochar to the pot soil (Experiment I) resulted in lower levels of *P. aegyptiaca* infection in the tomato plants, mainly through a decrease in the number of *P. aegyptiaca* attachments. This led to improved tomato plant growth. In Experiment II, *P. aegyptiaca* seed germination percentage decreased in the biochar-treated root zone as compared with the no-biochar control root zone; *P. aegyptiaca* radicle attachment numbers decreased accordingly. This experiment showed that biochar did not induce a systemic change in the activity of the stimulant molecules exuded by the tomato roots, toxicity to the radicles, or a change in the ability of the radicles to penetrate the tomato roots. The major cause for the decrease in germination percentage was physical adsorption of the stimulant molecule by the biochar (Experiment III). Adding biochar to soil to reduce infections by root parasitic weeds is an innovative means of control with the potential to become an important strategy both for non-chemical treatment of this family of pests, and for enhancing the economic feasibility of the pyrolysis/biochar platform. This platform is often viewed as one of a handful of credible strategies for helping to mitigate climate change.

Keywords: biochar, tomato, root parasitic weeds, broomrape, ecological control, climate change mitigation

INTRODUCTION

The pyrolysis/biochar platform has enjoyed considerable attention in recent years because it has the potential to convert organic wastes into renewable energy, sequester atmospheric carbon, and improve soil fertility. Yet, despite the intense and continually expanding scientific and industrial interest in this platform, it is still in an immature state, in part because much remains to be learned regarding its potential for restoring soil health and functioning. While the addition of biochar to soils has, on average, a positive influence on crop growth and productivity, some systems demonstrate negative or no effect of soil-added biochar (Jeffery et al., 2011, 2017; Biederman and Harpole, 2013; Crane-Droesch et al., 2013). The mechanisms responsible for such inconsistent results are far from understood. Adding biochar to soil is known to create changes in the plant/root zone/soil system that can affect plant performance and health by virtue of a number of inter-related physical, chemical and biochemical processes (Graber et al., 2014a), including (i) nutrient supply and balance; (ii) soil pH and redox, (iii) soil structure and functioning; (iv) microbial community structure; (v) plant–microbe signaling; and (vi) release/sequestration of phytotoxic, biotoxic, or bioactive compounds.

Biochar-elicited changes at the soil/root interface have been found to play a role in activating plants' innate defenses against disease-causing microbial pathogens along multiple hormone pathways (Meller Harel et al., 2012; Mehari et al., 2015). Frequently, impacts on both plant resistance to disease and on plant growth are related to biochar dose, with biochar exhibiting a hormone-like effect of low dose stimulation and high dose inhibition (Graber et al., 2014a; Jaiswal et al., 2015). Changes in microbial community structure, functioning and diversity caused by biochar additions are implicated in some of these effects (Kolton et al., 2017). One of the ways in which biochar can alter microbial community structure and dynamics is by adsorbing compounds involved in bacterial intercellular signaling (Masiello et al., 2013).

Generally, the adsorption ability of biochar greatly exceeds that of native adsorbing phases in soils (Graber and Kookana, 2015). Adding a strong adsorbent such as biochar to the soil thus can have considerable and varied impacts on the soil system. This has been documented with respect to soil contaminants, pest control products, and bacterial signaling molecules. The question arises whether soil-applied biochar can interfere with plant–plant or plant–bacteria communications, much the way it interferes with bacteria–bacteria signaling, via adsorption deactivation of signaling molecules.

One type of subsurface plant–plant interaction promoted by signaling molecules is germination of root parasitic weed seeds induced by chemicals released into the rhizosphere by the roots of the host plants (Yoneyama et al., 2010). Seed germination that relies on a host-specific signaling molecule precludes the possibility that the seeds will germinate in the absence of the host. This is crucial because the parasitic weed embryo will die if its radicle does not attach itself to a host root within a distance of only a few millimeters (Joel and Bar, 2013). Any process that interferes with host-parasite signaling, such as adsorption of the signaling molecule, or a change in the host production of the signaling molecule, could influence germination of parasitic weed seeds and hence, the intensity of infection.

To the best of our knowledge, the effect of biochar on weed growth has been studied only in autotrophic (Major et al., 2005; Adams et al., 2013; Soni et al., 2014) and hemiparasitic weed species (Smith and Cox, 2014). Neither of these types requires host-specific signaling molecules for seed germination. In contrast, root parasitic weed seeds can only germinate when they sense the presence of the host-exuded signaling molecules.

Lacking photosynthetic activity, root parasitic weeds rely wholly on the host plant for their germination, nutrition and water (Parker, 2013). They are a serious threat to many important crops, including tomato, potato, sunflower, canola, fava bean, pea, carrot and more. By and large, there are only a few effective treatment options for such weeds, and these options are neither economically viable nor ecologically sound. For example, *Phelipanche aegyptiaca* in open field tomatoes and *Orobanche cumana* in sunflower can be effectively controlled with herbicides delivered via drip irrigation system or by soil fumigation, but doing so requires advanced technologies (Eizenberg et al., 2012) and is expensive (700 to 7,000€ per ha). Today, no solutions for *P. aegyptiaca* management in organic farming and in greenhouse tomatoes are available, particularly in the Mediterranean basin (e.g., Turkey, Greece, Italy, Spain, Morocco and Israel), where infestation with *P. aegyptiaca* and *P. ramosa* is rampant (Parker, 2013). Both organic tomatoes and greenhouse tomatoes are attractive high cash crops for farmers.

The current study represents the first test of whether soil-applied biochar impacts the germination, attachment, and development of root parasitic weeds. *P. aegyptiaca* PERS. (Egyptian broomrape) in tomato (*Solanum lycopersicum*) was selected as the test case. Specifically, we evaluated three hypotheses, each of them involving a mechanism that could affect the extent of broomrape infection under biochar application: (i) biochar adsorbs host-exuded signaling molecules; (ii) biochar activates plants' innate system-wide defenses against invasion by the parasite; and (iii) biochar has a systemic influence on the amount of seed germination stimulant produced or released by the host plant.

MATERIALS AND METHODS

Biochars

Three types of biochar were tested: (i) biochar produced from wastes of greenhouse pepper plants in a modified slow pyrolysis unit (All Power Labs, San Francisco, CA, United States) at a highest treatment temperature (HTT) of c. 350°C, designated herein as GHW-350 (lab no. B8, GHW-350.5); (ii) biochar produced from the same wastes in the same unit at an HTT of c. 600°C; designated herein as GHW-600 (lab no. B9, GHW-600.2); and (iii) biochar produced from shredded date palm fronds using a home-made top-lit up-draft (TLUD) gasifier based on the ELSA design of Blucomb (Udine, Italy[1]) at an HTT of c.

[1] www.blucomb.com

650°C; designated herein as PALM-650 (lab no. D3, PALM-650.1). The two biochar feedstocks are common agricultural wastes in Israel and are candidates for pyrolysis solutions. These wastes are produced in the same geographical areas as are crops suffering from root parasitic weed infestations. Physical and chemical characteristics of the biochars [specific surface area (SSA), iodine number, inorganic matter content, solution pH, solution electrical conductivity, inorganic C, organic C, and element analysis] were determined according to methods in Graber et al. (2014b) and are reported in **Table 1**.

Tomato-Broomrape Pot Trial

Tomato plants were grown in soil infested with *P. aegyptiaca* seeds and treated with various levels of different biochars to determine if the biochars had any effect on *P. aegyptiaca* attachment and biomass, and tomato plant development. Tomato (*S. lycopersicum*) cv. 4343 (non-terminated type; Adama, Israel) was used. *P. aegyptiaca* inflorescences were collected from a broomrape-infested tomato field (Mevo Hama, Israel). Prior to the pot trial, *P. aegyptiaca* seed germination rates were confirmed to be 84% when treated with 10^{-6} M synthetic germination stimulant, GR-24.

Biochars were mixed with a naturally fertile, clay-rich soil (55% clay, 25% silt, 20% sand, 2% organic matter, pH 7.2) from the north of Israel in a cement mixer at 0, 0.15, 0.3, 0.9, and 1.5% by dry weight. *P. aegyptiaca* seeds were mixed with the dry soil-biochar mixtures at an infestation rate of 15 mg seeds/kg soil mixture. Three control treatments were included: (i) no biochar and no *P. aegyptiaca* infestation; (ii) no biochar and *P. aegyptiaca* infested; and (iii) biochar at 1.5% and no *P. aegyptiaca* infestation.

Forty-day old tomato seedlings were transplanted into 2 L pots filled with the various soil mixtures and grown between April and July 2014 in a net house with irrigation as required. No additional nutrients or chemicals were added. The pots were arranged randomly in a single block. Sixty days after the seedlings were transferred, the soil was removed, tomato roots were washed, and tomato and *P. aegyptiaca* dry above-ground biomass were determined gravimetrically after drying to a constant weight at 60°C. *P. aegyptiaca* attachments were gently removed from the tomato roots and counted. A repeat experiment was conducted between July and October 2014.

TABLE 1 | Physical and chemical characteristics of the biochars.

Characteristic	Units	GHW-350	GHW-600	PALM-650
Specific surface area (SSA)	m^2/g	1.3	2.9	9.0
Iodine number	mg/g	*n.a.*$^+$	93	108
Mineral content (Ash)	%	61.0	77.5	63.2
pH$^\#$	Units	8.5	10.5	8.5
EC$^\#$	mS/s	7.9	4.2	4.1
C inorganic	%	4.1	6.7	2.2
C organic (C_{org})*	%	44.0	24.4	62.9
H*	%	3.3	1.1	2.2
N*	%	3.5	1.6	1.4
S*	%	1.7	1.0	1.0
H/C_{org}	Molar ratio	0.90	0.55	0.42

*The biochars have high inorganic mineral contents because the feedstocks from which they originated were grown under intense fertilization and irrigation with saline water (3.5–4.5 mS/s). GHW refers to foliage and stems of greenhouse pepper plants, and PALM refers to date palm fronds. $^+$n.a. – not amenable to determination. $^\#$pH and EC determined in 1:20 biochar:deionized water suspension. *C_{org}, H, N, and S are reported on an ash-free basis.*

The experimental design and statistical analyses were carried out according to Onofri et al. (2010). Experiments were arranged in a two-factorial design (biochar type and concentration) with six replications, and subjected to ANOVA testing. Because of non-homogeneity of the variance, data were log transformed. There was no significant experiment by treatment interaction, therefore, results of the two repeat experiments were combined and analyzed as 12 replications for each treatment. Means comparison was carried out by Tukey–Kramer Honestly Significant Difference (HSD) test at an α level of 0.05 computed by JMP software (version 7).

Split Root Experiment in Polyethylene Bags

Tomato seedlings were cultivated in polyethylene bags (PEB) in a split-root type of arrangement that enabled evaluation of whether biochar had any systemic impact on seed germination or radicle attachment to the root. The PEBs were prepared according to Parker and Dixon (1983) and modified as required for this study. The root systems of 30-day-old tomato plants were cut and the stems re-rooted in sterile water for 7 days. *P. aegyptiaca* seeds were surface-sterilized for 3 min in 70% ethanol followed by 1% sodium hypochlorite for 10 min, and were then washed four times with sterile water. The upper half of a 10 cm × 22 cm sheet of glass fiber filter paper (Whatman, GF/A, Whatman International Ltd., Maidstone, England) was uniformly covered with 5 mL of a well-mixed biochar (GHW-350)/water suspension at concentrations of 0, 0.05, and 0.1% (w/v). One re-rooted plant was mounted onto each prepared filter paper, with the upper part of the root zone on the biochar-containing portion of the filter, and the lower part of the root zone on the biochar-free portion of the filter. Sterilized *P. aegyptiaca* seeds were then evenly scattered across the filter paper at a concentration of 50 seeds per cm^2. The plant-containing sheets were inserted into a clear polyethylene bag to which 20 ml of sterilized half-strength Hoagland nutrient solution (Hoagland and Arnon, 1950) was added. The experimental setup is depicted schematically in **Figure 1**. The PEBs were placed in a growth chamber under a 16/8 h day/night condition at 24°C and were nourished twice weekly from the top of the bag with 20 to 30 ml half strength Hoagland nutrient solution as required.

After 14 and 21 days, a binocular microscope (DZ, Zeiss, Germany) at 40× magnification was used to detect seed germination and root attachment, respectively. The percentage of germinated seeds and number of radicle attachments along the main tomato root were recorded.

Two identical experiments were performed in series; in both, the PEBs were arranged in completely randomized design in four replicates. No experiment–treatment interaction was detected; therefore, results of experiments were combined and analyzed as 8 replicates by 2-way ANOVA. Means comparison was carried

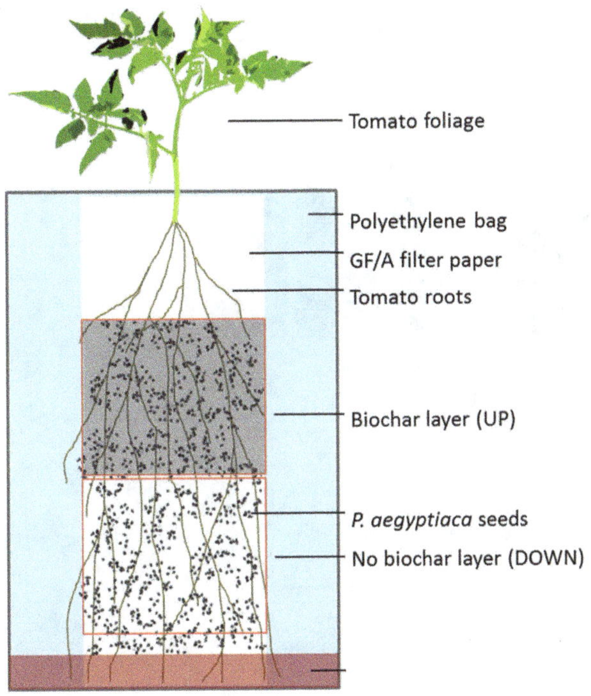

FIGURE 1 | Schematic representation of the split-root polyethylene bag experimental design.

out by Tukey–Kramer HSD test at an α level of 0.05 computed by JMP software (version 7).

Activity of the Synthetic Stimulant GR-24 as Affected by Adsorption on Biochar

GR-24 is a standard artificial stimulant used to evaluate parasitic weed seed germination under laboratory conditions. Typically, the germination rate in response to GR-24 is higher than the germination rate in the natural environment. The ability of biochar to adsorb the seed stimulant molecule, GR-24, and hence interfere with seed germination, was evaluated using a *P. aegyptiaca* seed germination test. *P. aegyptiaca* seeds were surface-sterilized as before. About 50–70 seeds were scattered on the surface of a GF/A disk (9 mm diameter). A total of 48 disks were prepared, the disks were placed in sterile plastic Petri dishes (50 mm diameter) at 4 disks per dish (12 total dishes). Each disk was wetted with 32 μl sterile water. The Petri dishes were closed and then sealed with Parafilm and covered completely in aluminum foil; the seeds were thusly conditioned at 22°C for 7 days. Following the conditioning phase, the seed-bearing disks were blotted and transferred to new Petri dishes, 4 per dish, for subsequent treatment by the test solutions.

Sixteen different treatment solutions were prepared by adding finely ground biochar (GHW-350) at different levels (0, 0.005, 0.01, and 0.05% by weight) to solutions of GR-24 prepared in Millipore water at desired initial nominal GR-24 concentrations (10^{-7}, 10^{-8}, 10^{-9}, and 10^{-10} M). The biochar/GR-24 water suspensions were kept at pH 7 and shaken at room temperature for 2 h, and then filtered using Millipore PVDF 0.2 μm filters to remove the biochar. By maintaining a neutral pH, any potential alkaline deactivation of the GR-24 stimulant was avoided. Aliquots (26 μl) of the filtered treatment solutions were applied to the preconditioned seed disks (4 per petri dish), the dishes were sealed and incubated as before. Each petri dish was arranged as a block containing one disk each of the four filtered test solutions (0, 0.005, 0.01, and 0.05% by weight); there were three replicate blocks for each GR-24 concentration.

Phelipanche aegyptiaca seed germination was quantified after 7 days exposure to the GR-24 solutions by counting total and germinated seeds under a binocular microscope. Two identical experiments were performed consecutively, both arranged in block design in three replicates. No experiment–treatment interaction was detected; therefore, results of experiments were combined and analyzed as 6 replications by 2-way ANOVA in blocks. Means comparison was carried out by Tukey–Kramer HSD test at an α level of 0.05, computed by JMP software (version 7).

RESULTS

Pot Trials

Under non-infested conditions, the addition of GHW-600 biochar at 1.5% resulted in a significant increase in tomato plant biomass as compared with the no-biochar, non-infested control (**Figure 2**). The addition of the other two biochars (GHW-350 and PALM-650) at 1.5% had no significant impact on tomato plant growth compared with the no-biochar, non-infested control (**Figure 2**).

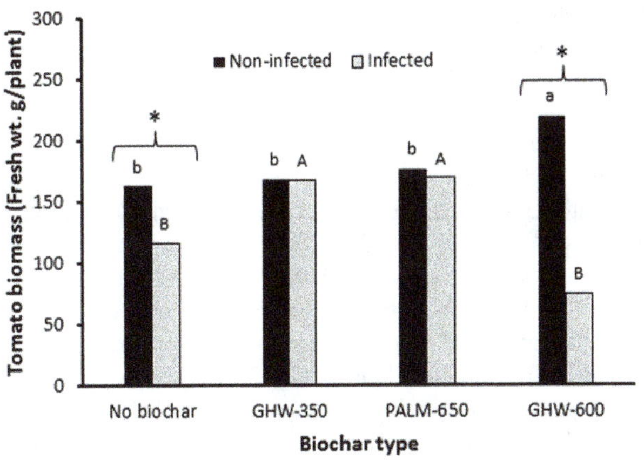

FIGURE 2 | Effect of biochar type (*X*-axis) and *P. aegyptiaca* (Egyptian broomrape) infection on tomato plant biomass (*Y*-axis). Histogram bars give the mean tomato plant wet weight in g per plant at biochar dose of 1.5 wt%. Means labeled by different lowercase letters (non-infected) are significantly different according to the Tukey HSD test (α = 0.05). Means labeled by different uppercase letters (infected) are significantly different according to the Tukey HSD test (α = 0.05). The asterisk (*) indicates a significant difference at α = 0.05 between the biomass of infected and non-infected tomato plants within a given biochar treatment.

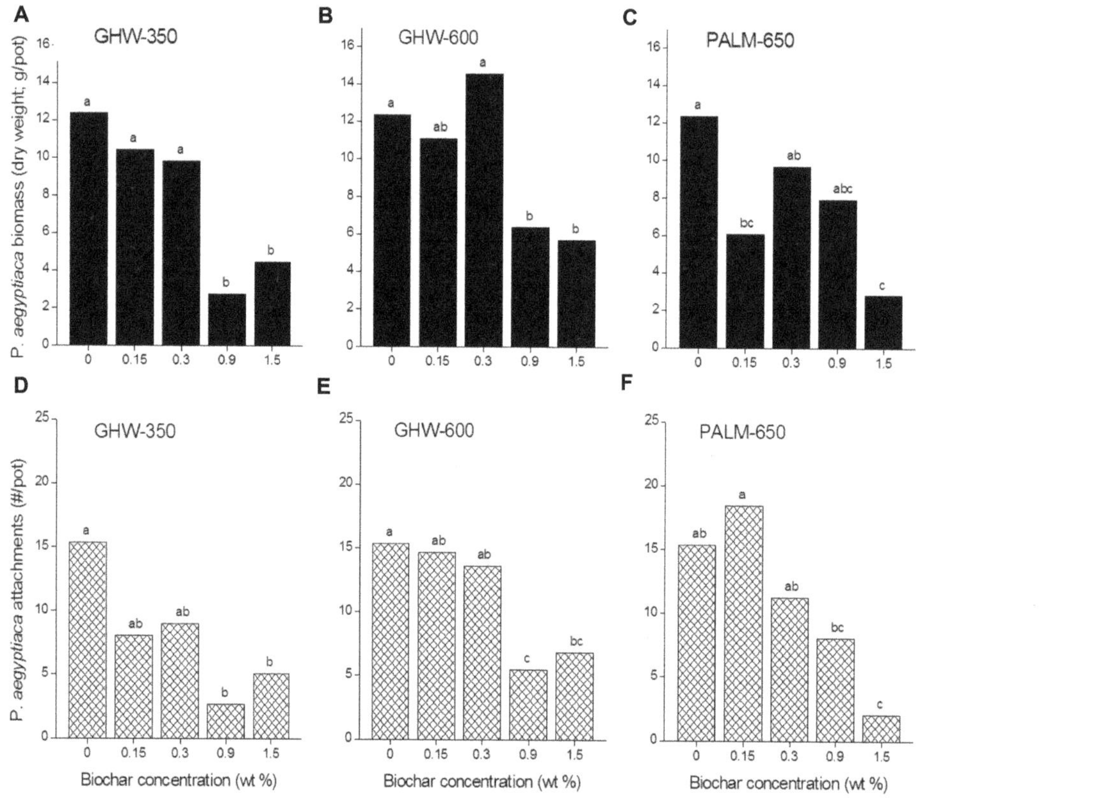

FIGURE 3 | *Phelipanche aegyptiaca* (Egyptian broomrape) infection parameters [biomass in dry weight in g/pot, **(A–C)**, and attachments in number/pot, **(D–F)**] versus biochar concentration (wt%) for the three different biochars. Experiments were arranged in a two-factorial design (biochar type and concentration) with six replications and subjected to ANOVA testing. Means comparison was carried out by Tukey–Kramer Honestly Significant Difference (HSD) test at an α value of 0.05 computed by JMP software (version 7). Means labeled by different letters are significantly different according to the HSD test.

Under *P. aegyptiaca* infestation, additions of GHW-350 and PALM-650 biochars resulted in increases in tomato plant biomass as compared with the no-biochar, infested control (**Figure 2**). In contrast, addition of the GHW-600 biochar did not improve tomato plant growth compared with the no-biochar, infested control treatment (**Figure 2**).

Compared with the no-biochar controls, all three biochars added at 1.5% caused a reduction in *P. aegyptiaca* biomass (**Figures 3A–C**), while both GHW biochars also reduced *P. aegyptiaca* biomass when added to the soil at 0.9% (**Figures 3A,B**). *P. aegyptiaca* attachments were reduced at 0.9 and 1.5% biochar doses for all three biochar types compared with the no-biochar controls (**Figures 3D–F**).

A negative relationship between number of attachments and biochar dose, pooled for all three biochars, was revealed (**Figure 4**). The higher biochar doses (0.9 and 1.5% by weight) effected a reduction in number of attachments to circa 35% of the number in the biochar-less control. Moreover, a positive linear relationship between number of attachments and *P. aegyptiaca* biomass was detected, with higher biochar dose treatments having fewer attachments and lower biomass (**Figure 5**). There was no general trend in efficacy amongst the three biochars (**Figure 5**).

Split-Root PEB Experiments

In the split root experiment, the percentage of *P. aegyptiaca* seeds that germinated was the same in both the upper and lower parts of the negative control treatment (no biochar on either upper or lower parts; **Figure 6**). Moreover, the germination percentage in the negative control was indistinguishable from that in the no-biochar lower parts of the test treatments (**Figure 6**). In the biochar-exposed upper parts, however, the germination percentage was strongly reduced compared with that in the no-biochar lower parts, and as compared with the negative control (no biochar on either upper or lower part; **Figure 6**). This same effect was observed at both levels of biochar addition (**Figure 6**). The number of radicle attachments (parasites per plant) was linearly related to the seed germination percentage across the whole root, including biochar-exposed and biochar-free parts (**Figure 7**). This linear relationship demonstrates that the biochar was not toxic to the sensitive radicle. The results also demonstrate there was no direct or systemic effect of biochar on the penetrability of the root membrane. Together, these results show that addition of biochar reduced seed germination, and it is the reduction in seed germination that caused the reduction in number of radicle attachments.

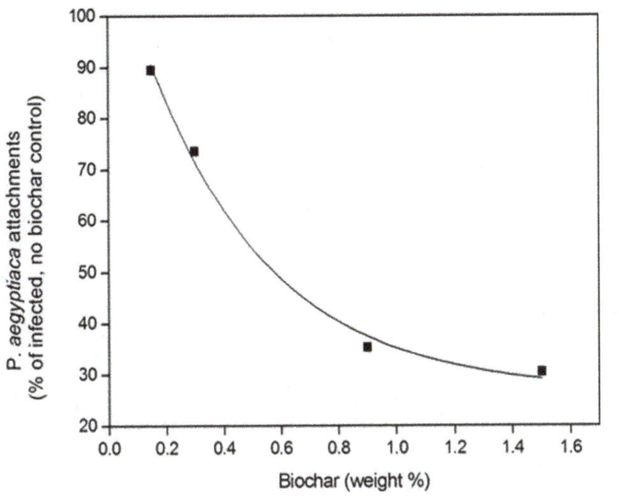

FIGURE 4 | *Phelipanche aegyptiaca* (Egyptian broomrape) attachments (number/pot) in the root zone relative to the no-biochar, infested control as a function of biochar dose. Results for the three biochar types are pooled. The fitted line is a first order exponential decay curve of the form $y = y_0 + Ae^{-x/t_1}$, where y_0 is 26.4, A is 91.1, and t_1 is 0.43, $R^2 = 0.995$.

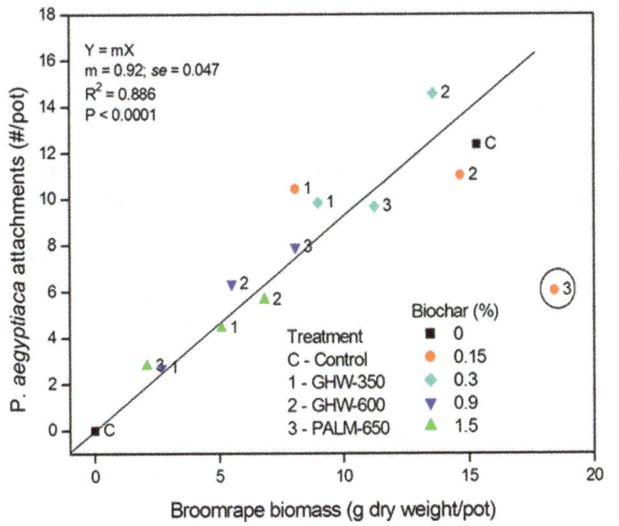

FIGURE 5 | *Phelipanche aegyptiaca* (Egyptian broomrape) attachments (number/pot) versus *P. aegyptiaca* biomass (dry weight in g/pot). Biochar concentrations are depicted in symbols of different colors and shapes (as given in legend), and each data point is labeled according to its biochar type (given in legend). One outlier (encircled red dot labeled with the number 3) is excluded from the regression.

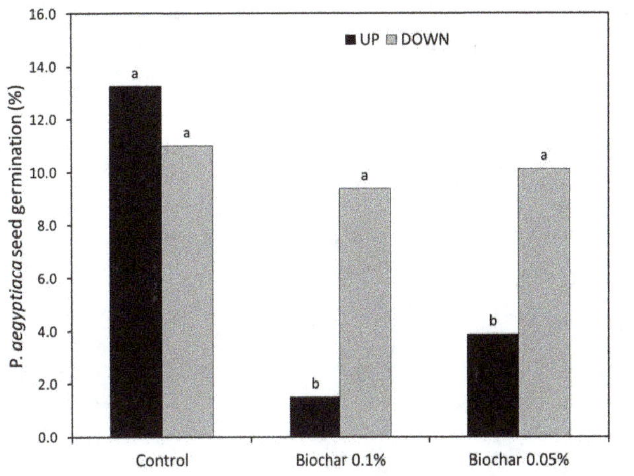

FIGURE 6 | Split-root experimental results. 'UP' signifies the upper portion of the root zone and 'DOWN' signifies the bottom portion of the root zone. In the Control treatment, neither UP nor DOWN were exposed to biochar. In the two biochar treatments (Biochar 0.1% and biochar 0.05%), the UP part was exposed to the specified concentration of biochar; the DOWN part was free of biochar. The *Y*-axis gives the percentage (%) of *P. aegyptiaca* (Egyptian broomrape) seeds that germinated. Results of two experiments were combined and analyzed as 8 replicates by 2-way ANOVA. Means comparison was carried out by Tukey-Kramer Honestly Significant Difference (HSD) test at an α level of 0.05 computed by JMP software (version 7). Means labeled by different letters are significantly different according to the HSD test.

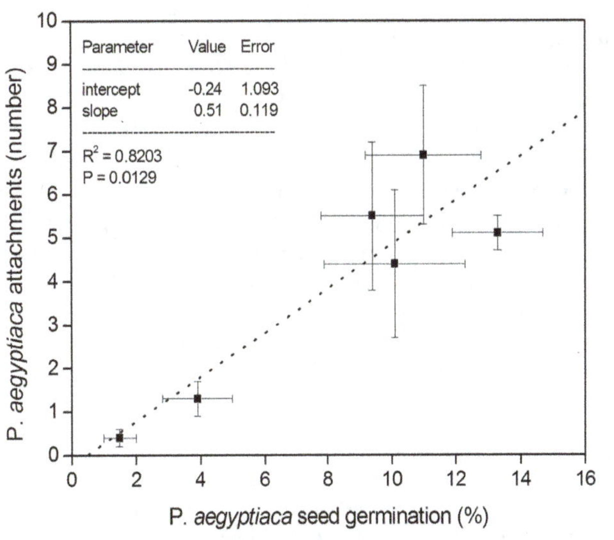

FIGURE 7 | Split-root experimental results. The *X*-axis presents the percentage (%) of *P. aegyptiaca* (Egyptian broomrape) seeds that germinated in both UP and DOWN portions of all three treatments; the *Y*-axis presents the number of attachments of *P. aegyptiaca* in both UP and DOWN portions of all three treatments. Error bars in both *X*- and *Y*-axis directions denote the standard error of the mean, represented by the data points.

Effect of GR-24 Sorption on Biochar on Germination of *P. aegyptiaca* Seeds

In the no-biochar control solutions of GR-24, the maximally effective concentrations of GR-24 were 10^{-8} and 10^{-7} M (78% of *P. aegyptiaca* seeds germinated; **Figure 8**). In the GR-24 solutions to which biochar had been added at the two highest doses (0.01 and 0.05% biochar), *P. aegyptiaca* seed germination was substantially reduced compared with the no-biochar control solutions (**Figure 8**). Indeed, at the highest biochar dose (0.05%), germination percentage was decreased by the equivalent of two

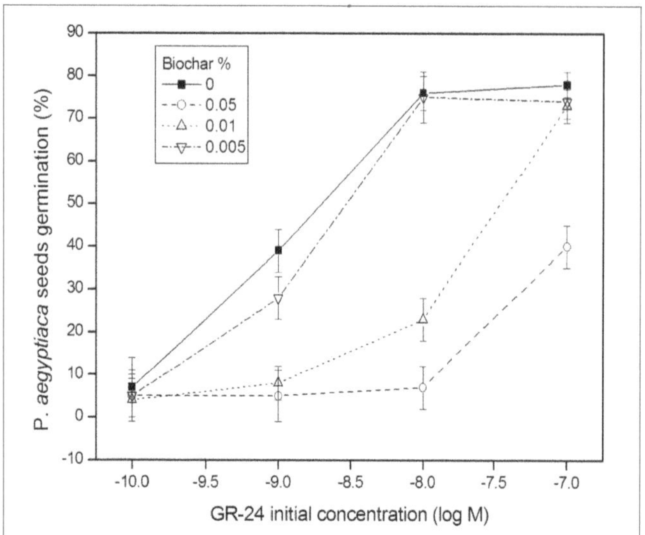

FIGURE 8 | *Phelipanche aegyptiaca* (Egyptian broomrape) seed germination percentage (*Y*-axis) as a function of GR-24 initial concentration (M) (*X*-axis) at different additions of GHW-350 biochar (%); mean and standard error (bars) are presented. Least significant difference (LSD) values (in %) based on the Tukey-Kramer HSD test ($P \leq 0.05$) are as follows: for GR24 concentration of 10^{-10} = 6.9; 10^{-9} = 11.4; 10^{-8} = 10.2; 10^{-7} = 14.4.

orders of magnitude in GR-24 concentration. This is apparent in **Figure 8**, where the same percentage of *P. aegyptiaca* seeds germinated in the solution having an initial concentration of 10^{-7} M GR-24 plus 0.05% biochar, as in the solution having an initial concentration of 10^{-9} M GR-24 and no biochar. Even the lowest biochar dose, 0.005%, resulted in a decrease in *P. aegyptiaca* seed germination by about 25% at an initial concentration of GR-24 of 10^{-9} M in comparison with the no-biochar added control. Insomuch as the pH in all the test solutions was held at a constant neutral value, it is apparent that the stimulant molecule was not alkaline-deactivated. The results show that reduced germination of *P. aegyptiaca* seeds in biochar-treated GR-24 solutions is a result of reduced GR-24 solution concentration due to its adsorption on the biochar.

DISCUSSION

In sum, the results shows that biochar added to the soil can reduce the extent of *P. aegyptiaca* infection in tomato, mainly by reducing *P. aegyptiaca* seed germination due to adsorption of the stimulant molecule on the biochar. The reduction in seed germination leads to a decrease in *P. aegyptiaca* attachments, which in turn leads to a decrease in *P. aegyptiaca* biomass. What is more, some types of biochar can contribute to improved tomato plant growth and development even in the event of *P. aegyptiaca* infestation. It was found that all three tested biochars exhibited similar abilities to reduce *P. aegyptiaca* attachments and biomass, which reflects their essential similarity in the physical and chemical attributes that are important for sorption, namely specific surface area and iodine number (**Table 1**).

In theory, there are a number of ways by which biochar could interfere with the germination of root parasitic weed seeds such as broomrape and *Striga* spp.: (1) Physically adsorb signaling molecules, preventing them from reaching the seeds; (2) Chemically inactivate the signaling molecules; (3) Release chemicals that are toxic to the seeds or interfere with stimulant receptor recognition; (4) Induce reduction in host production or exudation of the germination-signaling molecule; (5) Promote changes in the rhizosphere microbiome that affect root parasitic weed seed germination; and (6) Combinations of these factors. However, all but the first of these possible mechanisms can be eliminated from consideration in view of the results reported herein.

One of the outcomes of the split-root experiment is that biochar had no effect on the host production or exudation of the signaling molecule(s). This can be understood by virtue of the fact that in the no-biochar part of the split biochar/no-biochar systems, the extents of germination and attachment were identical to those in the negative control (no part of the root system exposed to biochar). Moreover, the linear relationship between germination percentage and number of radicle attachments across all the studied biochar concentrations (0, 0.05, and 0.1%) shows the biochar was not toxic to the radicle. Given that broomrape radicles are generally more sensitive to toxins than are broomrape seeds (Timko and Scholes, 2013), it can be assumed that the biochar also was not toxic to the seeds. It is also apparent from these results that the biochar had neither a systemic nor contact effect on the integrity of the root membrane.

Some rhizosphere microorganisms can affect root parasitic weed seed germination by competing for the signaling molecules, for example, arbuscular mycorrhizal fungi (AMF). Many of the compounds that activate root parasitic weed seed germination (such as flavonoids, sesquiterpenes, and strigolactones) are similar to or the same as those that stimulate growth and branching in AMF (Akiyama et al., 2005; Bucher et al., 2009). Biochar additions to soil are known to affect AMF abundance, and it was recently suggested that this is due to biochar adsorption of inhibitory or signaling compounds (Thies et al., 2015). While levels of microorganisms such as AMF that can compete with weed seeds for exuded stimulant molecules were not determined in these experiments, it is known that AMF do not develop in well-nourished semi-sterile hydroponic systems such as those employed (Gomez-Roldan et al., 2008). This means there is little likelihood that microorganisms such as AMF out-competed the seeds for the germination stimulant molecule.

The active chemical moiety of the synthetic stimulant GR-24 is the lactone ring, which can open at elevated pH and lose its stimulant activity (Yoneyama et al., 2013). However, pH was maintained at 7 in the adsorption/germination experiment. Thus, the totality of the experiments points toward physical adsorption of the signaling molecules by the biochar as being the major factor responsible for the reduction in germination of the root parasitic weed seeds.

Adding biochar to soil represents the first-ever ecological approach to root parasitic weed management. It is also suitable for organic agriculture. The potential of this discovery may be far-reaching. For example, the use of biochar may make it unnecessary to develop root parasitic weed resistant crops such as those based on the absence of *in planta* production of germination stimulant chemicals (Dor et al., 2009), particularly as such mutant plants often have irregular branching and growth patterns. It is possible that adding biochar can help offset temperature-dependent losses of natural resistance observed in various broomrape resistant varieties (Eizenberg et al., 2003). Reducing broomrape germination and parasitism by using biochar could potentially minimize the development of new virulent broomrape races (Pérez-de-Luque et al., 2009).

It is worthwhile noting that the use of biochar for ecological weed control will be effective only for root parasitic weeds that need chemical stimulation for germinating their seeds. When a stimulant is not required for seed germination, such as for autotrophic plants, adding biochar has not been an effective approach for weed control (Major et al., 2005). This is because, like with other non-parasitic plants, biochar mainly has a neutral or stimulating effect on autotrophic weed growth. Nor was biochar found to be an effective treatment for reducing the biomass of the hemiparasitic yellow rattle weed (Smith and Cox, 2014). The yellow rattle also does not require a host-derived stimulant molecule for germination of its seeds.

CONCLUSION

The finding that small additions of various biochars to soil can significantly reduce infection by the root parasitic weed, *P. aegyptiaca*, in tomato, portends an innovative means of ecological control over such pests and justifies testing additional host-parasite systems under horticultural and field conditions. The major mechanism responsible for biochar interference with parasitic weed seed germination is adsorption of the stimulant molecule. Hence, it will be straightforward to design biochars having high adsorption capacities specifically for this application. To date, the pyrolysis/biochar platform is widely viewed as a potentially important tool for global climate change mitigation; however, it still does not enjoy widespread implementation, in part because the benefits of biochar addition to soil are not yet well-understood, and in part because the costs of biochar use in agriculture are still too high. The ability of biochar to decrease infection by parasitic weeds in important crops may change treatment strategies for this family of parasites, and also enhance the economic feasibility of biochar use in agriculture, and as a result, of the pyrolysis/biochar platform in its entirety.

AUTHOR CONTRIBUTIONS

HE was the lead designer of the experiments, and an integral part of result analysis and interpretation, drafting, revising and approving the manuscript, and is accountable for the accuracy of the reported experimental data. DP was responsible for performing the split-root and seed germination experiments, HZ was responsible for performing the pot trials, and LT was responsible for performing the physical and chemical characterization of the biochars. DP, HZ, and LT acquired and analyzed the data for which they were responsible, are accountable for its accuracy, and reviewed the manuscript for accuracy. EG was responsible for conceiving the research project, results analysis and interpretation, drafting, revising and approving the manuscript, and is accountable for the data reported inside.

ACKNOWLEDGMENTS

This research was supported by grants from the Israel–Italy Program 2011 (project 301-750-11) funded by the Chief Scientist of the Israel Ministry of Agriculture and Rural Development, and jointly by the Plant Production and Marketing Board and Chief Scientist of the Israel Ministry of Agriculture and Rural Development (grant no. 132-1499-11).

REFERENCES

Adams, M. M., Benjamin, T. J., Emery, N. C., Brouder, S. J., and Gibson, K. D. (2013). The effect of biochar on native and invasive prairie plant species. *Invasive Plant Sci. Manag.* 6, 197–207. doi: 10.1614/IPSM-D-12-00058.1

Akiyama, K., Matsuzaki, K.-I., and Hayashi, H. (2005). Plant sesquiterpenes induce hyphal branching in arbuscular mycorrhizal fungi. *Nature* 435, 824–827. doi: 10.1038/nature03608

Biederman, L. A., and Harpole, W. S. (2013). Biochar and its effects on plant productivity and nutrient cycling: a meta-analysis. *GCB Bioenergy* 5, 202–214. doi: 10.1111/gcbb.12037

Bucher, M., Wegmüller, S., and Drissner, D. (2009). Chasing the structures of small molecules in arbuscular mycorrhizal signaling. *Curr. Opin. Plant Biol.* 12, 500–507. doi: 10.1016/j.pbi.2009.06.001

Crane-Droesch, A., Abiven, S., Jeffery, S., and Torn, M. S. (2013). Heterogeneous global crop yield response to biochar: a meta-regression analysis. *Environ. Res. Lett.* 8:044049. doi: 10.1088/1748-9326/8/4/044049

Dor, E., Alperin, B., Wininger, S., Ben-Dor, B., Somvanshi, V. S., Koltai, H., et al. (2009). Characterization of a novel tomato mutant resistant to the weedy parasites *orobanche* and *phelipanche* spp. *Euphytica* 171, 371–380. doi: 10.1007/s10681-009-0041-2

Eizenberg, H., Aly, R., and Cohen, Y. (2012). Technologies for smart chemical control of broomrape (*Orobanche* spp. and *Phelipanche* spp.). *Weed Sci.* 60, 316–323. doi: 10.1614/WS-D-11-00120.1

Eizenberg, H., Plakhine, D., Hershenhorn, J., Kleifeld, Y., and Rubin, R. (2003). Resistance to broomrape (*Orobanche* spp.) in sunflower (*Helianthus annuus* L.) is temperature-dependent. *J. Exp. Bot.* 54, 1305–1311. doi: 10.1093/jxb/erg129

Gomez-Roldan, V., Fermas, S., Brewer, P. B., Puech-Pages, V., Dun, E. A., Pillot, J.-P., et al. (2008). Strigolactone inhibition of shoot branching. *Nature* 455, 189–194. doi: 10.1038/nature07271

Graber, E. R., Frenkel, O., Jaiswal, A. K., and Elad, Y. (2014a). How may biochar influence severity of diseases caused by soilborne pathogens? *Carbon Manag.* 5, 169–183. doi: 10.1080/17583004.2014.913360

Graber, E. R., and Kookana, R. (2015). "Biochar and retention/efficacy of pest control products," in *Biochar for Environmental Management: Science and Technology*, eds J. Lehmann and S. Joseph (Abingdon: Routledge), 655–678.

Graber, E. R., Tsechansky, L., Lew, B., and Cohen, E. (2014b). Reducing capacity of water extracts of biochars and their solubilization of soil Mn and Fe. *Eur. J. Soil Sci.* 65, 162–172. doi: 10.1111/ejss.12071

Hoagland, D. R., and Arnon, D. I. (1950). *The Water-Culture Method for Growing Plants without Soil.* Berkeley, CA: University of California.

Jaiswal, A. K., Frenkel, O., Elad, Y., and Graber, E. R. (2015). Non-monotonic influence of biochar dose on bean seedling growth and susceptibility to rhizoctonia solani: the shifted Rmax-effect. *Plant Soil* 395, 125–140. doi: 10.1007/s11104-014-2331-2

Jeffery, S., Abalos, D., Prodana, M., Bastos, A. C., van Groenigen, J. W., Hungate, B. A., et al. (2017). Biochar boosts tropical but not temperate crop yields. *Environ. Res. Lett.* 12:053001. doi: 10.1088/1748-9326/aa67bd

Jeffery, S., Verheijen, F. G. A., van der Velde, M., and Bastos, A. C. (2011). A quantitative review of the effects of biochar application to soils on crop productivity using meta-analysis. *Agric. Ecosyst. Environ.* 144, 175–187. doi: 10.1016/j.agee.2011.08.015

Joel, D. M., and Bar, H. (2013). "The seed and the seedling," in *Parasitic Orobanchaceae, Parasitic Mechanisms and Control Strategies*, eds D. M. Joel, J. Gressel, and L. J. Musselman (Berlin: Springer), 147–166.

Kolton, M., Graber, E. R., Tsehansky, L., Elad, Y., and Cytryn, E. (2017). Biochar-stimulated plant performance is strongly linked to microbial diversity and metabolic potential in the rhizosphere. *New Phytol.* 213, 1393–1404. doi: 10.1111/nph.14253

Major, J., Steiner, C., Ditommaso, A., Falcao, N. P. S., and Lehmann, J. (2005). Weed composition and cover after three years of soil fertility management in the central Brazilian Amazon: compost, fertilizer, manure and charcoal applications. *Weed Biol. Manag.* 5, 69–76. doi: 10.1111/j.1445-6664.2005.00159.x

Masiello, C. A., Chen, Y., Gao, X., Liu, S., Cheng, H.-Y., Bennett, M. R., et al. (2013). Biochar and microbial signaling: production conditions determine effects on microbial communication. *Environ. Sci. Technol.* 47, 11496–11503. doi: 10.1021/es401458s

Mehari, Z. H., Elad, Y., Rav-David, D., Graber, E. R., and Meller Harel, Y. (2015). Induced systemic resistance in tomato (*Solanum lycopersicum*) against botrytis cinerea by biochar amendment involves jasmonic acid signaling. *Plant Soil* 395, 31–44. doi: 10.1007/s11104-015-2445-1

Meller Harel, Y., Elad, Y., Rav-David, D., Borenshtein, M., Schulcani, R., Lew, B., et al. (2012). Biochar mediates systemic response of strawberry to foliar fungal pathogens. *Plant Soil* 357, 245–257. doi: 10.1007/s11104-012-1129-3

Onofri, A., Carbonell, E. A., Piepho, H. P., Mortimer, A. M., and Cousens, R. D. (2010). Current statistical issues in weed research. *Weed Res.* 50, 5–24. doi: 10.1111/j.1365-3180.2009.00758.x

Parker, C. (2013). "The parasitic weeds of the orobanchaceae," in *Parasitic Orobanchaceae, Parasitic Mechanisms and Control Strategies*, eds D. M. Joel, J. Gressel, and L. J. Musselman (Berlin: Springer), 313–344.

Parker, C., and Dixon, N. (1983). The use of polyethylene bags in the culture and study of *Striga* spp. and other organisms on crop roots. *Ann. Appl. Biol.* 103, 485–488. doi: 10.1111/j.1744-7348.1983.tb02787.x

Pérez-de-Luque, A., Fondevilla, S., Pérez-Vich, B., Aly, R., Thoiron, S., Simier, P., et al. (2009). Understanding broomrape-host plant interaction and developing resistance. *Weed Res.* 49, 8–22. doi: 10.1111/j.1365-3180.2009.00738.x

Smith, R. G., and Cox, D. A. (2014). Effects of soil amendments on the abundance of a parasitic weed, yellow rattle (*Rhinanthus minor*) in hay fields. *Weed Sci.* 62, 118–124. doi: 10.1614/WS-D-13-00106.1

Soni, N., Leon, R. G., Erickson, J. E., Ferrell, J. A., Silveira, M. L., and Giurcanu, M. C. (2014). Vinasse and biochar effects on germination and growth of palmer amaranth (*Amaranthus palmeri*), sicklepod (*Senna obtusifolia*), and southern crabgrass (*Digitaria ciliaris*). *Weed Technol.* 28, 694–702. doi: 10.1614/WT-D-14-00044.1

Thies, J., Rilling, M., and Graber, E. R. (2015). "Biochar effects on the abundance, activity and diversity of the soil biota," in *Biochar for Environmental Management: Science and Technology*, eds J. Lehmann and S. Joseph (Abingdon: Routledge), 327–390.

Timko, M. P., and Scholes, J. D. (2013). "Host reaction to attack by root parasitic plants," in *Parasitic Orobanchaceae: Parasitic Mechanisms and Control Strategies*, eds D. M. Joel, J. Gressel, and L. J. Musselman (Berlin: Springer), 115–141. doi: 10.1007/978-3-642-38146-1_7

Yoneyama, K., Awad, A. A., Xie, X., Yoneyama, K., and Takeuchi, Y. (2010). Strigolactones as germination stimulants for root parasitic plants. *Plant Cell Physiol.* 51, 1095–1103. doi: 10.1093/pcp/pcq055

Yoneyama, K., Ruyter-Spira, C., and Bouwmeester, H. (2013). "Induction of germination," in *Root Parasitic Orobanchaceae: Parasitic Mechanisms and Control Strategies*, eds D. Joel, J. Gressel, and L. J. Musselman (New York, NY: Springer), 166–194.

Conflict of Interest Statement: The authors declare that the research was conducted in the absence of any commercial or financial relationships that could be construed as a potential conflict of interest.

13

The Effect of *Orobanche crenata* Infection Severity in Faba Bean, Field Pea, and Grass Pea Productivity

Edited by:
Edmundo Acevedo,
University of Chile, Chile

Reviewed by:
Grama Nanjappa Dhanapal,
University Of Agricultural Sciences,
Bangalore, India
Ilias Travlos,
Agricultural University of Athens,
Greece

***Correspondence:**
Mónica Fernández-Aparicio
monica.fernandez@ias.csic.es

Mónica Fernández-Aparicio[1,2], Fernando Flores[3] and Diego Rubiales[1]*

[1] Institute for Sustainable Agriculture, Consejo Superior de Investigaciones Científicas, Córdoba, Spain, [2] INRA, UMR1347 Agroécologie, Dijon, France, [3] Escuela Técnica Superior de Ingeniería – Universidad de Huelva, Palos de la Frontera, Spain

Broomrape weeds (*Orobanche* and *Phelipanche* spp.) are root holoparasites that feed off a wide range of important crops. Among them, *Orobanche crenata* attacks legumes complicating their inclusion in cropping systems along the Mediterranean area and West Asia. The detrimental effect of broomrape parasitism in crop yield can reach up to 100% depending on infection severity and the broomrape-crop association. This work provides field data of the consequences of *O. crenata* infection severity in three legume crops, i.e., faba bean, field pea, and grass pea. Regression functions modeled productivity losses and revealed trends in dry matter allocation in relation to infection severity. The host species differentially limits parasitic sink strength indicating different levels of broomrape tolerance at equivalent infection severities. Reductions in host aboveground biomass were observed starting at low infection severity and half maximal inhibitory performance was predicted as 4.5, 8.2, and 1.5 parasites per faba bean, field pea, and grass pea plant, respectively. Reductions in host biomass occurred in both vegetative and reproductive organs, the latter resulting more affected. The increase of resources allocated within the parasite was concomitant to reduction of host seed yield indicating that parasite growth and host reproduction compete directly for resources within a host plant. However, the parasitic sink activity does not fully explain the total host biomass reduction because combined biomass of host–parasite complex was lower than the biomass of uninfected plants. In grass pea, the seed yield was negligible at severities higher than four parasites per plant. In contrast, faba bean and field pea sustained low but significant seed production at the highest infection severity. Data on seed yield and seed number indicated that the sensitivity of field pea to *O. crenata* limited the production of grain yield by reducing seed number but maintaining seed size. In contrast, the size of individual parasites was not genetically determined but dependent on the host species and resource availability as a consequence of competition between parasites at increasing infection severities.

Keywords: parasitic weed damage, legume, broomrape, resource allocation, weed threshold density

INTRODUCTION

Broomrape weeds (*Orobanche* and *Phelipanche* species) are root-holoparasitic plants that possess extreme competitive ability against the crop. Rather than to compete with crops for field resources, their haustorial cells penetrate crop roots to directly divert water and nutritive resources (Parker and Riches, 1993). Broomrape weeds attack dicotyledonous crops along Mediterranean, central and Eastern Europe, and Asia (Parker, 2009). *Orobanche crenata* Forsk is a major constraint for grain and forage legume on over 4 Mha of the Mediterranean area (Parker, 2009). The increasing interest on sustainable agriculture promotes the cultivation of legumes as a tool of ecological optimization of resource use and promotion of pest resilience in cropping systems. Due to the severe effects of *O. crenata* parasitism in the host crop, and the high persistence of parasitic seedbank in agricultural soils it has been the cause of abandonment of legume cultivation in important cropping areas (Mesa-Garcia and Garcia-Torres, 1984; Sauerborn, 1991; Parker and Riches, 1993; Rubiales et al., 2009).

Many of the broomrape traits such as achlorophyllous nature, underground parasitism, the physical and metabolic overlap with the crop, or lack of functional roots, reduce the efficiency of conventional programs in weed management aimed to their control (Fernández-Aparicio et al., 2016). In order to achieve efficient control, broomrape parasitism should be targeted at different fronts by an integrated management strategy (Kebreab and Murdoch, 2001). Infection severity in broomrape strongly depends on environmental factors such as temperature and parasitic seedbank density (Linke et al., 1991; Eizenberg et al., 2005). For the determination of thresholds of weed density above which it is profitable to apply control measurements, equations correlating yield losses to broomrape infection severity are needed (Garcia-Torres et al., 1996) but their determination is spare in the majority of crops affected. No previous work has estimated the effect of *O. crenata* infection in field pea and grass pea. The *O. crenata* effect on faba bean productivity was estimated in pots by Manschadi et al. (1996) at three successively increased seedbank densities, in which even at the lowest seed density, a high number of broomrapes was attached per host plant, while at the other two increased densities the parasite number was observed so high that lead to faba bean death before maturity. In contrast, a field study by Mesa-Garcia and Garcia-Torres (1984) provided an accurate prediction system for expected damage by studying the range of infections normally found in the field. Caution should be exercised in the extrapolation of host and parasite growth in pots to real farming conditions, as the rapid saturation in parasite-carrying capacity occurring in pots complicates the determination of the host response curves related to infection severity.

Our study was conducted to determine the consequences in crop productivity of *O. crenata* parasitism at successively increasing infection severities in three highly susceptible legume crops: faba bean (*Vicia faba* L.), field pea (*Pisum sativum* L.), and grass pea (*Lathyrus sativus* L.). The experimentation was carried out in field conditions with a natural parasitic seed bank in south of Spain, an area where *O. crenata* is endemic attacking legume crops.

MATERIALS AND METHODS

Plant Material

One susceptible cultivar of each of the three species faba bean (cv. Prothabon), field pea (cv. Messire), and grass pea (cv. Lisa) were chosen to evaluate their response to increasing levels of *O. crenata* parasitism. A natural occurring parasitic seedbank of *O. crenata* was infesting the field used in the experiments.

Site and Experimental Design

The experimental work was carried out in Córdoba, (Alameda del Obispo Farm), southern Spain, on a deep loam soil (typic xerofluvent). The precrop on the experimental site was faba bean-wheat- field pea rotation. Average (30 years) annual precipitation and air temperature in the area were 536.0 mm and 17.6°C, respectively, with maximum and minimum daily air temperature of 46.6°C and −7.8°C. Each host species was laid out in 15×4 m^2 experimental plots, in a complete randomized block design with three replicates. Each crop was hand sown in November, each plant separated 50 cm apart inside rows with 50 cm distance between rows and 5 cm sowing depth. Legumes were rain feed, and hand weeding of weeds other than broomrape was carried out when required.

The field was deliberately chosen by its heterogeneous distribution of *O. crenata* seed bank indirectly observed in previous seasons by a patchy distribution of the parasite in susceptible legume plants. Consequently, host plants inside each plot suffered variation of infection severity measured as number of emerged broomrapes per host plant. At the end of cultivation cycle, data on emerged broomrapes ranging from 0 to 11 parasites per host plant was taken in 750 host plants. All broomrapes emerged per host plant were extracted from the host root by a gentle pull and counted. Host and parasitic tissue was collected per each host plant and carried to the laboratory. Samples were dried at 80°C during 48 h and each biomass compartment weighed independently per host plant.

Calculation and Statistics

O. crenata infection severity was estimated per individual host plant in each crop species as **number of emerged broomrapes per plant**. The distribution of dry matter within the host–parasite complex was determined by recording five parameters per host plant: **combined biomass** (host and broomrape biomass), **aboveground host dry matter, host reproductive dry matter** (host seeds), **host vegetative dry matter** (aboveground minus reproductive dry matter), and **cumulative broomrape dry matter** (total broomrape dry biomass per host plant). In addition, for field pea plants, **number of seeds** per plant was measured in each sampled plant.

Several parameters were calculated to characterize the effect of *O. crenata* parasitism in each crop. First, host **reproductive index** was determined taking into account only host tissue and it was calculated for each plant as the ratio between host reproductive dry matter and total aboveground host dry matter. Additional parameters were calculated to determine the biomass partitioning within the host–parasite complex: **relative host reproductive weight** (percentage of the combined biomass allocated into the host seeds); **relative parasitic weight** (percentage of the combined biomass allocated into the parasite); **relative weight of total sinks** (percentage of the combined biomass allocated into host seeds and in the parasite). In addition the **individual *O. crenata* weight** was estimated as parasitic biomass sustained by each host plant averaged by the number of parasites per plant. For field pea, the **individual seed weight** was calculated as host reproductive biomass averaged by the number of seeds.

Statistical analysis was performed using SAS (R) 9.3 (SAS Institute Inc.). Arcsine square root transformations of the data which did not meet the conditions of normality and homogeneous variance were performed to conform to the assumptions of analysis of variance (ANOVA). ANOVA was conducted on biomass data using a randomized design, to test for the significance of the infection severity.

In each host species, infection severity was regressed against all components in the system response to parasitism: total aboveground host biomass, host vegetative biomass, host reproductive biomass, number of seeds, average seed weight, combined biomass, relative parasitic weight, relative host reproductive sink, and relative weight of total sinks. In addition, number of parasites was regressed against individual parasitic weight average.

RESULTS AND DISCUSSION

In parasite-free plants, the average **aboveground dry matter** and **reproductive index** (ratio host seeds to aboveground host dry matter) were, respectively, 110.5 g and 53.6% for faba bean, 28.0 g and 53.5% for field pea and 43.2 g and 35.0% for grass pea plant (**Figures 1** and **2**). Our results are in agreement with previous studies showing that low reproductive index is characteristic of grass pea species (Siddique et al., 1996). During parasitism, broomrape plant acts as additional sink withdrawing water and assimilates from host vascular system. In the three legume species studied, aboveground host dry matter was consistently reduced by *O. crenata* parasitism in an infection severity-dependent fashion. Correlations between aboveground dry matter in the host and infection severity have been previously observed high in faba bean-*O. crenata* (Mesa-Garcia and Garcia-Torres, 1984), while was low or not significantly correlated in other crop-broomrape associations such as sunflower-*O. cernua* (Garcia-Torres et al., 1996), tobacco-*O. cernua* (Hibberd et al., 1998), and tomato-*P. ramosa* (Mauromicale et al., 2008). In our field data, the relation between aboveground host dry matter reduction and infection severity was best fitted to exponential functions where the downward slide began at low parasite densities and intensified at a biomass inhibition rate of 0.16, 0.09, and 0.82 g/g indicating that parasitism reduced aboveground host dry weight in faba bean, field pea, and grass pea by 16, 9, and 82% for weight unit, respectively. The severe reduction in grass pea biomass was linear up to three parasites per plant, after which there was not further decrease in host dry matter with increased parasite densities. Similar pattern was previously described in tomato infected with increased densities of the broomrape species *P. aegyptiaca* (Barker et al., 1996). In our field study, the infection severity with half maximal inhibitory performance was predicted as 4.5, 8.2, and 1.5 parasites per faba bean, field pea, and grass pea plant, respectively. Tolerance to broomrape weeds is defined by the capacity of a given crop genotype to endure broomrape infection with low productivity losses (Fernández-Aparicio et al., 2016). Our field data show that at any given level of infection severity, grass pea is the species less tolerant to *O. crenata* parasitism except at the highest parasite density, in which each legume species only achieved 30.0, 38.0, and 33.0% of the total aboveground dry matter of their respective faba bean, field pea, and grass pea uninfected counterparts (**Figures 1A–C**).

Figure 2 shows that the *O. crenata*-induced inhibition rate was more intense in reproductive than in vegetative dry matter measured at final harvest. Reductions in the reproductive index were previously observed in *O. minor*-red clover and *O. crenata*-faba bean by Mesa-Garcia and Garcia-Torres (1984), Manschadi et al. (1996), and Lins et al. (2007). The parasite density with half maximal inhibitory performance in faba bean, field pea, and grass pea was, respectively, predicted as 7.6, 8.2, and 4.8 parasites per plant for host vegetative dry matter and predicted as 3.5, 3.7, and 1.7 parasites per plant for host reproductive dry matter. Mesa-Garcia and Garcia-Torres (1984) predicted that four *O. crenata* per plant was the average infection severity responsible for the reduction of faba bean seed yield by half, being similar to our predictive system for faba bean seed reduction. It has been previously suggested that seed yield losses induced by *O. crenata* can reach up to 100% (Sauerborn, 1991). In grass pea, the seed yield was negligible at severities higher than four parasites per plant. In contrast, at the highest infection severity (11 broomrapes per plant) faba bean and field pea sustained a low but significant seed production.

The **cumulative dry biomass** of all broomrapes infecting one single plant was highly related with infection severity (**Figure 1**). Exponential rise functions were fitted to these data suggesting that biomass accumulation in parasites rise quickly at low parasite densities and the rate of increase starts to slow at densities higher than five parasites per plant regardless host species. High correlation between broomrape number and broomrape dry weight accumulated per host plant was previously observed by Mesa-Garcia and Garcia-Torres (1984), Barker et al. (1996), Garcia-Torres et al. (1996) but differed from that reported by Manschadi et al. (1996) and Hibberd et al. (1998). In faba bean, the maximum accumulation of parasitic biomass (55.18 ± 12.27) was higher than that observed in field pea (7.86 ± 1.04) and in grass pea (11.6 ± 1.67). For faba bean and grass pea the maximum accumulated parasitic dry weight was not significantly different from the seed yield produced by uninfected plants. On the contrary, for field pea the maximum parasitic dry matter was significantly lower than the dry matter of seeds in uninfected field pea plants.

Living as obligated holoparasites, broomrapes are deprived of autotrophy and in consequence infected host plants are

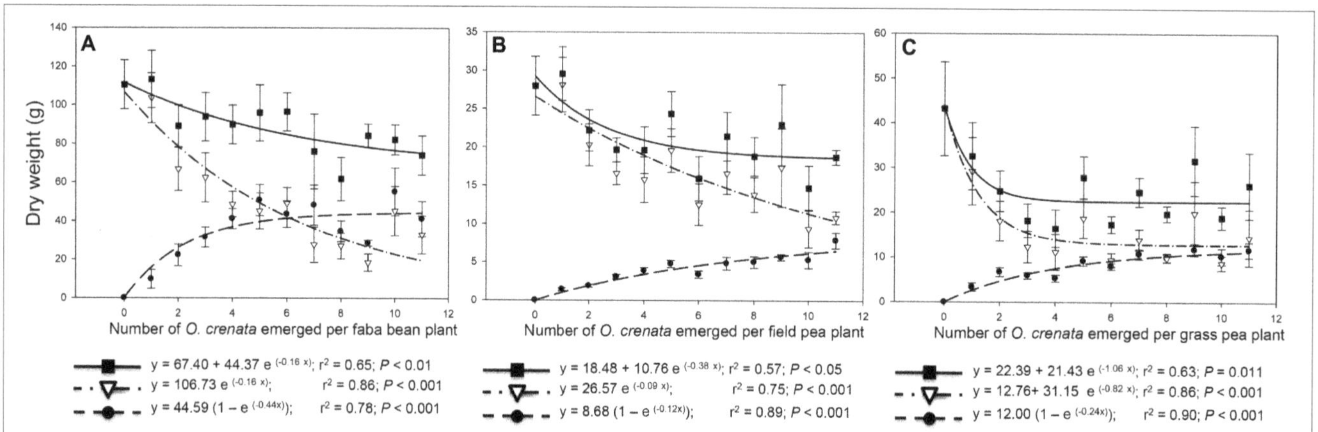

FIGURE 1 | Relationship between total dry weight and infection severity measured as *O. crenata* emergence per host plant. Dry weight of infected system (combined host+ *O. crenata*; ■), host dry weight (▽), cumulative *O. crenata* dry weight (●). **(A)** faba bean, **(B)** field pea; **(C)** grass pea.

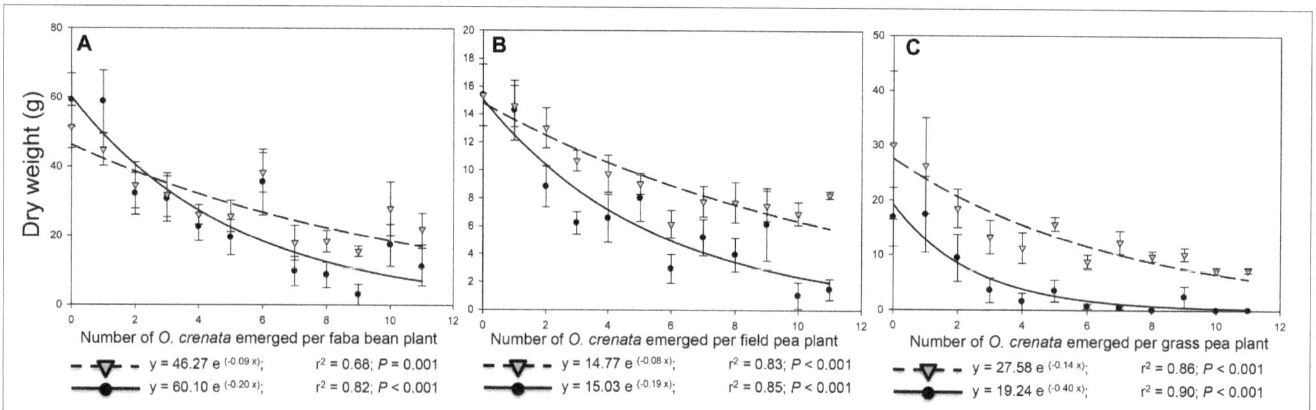

FIGURE 2 | Relationship between total dry weight and infection severity measured as *O. crenata* emergence per host plant. Host vegetative dry matter (leaves and stems; ▽), Host reproductive dry matter (host seeds; ●). **(A)** faba bean, **(B)** field pea; **(C)** grass pea.

responsible for capture and synthesis of resources used for the biomass construction of the overall parasitic plant-host plant complex. For some crop-broomrape associations, combined biomass is equivalent to that developed by uninfected crops being the biomass loss in the host equivalent to that accumulated by the parasite. This performance suggests that damage in the crop is directly attributed to the parasitic sink activity for nutrient withdrawal. However, the damage induced by parasitic weeds in other crops extends beyond assimilate diversion. In those cases, the parasitic weed displays a pathogenic-like nature promoting negative effects on the crop photosynthetic machinery and hormonal balance (Stewart and Press, 1990; Barker et al., 1996; Manschadi et al., 1996; Hibberd et al., 1998; Mauromicale et al., 2008). Therefore, we studied whether in our field conditions, *O. crenata* parasitism impairs the capacity of faba bean, field pea, and grass pea as autotrophic sources of energy and nutrients regardless the final sink allocation within the plant-parasitic plant complex. The productivity of the infected system (**total host-parasite combined dry weight)** was slightly lower than the total biomass developed by uninfected plants (**Figure 1**). Similar results have been reported for *O. cernua*-tobacco association (Hibberd et al., 1998). The inhibition rate of combined biomass was more intense in grass pea (1.06 g/g) than in field pea (0.38 g/g) and faba bean (0.16 g/g), reaching asymptotic values in which each legume species achieved 60, 67, and 67% of the total biomass of their respective uninfected counterparts. These results suggest that the reduction of biomass allocated into the host was not fully explained by the *O. crenata* sink activity.

In order to determine whether there is a trade-off between the resources allocated into reproductive versus parasite sink through the range of infection severity observed, we studied the **combined weight of sinks** (host seed plus parasite dry weights) and the **relative weight of combined sink** (ratio between combined sink weight to total aboveground biomass) at increased infection severities. In faba bean and grass pea, the combined weight of sinks equals the reproductive sink (seed weight) in uninfected plants regardless the level of infection (data not shown). Similar results were reported by Manschadi et al. (1996). By contrast, in field pea the combined sink weight was reduced up to 44.9% with respect

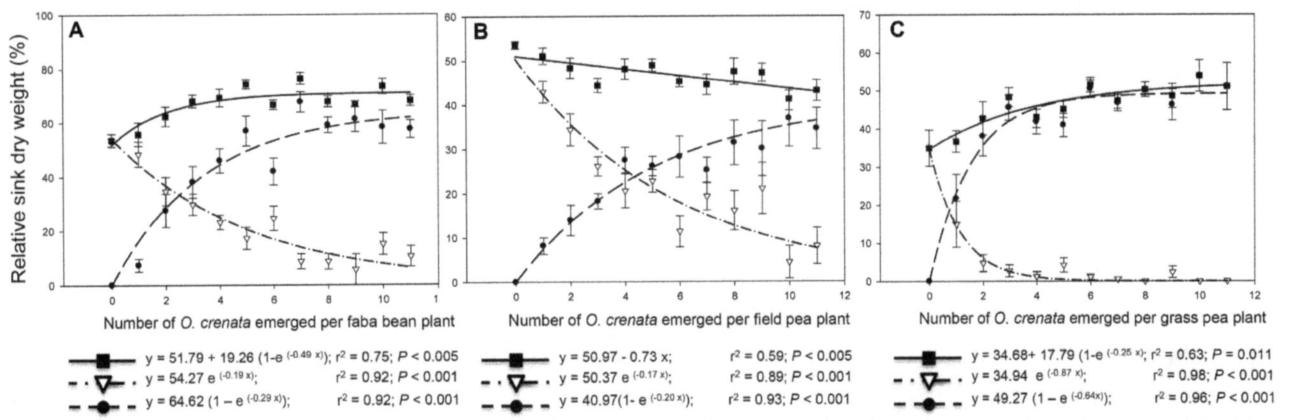

FIGURE 3 | Relationship between relative sink weight (ratio of sink weight to total complex dry weight) and infection severity measured as *O. crenata* emergence per host plant. Combined sink (host seed + *O. crenata*) relative weight (■), host seed relative weight (▽), *O. crenata* relative weight (●). **(A)** faba bean, **(B)** field pea; **(C)** grass pea.

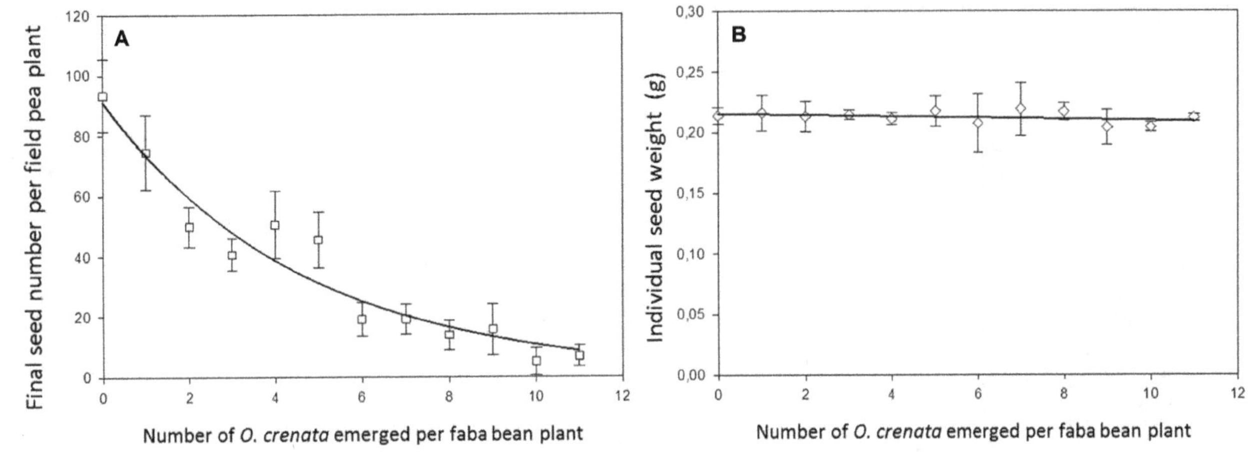

FIGURE 4 | Relationship between seed yield components in faba bean and infection severity. (A) Number of harvested seeds per host plant ($y = 9107\ e^{(-0.21x)}$, $r^2 = 0.93$, $P < 0.0001$), **(B)** Individual seed weight ($y = 0.21{-}0.0006x$), $r^2 = 0.16$, $P = 0.19$).

to reproductive weight in uninfected plants. This reduction was a consequence of relatively larger host seed inhibition than the concomitant gain in parasitic dry weight as described above. This reduction occurred at a rate of 0.64 g/g up to infection severity of two parasites per plant but remained constant thereafter. **Figure 3** shows the relative distribution of dry matter into sinks within the host-*O. crenata* complex. In faba bean and grass pea but not in field pea, the relative weight of combined sinks was significantly higher than the reproductive index and in the uninfected system and this difference was related with infection severity (ANOVA, $p < 0.001$; **Figure 2**). This performance was due to a larger decrease in dry weight of vegetative compartment at final harvest in relation to a constant combined sink weight suggesting that as the infection severity increases, higher levels of leaf reserves needs to be remobilized to attend the cost of combined sink biomass construction versus that occurred to build the equivalent seed biomass in uninfected plants. It was previously described that in the system *O. cernua*-tobacco, the host leaf area is not altered during *Orobanche* infection but the specific leaf area increases at late stages in the infection process indicating remobilization of leaf reserves (Hibberd et al., 1998).

In order to compare the strength of the parasitic sink for assimilate allocation in each host species, we studied the **relative weight of parasites** (ratio of accumulation of parasitic dry weight relative to combined biomass). For faba bean and grass pea the relative parasitic weight increased rapidly at low infection severity but it found an upper limit established around five parasites per plant beyond which it did not change significantly suggesting that a maximum percentage of dry matter within the host–parasite complex can be derived to the parasite regardless the infection severity (**Figures 3A,C**). However, in field pea, the increase rate in relative parasitic weight was slightly lower but more sustained

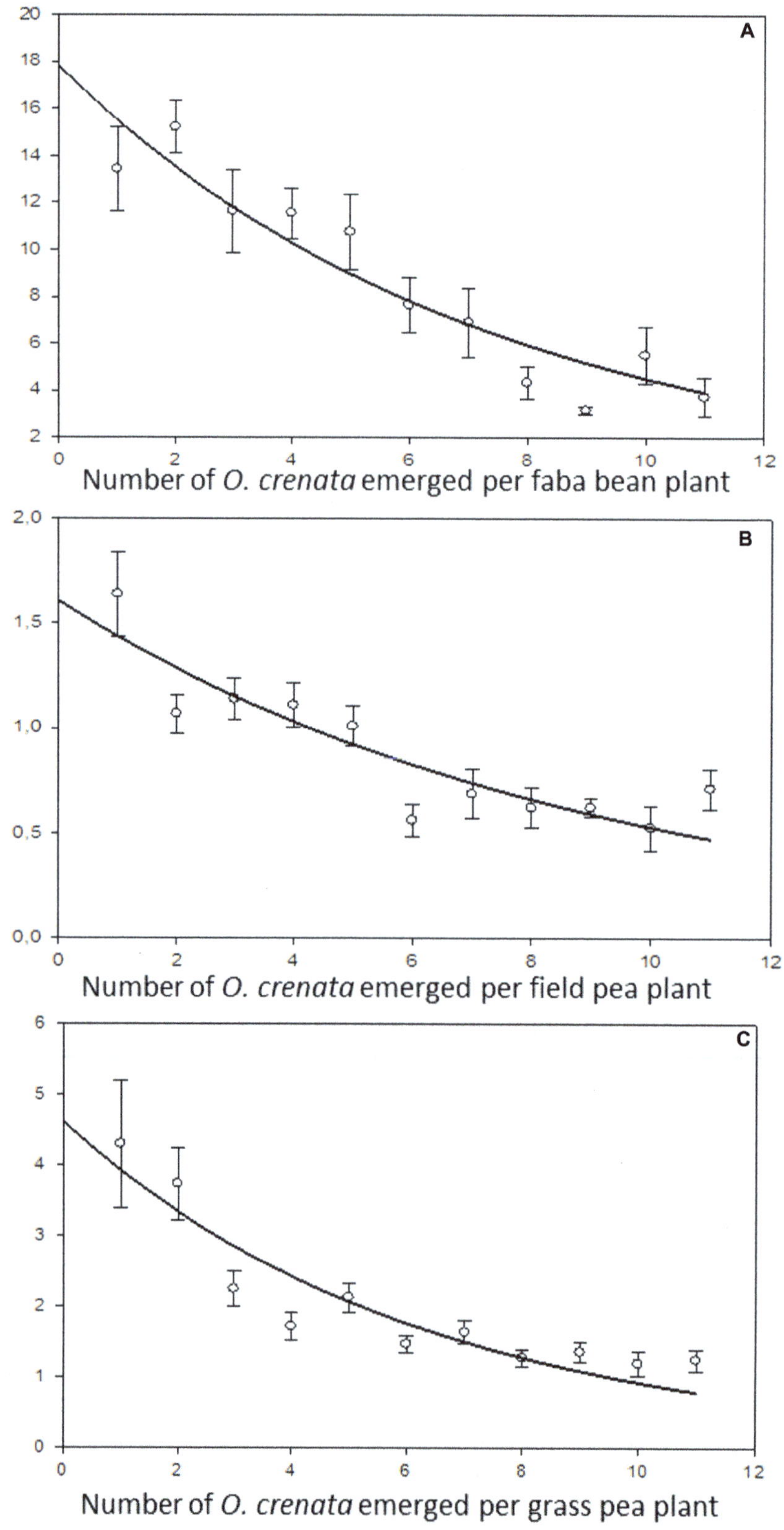

FIGURE 5 | Relationship between average dry weight of individual *O. crenata* plant and infection severity measured as *O. crenata* emergence per host plant. (A) Faba bean ($y = 17.82\ e^{(-0.14x)}$, $r^2 = 0.89$, $P < 0.0001$), **(B)** field pea ($y = 1.61e^{(-0.11x)}$, $r^2 = 0.80$, $P = 0.0002$), **(C)** grass pea ($y = 4.62\ e^{(-0.16x)}$, $r^2 = 0.85$, $P < 0.0001$).

through a broader range of parasite severity, reaching the upper limit at higher levels of infection (**Figure 3B**). The parasite relative weight ranged from 7.4 (1 parasite per plant) to 57.6% (11 parasites per plant) in faba bean; from 8.2 (1 parasite per plant) to 41.6% (11 parasites per plant) in field pea; and from 21.8 (1 parasite per plant) to 51.0% (11 parasites per plant) in grass pea. These data seems to indicate that less assimilates is available to *O. crenata* plants when they are attached to field pea roots. In the case of grass pea, parasitic relative weight equals combined sink relative weight at parasites densities higher than 4, due to the total substitution of seed yield by parasite dry matter. This effect was not observed in faba bean and field pea due to low but sustained seed production in the host. The proportion of resources allocated within the parasite was concomitant to reduction of host seed yield indicating that parasite growth and host reproduction compete directly for resources within a host plant. Host seed relative weight decreased at greater parasite densities according to exponential decay functions from 48.5 (1 parasite per plant) to 10.6% (11 parasites per plant) in faba bean, from 42.9 (1 parasite per plant) to 8.1% (11 parasites per plant) in field pea; and from 15.0 (1 parasite per plant) to 0.1% (11 parasites per plant) in grass pea (**Figure 3**).

Data recorded in field pea on seed yield and number of seeds per plant appears to suggest that field pea adjusts grain yield components in response to *O. crenata* parasitism by reducing number of seeds per host plant but maintaining constant the individual seed weight (**Figure 4**). Similarly *O. crenata* infection reduced the seed number but not the average seed unit weight in faba bean (Mesa-Garcia and Garcia-Torres, 1984). In our work, uninfected field pea plants produced an average of 93.4 ± 11.9 seeds per plant which was inhibited at a rate of 21% by increasing levels of *O. crenata* infection. The parasite density with half maximal inhibitory performance for seed number was predicted as 3.3 parasites per plant and at the highest parasite density, pea plants only produced 7.3% seeds of their uninfected counterparts (**Figure 4A**). The average weight of individual field pea seeds was 0.22 g per seed and was independent on the level of infection indicating that it was genetically controlled regardless the availability of resources (**Figure 4B**). This may be part of a host survival strategy to parasitic plant infection, where field pea concentrates its resources to a smaller number of viable seeds being this strategy also observed in legumes to survive drought (Gusmao et al., 2012). Those results are in agreement with the Smith and Fretwell (1974) model which establishes that seed individual weight is genetically determined and less variable in response to changes in resource availability than seed number which is directly regulated by the resource availability averaged by the genetically determined seed individual weight. Competition for pollinators between host and *O. crenata* inflorescences could not have a role in the reduction of number of seeds in field pea infected plants, because field pea is an autogamous species, however, it should be considered in other parasitic plant-host plant interactions when both partners in the host–parasite complex are allogamous and their flowering stage is synchronized.

The regulation model of seed production in response to differing resource availability contrasted with the regulation model of resource distribution observed in the parasites attached to the same host plant. The size of individual parasites (indirectly estimated as **average weight of individual parasite)** was not genetically controlled but dependent on the host species. Individual parasite weight was bigger in faba bean (at low parasite densities the individual weight per parasite was 10.34 ± 0.22 g) than in grass pea (2.36 ± 0.41 g) followed by field pea (1.06 ± 0.09 g). In addition, parasite individual weight was very plastic in response to resource availability. A negative relation between *O. crenata* individual weight and parasite density was observed, suggesting competition between individual parasites feeding in a single host plant at high parasite densities (**Figure 5**). Our observations agreed to those observed by Barker et al. (1996), Manschadi et al. (1996), and Hibberd et al. (1998) in various crop-broomrape associations.

CONCLUSION

Studies of competitive relations between parasitic weeds and its host crops are important in order to calculate intervention thresholds of control measurements essential in integrated pest management programs. Data of how *Orobanche* affects the growth of attacked crops is not always available or applicable. Most available studies in broomrape-crop associations have been performed in pots which reflect imperfectly the consequences of broomrape parasitism in the crop. We have contributed to breach this gap of knowledge in legume crops by characterizing in field *O. crenata* growth and the consequences of its parasitism in faba bean, field pea, and grass pea, three of its most preferred host crops.

AUTHOR CONTRIBUTIONS

MF-A designed, implemented the study and collected the data. FF analyzed the data. MF-A and FF interpreted the data. MF-A wrote the manuscript. FF and DR revised the manuscript. DR contributed with materials and laboratory equipment.

ACKNOWLEDGMENTS

This work was funded by the project AGL2014-52871-R. Additional support was obtained from Marie-Curie FP7 COFUND People Programme, through the award of an AgreenSkills' fellowship (under grant agreement n° PCOFUND-GA-2010-267196) to MFA.

REFERENCES

Barker, E. R., Press, M. C., Scholes, J. D., and Quick, W. P. (1996). Interactions between the parasitic angiosperm *Orobanche aegyptiaca* and its tomato host: growth and biomass allocation. *New Phytol.* 133, 637–642. doi: 10.1111/j.1469-8137.1996.tb01932.x

Eizenberg, H., Shtienberg, D., Silberbush, M., and Ephrath, J. E. (2005). A new method for monitoring early stages of *Orobanche cumana* development in sunflower (*Helianthus annuus*) with minirhizotron. *Ann. Bot.* 96, 137–140. doi: 10.1093/aob/mci252

Fernández-Aparicio, M., Reboud, X., and Gibot-Leclerc, S. (2016). Broomrape weeds. Underground mechanisms of parasitism and associated strategies for their control: a review. *Front. Plant Sci.* 7:135. doi: 10.3389/fpls.2016.00135

Garcia-Torres, L., Castejon-Muñoz, M., Jurado-Expósito, M., and López-Granados, F. (1996). Modelling the economics of controlling nodding broomrape (*Orobanche cernua*) in sunflower (*Helianthus annuus*). *Weed Sci.* 44, 591–595.

Gusmao, M., Siddique, K. H. M., Flower, K., Nesbitt, H., and Veneklaas, E. J. (2012). Water deficit during the reproductive period of grass pea (*Lathyrus sativus* L.) reduced grain yield but maintained seed size. *J. Agron. Crop Sci.* 198, 430–441. doi: 10.1111/j.1439-037X.2012.00513.x

Hibberd, J. M., Quick, W. P., Press, M. C., and Scholes, J. D. (1998). Can source-sink relations explain responses of tobacco to infection by the root holoparasitic angiosperm *Orobanche cernua*? *Plant Cell Environ.* 21, 333–340. doi: 10.1046/j.1365-3040.1998.00272.x

Kebreab, E., and Murdoch, A. J. (2001). Simulation of integrated control strategies for *Orobanche* spp. based on a life cycle model. *Exp. Agric.* 37, 37–51. doi: 10.1017/S001447970100401X

Linke, K. H., Sauerborn, J., and Saxena, M. C. (1991). Host-parasite relationships: effect of *Orobanche crenata* seed banks on development of the parasite and yield of faba bean. *Angew. Bot.* 65, 229–238.

Lins, R. D., Colquhoun, J. B., and Mallory-Smith, C. A. (2007). Effect of small broomrape (*Orobanche* minor) on red clover growth and dry matter partitioning. *Weed Sci.* 55, 517–520. doi: 10.1614/WS-07-049.1

Manschadi, A. M., Kroschel, J., and Sauerborn, J. (1996). Dry matter production and partitioning in the host-parasite association *Vicia faba–Orobanche crenata*. *J. Appl. Bot.* 70, 224–229.

Mauromicale, G., Lo Monaco, A., and Longo, M. G. A. (2008). Effect of branched broomrape (*Orobanche* ramosa) infection on the growth and photosynthesis of tomato. *Weed Sci.* 56, 574–581. doi: 10.1614/WS-07-147.1

Mesa-Garcia, J., and Garcia-Torres, L. (1984). A competition index for *Orobanche* creanta Forsk effecs on broad vean (*Vicia faba* L.). *Weed Res.* 24, 379–382. doi: 10.1111/j.1365-3180.1984.tb00600.x

Parker, C. (2009). Observations on the current status of *Orobanche* and *Striga* problems worldwide. *Pest Manag. Sci.* 65, 453–459. doi: 10.1002/ps.1713

Parker, C., and Riches, C. R. (1993). *Parasitic Weeds of the World: Biology and Control.* Wallingford: CAB International.

Rubiales, D., Fernández-Aparicio, M., Wegmann, K., and Joel, D. (2009). Revisiting strategies for reducing the seedbank of *Orobanche* and *Phelipanche* spp. *Weed Res.* 49, 23–33. doi: 10.1111/j.1365-3180.2009.00742.x

Sauerborn, J. (1991). "The economic importance of the phytoparasites *Orobanche* and *Striga*," in *Proceedings of the Fifth Symposium on Parasitic Weeds, Nairobi, Kenya; June 24–30, 1991*, eds J. K. Ransom, L. J. Musselman, A. D. Worsham, and C. Parker (Nairobi: CIMMYT), 137–143.

Siddique, K. H. M., Loss, S. P., Herwig, S. P., and Wilson, J. M. (1996). Growth, yield and neurotoxin (ODAP) concentration of three *Lathyrus* species in Mediterranean-type environments of Western Australia. *Aust. J. Exp. Agric.* 36, 209–218. doi: 10.1071/EA9960209

Smith, C. C., and Fretwell, S. D. (1974). The optimal balance between size and number of offspring. *Am. Nat.* 108, 499–506. doi: 10.1086/282929

Stewart, G. R., and Press, M. C. (1990). The physiology and biochemistry of parasitic angiosperms. *Annu. Rev. Plant Physiol. Plant Mol. Biol.* 41, 127–151. doi: 10.1146/annurev.pp.41.060190.001015

Conflict of Interest Statement: The authors declare that the research was conducted in the absence of any commercial or financial relationships that could be construed as a potential conflict of interest.

14

Aligning Microtomography Analysis with Traditional Anatomy for a 3D Understanding of the Host-Parasite Interface – *Phoradendron* spp. Case Study

Edited by:
Hanan Eizenberg,
Agricultural Research Organization, Israel

Reviewed by:
Rachid Lahlali,
Canadian Light Source Inc., Canada
Maoteng Li,
Huazhong University of Science and Technology, China

***Correspondence:**
Gregório C. T. Ceccantini
gregorio@usp.br

Luíza Teixeira-Costa and Gregório C. T. Ceccantini*

Group of Ecological Wood Anatomy and Parasitic Plant Biology, Department of Botany, Institute of Biosciences, University of São Paulo, São Paulo, Brazil

The complex endophytic structure formed by parasitic plant species often represents a challenge in the study of the host-parasite interface. Even with the large amounts of anatomical slides, a three-dimensional comprehension of the structure may still be difficult to obtain. In the present study we applied the High Resolution X-ray Computed Tomography (HRXCT) analysis along with usual plant anatomy techniques in order to compare the infestation pattern of two mistletoe species of the genus *Phoradendron*. Additionally, we tested the use of contrasting solutions in order to improve the detection of the parasite's endophytic tissue. To our knowledge, this is the first study to show the three-dimensional structure of host-mistletoe interface by using HRXCT technique. Results showed that *Phoradendron perrottetii* growing on the host *Tapirira guianensis* forms small woody galls with a restricted endophytic system. The sinkers were short and eventually grouped creating a continuous interface with the host wood. On the other hand, the long sinkers of *P. bathyoryctum* penetrate deeply into the wood of *Cedrela fissilis* branching in all directions throughout the woody gall area, forming a spread-out infestation pattern. The results indicate that the HRXCT is indeed a powerful approach to understand the endophytic system of parasitic plants. The combination of three-dimensional models of the infestation with anatomical analysis provided a broader understanding of the host-parasite connection. Unique anatomic features are reported for the sinkes of *P. perrottetii*, while the endophytic tissue of *P. bathyoryctum* conformed to general anatomy observed for other species of this genus. These differences are hypothesized to be related to the three-dimensional structure of each endophytic system and the communication stablished with the host.

Keywords: haustorium, microtomography, endophytic system, Santalaceae, mistletoe

INTRODUCTION

The complex endophytic structure formed by parasitic plant species renders a rather difficult three dimensional comprehension. Traditional techniques in plant anatomy have been extensively employed to describe the interface formed between parasites and their hosts. However, due to difficulties in imbedding and cutting lignified and large materials, these techniques are usually applied to small samples, which provide a detailed but restricted view of the host-parasite interface. Thus, in order to gain a three-dimensional comprehension of the structure, a huge series of sequential anatomical sections are necessary.

This issue becomes even more relevant when analyzing stem hemiparasites, which often form large woody infestation structures that comprise the interface between these plants and their hosts. In his work on the anatomy of the endophytic system of *Phoradendron flavescens*, Calvin (1967) mentions the analysis of hundreds of anatomical sections in order to follow the endophyte's extension. In this context, the use of High Resolution X-ray Computed Tomography (HRXCT) arises as a technique that allows general three-dimensional analysis of these complex structures, pushing forward the traditional field of plant anatomy.

This technique consists of using a micro-focused X-ray to illuminate the object while an X-ray detector collects the magnified projection images. As the object rotates inside the equipment, hundreds of images are generated. A powerful computer is than used to stack these images creating a virtual three-dimensional object allowing the analysis of its internal structure[1]. Despite its original development for medical diagnosis (Hounsfield, 1976), computed tomography have long been used for several other scientific purposes (Staedler et al., 2013), including studies in the broad field of Plant Sciences (Heeraman et al., 1997; Kaestner et al., 2006; Brodersen et al., 2012; Gee, 2013; Breteron et al., 2015; Cochard et al., 2015 and others).

Considering these matters, we combined traditional plant anatomy with the HRXCT technique in order to broadly and deeply analyze the endophytic system of *Phoradendron* species. Although several studies have dealt with the haustorial anatomy of this genus (Cannon, 1901; Calvin, 1967; Fineran and Calvin, 2000; Schmid et al., 2011; and others), the large number of species it comprises is reflected in the structural diversity of the endophytic system (Kuijt, 1964, 2003). Based on this proposition, we chose to analyze the endophytic system of two morphologically similar *Phoradendron* species of the Neotropical region.

[1]Information on the HRXCT technique available at: www.bruker-microct.com/company/methods.htm

MATERIALS AND METHODS

Plant Material Sampling

The mistletoe *Phoradendron perrottetii* Nutt. (Santalaceae) growing on branches of the host tree *Tapirira guianensis Aubl.* (Anacardiaceae; **Figure 1A**) was sampled in a riparian forest in Campanha municipality (Minas Gerais state, Brazil). The

FIGURE 1 | Two mistletoes of the genus *Phoradendron* and their respective host species. (A,B) *Phoradendron perrottetii* on *Tapirira guianensis*. **(C,D)** *Phoradendron bathyoryctum* on *Cedrela fissilis*. **(A,C)** Parasitic plants on host branches. **(B,D)** Highly infested trees.

other mistletoe *P. bathyoryctum* Eichler parasitizing the host tree *Cedrela fissilis* Vell. (Meliaceae; **Figure 1C**) was sampled in a wooded area inside the main campus of the University of Sao Paulo (Sao Paulo state, Brazil). Both parasitic species were capable of colonizing the host canopies, forming moderate to heavy infestations (**Figures 1B,D**).

After sampling we measure the length of all host-mistletoe interface structures - henceforth called "woody galls." Half of the material was then fixed in a 50% solution of formaldehyde-ethanol-acetic acid (FAA). The other half was air-dried.

An additional mistletoe species - *P. affine* (Pohl ex DC.) Engl. and K. Krause (**Figure 2A**) - was cultivated upon young specimens of the host tree *Melia azedarach* L. (Meliaceae; **Figure 2B**). This material was used for the testing of contrasting agents as explained in the next subsection.

Considering the three mistletoe species, we sampled a total amount of 60 woody galls. Several intact woody galls were added to the collection preserved in the Xylarium Nanuza Luiza de Menezes, at Department of Botany of the University of São Paulo (SPFw).

High Resolution X-ray Computed Tomography Analysis

In order to compare the three-dimensional structure and the infestation patterns of *P. perrottetii* and *P. bathyoryctum* three woody galls formed by each species were scanned using a high performance *in vivo* X-ray microtomography scanner (Skyscan, 1176). Prior to the scanning these woody galls were fixed in a 50% solution of FAA. The fixation aimed on preserving the structure of the endophyte, which are usually composed by abundant parenchyma cells (Thoday, 1957; Calvin, 1967; Kuijt, 2003).

Previous trials were also carried out in order to evaluate other preservation methods, such as air-drying and embedding of the samples. Indeed, the first method was not efficient in preserving the structure of the endophyte. On the other hand, the embedding medium (polyethyleneglycol) was observed to severely interfere with the X-ray penetration into the samples.

Additionally, two woody galls formed by *P. affine* were used to test whether the use of contrasting agents could improve the visualization of the parasite's endophytic system within the host branches. Based on the work of Staedler et al. (2013) two

FIGURE 2 | (A) *Phoradendron affine* on *Melia azedarach*. **(B)** Mistletoe species (*P. affine*) cultivated on young host tree (*M. azedarach*). **(C)** Schematic drawing of the method employed to perfuse contrasting solutions into the parasitic endophyte. H, host branch; P, parasitic plant; C, contrasting solution.

contrasting solutions were tested – Lugol's solution (0.1%) and a lead nitrate ($PbNO_3$) solution (0.2%). A third sample was not perfused with contrasting agents, thus serving a test control.

Contrasting solutions were applied to the woody galls according to the method described by Sperry et al. (1988) (**Figure 2C**) for vascular infusions. Briefly, one end of a rubber tube was fitted at one of the branches of the parasite close to woody gall. The other end of the tube was connected to an elevated reservoir (0.5 – 1 m) containing the contrasting solution. This set up was designed so that the flexibletube would be filled with the contrasting solution forming a liquid column with enough pressure to force the entry of the solution into the host branch. Special care was taken to avoid the presence of air bubbles in the tube that could be pushed into the xylem and clog the system. Fine surfacing of the wood with a razor blade was also needed to assure that vessels were open and permeable. After being perfused with contrasting solutions for ca. 8 h the material was disconnected from the apparatus and scanned immediately.

The scanning of each woody gall generated hundreds of X-ray images which were subsequently reconstructed using the NRecon software in order to provide a three-dimensional visualization. Analysis and image acquisition were carried out by using the software Dataviewer (two-dimensional analysis of internal structures) and the software Dataviewer (three-dimensional analysis).

Morphological and Anatomical Analyses

The same woody galls used for HRXCT analyses were also used for morphological and anatomical analyses as a way to provide a detailed understanding of host-mistletoe interface. One material of each host-mistletoe pair was air-dried, cut in transversal and longitudinal sections, and sanded using sand papers of ascending grifts until a smooth surface was obtained. The material was studied and photographed using a stereo-photomicroscope (Leica DML and camera DFC 310FX).

The two remaining woody galls of each mistletoe were used for anatomical analysis. Samples were imbedded in polyethyleneglycol (PEG) and sectioned in a sliding microtome (Leica SM 2000R) to produce 20 μm thick transverse, radial longitudinal, and tangential longitudinal sections, which were stained with safranin/Astra-blue (Johansen, 1940; Bukatsch, 1972 adapted by Kraus and Arduin, 1997).

Additional material of host-mistletoe pair was used to complement the analysis. A total amount of 10 woody galls was studied through stereo-photomicroscope, while five woody galls were studied through plant anatomy techniques.

RESULTS

Optimizing the Parameters for HRXCT at the Host-Mistletoe Interface

The use of HRXCT technique for analyzing the three-dimensional host-mistletoe interface began by testing a total of nine scanning parameters. Among these, the voltage and the current had to be adjusted for each sample. **Table 1** lists the adjustment used for each sample along with sample size, duration of each scan, and information about the resulting files.

After the initial test other relevant parameters, such as resolution, filter, and rotation step were fixed for all samples. Thus, all samples were scanned with maximum resolution (9 μm), using a 0.2 mm aluminum filter and a 0.3° rotation step. Despite the increase on scan duration, maximum resolution, and low rotation steps were used to allow a detailed observation of the parasite's endophytic system. Due to longer scan duration all samples were previously wrapped in thin plastic film to avoid wood desiccation.

TABLE 1 | Scanning parameters adjustment for each sample scanned using the High Resolution X-ray Computed Tomography technique.

Mistletoe-host pair	Species ID	Sample size (cm)	Voltage (kV)	Current (μA)	Scan duration	Number of files	File size (GB)
Phoradendron perrottetii on *Tapirira guianensis*	1	4.70	45	397	00:41:40	515	58
	2	5.80	55	630	01:04:31	721	37
	3	5.43	35	456	01:08:55	730	52
Phoradendron bathyoryctum on *Cedrela fissilis*	1	10.56	50	500	03:19:31	1201	402
	2	7.30	62	403	02:27:24	983	108
	3	7.02	46	540	02:31:25	990	103
Phoradendron affine on *Melia azedarach*				Control			
	1	6.00	45	397	01:38:38	815	39
				Lugol's solution			
	2	5.90	37	396	01:26:21	802	39
				$PbNO_3$ solution			
	3	5.75	35	375	01:26:16	711	33

Three-Dimensional Structure of the Endophytic System

During the sampling and the measurement of the woody galls *P. bathyoryctum* was observed to form larger infestation structures on branches of *C. fissilis* when compared to *P. perrottetii* on branches of *T. guianensis*. These observations served as an initial clue to understanding the extension of the endophytic system of each mistletoe species. The HRXCT technique allowed us to assess not only the endophyte's extension but also its spread pattern within the host branch.

A series of six internal images of the woody gall formed by *P. perrottetii* illustrates the interface between this mistletoe and the host *T. guianensis* (**Figure 3**). The images taken from proximal and distal ends of sample showed the host branch with no sign of infestation (**Figures 3A,F**, respectively). In the actual region comprising the endophyte (**Figures 3B–E**) parasitic tissue was observed to spread both within the wood and the bark of the host. Initially, few wedge-shape sinkers were observed (**Figure 3B**). As the number of sinkers increased, the amount of cortical strands also increased (**Figure 3C**). The sinkers progressively clustered into the parasite's wood (**Figure 3D**), which then occupied half of the branch circumference (**Figure 3E**).

A similar series of images was obtained for the woody gall of *P. bathyoryctum* on *C. fissilis* (**Figure 4**). In this case, the parasite's endophyte extended within the host branch from one end to the other (**Figures 4A–F**). Large cortical

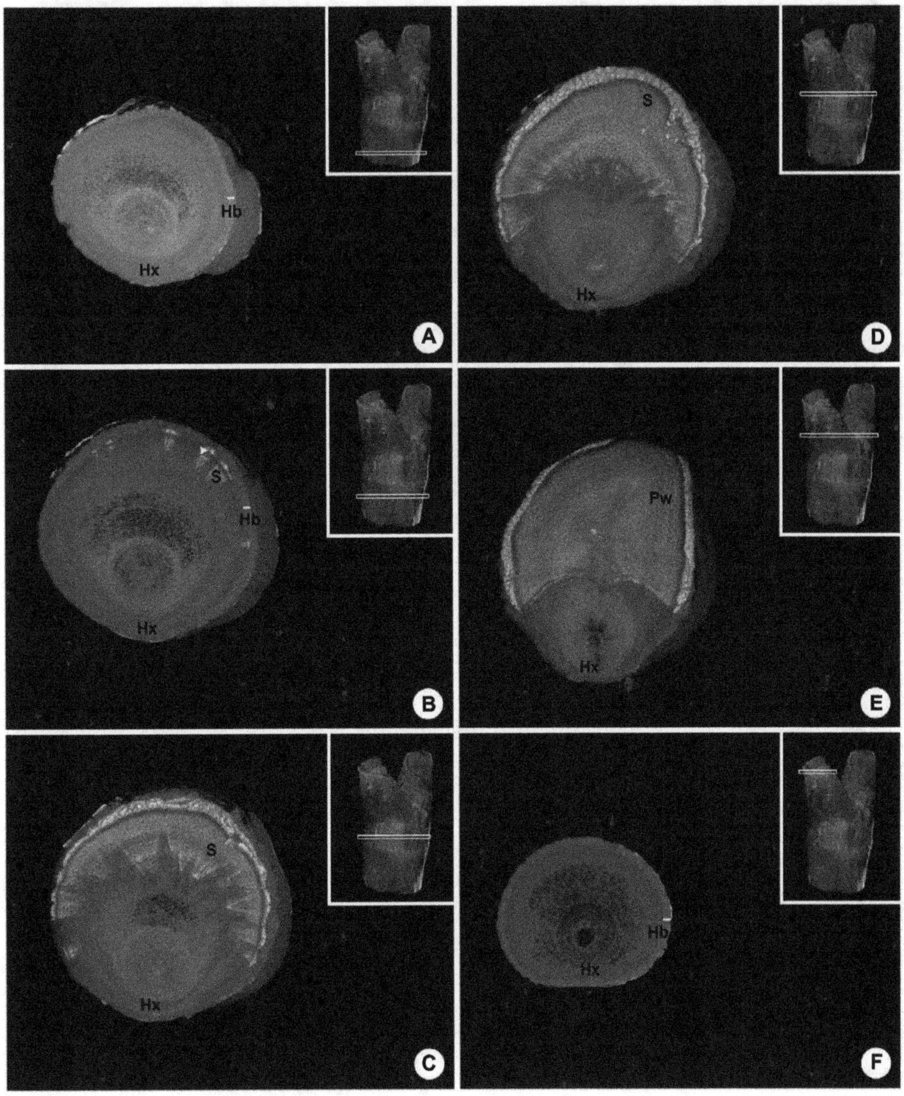

FIGURE 3 | Sequence of six internal images of the woody gall formed by *P. perrottetii* on the host *T. guianensis*. **(A)** Proximal end of the host branch free of parasitic tissue. **(B)** Wedge-shape sinkers penetrating the host wood. **(C)** Increased amount of sinkers and cortical strands. **(D)** Grouping of sinkers. **(E)** Parasitic tissue occupying the upper half of the branch circumference. **(F)** Distal end of the host branch free of parasitic tissue. Hx, host xylem; Hb, host bark; S, sinker; Pw, parasitic wood; white arrow-head, cortical strands.

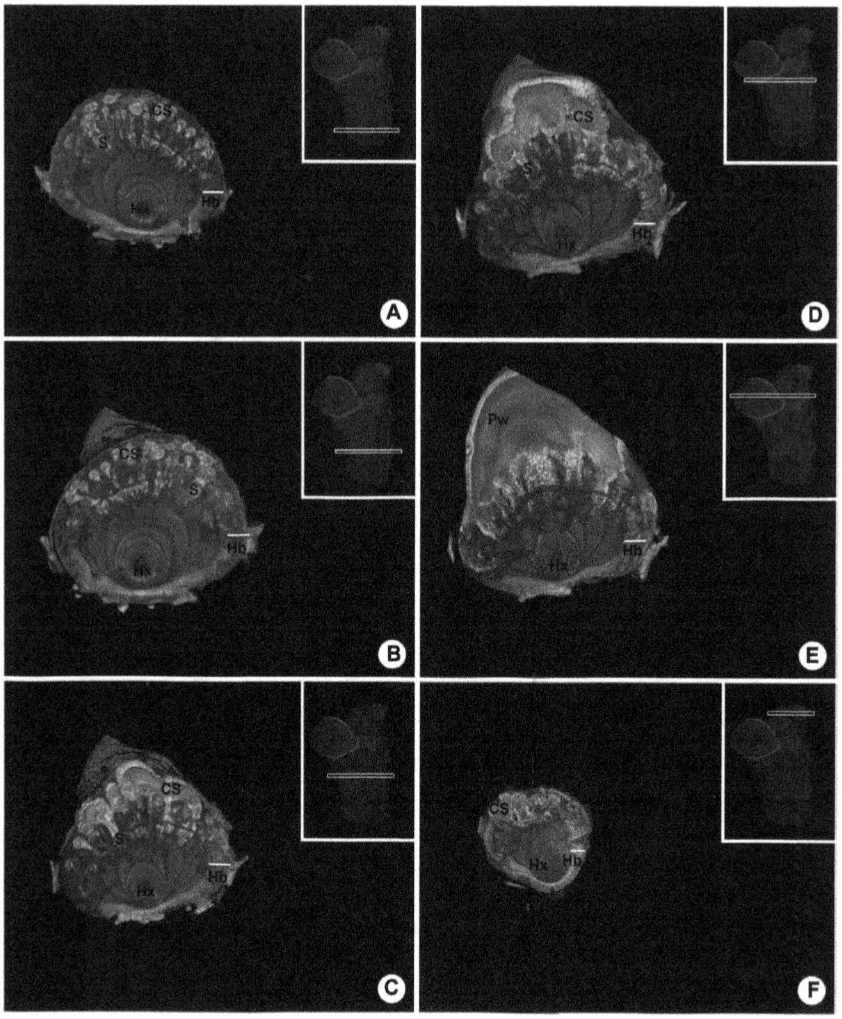

FIGURE 4 | Sequence of six internal images of the woody gall formed by *P. bathyoryctum* on the host *C. fissilis*. (A) Proximal end of the host branch already showing infestation by parasitic tissue. **(B)** Large cortical strands associated with elongated sinkers. **(C)** Initial clustering of cortical strands. **(D)** Further clustering of cortical strands. **(E)** Parasitic tissue associated to the parasite's wood. **(F)** Distal end of the host branch still showing infestation by parasitic tissue. Hx, host xylem; Hb, host bark; CS, cortical strands; S, sinker; Pw, parasitic wood.

strands were associated with elongated sinkers (**Figure 4A**). The cortical strands grew larger (**Figure 4B**) and began to cluster (**Figures 4C,D**) until the parasite's wood becomes visible (**Figure 4E**). Unlike the endophyte of *P. perrottetti*, only the cortical strands of *P. bathyoryctum* are eventually grouped into its wood as the sinkers remain separated.

The analysis of the three-dimensional internal structure of both parasites' endophytic systems revealed two opposite patterns. The sinkers of *P. perrottetii* did not penetrate deeply into the host wood and the endophytic tissue was usually restricted to half of the branch circumference forming a concise endophytic pattern. As for *P. bathyoryctum*, its longer sinkers projected throughout the woody gall area in all directions forming a spread out pattern. Please refer to **Supplementary Material** for short videos that provide a three-dimensional internal view of the structure formed by each parasite.

Anatomy of the Host-Parasite Interface

Despite their morphological similarities, *P. perrottetii* and *P. bathyoryctum* showed remarkable differences regarding their endophytic structure and anatomy (**Figure 5**). A transversal section of the woody gall shows the haustorium of the parasitic species *P. perrottetii* forming a continuous interface with the host branch of *T. guianensis* (**Figure 5A**). In this old infestation, the parasite encircled the host branch almost completely forming sinkers around the whole circumference of the host branch. The endophyte was composed of small cortical strands of the parasite, which were observed to give rise to short wedge-shaped sinkers that penetrate into the host xylem (**Figures 5B,C**).

On the other hand, the transverse section of the woody gall formed by *P. bathyoryctum* on branches of *C. fissilis* showed a fragmented interface due to the diffuse organization

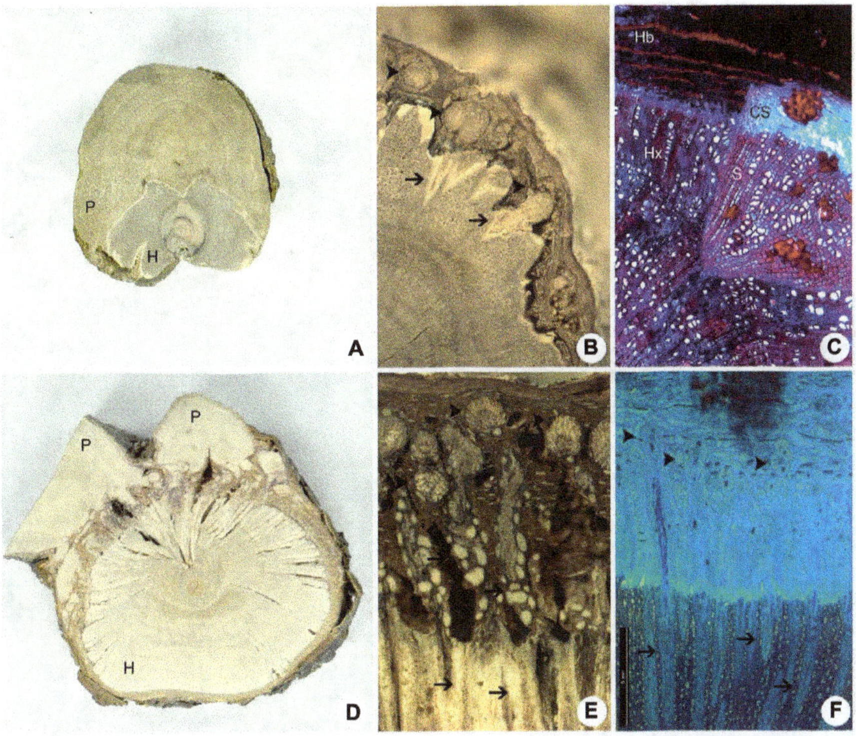

FIGURE 5 | Comparative cross-sections of the host-parasite interfaces. (A–C) *P. perrottetii* on *T. guianensis*. **(D–F)** *P. bathyoryctum* on *C. fissilis*. **(A,D)** Macroscopic cross-sections showing the maximum size observed for the parasite's primary haustorium. **(B,E)** Detail of the macroscopical cross-section showing sinkers arising from cortical strands. **(C,F)** Microscopy cross-section showing the general anatomy of cortical strands and sinkers. Arrow-heads, cortical strands; Arrows, sinkers; CS, cortical strand; H, host; Hb, host bark; Hx, host xylem; P, parasite; S, sinker.

of the endophytic system (**Figure 5D**). The contact between the mistletoe and the host tissue is accomplished by the multiple cortical strands that spread out through the thick host bark of *C. fissilis* (**Figure 5E**) and give rise to a large number of long needle-shaped sinkers (**Figure 5F**).

Despite being smaller than the cortical strands of *P. bathyoryctum*, the strands formed by *P. perrottetii* on the host tree *T. guianensis* caused morphological alterations to surface of host xylem (**Figure 6A**). At the beginning of its penetration into the host tissues, the cortical strand is composed almost exclusively of parenchyma cells (**Figure 6B**). Later on during its growth, tick-walled fibers and vessels elements are developed in the strands (**Figure 6C**). The vessel elements in the sinker are connected to vessel coming from the cortical strand (**Figure 6D**). A reduced number of cortical strands were observed to undergo secondary growth in *P. perrottetii* (**Figure 6E**). Direct vessel contact between the parasite and the host was also observed (**Figure 6F**).

The large cortical formed by *P. bathyoryctum* on the host tree *C. fissilis* were observed to cause severe alterations to the host wood (**Figure 7A**). The large number of cortical strands allowed us to observe these structures in different growth stages (**Figure 7B**), from primary to secondary growth (**Figure 7C**).

The needle-shaped sinkers were composed of lignified parenchyma which encircled short vessel elements (**Figure 7D**). Few lignified vascular parenchyma cells were also present. Direct vessel contact between the parasite and the host was also observed (**Figure 7E**). The formation of cross-grained wood was observed in some areas of the host xylem close to the sinkers (**Figure 7F**).

Additional images showing the morphology and anatomy of the host-mistletoe interface were compared with images obtained via HRXCT analysis, as suggested by Steppe et al. (2004). These figures are provided as **Supplementary Material**.

The Use of Contrasting Agents in HRXCT

The endophytic tissue of *P. perrottetii* and *P. bathyoryctum* showed a natural contrast against the wood of *T. guianensis* and *C. fissilis*, respectively. In both cases the entire parasitic endophyte was observed in white, while the host wood showed dark shades of gray. However, this was not entirely the case for *P. affine* parasitizing *M. azedarach*. The cortical strands of *P. affine* actually showed natural contrast against the host wood, but the sinkers was seen as light gray structures penetrating the wood of *M. azedarach* (**Figures 8A,B**).

The use of contrasting agents was observed to improve the contrast between the endophytic tissue of the parasite and the wood of the host. In the sample perfused with Lugol's solution both the cortical strands and the sinker showed a higher contrast (**Figure 8C**). When lower density tissue was removed from the image, the whole extension of the cortical strand became visible in bright white along with short parasitic sinkers, host vessels and a wood knot (**Figure 8D**).

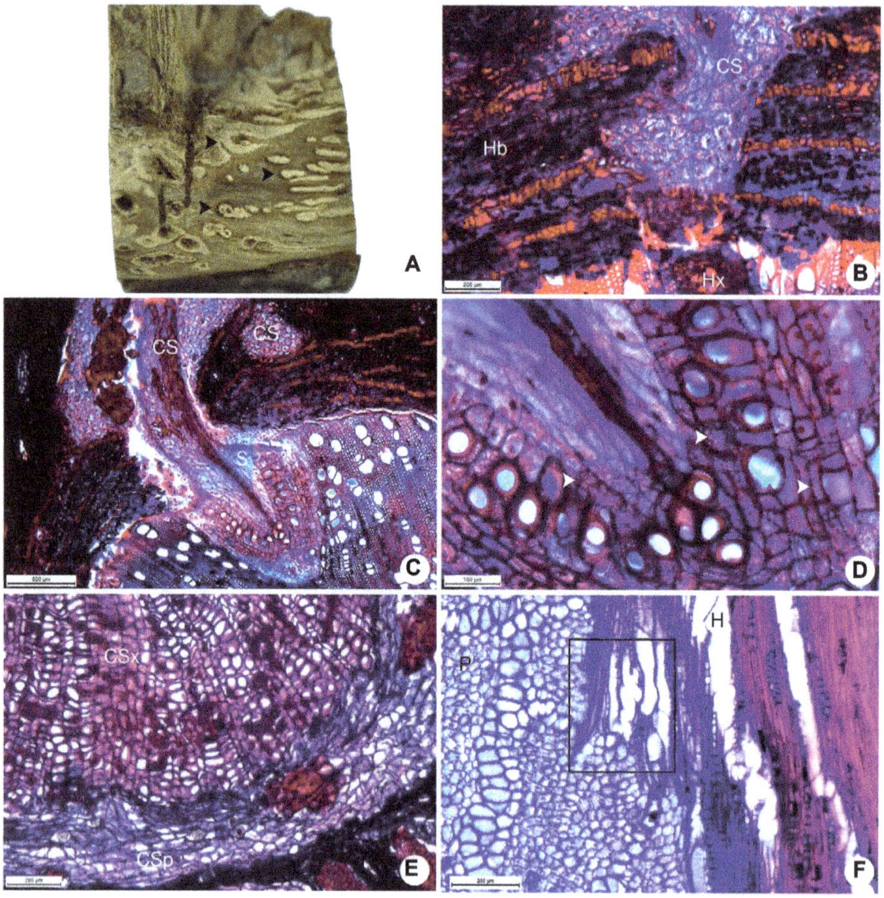

FIGURE 6 | Aspects of the host-parasite interface between *P. perrottetii* and *T. guianensis*. (A) Macroscopic view of the woody gall after removal of host bark, showing cortical strands on host xylem. **(B)** Young cortical strand penetrating the host xylem (scale bar = 200 μm). **(C)** Developed cortical strand giving rise to a sinker (scale bar = 500 μm). **(D)** Detail of the sinker showing lignified parenchyma cells (scale bar = 100 μm). **(E)** Cortical strand showing secondary growth (scale bar = 200 μm). **(F)** Direct vessel connection between parasite and host (scale bar = 200 μm). Black arrow-heads, cortical strands; white arrow-heads, lignified parenchyma cells; CS, cortical strand; CSp, phloem tissue of the cortical strand; CSx, xylem tissue of the cortical strand; H, host; Hb, host bark; Hx, host xylem; P, parasite; S, sinker.

The use of lead nitrate solution also provided a higher contrast to the parasitic tissues and to some of the host's vessels as well (**Figure 8E**). Nevertheless, the contrasting solution was observed only within the vessels as the surrounding parasitic tissue was again observed in shades of gray (**Figure 8F**).

It's noteworthy that the control sample had longer and more developed sinkers when compared to the other two samples perfused with contrasting agents. Although all mistletoe seeds were applied to host branches on the same day, their morphological and anatomical development rates varied greatly. Still, it was possible to compare the structure endophytic tissue among the samples.

DISCUSSION

As other member of the "Viscaceae" clade (currently within the Santalaceae family; Stevens, 2001), *Phoradendron* species attach to their hosts via a single connection (primary haustorium) that frequently form swollen regions referred by Many (1964) as woody galls. Their endophytic system is conspicuous and extensive, which is considered to be a feature of evolutionary derived species due to an increased protection conferred to the parasitic tissue (Heide-Jorgensen, 2008). The endophytic tissue is generally organized in cortical strands that grow longitudinally within the host bark and give rise to sinkers, which then penetrate the host xylem (Calvin, 1967).

However, minor variations to this general organization are observed in some species, such as *P. perrottetii*. In this species an eventual fusion of sinkers occurs and the distinction between sinkers and cortical strands becomes unclear, forming a continuous endophytic tissue and connects to the host. This peculiarity was also observed by Thoday (1957), who analyzed *P. perrottetii* growing on the host *Protium insigne* (Burseraceae). The uniqueness of this endophytic system could be related to preferred, although not restricted, use of a host species. At the

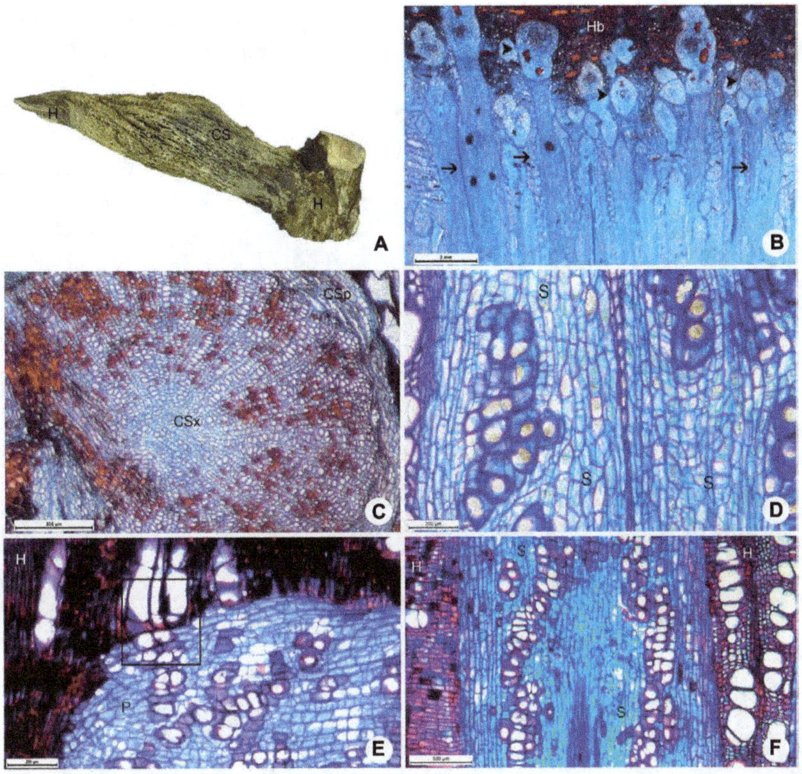

FIGURE 7 | Host-parasite interface between *P. bathyoryctum* and *C. fissilis*. (A) Macroscopic view of the woody gall after removal of host bark, showing cortical strands on host xylem. **(B)** Multiple cortical strands in different developmental stages within the host bark (scale bar = 2 mm). **(C)** Cortical strand showing secondary growth (scale bar = 500 μm). **(D)** Detail of the sinker showing lignified parenchyma cells and vessel elements (scale bar = 200 μm). **(E)** Direct vessel connection between parasite and host (scale bar = 200 μm). **(F)** Region of the host xylem with cross-grained wood close to the sinker (scale bar = 500 μm). Arrows, sinkers; arrow heads, cortical strands; CS, cortical strand; CSp, phloem tissue of the cortical strand; CSx, xylem tissue of the cortical strand; H, host; P, parasite; S, sinker.

study site *P. perrottetti* was mostly observed to use *T. guianensis* as its host despite the presence of other tree species. This host preference was also recorded[2] for other areas of the Cerrado (Brazilian savanna).

On the other hand, the endophytic anatomy of *P. bathyoryctum* showed more generalized features commonly reported for other *Phoradendron* species (Kuijt, 2003). A great number of mature cortical strands are also observed. Regarding host selection, *P. bathyoryctum* has been reported to grow on species from a total of 14 different plant families showing little predilection for hosts within the Sapindaceae and Meliaceae family[2].

Not only had the anatomy of the endophytic tissue differed between the studied species, the three-dimensional internal pattern of each parasite's endophytic system was also diverse. To our knowledge, this is the first study to show the three-dimensional structure of host-mistletoe interface by using HRXCT technique.

While *P. perrottetii* formed small woody galls with a concise endophytic restricted to one area of the host branch, *P. bathyoryctum* had a spread out pattern of the endophyte with longer sinkers that grew throughout the area of the woody galls. The understanding of these internal patterns was readily achieved through the X-ray tomography analysis.

Given its non-destructive method, we chose to begin our study by employing the tomography technique. The high-resolution images and three-dimensional models obtained for the woody galls allowed us to select the most interesting area of the host-parasite interface to be used in the following anatomical analysis eliminating the need for a long series of anatomical sections.

Higher resolutions are also possible, as showed by Dhondt et al. (2010) who made images of *Arabidopsis thaliana* at the cellular level. Nevertheless, due to scan duration and size of generated files, this increased resolution would require a previous reduction of the host-mistletoe interface. This reduction would result in loss of the whole three-dimensional pattern.

Additionally to guiding subsequent anatomical studies, the understanding of the endophytic spread allowed us to better understand the host-parasite communication. Although lumen-to-lumen contact was observed for both host-mistletoe pairs, the sinkers of *P. perrottetii* had a greater amount of vessel elements when compared to the sinkers of *P. bathyoryctum*. The

[2]According to herbaria data available at http://inct.splink.org.br (INCT – Virtual Herbaria of Flora and Fungi).

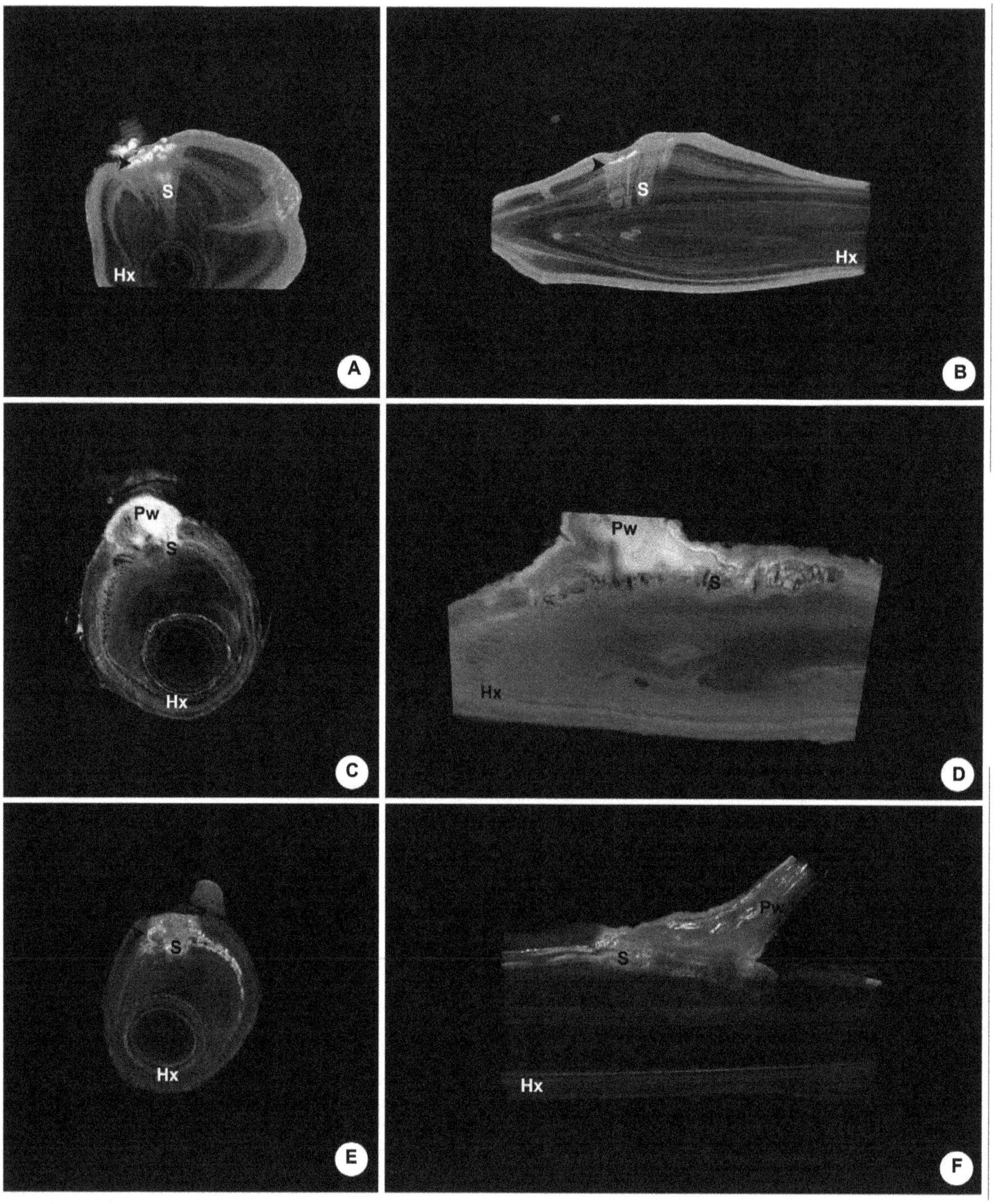

FIGURE 8 | Comparative internal images of the woody gall formed by *P. affine* on the host *M. azedarach* showing the test of contrasting solutions. **(A,B)** Control sample (no contrasting solution) showing cortical the natural contrast of the cortical strands as opposed to the sinkers. **(C,D)** Sample perfused with Lugol's solution showing the parasitic tissue with a higher contrast. The whole extension of the cortical strand is visible, along with short parasitic sinkers, host vessels and a wood knot. **(E,F)** Sample perfused with lead nitrate solution showing the parasitic tissue with a higher contrast, especially within the vessels of the parasite and the host. Hx, host xylem; S, sinker; Pw, parasitic wood; black arrow-head, cortical strands.

observation may be related to spread of each endophytic system. The spread-out sinkers of *P. bathyoryctum* could compensate the reduced amount of vessel elements by increasing the contact area with the host xylem. Future *in vivo* analysis, such as those carried out by Brodersen et al. (2010) and Cochard et al. (2015) should help addressing this hypothesis.

The host-parasite communication was also indirectly assessed by the use of contrasting solutions. When analyzing the infestation pattern of each parasitic species the high contrast of the endophytic tissue against the host wood was an important factor during the analysis. This was due to the different density between the endophyte and the host tissue. Therefore, the low natural contrast between the sinkers of *P. affine* and the wood of *M. azedarach* was probably related to a similar density of these structures. As observed by other authors (Metscher, 2013; Staedler et al., 2013), the use of contrasting solutions proved to be an efficient way to improve the analysis of this situation.

Each contrasting solutions employed here had an interesting result. Lugol's solution perfused to the whole endophytic tissue of *P. affine* due to the presence of iodine, which bound to starch grains previously observed within the parasite's endophyte (data not shown). Host vessels close to the endophyte were also perfused with this solution highlighting the vascular communication between host and parasite. The same occurred in the specimen perfused with lead nitrate solution. However, this solution was only observed within vessels, thus indicating an apoplastic connection between the two plants.

In synthesis, the use of High Resolution X-ray Computed Tomography analysis is proven to be an important non-invasive tool for the understanding of the connection and organization between hosts and their parasites. However, up to the current technology and computing development the use of traditional methods of plant anatomy are still needed for a comprehensive understanding of these complex structures. We are currently testing the use of the HRXCT, as well as the use of contrasting solutions in other host-parasite models.

AUTHOR CONTRIBUTIONS

This work is part of the research conducted by LT-C during her Master's Degree course. Therefore, LT-C was responsible for: experimental planning; all the preparations mentioned in this manuscript; result analysis and discussion; manuscript writing. GC was the Advisor during the conduction of the research presented in this manuscript. Therefore, GC was responsible for: funding obtainment, experimental planning, discussion of results and manuscript writing.

FUNDING

This work was funded by the Brazilian Council for Superior Education (CAPES), the National Council for Scientific and Technological Development (CNPq - grant 307041/2014-0), and the São Paulo Research Foundation (FAPESP - grant 12/22833-1).

ACKNOWLEDGMENTS

The authors would like to thank technicians and colleagues at Department of Botany of the University of São Paulo: Gisele Costa, Erismaldo de Oliveira, Paula Alécio, Simone Ferreira, and Tássia dos Santos for the technical support provided during the execution of this work; MSc. Vitor Barão, Dr. Giuliano Locosseli, Fábio Coelho and MSc. Plácido Buarque for helping during samplings; and MSc. Yasmin Vidal Hirao for providing the schematic drawing used in **Figure 2C**. We would also like to acknowledge Prof. Dr. Claudenir Simões Cairés for identifying the parasitic plant species.

VIDEO S1 | Internal view of the host-parasite interface between *P. perrottetii* and *T. guianensis* using the HRXCT technique.

VIDEO S2 | Internal view of the host-parasite interface between *P. bathyoryctum* and *C. fissilis* using the HRXCT technique.

REFERENCES

Breteron, N. J. B., Ahmed, F., Skyes, D., Ray, M. J., Shield, I., Karp, A., et al. (2015). X-ray micro-computed tomography in willow reveals tissue patterning of reaction wood and delay in programmed cell death. *BMC Plant Biology.* 15:83. doi: 10.1186/s12870-015-0438-0

Brodersen, C. R., McElrone, A. J., Choat, B., Matthews, M. A., and Shackel, K. A. (2010). The dynamics of embolism repair in Xylem: in vivo visualizations using high-resolution computed tomography. *Plant Physiol.* 154, 1088–1095. doi: 10.1104/pp.110.162396

Brodersen, C. R., Roark, L. C., and Pittermann, J. (2012). The physiological implications of primary xylem organization in two ferns. *Plant Cell Environ.* 35, 1898–1911. doi: 10.1111/j.1365-3040.2012.02524.x

Bukatsch, F. (1972). Bemerkungen zur doppelfärbung: astrablau-Safranin. *Mikrokosmos* 61, 255.

Calvin, C. L. (1967). Anatomy of the endophytic system of the mistletoe, *Phoradendron flavescens. Bot. Gaz.* 128, 117–137. doi: 10.1086/336388

Cannon, W. A. (1901). The anatomy of *Phoradendron villosum* Nutt. *Bull. Torrey Bot. Club* 28, 374–390. doi: 10.2307/2478757

Cochard, H., Delzon, S., and Badel, E. (2015). X-ray microtomography (micro-CT): a reference technology for high-resolution quantification of xylem embolism in trees. *Plant Cell Environ.* 38, 201–206. doi: 10.1111/pce.12391

Dhondt, S., Vanhaeren, H., Van Loo, D., Cnudde, V., and Inzé, D. (2010). Plant structure visualization by high-resolution X-ray computed tomography. *Trends Plant Sci.* 15, 419–422. doi: 10.1016/j.tplants.2010.05.002

Fineran, B. A., and Calvin, C. L. (2000). Transfer cells and flange cells in sinkers of the mistletoe *Phoradendron macrophyllum* (Viscaceae), and their novel combination. *Protoplasma* 211, 76–93. doi: 10.1007/BF01279901

Gee, C. T. (2013). Applying microCT and 3D visualization to jurassic silicified conifer seed cones: a virtual advantage over thin-sectioning. *Appl. Plant Sci.* 1, apps.1300039. doi: 10.3732/apps.1300039

Heeraman, D. A., Hopmans, J. W., and Clausnitzer, V. (1997). Three dimensional imaging of plant roots in situ with X-ray Computed Tomography. *Plant Soil.* 189, 167–179. doi: 10.1023/A:1004258818538

Heide-Jorgensen, H. S. (2008). *Parasitic Flowering Plants.* Leiden: Brill.

Hounsfield, G. N. (1976). Historical notes on computerized axial tomography. *J. Can. Assoc. Radiol.* 27, 135–142.

Johansen, D. A. (1940). *Plant Microtecnique*. New York, NY: Mc Graw-Hill Book Company.

Kaestner, A., Schneebeli, M., and Graf, F. (2006). Visualizing three-dimensional root networks using computed tomography. *Geoderma* 136, 459–469. doi: 10.1016/j.geoderma.2006.04.009

Kraus, J. E., and Arduin, M. (1997). *Manual Básico de Métodos em Morfologia Vegetal*. Seropédica: EDUR.

Kuijt, J. (1964). Critical observations on the parasitism of new world mistletoes. *Can. J. Bot.* 42, 1243–1278. doi: 10.1139/b64-118

Kuijt, J. (2003). Monograph of *Phoradendron* (Viscaceae). *Syst. Bot. Monogr.* 66, 1–643. doi: 10.2307/25011253

Many, M. S. (1964). *Ecology of Plant Galls*. The Hague: Junk, 335–434. doi: 10.1007/978-94-017-6230-4

Metscher, B. D. (2013). Biological applications of X-ray microtomography: imaging microanatomy, molecular expression and organismal diversity. *Microsc. Anal.* 27, 13–16.

Schmid, R., Calvin, C. L., and Wilson, C. A. (2011). Sinker structure of *Phoradendron californicum* (Viscaceae) confounds its presumed close relationship to other acataphyllous species. *Aliso* 29, 13–23. doi: 10.5642/aliso.20112901.03

Sperry, J. S., Donnelly, J. R., and Tyree, M. T. (1988). A method for measuring hydraulic conductivity and embolism in xylem. *Plant Cell Environ.* 11, 35–40. doi: 10.1111/j.1365-3040.1988.tb01774.x

Staedler, Y. M., Masson, D., and Schönenberger, J. (2013). Plant tissues in 3D via X-ray tomography: simple contrasting methods allow high resolution imaging. *PLoS ONE* 8:e75295. doi: 10.1371/journal.pone.0075295

Steppe, K., Cnudde, V., Girard, C., Lemeur, R., Cnudde, J. P., and Jacobsb, P. (2004). Use of X-ray computed microtomography for non-invasive determination of wood anatomical characteristics. *J. Struct. Biol.* 148, 11–21. doi: 10.1016/j.jsb.2004.05.001

Stevens, P. F. (2001). [continuously updated since]. Angiosperm phylogeny website. Version 14, 2014. Available at: http://www.mobot.org/MOBOT/research/APweb/

Thoday, D. F. R. S. (1957). Modes of union and interaction between parasite and host in the Loranthaceae - II Phoradendreae. *Proc. R. Soc. Lond.* 146, 320–338. doi: 10.1098/rspb.1957.0014

Conflict of Interest Statement: The authors declare that the research was conducted in the absence of any commercial or financial relationships that could be construed as a potential conflict of interest.

15

Trophic Relationships between the Parasitic Plant Species *Phelipanche ramosa* (L.) and Different Hosts Depending on Host Phenological Stage and Host Growth Rate

Edited by:
Paul Christiaan Struik,
Wageningen University, Netherlands

Reviewed by:
Grama Nanjappa Dhanapal,
University of Agricultural Sciences,
Bengaluru, India
Aad Van Ast,
Wageningen University, Netherlands

***Correspondence:**
Delphine Moreau
delphine.moreau@dijon.inra.fr

Delphine Moreau*, Stéphanie Gibot-Leclerc, Annette Girardin, Olivia Pointurier, Carole Reibel, Florence Strbik, Mónica Fernández-Aparicio and Nathalie Colbach

Agroécologie, AgroSup Dijon, INRA, Université de Bourgogne Franche-Comté, Dijon, France

Phelipanche ramosa (L.) Pomel (branched broomrape) is a holoparasitic plant that reproduces on crops and also on weeds, which contributes to increase the parasite seed bank in fields. This parasite extracts all its nutrients at the host's expense so that host–parasite trophic relationships are crucial to determine host and parasite growth. This study quantified the intensity with which *P. ramosa* draws assimilates from its host and analyzed whether it varied with host species, host phenological stage and host growth rate. A greenhouse experiment was conducted on three host species: the crop species *Brassica napus* (L.) (oilseed rape) and two weed species, *Capsella bursa-pastoris* (L.) Medik. and *Geranium dissectum* (L.). Plants were grown with or without *P. ramosa* and under three light levels to modulate host growth rate. The proportion of host biomass loss due to parasitism by *P. ramosa* differed between host species (at host fructification, biomass loss ranged from 34 to 84%). *B. napus* and *C. bursa-pastoris* displayed a similar response to *P. ramosa*, probably because they belong to the same botanical family. The sensitivity to *P. ramosa* in each host species could be related to the precocity of *P. ramosa* development on them. Host compartments could be ranked as a function of their sensitivity to parasitism, with the reproductive compartment being the most severely affected, followed by stems and roots. The proportion of biomass allocated to leaves was not reduced by parasitism. The proportion of pathosystem biomass allocated to the parasite depended on host species. It generally increased with host stage progression but was constant across light induced-host growth rate, showing that *P. ramosa* adapts its growth to host biomass production. The rank order of host species in terms of sink strength differed from that in terms of host sensitivity. Finally, for *B. napus*, the biomass of individual parasite shoots decreased with increasing their number per host plant, regardless of host growth rate. Results will be incorporated into a mechanistic model in order to analyze the effect of parasitic plant species on weed community assembly and to design new cropping systems for controlling *P. ramosa*.

Keywords: *Phelipanche ramosa*, *Brassica napus*, *Geranium dissectum*, *Capsella bursa-pastoris*, weed, biomass, host, parasite

INTRODUCTION

Broomrapes are root-parasitic plant species of the *Phelipanche* and *Orobanche* genera that can cause severe yield losses to economically important crop species all over the world (Dhanapal et al., 1996; Parker, 2009). In France, branched broomrape – *Phelipanche ramosa* (L.) Pomel (syn. *Orobanche ramosa*) – is a frequent and harmful parasitic plant species with a large range of hosts including Solanaceae, Brassicaceae, and Fabaceae species. Among arable crops, *Brassica napus* (L.) (oilseed rape) is the favorite *P. ramosa* host with yield losses up to 90% (Gibot-Leclerc et al., 2012). *P. ramosa* reproduces not only on crop species but also on weed species (Boulet et al., 2001). Thus, even in the absence of a host crop, weeds can allow an increase in *P. ramosa* soil seed bank which can then infest a subsequent crop. To date, no efficient method is available for controlling broomrape in arable crops and integrated management practices are required (Dhanapal et al., 1996).

P. ramosa is a root-holoparasitic (i.e., chlorophyll-lacking) angiosperm that connects with the vascular system of its host's roots and extracts all its nutrients (carbohydrates and minerals) and water at the host's expense (Parker and Riches, 1993). Consequently, host–parasite trophic relationships are crucial to determine the growth and seed production of the two interacting plants and, therefore, the harmfulness of *P. ramosa*. For holoparasitic plants, host–parasite trophic relationships have never been compared for different host species under the same experimental conditions. Several studies dealing with trophic relationships in single broomrape–host combination pairs revealed that the deleterious effect of broomrape parasitism on host biomass production and distribution among host compartments (leaves, stems, and roots) varies across host and parasitic species, number of parasites and growing conditions (Barker et al., 1996; Dale and Press, 1998; Hibberd et al., 1998; Manschadi et al., 2001; Lins et al., 2007; Mauromicale et al., 2008). In the particular case of *P. ramosa*, host–parasite trophic relationships were studied for only one host species, *Solanum lycopersicum* L. (tomato) (Mauromicale et al., 2008). The trophic relationships have never been characterized for this parasite with either *B. napus* (*P. ramosa*'s favorite arable host crop in France) or weed species. Yet, such knowledge is crucial to better understand the impact of *P. ramosa* on *B. napus* production, both directly (by nutrient withdrawal from *B. napus* plants) and indirectly (via weed parasitism increasing *P. ramosa* soil seed bank) and thereby to parametrize simulation models allowing to identify cropping systems adapted to the control of *P. ramosa* (Colbach et al., 2011, 2014).

The trophic relationships can be characterized by, on the one hand, the response of the host plant to parasitism, referring to both number of parasite attachments and host growth reduction due to parasitism (Mauromicale et al., 2008) and, on the other hand, by the parasite sink strength, i.e., the proportion of the parasite biomass or growth rate relative to that of the host plant or the pathosystem (Lins et al., 2007; Hautier et al., 2010). Previous studies on host–parasite trophic relationships comparing different host species concerned hemiparasitic plant species (Matthies, 1996; Cameron et al., 2006; Hautier et al., 2010). In contrast to holoparasitic plant species like *P. ramosa*, hemiparasitic plant species are photosynthetically active, relying on the host for mineral nutrients, water and sometimes for carbohydrates (Watling and Press, 2001; Irving and Cameron, 2009). For these parasites, host growth reduction due to parasitism was shown to depend on host species (Cameron et al., 2006) while the analysis of the parasite sink strength showed that the parasite growth rate, and hence parasite biomass, increased with host growth rate (Hautier et al., 2010). Host species parasitized by a hemiparasitic plant were shown to be ranked the same, both in terms of host response to parasitism and parasite sink strength (Matthies, 1996).

Focusing on a holoparasitic plant species, the objective of the present study was to quantify the intensity with which *P. ramosa* draws assimilates from *B. napus*, its favorite arable host crop in France, and to analyze whether this intensity varies with host phenological stages and light intensity-induced host-growth rate. Our objective was also to compare the responses and effects of *P. ramosa* in two additional host species with differing levels of ability to promote parasite growth. The following research hypotheses were tested. Firstly, for a given host species, we assumed a constant response to parasitism, meaning that the growth of parasitized host plants will be proportional to the growth of healthy host plants, whatever the host stage, the host growth rate and the number of attached parasites. Secondly, for a given host species, we assumed a constant sink strength of the parasite, meaning that the growth of *P. ramosa* will be proportional to that of the pathosystem including host and parasite, irrespective of host stage, host growth rate and number of parasite attachments. Plant biomass will be considered as a proxy of assimilate fluxes to analyze host–parasite trophic relationships. These hypotheses were tested on host species grown in the presence or absence of *P. ramosa* in order to evaluate the impact of parasitism on host growth and under three light levels in order to generate different host growth rates. Oilseed rape (*B. napus*) was used as the preferred host crop species that supports numerous attachments and aboveground shoots of *P. ramosa* (Gibot-Leclerc et al., 2012). *Capsella bursa-pastoris* (L.) and *Geranium dissectum* (L.) were also used. They are both host weed species that support a limited number of viable attachments of *P. ramosa* (Boulet et al., 2001). Attachments generally result in aboveground parasite shoots on *C. bursa-pastoris* (Boulet et al., 2001) but more erratically on *G. dissectum*: no emergence was observed in greenhouse conditions (Boulet et al., 2001), while emerged shoots were reported in the field (Supplementary Data Sheet 1). Host reponse to parasitism was analyzed in terms of *P. ramosa* attachments, total host biomass production and biomass partitioning among host compartments (i.e., leaves, stems, roots, and reproductive compartments). The sink strength of the parasite was analyzed in terms of *P. ramosa* biomass production relative to that of the pathosystem as well as parasite shoot number per host plant and average parasite shoot biomass.

MATERIALS AND METHODS

Experimental Treatments

A greenhouse experiment was conducted in Dijon (France) from September 2013 to July 2014 with three host species, two parasite seed densities, three light conditions and three replicates at each harvest date. The three host species were *B. napus* (Aviso cultivar, seeds from SW Seeds), *C. bursa-pastoris* and *G. dissectum* (seeds from Herbiseed). There exist several pathovar for *P. ramosa* (Le Corre et al., 2014). Here, the pathovar specific to the most cultivated French arable crop species (i.e., oilseed rape) was chosen. Parasite seeds were collected in 2002 from severely infested oilseed rape fields and identified as belonging to the *B. napus* pathovar (Le Corre et al., 2014). Plants of the three host species were grown without and with *P. ramosa* seeds (165 mg of *P. ramosa* seeds per pot, corresponding to ca. 55000–60000 seeds). Hereafter, plants of both treatments are referred as healthy and parasitized plants, respectively. Plants were also grown under three light levels (100, 34, and 29% daylight). The 100% light level corresponded to natural light whereas the 34 and 29% light levels were obtained using two types of shading net allowing to artificially reduce light levels over the plots[1]. The experimental design was a complete factorial design with three replicates. For each of the 18 experimental treatments (3 host species × 2 *P. ramosa* seed densities × 3 light levels), 12 plants were grown (4 harvest dates × 3 replicates), resulting in a total of 216 plants in the experiment.

Cultural Conditions

Two L-pots were filled with 0.4 L of clay balls at the bottom and 1.6 L of a substrate made up of 1:3 of Biot sand and 2:3 of a disinfected soil (collected at Experimental station of Dijon-Epoisses, France) determined as 8% sand, 57% silt, 35% clay. For the treatment with *P. ramosa*, parasite seeds were mixed with all of the substrate moistened beforehand. Host plant seeds were sown 14 days later to allow preconditioning of *P. ramosa* seeds, i.e., the exposure of parasite seeds to temperate and moist conditions to make them more susceptible to host root exudates (Musselman, 1980). For each host species, one host seed was sown per pot on September 12th 2013. As *G. dissectum* seedlings did not emerge and grow correctly, a new batch of seeds was sown 4 months later. *G. dissectum* plants were grown during 6 weeks in a growth chamber to mimic greenhouse conditions similar to those experienced by the two other species, before transfer into the greenhouse in March.

Tap water was provided at frequencies and quantities that were adjusted to maintain humidity at about 70% of soil water-holding capacity. The water loss through evapotranspiration was estimated every 1–3 days by weighing reference pots (one per experimental treatment). Whenever the water content of the soil was below 70% of soil water-holding capacity, the pots were irrigated. In addition, 10 mL of a nutrient solution (10-10-10 corresponding to N-P-K) was provided once a week per pot. Our objective was to obtain environmental conditions as close as possible to those in the field. Thus, neither heating nor artificial light were used. Air temperature (PT100 sensors; Pyro-Contrôle, Vaulx-en-Velin, France) and incident photosynthetically active radiation (PAR; silicium sensors; Solems, Palaiseau, France) measurements were taken every 600 s and stored in a data logger (DL2e; Delta-T Devices, Cambridge, England). For *B. napus* and *C. bursa-pastoris*, the mean daily temperature (including night temperatures) was 13.3°C (ranging from 3.3 to 28.7°C), the mean daily incident photosynthetic active radiation for the 100% light level treatment was 8 mol m^{-2} day^{-1} (ranging from 1 to 33 mol m^{-2} day^{-1}) and day length was 12.0 h (ranging from 8.2 to 16.1 h). For *G. dissectum* during the whole growth period (including in growth chamber), the mean daily temperature was 16.7°C (ranging from 6.2 to 28.7°C), the mean daily incident photosynthetic active radiation for the 100% light level treatment was 8 mol m^{-2} day^{-1} (ranging from 2 to 11 mol m^{-2} day^{-1}) and day length was 14.3 h (ranging from 11.2 to 16.1 h).

Plant Measurements

Plants were harvested at four phenological stages of the host species: rosette, elongation, flowering, and fructification. As parasitism did not delay host phenology, parasitized and healthy plants were harvested at the same date for a given host species in a given light treatment (Supplementary Data Sheet 2). Biomass was determined after drying for 48 h at 80°C with a balance precision up to 10^{-5} g. Host biomass was measured separately for roots, stems, leaves, and reproductive organs (including flowers and fruits). For *P. ramosa*, the number of attached individual parasitic plants and the number of aboveground *P. ramosa* shoots per host plant were counted and aboveground and belowground biomass per host plant were determined separately.

Statistical Analyses

Analyzed variables were: total host biomass, host biomass per compartment (i.e., roots, stems, leaves, and reproductive organs), aboveground and belowground parasite biomass per host plant and number of aboveground shoots of *P. ramosa* per host plant. Analysis of variance was performed to study the effects of experimental factors, i.e., host species, *P. ramosa* parasitism, light level and host phenological stage, and their interactions on growth variables. Analysis of covariance was performed to analyze correlations between growth variables in relation to experimental factors. Significance was determined using $\alpha = 0.05$ and analyses were performed using the lm function of R x64 3.0.3 (R Development Core Team, 2014).

RESULTS

Parasite Number and Phenology Across Host Species

The number of attached *P. ramosa* plants per host plant differed between host species ($P < 0.001$), light levels ($P < 0.001$) and host phenological stages ($P = 0.002$; **Table 1**). *P. ramosa* infection success was the highest in *G. dissectum*, especially until flowering,

[1]We initially aimed at 60 and 30% light levels for the two shaded treatments but the continuous light measurements during the experiment showed the shading to be more similar than planned, despite preliminary tests

TABLE 1 | Number of attached parasites for the three host species, the four phenological stages and the three light levels.

Host species	Light level	Rosette	Elongation	Flowering	Fructification
Brassica napus	100%	27.7	25.0	8.7	35.3
	34%	10.7	10.7	17.7	17.7
	29%	0.0	4.3	0.7	4.7
Capsella bursa-pastoris	100%	0.3	0.3	1.3	2.0
	34%	1.0	1.0	0.7	3.3
	29%	0.0	0.3	0.0	0.3
Geranium dissectum	100%	47.0	108.7	21.7	30.7
	34%	20.0	34.3	7.7	6.3
	29%	14.0	19.7	14.7	3.3

followed by *B. napus* while it was low for *C. bursa-pastoris*. For *B. napus* and *G. dissectum*, the number of attachments decreased with decreasing light level ($P < 0.003$) and, for *G. dissectum* only, it varied with host phenological stage, being maximal at elongation ($P < 0.001$).

For the three host species, *P. ramosa* aboveground shoot number per host plant was maximal at the host fructification stage. It was much lower for *G. dissectum* (0.4 parasite shoots per host plant) and *C. bursa-pastoris* (1) than for *B. napus* (11) averaged over all light levels. So, despite the high number of *P. ramosa* attachments on *G. dissectum*, the success of *P. ramosa* to develop on this host species was low. Synchronization between host and parasite varied among species, with *P. ramosa* development, expressed relatively to host development, being later on *C. bursa-pastoris* (**Table 2**). Whatever the host species, the parasite did not have the time to fructify during the experiment.

Pathosystem Biomass Production

Host species, light level and host phenological stage all affected total pathosystem biomass production, i.e., the combined host and parasite biomass in parasitized plants, or the sole host biomass in healthy plants ($P < 0.001$). In addition, parasitism significantly decreased the pathosystem biomass ($P < 0.001$; **Figure 1**). This pathosystem biomass loss did not differ among host species ($P = 0.53$) but it varied with host phenological stage ($P < 0.001$). For *B. napus* and *C. bursa-pastoris* from rosette to flowering, pathosystem biomass was not or only little affected by parasitism. The effect was either non-significant or associated to a very low partial R^2 (**Table 3**) while parasitism had a significant effect on pathosystem biomass at fructification stage **Table 3**. At host fructification, pathosystem biomass loss was 36 and 26% for *B. napus* and *C. bursa-pastoris*, respectively (average over all light levels). For *G. dissectum*, parasitism decreased pathosystem biomass at all host phenological stages (**Table 3**), with the largest effect at host flowering and fructification stages where pathosystem biomass was respectively reduced by 78 and 79% compared to healthy host plants (average over all light levels). For the three species, the deleterious effect of parasitism on pathosystem biomass production also varied with light level ($P < 0.001$; **Figure 1**). It increased with light level for both *B. napus* and *G. dissectum*, while it was the strongest at the lowest light level for *C. bursa-pastoris*.

TABLE 2 | Synchronization between host and parasite phenology.

Host species	Light level	Host stage at parasite emergence	Host stage at parasite flowering
Brassica napus	100%	Flowering	Fructification
	34%	Flowering	Fructification
	29%	Flowering	Fructification
Capsella bursa-pastoris	100%	Fructification	–
	34%	Fructification	Fructification
	29%	–	–
Geranium dissectum	100%	Fructification	–
	34%	Flowering	Flowering
	29%	Flowering	Flowering

Host Biomass Production

Host species, light level, host phenological stage and parasitism by *P. ramosa* all affected total host biomass ($P < 0.001$; **Figure 1**). In addition, the effect of parasitism differed among host species ($P < 0.001$) and host phenological stages ($P < 0.001$). For *B. napus* and *G. dissectum*, parasitism affected host biomass at all phenological stages except rosette stage while for *C. bursa-pastoris* it affected host biomass at fructification stage only (**Table 4**). At fructification, host biomass was reduced by ca. 84% for *G. dissectum*, 76% for *B. napus* and 34% for *C. bursa-pastoris* (average over all light levels). Finally, the deleterious effect of parasitism increased with light availability for both *B. napus* and *G. dissectum*, whereas it was strongest at the intermediate light level for *C. bursa-pastoris* ($P < 0.001$). For *B. napus*, the effect of the parasitism increased with host phenological stage while that of the light level diminished. This is shown by the increase of partial R^2-values related to parasitism and the concomitant decrease of the partial R^2-values related to the light level with progression of phenological stages (**Table 4**).

The biomass of the parasitized host plants was positively and linearly related to that of the healthy host plants ($P < 0.001$). The regression parameters did not depend on light availability ($P = 0.55$) but varied with the host species ($P < 0.001$) and its phenological stage ($P < 0.001$). In *B. napus* (except at fructification) and *C. bursa-pastoris*, healthy and parasitized plant biomasses were correlated according to a single relationship ($P < 0.001$) that was valid whatever the host species ($P = 0.15$), the light level ($P = 0.63$) and the phenological stage ($P = 0.09$;

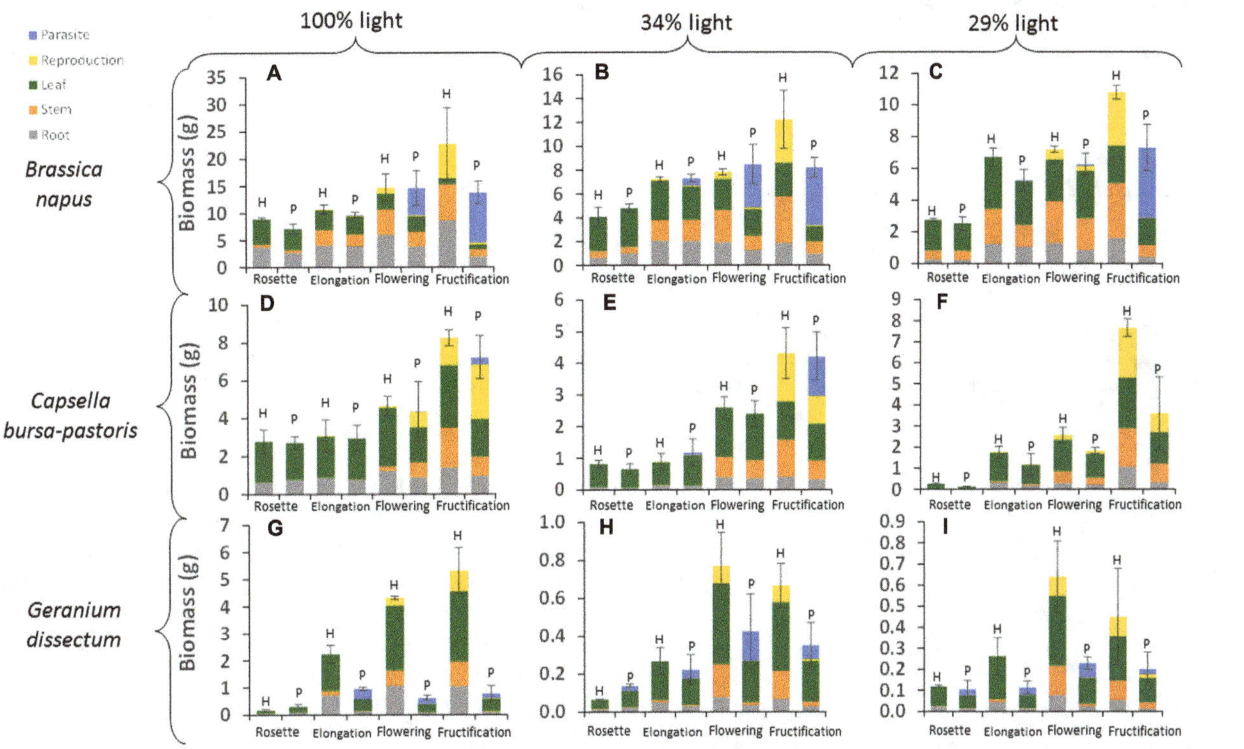

FIGURE 1 | Absolute values of biomass per compartment for each host species × host phenological stage × light level combination. H is for healthy plants and P for parasitized plants. Data are means for three independent replicates. SE is represented for pathosystem biomass values (i.e., the combined host and parasite biomass in parasitized plants, or the sole host biomass in healthy plants). The part-labels from **(A)** to **(I)** show the different host species per light level combinations.

Figure 2). The slope value of 0.82 g/g indicates that parasitism reduced host biomass by ca. 18% for *C. bursa-pastoris* whatever the phenological stage and *B. napus* from rosette to flowering stages. Parasitized plant biomass of *B. napus* at fructification stage was approximately 4 g, whatever light availability ($P = 0.17$). Biomasses of healthy and parasitized *G. dissectum* plants were also correlated ($P < 0.001$), with an average slope of 0.10 g/g showing that parasitism reduced host biomass by ca. 90% (average over all light levels) which is much higher than for the other host species. The regression slope varied with light level ($P = 0.004$) and host phenological stage ($P = 0.007$) but these effects were negligible (partial $R^2 = 0.09$ and 0.10, respectively). A similar regression analysis was performed on the pathosystem biomass, i.e., the sum of both host and parasite biomass, giving similar conclusions (Supplementary Data Sheet 3).

To analyze the tolerance of host species to parasitism, biomass loss per host plant was expressed relatively to parasite biomass. Averaged over the light levels at host fructification, values were

TABLE 3 | Effects of parasitism by *Phelipanche ramosa*, light level and interaction on pathosystem biomass for each host species and phenological stage.

Host species	Factors	Rosette	Elongation	Flowering	*Fructification*
Brassica napus	Light level	0.89***	0.85***	0.97***	0.48**
	Parasitism	ns	0.04*	ns	0.21**
	Interaction	0.05*	ns	ns	ns
Capsella bursa-pastoris	Light	0.93***	0.69***	0.64**	0.45**
	Parasitism	ns	ns	ns	0.17*
	Interaction	ns	ns	ns	0.16*
Geranium dissectum	Light level	0.47**	0.73***	0.42***	0.45***
	Parasitism	0.15*	0.10***	0.27***	0.21***
	Interaction	ns	0.13***	0.30***	0.29***

*Partial R^2-values determined by analysis of variance. ns for a non-significant effect, *$P < 0.05$; **$P < 0.01$; ***$P < 0.001$.*

TABLE 4 | Effects of parasitism by *Phelipanche ramosa*, light level and interaction on total host biomass for each host species and phenological stage.

Host species	Factors	Rosette	Elongation	Flowering	*Fructification*
Brassica napus	Light level	0.89***	0.83***	0.61***	0.16*
	Parasitism	ns	0.07**	0.19**	0.60***
	Interaction	ns	ns	ns	ns
Capsella bursa-pastoris	Light level	0.93***	0.70***	0.64**	0.50***
	Parasitism	ns	ns	ns	0.26***
	Interaction	ns	ns	ns	ns
Geranium dissectum	Light level	0.55***	0.56***	0.37***	0.42***
	Parasitism	ns	0.18***	0.31***	0.23***
	Interaction	ns	0.22***	0.31***	0.31***

*Partial R^2-values determined by analysis of variance. ns for a non-significant effect, *$P < 0.05$; **$P < 0.01$; ***$P < 0.001$.*

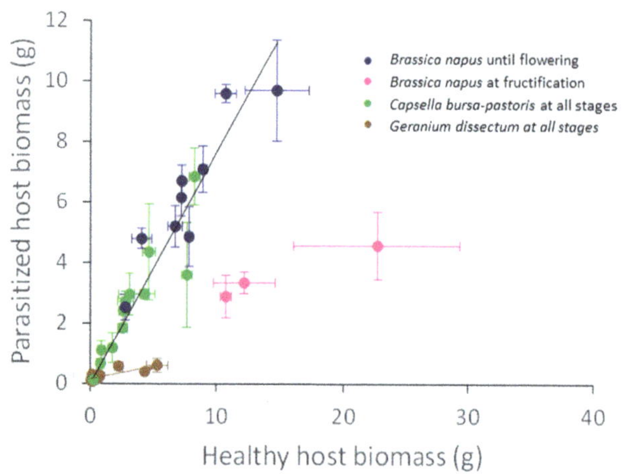

FIGURE 2 | Correlation between parasitized (P) and healthy (H) host biomass per host species. Data are mean ± SE for three independent replicates ($n = 3$). The black line is for both *Capsella bursa-pastoris* at all stages and *Brassica napus* until flowering only ($y = 0.82x$; $R^2 = 0.97$). The brown line is for both *Geranium dissectum* at all stages ($y = 0.10x$; $R^2 = 0.77$).

1.9 ± 0.1 g/g for *B. napus*, 2.4 ± 1.3 g/g for *C. bursa-pastoris*, and 14.4 ± 8.8 g/g for *G. dissectum* showing that *G. dissectum* was much less tolerant to *P. ramosa* than the other two species.

Host Biomass Distribution per Compartment

The biomass distribution per compartment of the host plant (leaves, stems, roots, and reproductive organs) was calculated by dividing the biomass of a given compartment by the biomass of the pathosystem, including both host biomass and parasite biomass if any (**Figure 3**). Whatever the compartment, host species ($P < 0.001$), light level ($P < 0.024$), host phenological stage ($P < 0.001$), and parasitism ($P < 0.004$) all affected biomass distribution. Except for leaves, the effect of the parasitism also varied with host species ($P < 0.001$), phenological stage ($P < 0.006$), and light availability ($P < 0.01$). The effect of parasitism on leaf biomass distribution varied with host phenological stage ($P = 0.002$) but not with host species ($P = 0.08$) and light level ($P = 0.17$).

The biomass distribution per compartment was analyzed per host species (**Table 5**). For *B. napus*, the biomass distribution in leaves was affected by parasitism at the two lowest light levels. However, this effect was minor compared to that of the other factors (**Table 5**). For stem, root and reproductive compartments, the adverse effect of parasitism increased with the host phenological stages progression (**Figure 3**). The compartments could be ranked according to their sensitivity to parasitism. At *B. napus* fructification, the reproductive compartment was the most severely affected (averaged over all light levels, reproductive-biomass proportion was reduced by 97% for parasitized vs. healthy plants), followed by stems (reduced by 66%), roots (reduced by 50%) and leaves (not affected).

Generally, in *C. bursa-pastoris*, parasitism did not affect the distribution of biomass, whatever the compartment (**Figure 3**; **Table 5**). Biomass allocation to leaves and reproductive compartment was though affected by parasitism at the highest light level whereas biomass allocation to stems was reduced at host fructification. However, these effects were minor compared to the direct effects of the other factors (low partial R^2-values associated to the effects of interactions between factors in **Table 5**).

In *G. dissectum*, parasitism generally decreased biomass allocation to host compartments to the benefit of the parasite compartment (**Figure 3**, **Table 5**). The magnitude of the parasitism effect varied for all compartments (except roots) with host phenological stage (**Table 5**). At host fructification, the ranking of the host compartments according to their sensitivity to parasitism was identical to that for *B. napus*: the reproductive compartment was the most reduced (averaged over all light levels, reproductive-biomass proportion was reduced by 75% for parasitized vs. healthy plants), followed by stems (reduced by 66%), roots (reduced by 33%), and leaves (increased by 20%).

Phelipanche ramosa Biomass

Host species, phenological stage and light level all affected absolute values of total *P. ramosa* biomass (including aboveground and belowground biomass) per host plant (**Figure 1**; **Table 6**). The main effects were host species and phenological stage. Parasite biomass was larger on *B. napus* than on the other two species. For *B. napus*, host phenological

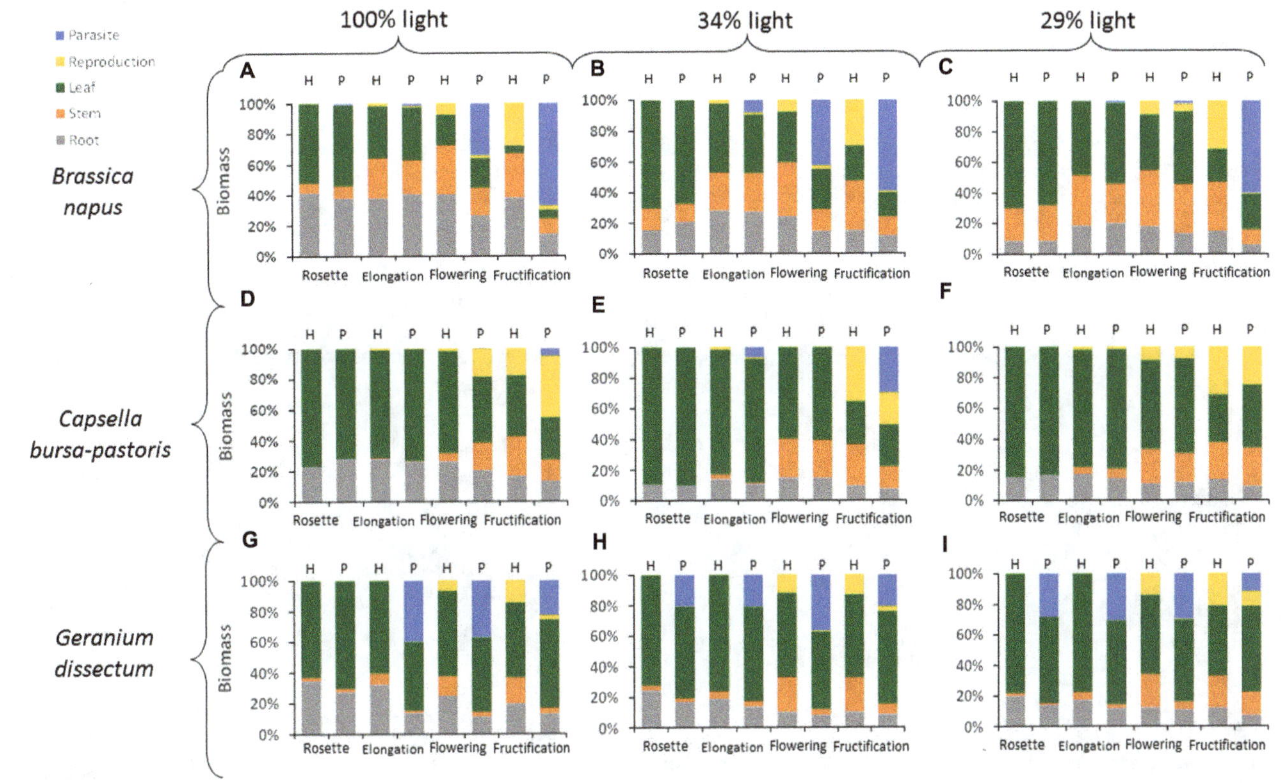

FIGURE 3 | Distribution of biomass per compartment for each host species × host phenological stage × light level combination. H is for healthy plants and P for parasitized plants. Data are means for three independent replicates. The part-labels from **(A)** to **(I)** show the different host species per light level combinations.

stage was the main factor affecting *P. ramosa* biomass per host plant and, for *C. bursa-pastoris*, it was the only significant factor (**Table 6**). For both species, *P. ramosa* biomass increased with host stage progression (**Figure 1**). For *G. dissectum*, light level had a greater impact (**Table 6**): depending on the light level, parasite biomass was the largest either at host elongation or flowering stage (**Figure 1**).

Parasite biomass per host plant was positively and linearly related to the biomass of the pathosystem including both host and parasite biomasses ($P < 0.001$). Host phenological stage ($P < 0.001$), host species ($P < 0.001$), and light level ($P < 0.001$) all influenced the regression parameters (**Figure 4**). For *B. napus*, the parameters of the regression varied with host stage ($P < 0.001$) but not with the light level ($P = 0.23$) and the interaction between both factors ($P = 0.17$). Parasite biomass was significantly correlated to pathosystem biomass at host flowering ($P = 0.006$) and fructification ($P < 0.001$). The slopes of the correlations (0.62 and 0.70 g/g at flowering and fructification, respectively) did not differ between both stages ($P = 0.65$), showing that *P. ramosa* biomass increased in similar proportions (around 66%) with increasing pathosystem biomass (**Figure 4A**). Y-intercepts differed between host stages ($P = 0.003$) because of the larger parasite biomass at host fructification than at flowering for a given pathosystem biomass. The difference in y-intercept suggests that, at flowering stage, a larger minimum amount of pathosystem biomass is needed before any measurable parasite biomass is being produced.

For *G. dissectum*, as for *B. napus*, the regression parameters varied with the host stage ($P < 0.001$) but not with the light level ($P = 0.96$) and the interaction between both factors ($P = 0.22$). Parasite biomass was significantly correlated to that of the pathosystem at all host stages ($P < 0.001$), except rosette ($P = 0.19$). Slopes ($P = 0.49$) and y-intercepts ($P = 0.99$) were not different between elongation and flowering stages (slope value at 0.43 g/g) while the slope was lower (slope value at 0.25 g/g) and y-intercept was higher at fructification stage ($P < 0.001$; **Figure 4C**). The lower slope value at fructification indicated that the increase in *P. ramosa* biomass with increasing pathosystem biomass was less than at elongation and flowering.

For *C. bursa-pastoris* with its small number of attached *P. ramosa* plants (see Parasite Number and Phenology Across Host Species), most of the *P. ramosa* biomass values were close to nil, except at host fructification (**Figure 4B**). Parasite biomass was not correlated to pathosystem biomass, whatever the host phenological stage.

Aboveground *P. ramosa* Shoot Number and Biomass

The number of *P. ramosa* aboveground shoots per host plant was analyzed in relation to the aboveground biomass per shoot

TABLE 5 | Effects of parasitism by *Phelipanche ramosa*, light level, host stage, and interactions on biomass distribution per compartment for each host species.

Host species	Factors	Leaf	Stem	Root	Reproduction
Brassica napus	Light	0.15***	0.11***	0.55***	ns
	Stage	0.80***	0.30***	0.14***	0.41***
	Parasitism	ns	0.22***	0.03**	0.19***
	Parasitism:light	0.01***	ns	ns	ns
	Parasitism:stage	ns	0.18***	0.05**	0.37***
	Light:stage	0.01**	0.07***	0.05*	ns
	Parasitism:light:stage	ns	0.04*	ns	ns
Capsella bursa-pastoris	Light	0.01**	0.04***	0.55***	ns
	Stage	0.87***	0.79***	0.14***	0.79***
	Parasitism	ns	ns	ns	ns
	Parasitism:light	0.02**	ns	ns	0.04***
	Parasitism:stage	ns	0.04***	ns	ns
	Light:stage	0.03**	0.04***	0.09***	0.02*
	Parasitism:light:stage	ns	0.04***	ns	0.06***
Geranium dissectum	Light	0.05*	0.02**	0.26***	0.03*
	Stage	0.25***	0.41***	0.33***	0.43***
	Parasitism	0.05**	0.25***	0.18***	0.18***
	Parasitism:light	ns	ns	0.06***	ns
	Parasitism:stage	0.19***	0.17***	ns	0.18***
	Light:stage	0.09*	0.04**	ns	0.04*
	Parasitism:light:stage	ns	0.03*	0.03*	ns

*Partial R^2-values determined by analysis of variance. ns for a non-significant effect, *P < 0.05; **P < 0.01; ***P < 0.001.*

of *P. ramosa*. The analysis was performed for *B. napus* which was the only host species on which *P. ramosa* produced several aboveground shoots (see Parasite Number and Phenology Across Host Species). Light level did not significantly affect either the number of shoots per host plant ($P = 0.12$) nor the biomass per shoot ($P = 0.20$) nor the total aboveground parasitic biomass per host plant ($P = 0.21$) at 2.8 ± 0.5 g/g. Nonetheless, as the number of *P. ramosa* shoots increased, the biomass of each shoot decreased following a negative power relation (**Figure 5**), regardless of the light level.

TABLE 6 | Effects of light level, host stage and interaction on *Phelipanche ramosa* total biomass for each host species.

Host species	Factors	*F*-values
Brassica napus	Light level	0.10**
	Host stage	0.62***
	Interaction	0.12*
Capsella bursa-pastoris	Light level	ns
	Host stage	0.26**
	Interaction	ns
Geranium dissectum	Light level	0.33***
	Host stage	0.22***
	Interaction	0.29***
All species	Host species	0.24***
	Host stage	0.18***
	Light level	0.03***
	Species:stage	0.28***
	Species:light	0.05***
	Stage:light	0.03*
	Species:stage:light	0.07***

*Partial R^2-values determined by analysis of variance. ns for a non-significant effect, *P < 0.05; **P < 0.01; ***P < 0.001.*

DISCUSSION

Response to *P. ramosa* Differed between Host Species

The effect of *P. ramosa* on host growth varied across host species, with host biomass losses ranging from 34 to 84% at host fructification. In the literature, the values of host biomass loss due to parasitism by broomrape species range from 15 to 72% (Dale and Press, 1998; Hibberd et al., 1998; Manschadi et al., 2001; Lins et al., 2007; Mauromicale et al., 2008). These values stem from independent studies differing on experimental design in terms of parasitic and host plant species, phenological stages at data collection, environmental conditions, and parasitic seed density. Even though our study considered only one parasitic plant species at a constant seed bank density and measurements were made at similar phenological stages for all host species, the range of values of biomass loss due to parasitism was as large as that found in literature.

B. napus and *C. bursa-pastoris* displayed similarities in their response to *P. ramosa*. Firstly, their tolerance to *P. ramosa*, i.e., their capacity to endure parasitism with minor losses of productivity, was similar. Secondly, the analysis of the correlation between the biomasses of parasitized and healthy hosts showed

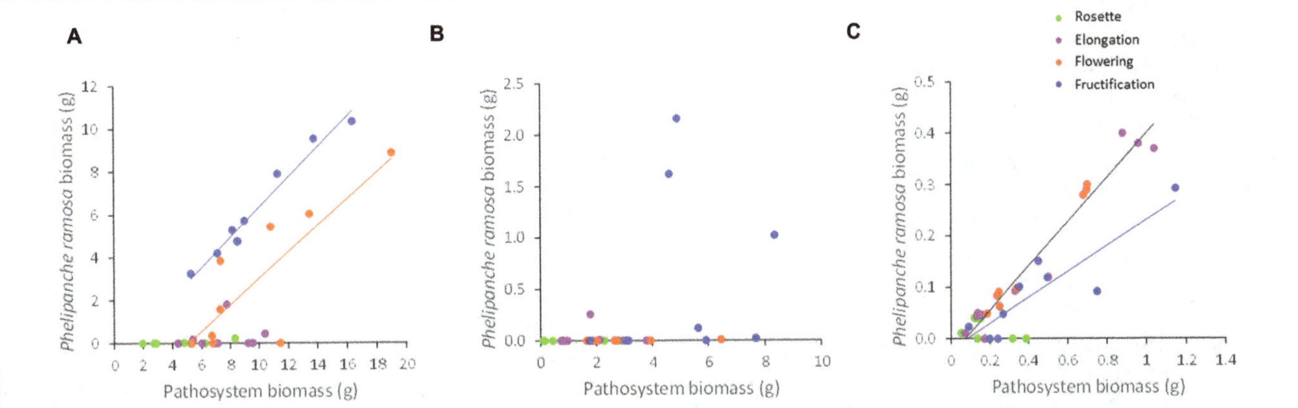

FIGURE 4 | Correlation between *Phelipanche ramosa* biomass and pathosystem biomass (including host and parasite biomasses) for the three host species, namely (A) *Brassica napus*, (B) *Capsella bursa-pastoris*, and (C) *Geranium dissectum*. Each point on the figure is a replicate. For *Brassica napus* the red and blue lines are for flowering stage (y = 0.62x-3.15; $R^2 = 0.68$) and fructification stage (y = 0.70x-0.70; $R^2 = 0.96$), respectively. For *Geranium dissectum*, the black line is for both elongation and flowering stages (y = 0.43x-0.03; $R^2 = 0.95$) and the blue line is for fructification stage (y = 0.25x-0.02; $R^2 = 0.80$).

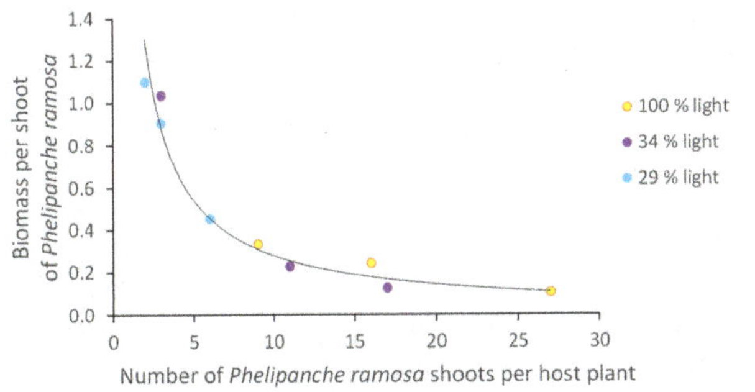

FIGURE 5 | Correlation between the biomass per shoot of *Phelipanche ramosa* and the number of *Phelipanche ramosa* shoots for *Brassica napus* at three different light levels. Each point on the figure is a replicate. The line is a power function ($y = 2.53x^{-0.96}$; $R^2 = 0.96$).

that the proportion of biomass loss due to parasitism was similar, except at host fructification when *B. napus* was more sensitive than *C. bursa-pastoris*. Thirdly, for both host species, the effect of parasitism on host biomass was maximal toward the end of the host life (fructification). Fourthly, the proportion of biomass loss due to parasitism was independent of the light level. These similarities between *B. napus* and *C. bursa-pastoris* could be due to their affiliation to the same botanical family. The lower resistance (characterized in our study by the number of parasite attachments) of *B. napus* compared to *C. bursa-pastoris* could be due to the *P. ramosa* pathovar that was used in this study, corresponding to a *B. napus* pathovar.

In contrast, *G. dissectum* was much more responsive to *P. ramosa*, with many more *P. ramosa* attachments reflecting a lower resistance of this host species. In spite of this, very few *P. ramosa* shoots emerged on this host. This could be due to (i) a genetic component of late-resistance in the host and/or (ii) an effect of high competition for host-derived nutritive resources between parasite individuals. Moreover, this species showed much more host biomass loss per g of parasite biomass, reflecting a lower tolerance to *P. ramosa*. Finally, for this host species, the maximal effect of parasitism occurred earlier in the host life-cycle (i.e., flowering). The dissimilarities between *G. dissectum* and the other two species could be linked to their different botanical families (Geraniaceae vs. Brassicaceae). Part of differences could though also be explained by the fact that this species was not grown at exactly the same period of time as the two other host species, though with identical parasite seed densities.

The relative timing of host and parasite stages seems to be a major factor for explaining differences in host tolerance among host species. The parasite was earlier on *G. dissectum* and consequently probably disrupted host plant functioning more. Parasite phenology was delayed on *C. bursa pastoris*, and therefore competition for assimilates tipped in favor of the host, with a lower impact on host growth. Considering a single host species, Manschadi et al. (1996) showed that the impact of the parasite *Orobanche crenata* on its host *Vicia faba* depended on the synchronization between parasite and host phenological stages. In our study, it is difficult to determine whether the relationship between host sensitivity and parasite precocity stem strictly from differences between plant species or also from differences in growth conditions (as *G. dissectum* was grown at a period of time different from that of the other two species). Nonetheless, both our results and Manschadi et al. (1996), suggest that this relationship could exist both at the intra- and inter-specific level.

Host Reproductive Compartment was the Most Severely Affected by Parasitism

Broomrapes plants are known to compete strongly for resources within host plants, causing significant changes in biomass distribution (Barker et al., 1996; Hibberd et al., 1998; Lins et al., 2007). Here, for *B. napus* and *G. dissectum*, parasitism by *P. ramosa* strongly affected biomass partitioning among compartments, namely leaf, stem, root and reproductive compartments. For *C. bursa-pastoris*, no direct effect of parasitism was observed on biomass partitioning, probably because this host species was not very sensitive to *P. ramosa* (see Response to *P. ramosa* Differed between Host Species) and because the parasite produced little biomass on this host (see Host Biomass Loss Due to Parasitism is Not Systematically Invested in *P. ramosa* Biomass). For *B. napus* and *G. dissectum*, the most severely affected compartment was the reproductive compartment, particularly for *B. napus*. The stem compartment was the second most affected, followed by roots, whereas the leaf compartment was not reduced by parasitism whatever the host species. A similar ranking of the host compartments was identified for other broomrape species (Barker et al., 1996; Manschadi et al., 1996; Hibberd et al., 1998; Grenz et al., 2008). Broomrape species act as an additional sink that competes with host compartments for assimilates. The strongest impact of parasitic plants on host reproductive growth was shown by Manschadi et al. (2001, 2004). It could be related to the synchronism between the periods of host reproductive growth and parasite shoot growth. Conversely, the other host compartments (roots, stems, and leaves) are already well-established at the time when parasite shoots start to require large amounts of assimilates for their growth. Even if the proportion of leaves was little affected by parasitism, plant photosynthesis was probably reduced as suggested by the substantially lower total biomass production of the pathosystem compared to total biomass production of healthy plants for most of our experimental treatments. In literature, the effect of parasitism on host photosynthesis is pathosystem dependent, with studies showing that photosynthesis is unaffected (example of *Orobanche minor* in Dale and Press, 1998) whereas others report a reduced photosynthesis, in particular caused by *P. ramosa* (Mauromicale et al., 2008).

Host Biomass Loss Due to Parasitism is not Systematically Invested in *P. ramosa* Biomass

Parasite sink strength was evaluated by the slope of the linear regression of parasite biomass vs. pathosystem biomass. It could not be evaluated for *C. bursa-pastoris* on which parasite biomass was negligible. On the two other species, *P. ramosa* was shown to adapt its sink strength to the host species and the host stage with, for a given host species and host stage, *P. ramosa* extracting a constant proportion of host assimilates whatever the host growth rate. To our knowledge, the present study is the first to analyze the effects of the host species and phenological stages on biomass allocation to the parasite. It allows ranking host species as a function of *P. ramosa* sink strength, with the lowest sink strength on *C. bursa-pastoris*, followed by *G. dissectum* and then *B. napus*. The highest sink strength on *B. napus* is consistent with the highest number of *P. ramosa* attachments and emerged shoots on this host species. The use of *P. ramosa* seeds from the *B. napus* pathovar is probably the explanation. The lowest sink strength on *C. bursa-pastoris* is consistent with the lowest sensitivity of this host species to *P. ramosa* (see Response to *P. ramosa* Differed between Host Species). However, the host species with the highest sensitivity to *P. ramosa* (*G. dissectum*) is not the species generating the highest *P. ramosa* sink strength (*B. napus*). In other words, *G. dissectum* was very sensitive to *P. ramosa* parasitism but *P. ramosa* was poorly efficient on this host, meaning that the host biomass loss due to parasitism was only poorly invested in *P. ramosa* biomass, contrary to what happened for *B. napus*.

Rethinking the Host Concept

While parasite biomass was significant on the three species, the parasite did not reproduce, questioning the host status the studied species. A host is defined as a species which allows the parasite to attach and which confers benefits to the parasite in terms of growth and reproduction. However, it is established that a greater spectrum of responses are displayed by potential hosts (from true host to non-host and all stages in between) especially in the case of parasitic plants that are not host-specific, i.e., able to parasitize different host species, such as *P. ramosa* (Cameron et al., 2006). Our results show that, even though *P. ramosa* does not reproduce, it can generate large biomass losses for the plant on which it is attached. Thus, the three studied species cannot be considered as non-host. Moreover, *P. ramosa* was previously shown to reproduce on these species (Supplementary Data Sheet 1; Boulet et al., 2001; Gibot-Leclerc et al., 2012). Altogether, these findings indicate that variations in the host status are possible for a given species. As discussed above, variations in synchronization between parasite and host phenological stages, due to variations in environmental conditions, could explain why *P. ramosa* is not always able to reproduce on a given host species. An improper synchronization for the parasite could result in not enough time for the parasite to complete its biological cycle, and/or a too large nutrient flow extracted by the parasite, causing host death.

Parasitic Aboveground Biomass per Shoot Decreased with Shoot Number per Host Plant

For *B. napus*, the aboveground biomass of individual *P. ramosa* shoots decreased with increasing shoot number per host plant, following a single relationship. Similar relationships were observed for two broomrape species, including *P. ramosa* (Hibberd et al., 1998; Mauromicale et al., 2008). Our study is the first to analyze the effect of host growth rate, modulated by the light level, on this relationship, showing that the relationship remained unchanged, whatever the host growth rate. Our data are consistent with Hibberd et al. (1998) in concluding that, as the number of parasite shoots increased, competition for assimilates between individual shoots decreased their individual biomass. Without such a regulation, i.e., if individual shoot biomass did not decrease with shoot number, a very high resource extraction by the parasite could potentially severely reduce host growth and consequently compromise parasite biomass and survival (Lins et al., 2007; Hautier et al., 2010).

Agronomic Implications

This study provides clues on the effects of the trophic relationships between *P. ramosa* and the weed flora on parasite soil seed bank in cropping systems. It is established that each of the studied host species can induce *P. ramosa* reproduction (see Gibot-Leclerc et al., 2012 for *B. napus*; Boulet et al., 2001 for *C. bursa*-pastoris; Supplementary Data Sheet 1 for *G. dissectum*). Even though *P. ramosa* did not reproduce in our study, our findings allow comparing the three host species. Parasite biomass was much lower on the two weed species than on *B. napus*. As seed production is known to be correlated to plant biomass (e.g., Lutman et al., 2011), our results suggest that these two weed species would contribute little to parasite seed bank. They could possibly even contribute to reduce rather than increase *P. ramosa* seed bank. Indeed, even though more than 50 *P. ramosa* individuals were fixed per *G. dissectum* plant, only few of them emerged and none produced seeds. This could be particularly true during summer fallow where the sheer abundance of weeds could compensate their small root system (and thus the resulting lower stimulation of fatal parasite germination) without hindering cash crop development. In the future, conducting new studies on a range of potential host weed species will be helpful to discriminate the weed species promoting from those impeding *P. ramosa* dissemination particularly checking parasite seed production ability. In order to characterize possible variations in the synchronization between host and parasite phenology for a given host species, these studies will have to consider different cohorts for the weed species which are able to germinate at different periods during the year (e.g., *C. bursa-pastoris* and *G. dissectum*; Mamarot and Rodriguez, 2014). Such studies should help to identify the host weed species that should be controlled in priority in order to manage *P. ramosa* seed bank in cropping systems.

Parasitism reduced host growth less in *C. bursa-pastoris* than in *G. dissectum*, suggesting that *P. ramosa* can affect plant species differently and therefore could modify weed community assembly in cropping systems. This result is supported by studies at the community level on other parasitic plants: some species (i.e., highly sensitive host species) were shown to be penalized by the presence of the parasite, leading to a lower resources uptake to the benefit of other species in the community (Gibson and Watkinson, 1991; Ameloot et al., 2005).

Our study tested only three host species. Considering a wider range of species, especially weed species, would be necessary to go further in the analysis of a host species effect. Our next step will be to integrate our data into a mechanistic model of the effects of cropping systems on parasite population dynamics. While previous models integrated interactions with the host crop species only (Grenz et al., 2006), a model that integrates parasite population dynamics in interaction with non-parasitic weed hosts will be considered (Colbach et al., 2011). Mechanistic models are particularly useful to predict and understand complex systems (Rossing et al., 1997; Colbach et al., 2014). They will thus help to further analyze the effect of parasitic plant species on weed growth and community assembly and to design cropping systems for controlling *P. ramosa*.

CONCLUSION

The intensity of the response to *P. ramosa* differed among host species and, depending on the host species, host biomass loss due to parasitism was not systematically invested in *P. ramosa* biomass. The parasite adapted its growth to host biomass production and the proportion of pathosystem biomass allocated to the parasite generally increased with host stage progression. Results suggests that some host weed species could contribute to increase while others could contribute to reduce *P. ramosa* soil seed bank. If confirmed, weed management may have to be rethought in order to restrain *P. ramosa* dissemination in infested fields and in fields with risks of infestations.

AUTHOR CONTRIBUTIONS

DM, SG-L, and NC designed the study. AG, CR, and FS collected the data. DM, AG, OP, MF-A, and NC analyzed and interpreted the data. All the authors contributed to manuscript writing.

FUNDING

The present work was supported by INRA (French National Institute for Agricultural Research), the research program "Assessing and reducing environmental risks from plant protection products" funded by the French Ministries in charge of Ecology and Agriculture, and the Marie-Curie FP7 COFUND People Program, through the award of an AgreenSkills' fellowship (under grant agreement n° PCOFUND-GA-2010-267196) to MF-A.

ACKNOWLEDGMENT

The authors are grateful to the greenhouse staff for their technical assistance.

REFERENCES

Ameloot, E., Verheyen, K., and Hermy, M. (2005). Meta-analysis of standing crop reduction by *Rhinanthus* spp. and its effect on vegetation structure. *Folia Geobot.* 40, 289–310. doi: 10.1007/BF02803241

Barker, E., Press, M., Scholes, J., and Quick, W. (1996). Interactions between the parasitic angiosperm *Orobanche aegyptiaca* and its tomato host: growth and biomass allocation. *New Phytol.* 133, 637–642. doi: 10.1111/j.1469-8137.1996.tb01932.x

Boulet, C., Labrousse, P., Arnaud, M. C., Zehhar, N., and Fer, A. (2001). "Weed species present various responses to *Orobanche ramosa* L. attack," in *Proceedings of the 7th International Parasitic Weed Symposium*, eds A. Fer, P. Thalouarn, D. M. Joel, L. J. Musselman, C. Parker, and J. A. C. Verkleij, (Nantes: Faculte des sciences, Université de Nantes), 228–231. doi: 10.1016/S1161-0301(00)00058-7

Cameron, D. D., Coats, A. M., and Seel, W. E. (2006). Differential resistance among host and non-host species underlies the variable success of the hemi-parasitic plant *Rhinanthus minor*. *Ann. Bot.* 98, 1289–1299. doi: 10.1093/aob/mcl218

Colbach, N., Abdennebi-Abdemessed, N., and Gibot-Leclerc, S. (2011). A preliminary approach for modelling the effects of cropping systems on the dynamics of broomrape (*Phelipanche ramosa*) in interaction with the non-parasitic weed flora. *Oléagineux Corps gras Lipides* 18, 39–45. doi: 10.1051/ocl.2011.0360

Colbach, N., Biju-Duval, L., Gardarin, A., Granger, S., Guyot, S., Mézière, D., et al. (2014). The role of models for multicriteria evaluation and multiobjective design of cropping systems for managing weeds. *Weed Res.* 54, 541–555. doi: 10.1111/wre.12112

Dale, H., and Press, M. C. (1998). Elevated atmospheric CO_2 influences the interaction between the parasitic angiosperm *Orobanche minor* and its host *Trifolium repens*. *New Phytol.* 140, 65–73. doi: 10.1046/j.1469-8137.1998.00247.x

Dhanapal, G. N., Struik, P. C., Udayakumar, M., and Timmermans, P. C. J. M. (1996). Management of broomrape (*Orobanche* spp.)-a review. *J. Agron. Crop Sci.* 175, 335–359. doi: 10.1111/j.1439-037X.1996.tb00479.x

Gibot-Leclerc, S., Sallé, G., Reboud, X., and Moreau, D. (2012). What are the traits of *Phelipanche ramosa* (L.) Pomel that contribute to the success of its biological cycle on its host *Brassica napus* L? *Flora* 207, 512–521. doi: 10.1016/j.flora.2012.06.011

Gibson, C., and Watkinson, A. (1991). Host selectivity and the mediation of competition by the root hemiparasite *Rhinanthus minor*. *Oecologia* 86, 81–87. doi: 10.1007/BF00317393

Grenz, J., Iştoc, V., Manschadi, A., and Sauerborn, J. (2008). Interactions of sunflower (*Helianthus annuus*) and sunflower broomrape (*Orobanche cumana*) as affected by sowing date, resource supply and infestation level. *Field Crops Res.* 107, 170–179. doi: 10.1016/j.fcr.2008.02.003

Grenz, J., Manschadi, A., Meinke, H., and Sauerborn, J. (2006). Simulating crop–parasitic weed interactions using APSIM: model evaluation and application. *Eur. J. Agron.* 24, 257–267. doi: 10.1016/j.eja.2005.10.002

Hautier, Y., Hector, A., Vojtech, E., Purves, D., and Turnbull, L. A. (2010). Modelling the growth of parasitic plants. *J. Ecol.* 98, 857–866. doi: 10.1111/j.1365-2745.2010.01657.x

Hibberd, J., Quick, W., Press, M., and Scholes, J. (1998). Can source–sink relations explain responses of tobacco to infection by the root holoparasitic angiosperm *Orobanche cernua*? *Plant Cell Environ.* 21, 333–340. doi: 10.1046/j.1365-3040.1998.00272.x

Irving, L. J., and Cameron, D. D. (2009). You are what you eat: interactions between root parasitic plants and their hosts. *Adv. Bot. Res.* 50, 87–138. doi: 10.1016/S0065-2296(08)00803-3

Le Corre, V., Reibel, C., and Gibot-Leclerc, S. (2014). Development of microsatellite markers in the branched broomrape *Phelipanche ramosa* L. (Pomel) and evidence for host-associated genetic divergence. *Int. J. Mol. Sci.* 15, 994–1002. doi: 10.3390/ijms15010994

Lins, R. D., Colquhoun, J. B., and Mallory-Smith, C. A. (2007). Effect of small broomrape (*Orobanche minor*) on red clover growth and dry matter partitioning. *Weed Sci.* 55, 517–520. doi: 10.1614/WS-07-049.1

Lutman, P. J. W., Wright, K. J., Berry, K., Freeman, S. E., and Tatnell, L. (2011). Estimation of seed production by *Myosotis arvensis*, *Veronica hederifolia*, *Veronica persica* and *Viola arvensis* under different competitive conditions. *Weed Res.* 51, 499–507. doi: 10.1111/j.1365-3180.2011.00863.x

Mamarot, J., and Rodriguez, A. (2014). *Mauvaises Herbes Des Cultures*. Paris: ACTA.

Manschadi, A., Kroschel, J., and Sauerborn, J. (1996). Dry matter production and partitioning in the host-parasite association *Vicia faba-Orobanche crenata*. *J. Appl. Bot.* 70, 224–229.

Manschadi, A., Sauerborn, J., and Stützel, H. (2001). Quantitative aspects of *Orobanche crenata* infestation in faba beans as affected by abiotic factors and parasite soil seedbank. *Weed Res.* 41, 311–324. doi: 10.1046/j.1365-3180.2001.00240.x

Manschadi, A. M., Hargreaves, J. N. G., Grenz, J., DeVoil, P., and Meinke, H. (2004). "Simulating damage effects of parasitic weeds in APSIM: a generic cohort-based approach," in *Proceedings of the Fourth International Crop Science Conference*, Brisbane, QLD.

Matthies, D. (1996). Interactions between the root hemiparasite *Melampyrum arvense* and mixtures of host plants: heterotrophic benefit and parasite-mediated competition. *Oikos* 75, 118–124. doi: 10.2307/3546328

Mauromicale, G., Monaco, A. L., and Longo, A. M. (2008). Effect of branched broomrape (*Orobanche ramosa*) infection on the growth and photosynthesis of tomato. *Weed Sci.* 56, 574–581. doi: 10.1614/WS-07-147.1

Musselman, L. J. (1980). The biology of *Striga*, *Orobanche* and other root-parasitic weeds. *Annu. Rev. Phytopathol.* 18, 463–489. doi: 10.1146/annurev.py.18.090180.002335

Parker, C. (2009). Observations on the current status of *Orobanche* and *Striga* problems worldwide. *Pest Manag. Sci.* 65, 453–459. doi: 10.1002/ps.1713

Parker, C., and Riches, C. R. (1993). *Parasitic Weeds of the World: Biology and Control*. Wallingford: CAB International.

R Development Core Team (2014). *R: A Language and Environment for Statistical Computing*. Vienna: R Foundation for Statistical Computing.

Rossing, W., Meynard, J., and Van Ittersum, M. (1997). Model-based explorations to support development of sustainable farming systems: case studies from France and the Netherlands. *Dev. Crop Sci.* 25, 339–351. doi: 10.1016/S0378-519X(97)80033-5

Watling, J., and Press, M. (2001). Impacts of infection by parasitic angiosperms on host photosynthesis. *Plant Biol.* 3, 244–250. doi: 10.1055/s-2001-15195

Conflict of Interest Statement: The authors declare that the research was conducted in the absence of any commercial or financial relationships that could be construed as a potential conflict of interest.

The reviewer AVA and handling Editor declared their shared affiliation, and the handling Editor states that the process nevertheless met the standards of a fair and objective review.

16

Metabolic Investigation of *Phelipanche aegyptiaca* Reveals Significant Changes during Developmental Stages and in its Different Organs

Edited by:
Monica Fernandez-Aparicio,
Institut National de la Recherche Agronomique (INRA), France

Reviewed by:
Philippe Simier,
Nantes University, France
Jesus V. Jorrin Novo,
University of Cordoba, Spain

***Correspondence:**
Rachel Amir
rachel@migal.org.il

[†] These authors have contributed equally to this work.

Noam Nativ[1,2†], *Yael Hacham*[1,2†], *Joseph Hershenhorn*[3], *Evgenia Dor*[3] *and Rachel Amir*[1,2*]

[1] Migal Galilee Technology Center, Kiryat Shmona, Israel, [2] Biotechnology Department, Tel-Hai College, Upper Galilee, Israel, [3] Weed Research Department, Newe Ya'ar Research Center, Ramat-Yishay, Israel

Phelipanche aegyptiaca Pers. is a root holoparasitic plant considered to be among the most destructive agricultural weeds worldwide. In order to gain more knowledge about the metabolic profile of the parasite during its developmental stages, we carried out primary metabolic and lipid profiling using GC-MS analysis. In addition, the levels of amino acids that incorporate into proteins, total protein in the albumin fraction, nitrogen, reduced sugars, and phenols were determined. For the assays, the whole plants from the four developmental stages—tubercle, pre-emergent shoot, post-emergent shoot, and mature flowering plants—were taken. Thirty-five metabolites out of 66 differed significantly between the various developmental stages. The results have shown that the first three developmental stages were distinguished in their profiles, but the latter two did not differ from the mature stage. Yet, 46% of the metabolites detected did not change significantly during the developmental stages. This is unlike other studies of non-parasitic plants showing that their metabolic levels tend to alter significantly during development. This implies that the parasite can control the levels of these metabolites. We further studied the metabolic nature of five organs (adventitious roots, lower and upper shoot, floral buds, and flowers) in mature plants. Similar to non-parasitic plants, the parasite exhibited significant differences between the vegetative and reproductive organs. Compared to other organs, floral buds had higher levels of free amino acids and total nitrogen, whereas flowers accumulated higher levels of simple sugars such as sucrose, and the putative precursors for nectar synthesis, color, and volatiles. This suggests that the reproductive organs have the ability to accumulate metabolites that are required for the production of seeds and as a source of energy for the reproductive processes. The data contribute to our knowledge about the metabolic behavior of parasites that rely on their host for its basic nutrients.

Keywords: developmental stages, GS-MS analysis, parasite's organs, *Phelipanche aegyptiaca*, primary metabolic profiling, total nitrogen

INTRODUCTION

Orobanche and *Phelipanche* spp. are obligate plant-parasitic plants in the Orobanchaceae family (Joel et al., 2007). Due to their achlorophyllous nature, they are constrained to obtain their nutritional resources by feeding off broad-leaf plants using the haustorium, a unique organ in parasitic plants that links the parasite to the root of the hosts. Through the haustorium, the parasite diverts water, and nutrients from the host (Westwood, 2013), leading to severe yield loss, and quality in numerous vegetables crops (Joel et al., 2007; Pérez-de Luquea et al., 2010; Fernández-Aparicio et al., 2016). It was suggested that once the connection between the parasite and the host roots is established, the parasite functions as an active sink, redirecting solutes away from autotrophic sink tissues and subsequently leading to a decrease in the accumulation of host biomass (Péron et al., 2016).

Despite accumulated knowledge about the parasite's agricultural damage, lifecycle and mode of action, our knowledge about the primary metabolic profiling and its metabolism is mostly unknown. Several studies previously measured the levels of a few amino acids, sugars, or polyols in several parasitic plants (e.g., Aber et al., 1983; Abbes et al., 2009a,b; Delavault, 2015), however, they usually concentrated on one group of metabolites, or even on one metabolite. Metabolic profiling or metabolomics is a rapidly developing technology of the post-genomics era that focuses on global changes in biological samples and enables determining metabolic differences in plants during various developmental stages, growth conditions and stresses (Nakabayashi and Saito, 2015).

We recently carried out a primary metabolic profiling analysis of the tips of young shoots of *Phelipanche aegyptiaca* Pers. that emerged from the "spiders" stage. In addition, the parasite's infected tomato roots and non-parasitized roots were analyzed (Hacham et al., 2016). Major differences in metabolites contents between the parasite and the tomato roots were defined, suggesting that *P. aegyptiaca* has its own metabolism that differs significantly in its regulation from those found in their host. These findings support previous lines of evidence showing significant accumulations of metabolites such as mannitol in the parasite (Delavault et al., 2002). In addition to primary metabolites, we also determined the total phenolic compounds that mostly belong to secondary metabolites, revealing that their levels were about three-fold higher compared to the host-infected roots (Hacham et al., 2016). The results have also shown that the levels of most of the metabolites in the infected roots were similar to the levels detected in the non-infected roots, except for seven metabolites whose levels increased in the infected vs. the non-infected roots (Hacham et al., 2016).

The motivation of the current study is to enhance our knowledge about the *P. aegyptiaca* primary metabolic profiling. To achieve this, we defined the changes in the whole plant of *P. aegyptiaca* during four distinct developmental stages and in five different organs of the mature and flowering plant. Such data could not only contribute to our knowledge about the behavior of parasitic plants and their metabolic nature, but also provide more data about the similarity to non-parasitic plants.

Abbreviations: GC-MS, Gas chromatography-mass spectrometry; GABA, γ-aminobutyric acid TCA, tricarboxylic acid cycle; PCA, principal component analysis.

MATERIALS AND METHODS

Plant Material

Tomato (*Solanum lycopersicon*) cultivar M-82 seeds were obtained from Tarsis Agricultural Chemicals Ltd., Israel. Seeds were collected from inflorescences of *P. aegyptiaca* parasitizing tomato grown on Kibbutz Beit Ha'shita in the 2015 season. Seeds were stored and held in the dark at 4°C until use. M82 tomato seedlings were planted in 2-L pots (Tefen Nachsholim, Israel) using medium-heavy clay-loam soil containing a dry weight basis of 55% clay, 23% silt, 20% sand, 2% organic matter, pH 7.1 (one plant per pot). Slow-release fertilizer at a concentration of 0.6% (w/v) (Osmocote, ScottsMiracle-Gro, Marysville, OH) and *P. aegyptiaca* seeds at a concentration of 15 ppm (15 mg seeds kg^{-1} soil ~2,250 seeds kg^{-1}) were added to the soil. The above-mentioned components were mixed to homogeneity in a cement mixer for 10 min. The pots were placed in a net house and drip-irrigated as needed.

The different developmental stages and different organs were collected 12 weeks after planting. Four different developmental stages were collected (**Figure 1**): Tubercle carrying adventitious roots; pre-emergent shoots that also include adventitious roots; post-emergent vegetative shoots that also include very small and condense floral buds; and whole mature *P. aegyptiaca* that

FIGURE 1 | Comparative illustration of the four key life stages and five organs collected from mature flowering plants of *Phelipanche aegyptiaca*. The developmental stages are: Tub, Tubercle; PeES, Pre-emergent shoot; PoES, Post-emergent shoot; and Mat, Mature flowering plant. The different organs are: Rt, Adventitious roots; LS, underground pre-emergent shoot; St, post-emergent shoot; FB, Floral buds; and Fwr, Flowers. Samples were collected 12 weeks after planting.

have adventitious roots, pre- and post-emergent shoots and reproductive tissue, including floral buds and flowers (according to Westwood et al., 2012). Five organs were collected from mature adult plants that have all the organs: Adventitious roots; underground pre-emergent shoot; post-emergent stem; floral buds; and flowers (**Figure 1**). The whole parasitic plants during the four developmental stages and the five organs were frozen in liquid nitrogen and dried by lyophilization. The samples were then ground to dried powder using TissueLyser (Qiagen Retsch MM301) for the various analyses.

Extraction and Analysis of Primary Metabolites Using Gas Chromatography-Mass Spectrometry (GC-MS)

Primary metabolites were extracted from 20 mg *P. aegyptiaca* dried powder in 1 mL of pre-cooled extraction buffer containing HPLC-grade methanol:chloroform:H_2O (2.5:1:1 v/v/v) and transferred into new 2 mL lock-cap Eppendorf tubes. Each sample was added with 4.6 μL of norleucine internal standard (2 mg mL^{-1} in HPLC-grade H_2O), vortexed for 10 s, and centrifuged for 10 min at 20,817 g at 4°C pre-cooled table-top centrifuge to permit the separation of polar (methanol/water) and non-polar (chloroform) phases. The methanol-water fractions were treated and derivatized as previously described (Matityahu et al., 2013; Cohen et al., 2014). The single-ion mass method was used for soluble and protein-incorporated amino acid determination with the RXI-5-Sil MS capillary column (RESTEK; 30 m, 0.25-mm i.d., and 0.25-mm thickness), while the total-ion-count method was used for metabolic profiling and separation using the VF-5 ms capillary column (Agilent; 30 m + 10 m EZ-guard, 0.25-mm i.d., and 0.25-mm thicknesses). All analyses were carried out on a GC-MS system (Agilent 7890A) coupled with a mass selective detector (Agilent 5975c) and a Gerstel multipurpose sampler MPS2 (Cohen et al., 2014). Peak finding, peak integration and retention time correction were performed with the Agilent GC/MSD Productivity ChemStation package (http://www.agilent.com). Areas of the peaks were normalized to integral standard (norleucine) signal. The identification of these metabolites was based on standards (amino acids, fatty acids, and most of the primary metabolites), and the spectra of all the peaks were compared with commercially available electron mass spectrum libraries, NIST, and WILEY.

Extraction of Total Amino Acids

For total amino acid determination including protein-bound amino acids, 20 mg of *P. aegyptiaca* dried powder from different developmental stages and plant organs were mixed with distilled water. After two centrifugation cycles (14,000 rpm for 20 min), 25 μl of protein extraction were placed in a glass tube with norleucine solution (60 μg/ml). The samples were frozen at −70°C and lyophilized, followed by 6N HCl acidic hydrolysis for 22 h at 110°C under vacuum (Frank et al., 2015). Hydrolysis products were suspended in 0.45 ml DDW and 0.3 ml chloroform. Following centrifugation (14,000 rpm, 20 min), 0.3 ml aliquots from the upper phase were collected, and samples were silylated, and analyzed as previously described (Matityahu et al., 2013).

Determination of Fatty Acid Contents Using GC-MS

Fatty acid composition was analyzed using an established GC-MS protocol (Bai et al., 2012). Three hundred microliters from the non-polar chloroform phases taken from the samples prepared for metabolic profiling were collected and transferred into new Eppendorf tubes. To each sample, 7 μL of heptadecanoic acid (C17:0) internal standard (5 mg mL^{-1} in HPLC-grade chloroform) was added. The chloroform phase was dried under a stream of nitrogen gas, and lipids were transmethylated with 2% (v/v) H_2SO_4 in methanol at 85°C for 1 h. The reaction was terminated by the addition of water, and fatty acid methyl esters (FAMEs) were extracted in hexane and transferred to 300 μL autosampler vials. FAMEs were quantified and identified on a GC-MS system (Agilent 7890A) coupled with a mass selective detector (Agilent 5975c) and a Gerstel multipurpose sampler MPS2. Helium was used as the carrier gas at a flow rate of 0.87531 mL/min. A 1 μL sample was injected into the split mode, the inlet temperature was 250°C and the pressure was 15.0 psi. Separation was achieved on a 70% cyanopropyl polysilphenylene-siloxane column (BPX70, SGE Analytical Science, 25 m, 0.22 mm, 0.25 μm) using the temperature gradient from 130°C (hold time 1 min) to 210°C (linear increase of 3°C min^{-1}; hold time 5 min) and finally to 240°C (linear increase of 10°C min^{-1}; hold time 2 min). FAMEs were identified by co-chromatography with authentic standards (Sigma-Aldrich).

Total Protein and Nitrogen Determinations

For total protein determination, 20 mg of *P. aegyptiaca* dried powder of tissues from the samples were ground in 400 μL buffer phosphate pH = 7.8 with a protease inhibitor cocktail (Sigma, P9599). After two centrifugation cycles (14,000 rpm for 20 min), total protein was determined using a Bradford reagent (Bio-Rad Hercules, Calif.) in three sample concentrations (Bradford, 1976). Bovine serum albumin was used as standard.

For total nitrogen measurement, 20 mg dried powder of the *P. aegyptiaca* tissues was digested with 0.4 mL of H_2SO_4 overnight. Afterwards 2 mL of H_2O_2 were added for 24 min at a heating block at 280°C. Samples were then cooled for 15 min at room temperature and left overnight in order to enable the evaporation of H_2O_2. Double-distilled water (DDW) was added to bring the total volume to 100 mL. The solution was diluted 10 times and used to determine the concentration of total N (Wolf, 1982; Prodhan et al., 2016). The total dissolved nitrogen was determined using a total organic N analyzer (Multi NC 2100S, Analytik Jena, Jena, Germany).

Total Reducing Sugars Determination

Reducing sugars were measured colorimetrically using the Sumner method following carbohydrates and starch hydrolysis (Sumner, 1924). Forty mg dry weights of each of the *P. aegyptiaca* samples from the different developmental stages and different organs were ground using a Restch MM 301 homogenizer and extracted in 40 ml HCl 1N at 70°C for 2 h. After cooling, the pH was adjusted to 7.3 and the volume to 50 ml with water. One half ml of the samples was then mixed with 1.5 ml water and 2 ml of Sumner reagent and left for 5 min at 90°C. After cooling, the reducing sugars were detected at 550 nm by the UV-vis spectrophotometer 1240 (Shimadzu) (Hacham et al., 2016). A standard curve was created using glucose in a range of 50–500 µg.

Total Phenolic Compounds Content Determination

For total phenolic compounds content determination, most of which are secondary metabolites, 20 mg dried powder from samples of *P. aegyptiaca* were ground in 0.5 ml water, and the colorimetric method, which modified the Ben Nasr method for small volumes (Ben Nasr et al., 1996), was employed. Ten microliters of the extraction sample were loaded on a 96-well ELISA plate. To each well, 50 µl of 10% Folin-Ciocalteu reagent and 40 µl of 7.5% (w/v) Na_2CO_3 were added. The plate was incubated for 40 min at 37°C and then read at 765 nm (Infinite M200PRO Tecan, Grodig Austria). A standard curve was created using gallic acid in a range of 0.5–10 µg.

Statistical Analyses

For the developmental stages, 5–10 parasitic plants were taken for each biological replicate, while for the different organs at least five plants for each biological replicate were taken. The data represent the mean of four independent replications. Statistical significance was evaluated using JMP software version 8.0 (SAS Institute Inc., Cary, NC). Significant differences between treatments were calculated according to the Turkey-Kramer HSD test ($p < 0.05$). Principal component analysis (PCA) and a heat-map of GC-MS data were conducted using the MetaboAnalyst 3.0 comprehensive tool (http://metaboanalyst.ca/; Xia et al., 2015) with Auto scaling (mean-centered and divided by the standard deviation of each variable) manipulations. Graphs were compiled using GraphPad Prism 5.01 scientific software (http://www.graphpad.com/).

RESULTS

Metabolic Profile of *P. aegyptiaca* during Four Developmental Stages

Four different developmental stages were taken from *P. aegyptiaca* for primary metabolic profiling analyses: Tubercle; pre-emergent shoot; post-emergent vegetative shoots; and mature flowering plants (**Figure 1**). The whole plants from each developmental stage were ground and used for global primary metabolic profile analyses, as well as a lipid profile analysis, using established GC-MS protocols (Bai et al., 2012; Cohen et al., 2014; Hacham et al., 2016). The analysis revealed 66 annotated metabolites having a significantly higher signal-to-noise value. The annotated metabolites belong to seven distinct biochemical groups: Amino acids (19); sugars (16 in total, 13 annotated and 3 non-annotated); polyols (6); tricarboxylic acid cycle (TCA) intermediates (4); organic acids (5); sugar acids (5); fatty acids (6); and others (5) (Supplemental Table S1).

The metabolite profiles of these four developmental stages were plotted onto a PCA (Xia et al., 2015). The two principal components account for 65% of the total variance (**Figure 2A**). The PCA analysis showed that the first three developmental stages—tubercle, pre-emergent shoot, and post-emergent vegetative shoot—were separated from one another. However, the pre-emergent shoot and post-emergent shoot stages were not separated from the mature stage (**Figure 2A**). A heat-map analysis showed that compared to the other developmental stages, the tubercle has the lowest contents of most of the metabolites detected (**Figure 2B**; Supplemental Table S1).

To gain a better resolution, each metabolite whose level changed significantly between the different stages was plotted separately on **Figure 3**. Examination of the tubercle metabolites revealed that trehalose showed significantly higher level at this stage compared to the other developmental stages (**Figure 3**). In addition, the level of γ-aminobutyric acid (GABA) was high at this stage compared to the last two developmental stages (**Figure 3**). The metabolic profile of the different developmental stages basically shows that the levels of nine metabolites gradually increased during the four developmental stages. These are methionine, isoleucine, valine, tryptophan, cysteine, proline, arabinose, ribose, and fumaric acid (**Figure 3**; Supplemental Table S1). In the pre-emergent shoots, only glutamate and aspartate were significantly higher compared to the other stages, however, the levels of 16 other metabolites, serine, homoserine, threonine, fructose, rahmnose, NA1, NA3, glycerol, myo-inositol, glyceric acid, erythronic acid, galactonic acid, pyroglutamic acid, phosphoric acid, ethanolamine, and C18:3, were higher than at least one of the other stages (Supplemental Table S1).

Among the 19 amino acids detected, leucine, serine, aspartate, glutamate, asparagine, and glutamine (the last two are amides that have an additional nitrogen molecule) were found at the highest levels, above 100 nmol/g DW during at least one developmental stage (**Figure 3**; Supplemental Table S1). Since most of the soluble amino acids were incorporated into proteins (Cohen et al., 2014), we next studied if the compositions of the protein-incorporated amino acids that changed during the developmental stages. We tested the albumin fraction, which is the aqueous-soluble protein, representing about 80% of leaf proteins (Galili and Hofgen, 2002). The levels of isoleucine, valine, and aspartate increased significantly throughout the development, having the highest levels in mature plants (**Table 1**).

We also studied fatty acid composition using another GC-MS method (Bai et al., 2012). The analysis detected six fatty, three saturated and three unsaturated acids (Supplemental Table S1). Except for α-linolenic (C18:3), whose levels significantly decreased from the pre-emergent shoot to the mature flowering stage (**Figure 3**), the levels of the other fatty acids did not change significantly between the four developmental stages. Notably, although the levels of most of the metabolites detected were relatively lower in the tubercle stage, the level of fatty acids did not decrease at this stage compared to the other developmental stages (Supplemental Table S1).

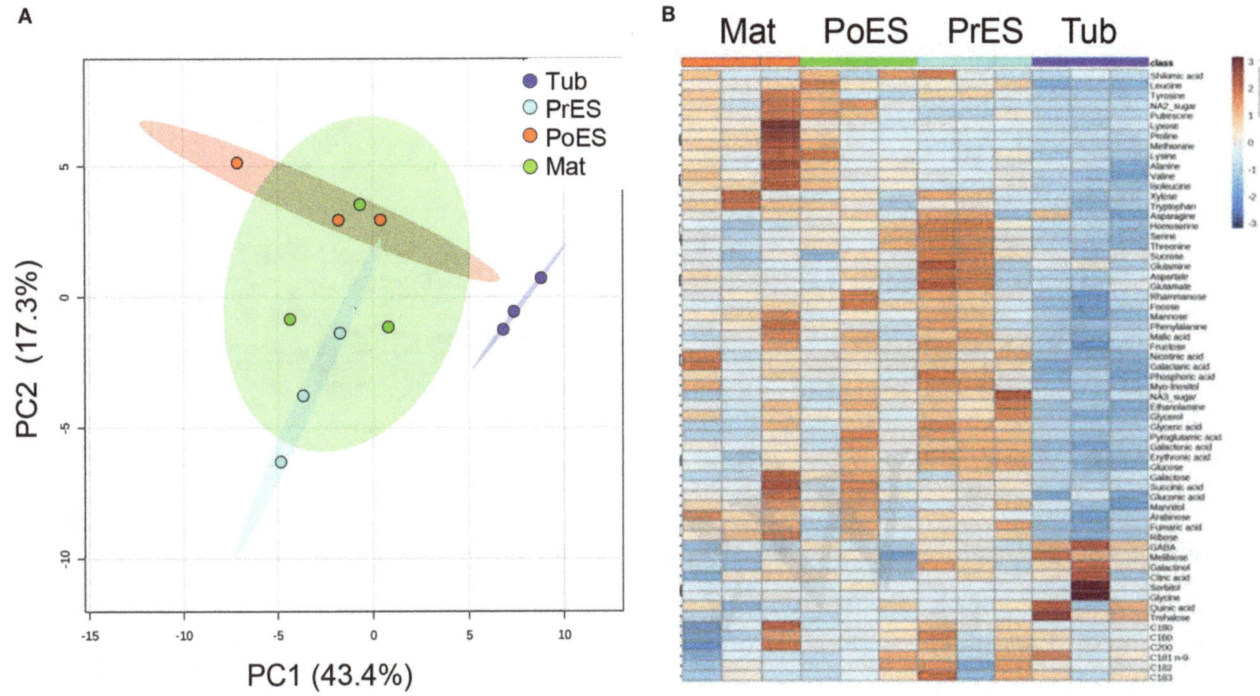

FIGURE 2 | Graphical representation of changes in *Phelipanche aegyptiaca* metabolite profiles at four developmental stages. (A) Principal component analyses (PCA) applied to the four developmental stages according to their entire primary metabolome set of 66 metabolites. The data points are displayed as projections onto the two primary axes (eigenvectors). Variances explained by the first two components (PC1 and PC2) appear in parentheses; **(B)** heat-map of these 66 primary metabolites detected by GC-MS. The data represent three replicates for each developmental stage.

Metabolic Profile of Five Different *P. aegyptiaca* Organs Taken from Mature Flowering Plants

The metabolic profiling of five different organs of the mature parasite was determined: Adventitious roots; underground pre-emergent shoot; post-emergent shoot; floral buds; and flowers (**Figure 1**). The metabolic profiles of these five organs were plotted onto a PCA to reveal that a separation exists between the three vegetative tissues, adventitious roots, pre- and post-emergent shoots to the reproductive organs, floral buds and flowers (**Figure 4A**). A heat-map analysis indeed determined that the two reproductive tissues (buds and flowers) have higher levels of most of the metabolites detected than the three vegetative organs (**Figure 4B**).

To gain a better resolution, each metabolite whose level changed significantly between the different organs was plotted separately on **Figure 5**. The results showed that the levels of the 15 soluble amino acids increased significantly in the floral buds compared to the three vegetative organs. The levels of 11 of them decreased significantly in the flowers (**Figure 5**; Supplemental Table S2). In addition, the levels of non-protein amino acids (homoserine and GABA) increased significantly in the floral buds. The levels of asparagine and the three aromatic amino acids (tyrosine, phenylalanine, and tryptophan) displayed similar high levels in both flowers and buds, while the level of cysteine was found to be higher in flowers. Notably, the levels of aspartate and glutamate did not differ significantly between the five organs, while those of asparagine and glutamine, which show relatively high levels, accumulated preferentially in the reproductive organs (**Figure 5**; Supplemental Table S2). The amino acids whose levels were the highest (over 100 nmol/g DW) in the floral buds were leucine, valine, serine, asparagine, glutamate, threonine, alanine, and proline.

The higher levels of soluble amino acids in the floral buds might be the result of lesser incorporation of these amino acids into proteins. To test this possibility, we studied the compositions of the protein-incorporated amino acids in the albumin fraction of the different organs. The analysis shows that five out the 13 amino acids changed significantly between the organs. Adventitious roots had higher levels (by means of mol%) of phenylalanine and tyrosine compared to the flowers, while the floral buds had significantly higher levels of the three branch chain amino acids, isoleucine, leucine, and valine, compared to the other organs (**Table 2**). Therefore, the results suggest that the total levels of branch chain amino acids (soluble and protein-incorporated) increased in the floral buds. The higher levels of other soluble amino acids did not significantly affect their incorporation to the proteins.

In addition to amino acids, the levels of other 28 metabolites changed significantly in at least in one organ compared to the others. The levels of the sugars sucrose and mannose were significantly higher in the flowers, while fructose, ribose, and NA2 had higher levels in flowers as well as in floral buds (Table S2). Flowers and/or floral buds had higher levels of

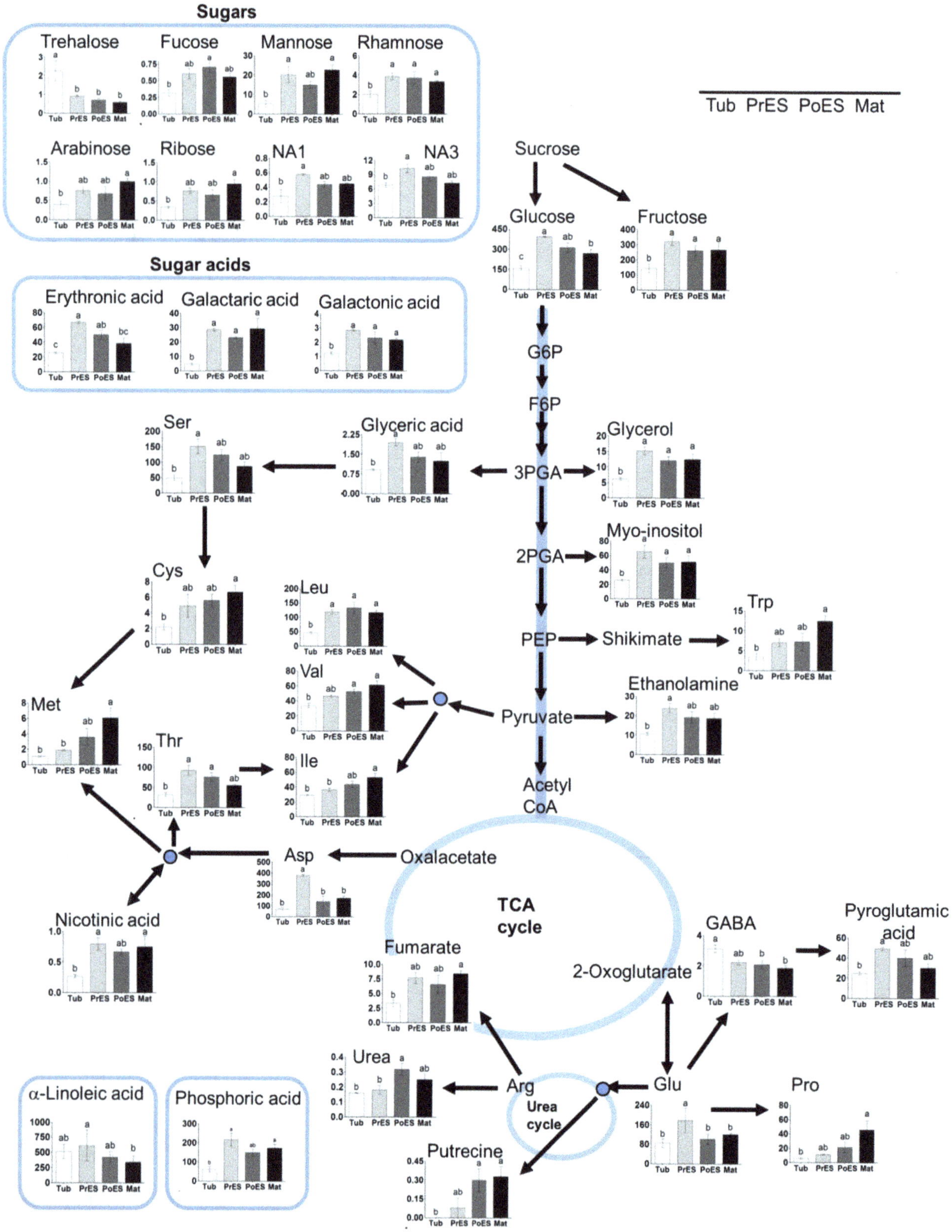

FIGURE 3 | The levels of primary metabolites and lipids of *Phelipanche aegyptiaca* during four developmental stages. Only metabolites whose levels changed significantly in at least one developmental stage were presented. The developmental stages are: Tub, Tubercle; PeES, Pre-emergent shoot; PoES,

(Continued)

FIGURE 3 | Continued
Post-emergent shoot; and Mat, Mature flowering plant. The y axis of amino acids is shown in nmol/g DW, for fatty acid in μg/g DW, while for the other metabolites it represents the area of relative m/z response of each metabolite following normalization to the norleucine internal standard. Data shown are means ± *SE* of three replicates for each type of developmental stage. Significance was calculated according to the Turkey-Kramer HSD test ($p < 0.05$) and is identified by different small letters. G6P, glucose 6-phosphate; F6P, fructose 6-phosphate; 3-PGA, 3-phosphoglyceric acid; 2-PGA, 2-phosphoglyceric acid; PEP, phosphoenolpyruvic acid; Glu (glutamic acid); Gln (glutamine); Asp, aspartic acid; Ser, serine; Ile, isoleucine; Val, valine; Leu, leucine; Trp, tryptophan; Cys, cysteine; Met, methionine; Thr, threonine; Glu, glutamic acid; Pro, proline; GABA, γ-aminobutyric acid.

TABLE 1 | The levels of total amino acids in the four developmental stages of *Phelipanche aegyptiaca* after protein hydrolysis of the soluble proteins fraction, given in mol%#.

Protein-bound amino acids (%)	Tubercle	Pre-emergent shoot	Post-emergent shoot	Mature
Glutamate	3.23 ± 0.31(a)	2.31 ± 0.22(a)	3.1 ± 0.14(a)	2.87 ± 0.22(a)
Aspartate	25.86 ± 1.5(a)	24.66 ± 1.24(a)	27.8 ± 0.95(a)	31.37 ± 0.28(a)
Lysine	1.48 ± 0.14(a)	2 ± 0.18(a)	0.94 ± 0.05(a)	1.24 ± 0.09(a)
Methionine	1.33 ± 0.08(a)	1.85 ± 0.09(a)	1.67 ± 0.08(a)	1.5 ± 0.08(a)
Threonine	5.23 ± 0.45(a)	4.94 ± 0.3(a)	4.44 ± 0.11(a)	4.21 ± 0.2(a)
Isoleucine	2.52 ± 0.2(b)	2.56 ± 0.16(b)	3.7 ± 0.14(ab)	5.21 ± 0.22(a)
Leucine	7.37 ± 0.44(a)	8.97 ± 0.53(a)	11.25 ± 0.44(a)	12.78 ± 0.25(a)
Valine	5.89 ± 0.26(b)	5.29 ± 0.23(b)	7.54 ± 0.31(ab)	10.05 ± 0.33(a)
Phenylalanine	10.9 ± 0.75(a)	8.14 ± 0.57(a)	8.31 ± 0.32(a)	8.7 ± 0.92(a)
Tyrosine	4.29 ± 0.34(a)	3.46 ± 0.29(a)	4.3 ± 0.54(a)	1.78 ± 0.2(a)
Glycine	10.37 ± 0.8(a)	8.05 ± 0.58(a)	10.72 ± 0.64(a)	9.84 ± 0.66(a)
Serine	7.77 ± 0.67(a)	8.24 ± 0.63(a)	7.39 ± 0.26(a)	7.53 ± 0.35(a)
Proline	1.68 ± 0.15(a)	1.89 ± 0.2(a)	1.94 ± 0.09(a)	2.92 ± 0.09(a)

 # The levels of amino acids were measured using GC-MS. The peak area was normalized to norleucine internal standard. For each biological replicate, the percentage (%) was calculated from the total of all amino acids that were detected at the specific sample, and then the average was calculated for all five biological replicates. Values are the mean ± standard deviation of five biological replicates, each taken from at least five plants. Significance was calculated according to the Turkey-Kramer HSD test ($p < 0.05$) and identified by different letters.

glycerol, xylitol, malic acid, succinic acid, glyceric acid, benzoic acid, pyroglutamic acid, putrescine, and ethanolamide. In total, 37 metabolites increased in one or both of these two reproductive organs. However, adventitious roots showed the highest levels of trehalose and arabinose, while the pre-emergent shoot had the highest levels of quininc and galactaric acids (**Figure 5**).

Next, we studied fatty acid composition. As shown for most of the other metabolites, floral buds have higher levels of three fatty acids, linoleic (C18:2), α-linolenic (C18:3), and palmitic (C16:0), compared to the previous stages. However, oleic acid (18:1) was higher in the pre-emergent shoot compared to the adventitious roots, post-emergent shoot, and flowers (**Figure 5**; Supplemental Table S2). The level of stearic acid (18:0) did not change significantly between the different organs.

The Contents of Total Nitrogen, Total Soluble Protein in the Albumin Fraction, and Total Reducing Sugars (Represented as Starch)

The higher levels of 11 out of the 18 amino acids in the last three developmental stages, and the relatively higher levels of the total amino acids (**Figure 6A**), encouraged us to determine the total nitrogen content in these stages. The analyses revealed that there were no significant differences between the different developmental stages (**Figure 6A**). Next, the total soluble proteins (albumin fraction) were examined using the Bradford analysis. The analyses again revealed that there were no significant differences between the different developmental stages (**Figure 6A**).

The levels of total soluble amino acids, total nitrogen, and soluble proteins were also examined in the five organs of the adult plants. Floral buds had significantly higher levels of total free amino acids and nitrogen compared to other organs (**Figure 6B**), while soluble proteins were higher in floral buds and flowers. This indicates that the reproductive organs, especially the floral buds, have higher levels of free amino acids, nitrogen, and soluble proteins. Adventitious roots, however, had about 5-, 2-, and 4-fold lower levels of total soluble amino acids, nitrogen and soluble proteins, respectively, than the floral buds.

Changes are also expected due to the finding that the levels of simple sugars changed significantly during the development and in the organs tested. To this end, we examined the levels of total reducing sugars (represented as starch) following carbohydrate hydrolysis. The reducing sugars did not change significantly during the developmental stages. However, their levels increased significantly in the underground pre-emergent shoot by 2.7-fold compared to the adventitious roots and flowers (**Figure 6B**).

Determination of Total Soluble Phenolic Compounds

We previously determined that *P. aegyptiaca* has about a three-fold higher level of total phenols (most of them secondary metabolites) compared to the host's roots (Hacham et al., 2016).

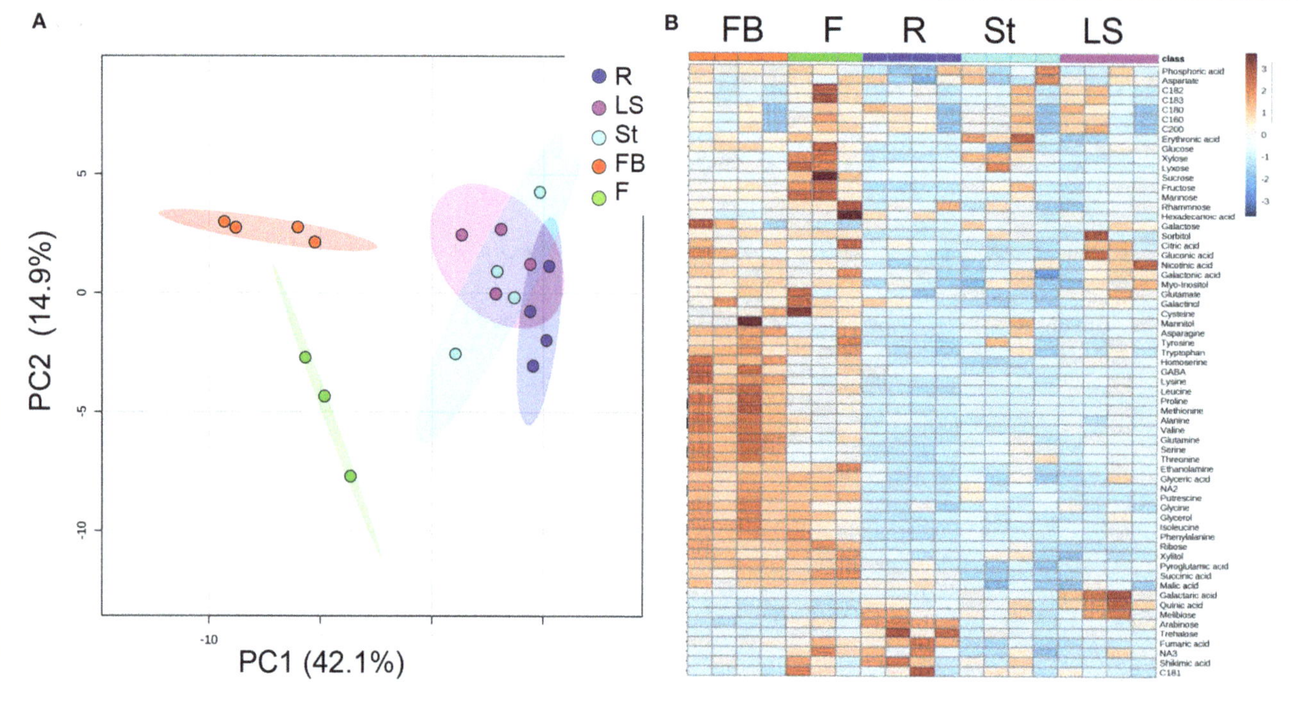

FIGURE 4 | Graphical representation of changes in metabolite profiles of five organs of mature *Phelipanche aegyptiaca*. (A) Principal component analyses (PCA) applied to the five organs according to their entire primary metabolome set of 66 metabolites. The data points are displayed as projections onto the two primary axes. Variances explained by the first two components (PC1 and PC2) appear in parentheses; **(B)** heat-map of these 66 primary metabolites detected by GC-MS. The data represent three to four replicates for each organ. The different organs are: Rt, adventitious roots; LS, underground pre-emergent shoot; St, post-emergent shoot; FB, Floral buds; and Fwr, Flowers.

Determination of total phenols content showed that their levels did not change significantly in the four developmental stages (**Figure 6A**). In plant organs, however, the floral buds had 45 and 34% more phenols than the adventitious roots and post-emergent shoot, respectively (**Figure 6B**).

DISCUSSION

Differences in Metabolic Profiling during the Four Developmental Stages

Determination of the metabolic profile of the four developmental stages revealed that the levels of 35 out of the 66 metabolites, which come from all determined biochemical groups, differed significantly in at least one of the developmental stages. However, the levels of 31 metabolites (constituting 45% of the total) still did not change significantly during development (Supplemental Table S1). This finding is unlike those reported for various developmental stages in different organs of non-parasitic plants. In the latter plants, the levels of most of the detected metabolites changed significantly during development. For example, significant changes were found in the levels of soluble and protein-bound amino acids during the development of all Arabidopsis aerial organs (rosette leaves, upper leaves, flowers, siliques, seeds) (Frank et al., 2015). In addition, the levels of all metabolites examined in the primary metabolic profiling of Arabidopsis, sunflower, and tobacco leaves changed significantly during their developmental stages (Watanabe et al., 2013; Li et al., 2016; Moschen et al., 2016), and also in the upper leaves and siliques in Arabidopsis (Watanabe et al., 2013). Although we compared between the whole *P. aegyptiaca* plants to the organs in non-parasitic plants, we assume that the metabolic behavior of the parasite is considerably less altered during development compared to non-parasitic plants. We hypothesize that the gap between *P. aegyptiaca* and non-parasitic plants in the levels of the different metabolites is one of the significant differences between the two types of plants. Although the reasons for such changes are not yet known, we assume that it is related to the life cycle of these two types of plants. For example, the leaves of non-parasitic plants served as a sink when they were young, as a source when they matured, and at senescence, most of their metabolites transferred toward the developing reproductive tissues including seeds. These led to significant changes in their metabolic profiling during development (Gregersen and Holm, 2007; Lim et al., 2007; Gregersen et al., 2008; Watanabe et al., 2013; Distelfeld et al., 2014; Moschen et al., 2016). However, *P. aegyptiaca*, as a parasitic plant, does not have functional leaves, and as a parasite, it relies at least for its carbon source on its hosts. This might be the reason for lesser change in the levels of their metabolites. The results also proposed that the parasite can control the levels of about half of its metabolites during its developmental stages.

Basically, the tubercle stage tends to have the lowest average level of amino acids, sugars and other metabolites compared to

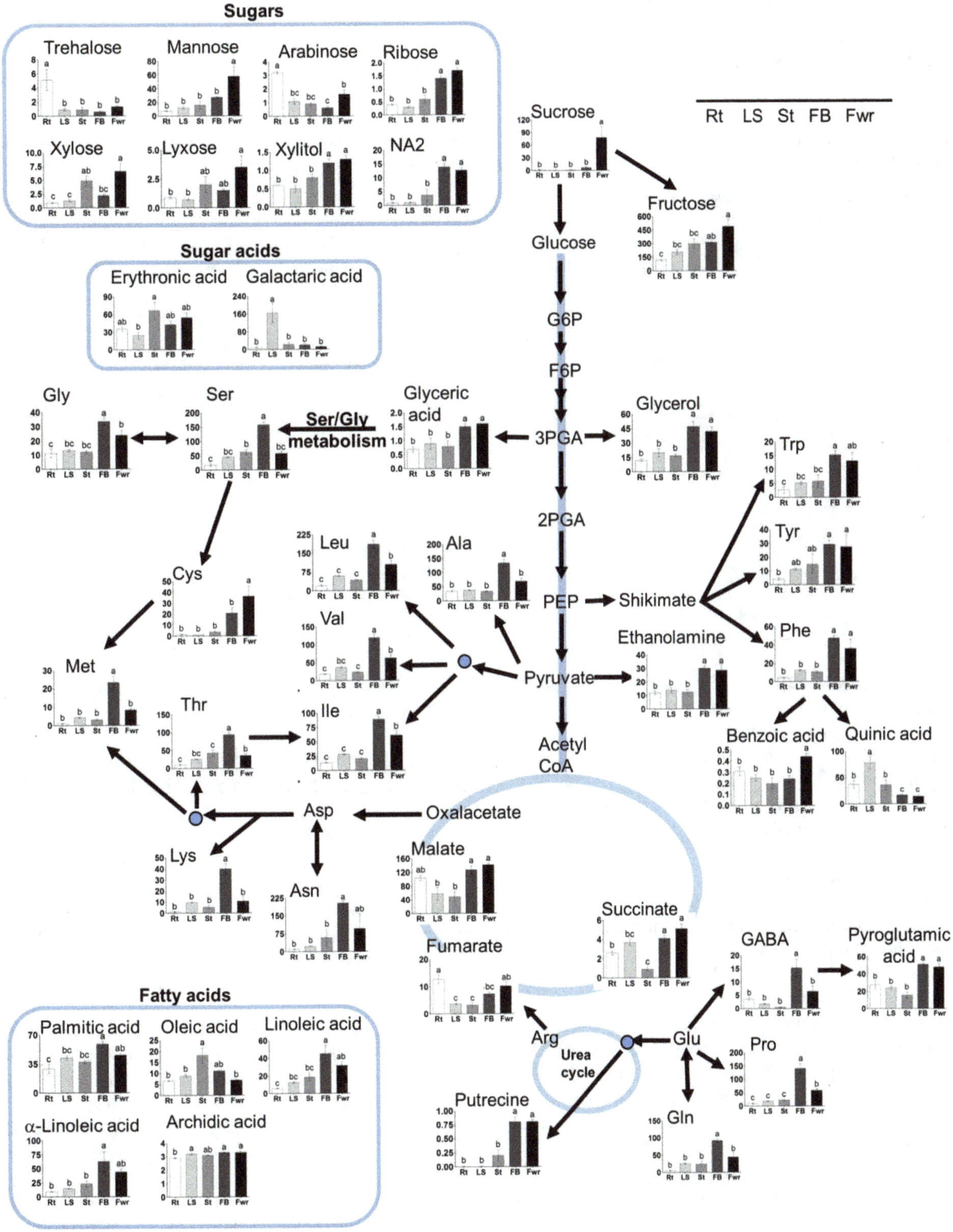

FIGURE 5 | The levels of primary metabolites and lipids of *Phelipanche aegyptiaca* in five organs of mature plants. Only metabolites whose levels changed significantly in at least one developmental stage were presented. The different organs are: Rt, adventitious roots; LS, underground pre-emergent shoot; St,

(Continued)

FIGURE 5 | Continued
post-emergent shoot; FB, Floral buds; and Fwr, Flowers. The y axis of amino acids is shown in nmol/g DW, for fatty acid in μg/g DW, while for the other metabolites it represents the area of relative m/z response of each metabolite following normalization to the norleucine internal standard. Data shown are means ± *SE* of three to four replicates for each type of plant's organ. Significance was calculated according to the Turkey-Kramer HSD test ($p < 0.05$) and is identified by different small letters. For abbreviations of metabolite names, see **Figure 3**, except for abbreviations of Asn, asparagine; Gly, glycine; Tyr, tyrosine; Phe, phenylalanine; Ala, alanine; Lys, lysine; Gln, glutamine.

TABLE 2 | The levels of total amino acids in five organs of mature flowering plants of *Phelipanche aegyptiaca* after protein hydrolysis of the soluble proteins fraction, given in mol%#.

Protein-bound amino acids (%)	Adventitious roots	Lower shoot	Stem	Floral buds	Flowers
Glutamate	4.68 ± 0.45(a)	4.44 ± 0.37(a)	2.88 ± 0.25(a)	1.15 ± 0.22(a)	2.16 ± 0.55(a)
Aspartate	22.71 ± 1.13(a)	18.92 ± 1.18(a)	22.52 ± 1.26(a)	16.94 ± 0.3(a)	23.31 ± 0.7(a)
Lysine	2.31 ± 0.83(a)	1.16 ± 0.08(a)	0.82 ± 0.04(a)	1.77 ± 0.09(a)	1.84 ± 0.13(a)
Methionine	1.76 ± 0.15(b)	3.08 ± 0.12(a)	1.59 ± 0.14(b)	2.42 ± 0.15(ab)	1.7 ± 0.12(b)
Threonine	3.84 ± 0.08(a)	5.07 ± 0.19(a)	4.34 ± 0.34(a)	6.25 ± 0.52(a)	5.26 ± 0.32(a)
Isoleucine	3.59 ± 0.08(b)	4.99 ± 0.13(b)	4.73 ± 0.3(b)	8.49 ± 0.21(a)	8.1 ± 0.25(a)
Leucine	7.64 ± 0.35(c)	10.7 ± 0.26(b)	9.32 ± 0.6(bc)	15.38 ± 0.23(a)	13.7 ± 0.23(ab)
Valine	8.76 ± 0.48(b)	10.6 ± 0.42(b)	10.25 ± 0.64(b)	17.2 ± 0.44(a)	12.4 ± 0.68(ab)
Phenylalanine	15.84 ± 0.44(a)	12.9 ± 1.08(ab)	11.3 ± 1.18(ab)	5 ± 0.51(b)	6 ± 0.32(b)
Tyrosine	3.75 ± 0.33(a)	1.13 ± 0.15(b)	2.19 ± 0.19(ab)	2.14 ± 0.25(ab)	1.27 ± 0.13(b)
Glycine	13.56 ± 0.55(a)	13.72 ± 0.54(a)	9.14 ± 0.85(a)	11.58 ± 1.25(a)	11.97 ± 0.77(a)
Serine	7.77 ± 0.17(a)	9.63 ± 0.34(a)	8.53 ± 0.61(a)	8.47 ± 0.36(a)	8.78 ± 0.44(a)
Proline	3.79 ± 0.32(a)	2.77 ± 0.09(a)	2.62 ± 0.17(a)	3.22 ± 0.18(a)	3.44 ± 0.32(a)

#The levels of amino acids were measured using GC-MS. The peak area was normalized to norleucine internal standard. For each biological replicate, the percentage (%) was calculated from the total of all amino acids that were detected at the specific sample, and then the average was calculated for all five biological replicates. Values are the mean ± standard deviation of five biological replicates, each taken from at least five plants. Significance was calculated according to the Turkey-Kramer HSD test ($p < 0.05$) and identified by different letters.

the other stages (**Figures 2B, 3**). This suggests that the parasite at this stage attracts metabolites relatively less from the host, or has a lesser ability to produce many of these metabolites compared to the more advanced developmental stages. Among the metabolites identified, only trehalose was significantly higher at the tubercle stage compared to the other stages. The level of trehalose increased significantly in *P. aegyptiaca* infected tomato roots compared to the non-infected roots (Hacham et al., 2016). Trehalose is known to be involved in stress responses, plant-microorganism interactions and defense response against pathogens (Zhang et al., 2016). Therefore, it might be that the significantly higher level of trehalose in the tubercle stage resulted from the infected roots that had higher levels of this metabolite that transferred to the parasite. Despite this assumption, we cannot exclude the possibility that they also form in the parasite.

The second stage (pre-emergent shoot) had the highest total hydrophilic metabolites (55% more than the tubercle and 33% more than in the two later developmental stages; Supplemental Table S1). This implies that the pre-emergent shoot became a much stronger sink than the tubercle, gained more abilities to synthesize these metabolites, and/or utilized these metabolites less. However, the levels of all of these metabolites tend to decrease in the last two developmental stages when the reproductive organs are produced (**Figure 3**). This implies that they are used as precursors for the synthesis of other metabolites, and/or as respiratory and energy substrates. We also assume that the plant at this pre-emergent stage accumulates metabolites to support emergence from the ground when relatively rapid growth occurred.

Differences in Metabolic Profiling in the Five Different Organs of the Mature Plant

Non-parasitic plants exhibit significant differences in their primary metabolites between different organs during development (Watanabe et al., 2013; Frank et al., 2015). However, only 44 metabolites out of 66 changed significantly between the different organs of the mature *P. aegyptiaca* plants (Supplemental Table S2). Yet, 22 metabolites (representing 32% of the total) were not significantly changed, suggesting that the parasite has the ability to control the levels of these unchanged metabolites in its organs.

PCA and heat-map analyses of the organs of the mature parasite showed a distinguished separation between the vegetative and reproductive organs. Indeed, it was observed that floral buds and flowers have the highest levels of most metabolites compared to vegetative organs (29 in buds and 19 in flowers). The gap in levels between floral buds and flowers proposed that several metabolites were used in the flowers to synthesize other metabolites that cannot be detected by the methods used in the current study, or they were used as an energy source, as proposed from the higher levels of malate and succinate in this organ (**Figure 5**). These results are similar to those reported for non-parasitic plants that exhibit diverse primary metabolic profiling in their different organs during

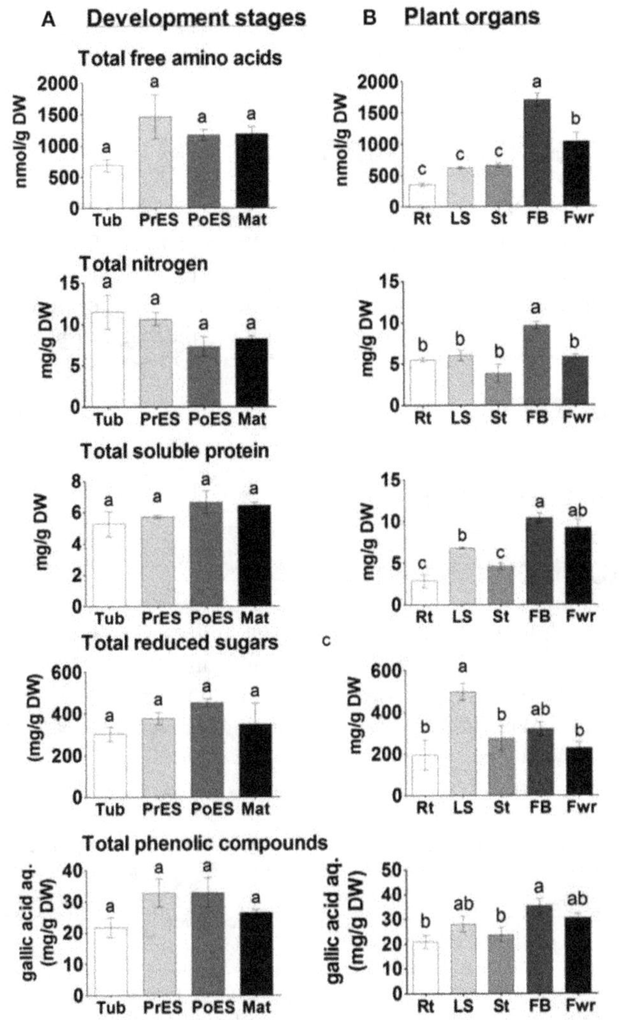

FIGURE 6 | The levels of total free amino acids, nitrogen, soluble proteins, reduced sugars and total phenols in *Phelipanche aegyptiaca* at four developmental stages (A); and in five organs from mature plants **(B)**. Total free amino acids contents measured by GC-MS analysis; total protein contents in the albumin fraction as measured using a Bradford assay; total reducing sugar contents analyzed after hydrolysis by the Sumner method; and total polyphenol content (represented as mg gallic acid equivalents per L of water extract). Data shown are means ± *SE* of three to four replicates for each sample. Significance between plants was calculated according to the Turkey-Kramer HSD test ($p < 0.05$) and is identified by different small letters. Tub, Tubercle; PeES, Pre-emergent shoot; PoES, Post-emergent shoot; Mat, Mature; Rt, adventitious roots; LS, underground pre-emergent shoot; St, post-emergent shoot; FB, Floral buds; and Fwr, Flowers.

development (Gregersen and Holm, 2007; Watanabe et al., 2013; Distelfeld et al., 2014). The major differences in the metabolom is largely shown during the transition between the vegetative to the reproductive stage (Gregersen et al., 2008; Watanabe et al., 2013; Li et al., 2016; Moschen et al., 2016). This metabolic shift is a key for the mobilization and recycling of nutrients from senescing leaves to sinks, such as young leaves, storage tissues, and/or developing seeds (Lim et al., 2007).

Out of the 16 soluble amino acids whose levels changed significantly in reproductive organs, 12 had the highest contents in floral buds (**Figure 5**; Supplemental Table S2). The total soluble amino acids in this organ were 4.8-fold higher than in adventitious roots, which exhibited the lowest accumulation (**Figure 6B**). Relatively higher levels were also detected in flowers. These elevations were in accordance with a higher level of soluble proteins, and higher total nitrogen content in these organs, especially in the floral buds (**Figure 6B**), which are required to prepare the flowers and the internal reproductive tissues. Additionally, higher levels of ethanolamine, putrescine, GABA, and pyroglutamic acid were detected in buds and/or flowers (**Figure 5**; Supplemental Table S2). This suggests that these organs required further nitrogen compounds (in addition to amino acids), since all of them have an amine group. They are proposed to contribute to the high levels of nitrogen, at least in the buds. However, they might also play a role in flowers, since putrescine synthesis is correlated metabolically during flower development in Citrus and Polianthes, and its changes are involved in floral initiation and development (Huang et al., 2004; Zierer et al., 2016). As suggested for non-parasitic plants, the higher soluble compounds in the reproductive organs decreased the water potential and attracted more metabolites toward this organ from the other organs, which were then used for flower and seed production.

The levels of three soluble aromatic amino acids increased in the flower buds compared to the other organs (**Figure 5**). In addition, the levels of phenylalanine decreased in the buds and flowers in the protein fraction. These two observations proposed that the pathways derived from phenylalanine (such as phenylprepnoinds) were enhanced in these two organs, which apparently contribute to the color (anthocyanins) and volatiles (like benzenoids) of flowers. Indeed, the high level of benzoic acid in flowers proposed that it is used for the synthesis of volatile benzenoids that are known to attract pollinators (Khurana and Cleland, 1992). This reduction of phenylalanine in the protein fraction is also apparently related to the significantly higher levels of total phenols found in the buds (**Figure 6**). Higher levels of aromatic amino acids and total phenols (about 3-fold) compared to the host's roots were previously detected in tubercle (Hacham et al., 2016), suggesting that these pathways are active in the parasite.

In addition to these compounds, flowers exhibit high levels of sucrose and mannose, while fructose, ribose and NA2 have higher levels in flowers and in floral buds (Supplemental Table S2). The level of sucrose, which was relatively very low in the vegetative organs of the parasite, increased in flowers by about 3.3-fold compared to the adventitious roots. Sucrose is proposed to be one of the main carbohydrates that are channeled from the host toward the *O. crenata* (Aber et al., 1983). However, the sucrose level in *P. aegyptiaca* was 77% lower than in the infected tomato roots, while the level of its two components, glucose and fructose, increased significantly in the parasite (Hacham et al., 2016). This is mainly due to the invertase activity that was indeed detected in *P. ramosa* (Draie et al., 2011). We assume that the high levels of sucrose and hexoses in flowers are due to the presence of nectar, which are the predominant floral nectar sugar in many species (Wolff, 2006).

Although most of the metabolites, including the total proteins, nitrogen, and phenols are accumulated in the reproductive organs, the levels of the reducing sugars representing starch increased significantly in the underground pre-emergent shoot (**Figure 6B**). The accumulation of reducing sugars proposes that this organ serves as a reservoir of carbohydrates for the upper organs. One possibility is that this reservoir can be used when the upper shoot is removed, which enables the synthesis of other shoots, an assumption that should be tested, and/or when the sullies from the host are reduced. We also assume that the less soluble sugars in this organ (**Figure 5**) enable the water coming from the host to be channeled through this organ to the upper organs at the mature stage.

This study provides more knowledge about changes occurring in the metabolic profiling during the developmental stages and in five organs of mature *P. aegyptiaca* plants. In the future, it would be worthwhile to reveal which metabolites are derived from the host, and which are mainly synthesized inside the parasite. Such data could contribute to our knowledge about the behavior of *P. aegyptiaca* and its metabolic nature. Differences in the metabolic pathways and their regulation between *P. aegyptiaca* and its hosts could enable additional approaches in the future to control this parasite.

AUTHOR CONTRIBUTIONS

YH and RA conceived and designed the research. NN and YH conducted all of the experiments and data analyses. JH and ED performed the growth and collection of the parasite, and assisted in the design part of the analyses. RA and YH prepared the manuscript. All authors read and approved the manuscript.

FUNDING

This work was supported by a grant from the JCA Foundation (Accelerator project).

ACKNOWLEDGMENTS

The authors would like to thank Evgeny Smirnov (Weed Research Dept., Newe Ya'ar Research Center, ARO), and Aviv Guy for their technical assistance in the growth and collection of the plant samples. We also thank Janet Covaliu for her assistance with the English editing of the manuscript.

REFERENCES

Abbes, Z., Kharrat, M., Delavault, P., Chaïbi, W., and Simier, P. (2009a). Nitrogen and carbon relationships between the parasitic weed *Orobanche foetida* and susceptible and tolerant faba bean lines. *Plant Physiol. Biochem.* 47, 153–159. doi: 10.1016/j.plaphy.2008.10.004

Abbes, Z., Kharrat, M., Delavault, P., Chaïbi, W., and Simier, P. (2009b). Osmoregulation and nutritional relationships between *Orobanche foetida* and faba bean. *Plant Signal. Behav.* 4, 336–338. doi: 10.4161/psb.4.4.8192

Aber, M., Fer, A., and Sallé, G. (1983). Transfer of organic substances from the host plant *Vicia faba* to the parasite *Orobanche crenata* Forsk. *Z. Pflanzenphysiol.* 112, 297–308.

Bai, B., Sikron, N., Gendler, T., Kazachkova, Y., Barak, S., Grafi, G., et al. (2012). Ecotypic variability in the metabolic response of seeds to diurnal hydration-dehydration cycles and its relationship to seed vigor. *Plant Cell Physiol.* 53, 38–52. doi: 10.1093/pcp/pcr169

Ben Nasr, C., Ayed, N., and Metche, M. (1996). Quantitative determination of the polyphenolic content of pomegranate peel. *Z. Lebensm. Unters. Forsch.* 203, 374–378.

Bradford, M. M. (1976). A rapid and sensitive method for the quantitation of microgram quantities of protein utilizing the principle of protein-dye binding. *Anal. Biochem.* 72, 248–254.

Cohen, H., Israeli, H., Matityahu, I., and Amir, R. (2014). Seed-specific expression of a feedback-insensitive form of CYSTATHIONINE-gamma-SYNTHASE in Arabidopsis stimulates metabolic and transcriptomic responses associated with desiccation stress. *Plant Physiol.* 166, 1575–1592. doi: 10.1104/pp.114.246058

Delavault, P. (2015). Knowing the parasite: biology and genetics of orobanche. *Helia* 38, 15–29. doi: 10.1515/helia-2014-0030

Delavault, P., Simier, P., Thoiron, S., Véronési, C., Fer, A., and Thalouarn, P. (2002). Isolation of mannose 6-phosphate reductase cDNA, changes in enzyme activity and mannitol content in broomrape (*Orobanche ramosa*) parasitic on tomato roots. *Physiol. Plant.* 115, 48–55. doi: 10.1034/j.1399-3054.2002.1150105.x

Distelfeld, A., Avni, R., and Fischer, A. M. (2014). Senescence, nutrient remobilization, and yield in wheat and barley. *J. Exp. Bot.* 65, 3783–3798. doi: 10.1093/jxb/ert477

Draie, R., Péron, T., Pouvreau, J. B., Véronési, C., Jégou, S., Delavault, P., et al. (2011). Invertases involved in the development of the parasitic plant *Phelipanche ramosa*: characterization of the dominant soluble acid isoform, PrSAI1. *Mol. Plant Pathol.* 12, 638–652. doi: 10.1111/j.1364-3703.2010.00702.x

Fernández-Aparicio, M., Flores, F., and Rubiales, D. (2016). The effect of *Orobanche crenata* infection severity in Faba Bean, field pea, and grass pea productivity. *Front. Plant Sci.* 7:1409. doi: 10.3389/fpls.2016.01409

Frank, A., Cohen, H., Hoffman, D., and Amir, R. (2015). Methionine and S-methylmethionine exhibit temporal and spatial accumulation patterns during the Arabidopsis life cycle. *Amino Acids* 47, 497–510. doi: 10.1007/s00726-014-1881-1

Galili, G., and Höfgen, R. (2002). Metabolic engineering of amino acids and storage proteins in plants. *Metab. Eng.* 4, 3–11. doi: 10.1006/mben.2001.0203

Gregersen, P. L., and Holm, P. B. (2007). Transcriptome analysis of senescence in the flag leaf of wheat (*Triticum aestivum* L.). *Plant Biotechnol. J.* 5, 192–206. doi: 10.1111/j.1467-7652.2006.00232.x

Gregersen, P. L., Holm, P. B., and Krupinska, K. (2008). Leaf senescence and nutrient remobilisation in barley and wheat. *Plant Biol.* 10(Suppl. 1), 37–49. doi: 10.1111/j.1438-8677.2008.00114.x

Hacham, Y., Hershenhorn, J., Dor, E., and Amir, R. (2016). Primary metabolic profiling of Egyptian broomrape (*Phelipanche aegyptiaca*) compared to its host tomato roots. *J. Plant Physiol.* 205, 11–19. doi: 10.1016/j.jplph.2016.08.005

Huang, C. K., Chang, B. S., Wang, K. C., Her, S. J., Chen, T. W., Chen, Y. A., et al. (2004). Changes in polyamine pattern are involved in floral initiation and development in Polianthes tuberosa. *J. Plant Physiol.* 161, 709–713. doi: 10.1078/0176-1617-01256

Joel, D. M., Hershenhorn, J., Eizenberg, H., Aly, R., Ejeta, G., Rich, P. J., et al. (2007). "Biology and management of weedy root parasites," in *Horticultural Reviews*, ed J. Janick (Hoboken, NJ: John Wiley & Sons), 267–349.

Khurana, J. P., and Cleland, C. F. (1992). Role of salicylic acid and benzoic acid in flowering of a photoperiod-insensitive strain, *Lemna paucicostata* LP6. *Plant Physiol.* 100, 1541–1546.

Li, L., Zhao, J., Zhao, Y., Lu, X., Zhou, Z., Zhao, C., et al. (2016). Comprehensive investigation of tobacco leaves during natural early senescence via multi-platform metabolomics analyses. *Sci. Rep.* 6:37976. doi: 10.1038/srep37976

Lim, P. O., Kim, H. J., and Nam, H. G. (2007). Leaf senescence. *Annu. Rev. Plant Biol.* 58, 115–136. doi: 10.1146/annurev.arplant.57.032905.105316

Matityahu, I., Godo, I., Hacham, Y., and Amir, R. (2013). Tobacco seeds expressing feedback-insensitive cystathionine gamma-synthase exhibit elevated content of methionine and altered primary metabolic profile. *BMC Plant Biol.* 13:206. doi: 10.1186/1471-2229-13-206

Moschen, S., Bengoa Luoni, S., Di Rienzo, J. A., Caro Mdel, P., Tohge, T., Watanabe, M., et al. (2016). Integrating transcriptomic and metabolomic analysis to understand natural leaf senescence in sunflower. *Plant Biotechnol. J.* 14, 719–734. doi: 10.1111/pbi.12422

Nakabayashi, R., and Saito, K. (2015). Integrated metabolomics for abiotic stress responses in plants. *Curr. Opin. Plant Biol.* 24, 10–16. doi: 10.1016/j.pbi.2015.01.003

Pérez-de Luquea, A., Eizenbergb, H., Grenzc, H. J., Sillerod, J. C., Ávilad, C., Sauerbornc, J., et al. (2010). Broomrape management in faba bean. *Field Crops Res.* 115, 319–328. doi: 10.1016/j.fcr.2009.02.013

Péron, T., Candat, A., Montiel, G., Veronesi, C., Macherel, D., Delavault, P., et al. (2016). New insights into phloem unloading and expression of sucrose transporters in vegetative sinks of the parasitic plant *phelipanche ramosa* L. (Pomel). *Front. Plant Sci.* 7:2048. doi: 10.3389/fpls.2016.02048

Prodhan, M. A., Jost, R., Watanabe, M., Hoefgen, R., Lambers, H., and Finnegan, P. M. (2016). Tight control of nitrate acquisition in a plant species that evolved in an extremely phosphorus-impoverished environment. *Plant Cell Environ.* 39, 2754–2761. doi: 10.1111/pce.12853

Sumner, J. B. (1924). The estimation of sugar in diabetic urine, using dinitrosalicylic acid. *J. Biol. Chem.* 62, 287–290.

Watanabe, M., Balazadeh, S., Tohge, T., Erban, A., Giavalisco, P., Kopka, J., et al. (2013). Comprehensive dissection of spatiotemporal metabolic shifts in primary, secondary, and lipid metabolism during developmental senescence in Arabidopsis. *Plant Physiol.* 162, 1290–1310. doi: 10.1104/pp.113.217380

Westwood, J. H. (2013). "The physiology of the established parasite-host association," in *Biology of Root Parasitic Orobanchaceae and Control Strategies*, eds D. M. Joel, L. J. Musselman, and J. Gressel (Springer), 87–114.

Westwood, J. H., dePamphilis, C. W., Das, M., Fernandez-Aparicio, M., Honaas, L., Timko, M. P., et al. (2012). The parasitic plant genome project: new tools for understanding the biology of Orobanche and Striga. *Weed Sci.* 60, 295–306. doi: 10.1614/WS-D-11-00113.1

Wolf, B. (1982). A comprehensive system of leaf analyses and its use for diagnosing crop nutrient status. *Commun. Soil Sci. Plant* 13, 1035–1059. doi: 10.1080/00103628209367332

Wolff, D. (2006). Nectar sugar composition and volumes of 47 species of Gentianales from a southern Ecuadorian montane forest. *Ann. Bot.* 97, 767–777. doi: 10.1093/aob/mcl033

Xia, J., Sinelnikov, I. V., Han, B., and Wishart, D. S. (2015). MetaboAnalyst 3.0-making metabolomics more meaningful. *Nucleic Acids Res.* 43, W251–W257. doi: 10.1093/nar/ gkv380

Zhang, H., Hong, Y., Huang, L., Liu, S., Tian, L., Dai, Y., et al. (2016). Virus-induced gene silencing-based functional analyses revealed the involvement of several putative trehalose-6-phosphate synthase/phosphatase genes in disease resistance against *Botrytis cinerea* and *Pseudomonas syringae* pv. tomato DC3000 in Tomato. *Front. Plant Sci.* 7:1176. doi: 10.3389/fpls.2016.01176

Zierer, W., Hajirezaei, M. R., Eggert, K., Sauer, N., von Wiren, N., and Pommerrenig, B. (2016). Phloem-specific methionine recycling fuels polyamine biosynthesis in a sulfur-dependent manner and promotes flower and seed development. *Plant Physiol.* 170, 790–806. doi: 10.1104/pp.15.00786

Conflict of Interest Statement: The authors declare that the research was conducted in the absence of any commercial or financial relationships that could be construed as a potential conflict of interest.

17

Enhanced Host-Parasite Resistance Based on Down-Regulation of *Phelipanche aegyptiaca* Target Genes is Likely by Mobile Small RNA

Edited by:
Soren K. Rasmussen,
University of Copenhagen, Denmark

Reviewed by:
Grama Nanjappa Dhanapal,
University of Agricultural Sciences,
Bangalore, India
Harro Bouwmeester,
University of Amsterdam, Netherlands

***Correspondence:**
Radi Aly
radi@volcani.agri.gov.il

*Neeraj K. Dubey[1], Hanan Eizenberg[1], Diana Leibman[2], Dalia Wolf[3], Menahem Edelstein[3], Jackline Abu-Nassar[1], Sally Marzouk[1], Amit Gal-On[2] and Radi Aly[1]**

[1] Department of Plant Pathology and Weed Research, Newe Ya'ar Research Center, Agricultural Research Organization, Volcani Center, Ramat Yishay, Israel, [2] Department of Plant Pathology and Weed Research, Agricultural Research Organization, Volcani Center, Rishon LeZion, Israel, [3] Department of Plant Science, Agricultural Research Organization, Volcani Center, Rishon LeZion, Israel

RNA silencing refers to diverse mechanisms that control gene expression at transcriptional and post-transcriptional levels which can also be used in parasitic pathogens of plants that Broomrapes (*Orobanche/Phelipanche* spp.) are holoparasitic plants that subsist on the roots of a variety of agricultural crops and cause severe negative effects on the yield and yield quality of those crops. Effective methods for controlling parasitic weeds are scarce, with only a few known cases of genetic resistance. In the current study, we suggest an improved strategy for the control of parasitic weeds based on trans-specific gene-silencing of three parasite genes at once. We used two strategies to express dsRNA containing selected sequences of three *Phelipanche aegyptiaca* genes *PaACS, PaM6PR,* and *PaPrx1* (pma): transient expression using *Tobacco rattle virus* (TRV:pma) as a virus-induced gene-silencing vector and stable expression in transgenic tomato *Solanum lycopersicum* (Mill.) plants harboring a hairpin construct (pBINPLUS35:pma). siRNA-mediated transgene-silencing (20–24 nt) was detected in the host plants. Our results demonstrate that the quantities of *PaACS* and *PaM6PR* transcripts from *P. aegyptiaca* tubercles grown on transgenic tomato or on *TRV*-infected *Nicotiana benthamiana* plants were significantly reduced. However, only partial reductions in the quantity of *PaPrx1* transcripts were observed in the parasite tubercles grown on tomato and on *N. benthamiana* plants. Concomitant with the suppression of the target genes, there were significant decreases in the number and weight of the parasite tubercles that grew on the host plants, in both the transient and the stable experimental systems. The results of the work carried out using both strategies point to the movement of mobile exogenous siRNA from the host to the parasite, leading to the impaired expression of essential parasite target genes.

Keywords: *Phelipanche*, root parasite, siRNA, VIGS, expression, trans-silencing

INTRODUCTION

Parasitic weeds such as broomrapes do not possess functional roots and do not have effective photosynthesis (Parker and Riches, 1993). Instead, they develop special intrusive organs (haustoria) that penetrate crop roots, directly connecting them to the vascular system of the crop plants that serve as their hosts (Joel and Losner-Goshen, 1994; Yoder, 1999; Westwood, 2000). The haustorium is the organ that distinguishes parasitic from non-parasitic plants (Kuijt, 1969). This organ forms the physical and physiological connection between parasite and host, and its interaction with tissues is important for the translocation of molecules and macromolecules (Aly, 2013). By acting as a strong sink relative to the host, broomrapes channel the flow of water, nutrients, and other molecules from the host to themselves, thereby damaging crop development and greatly reducing yields (Joel, 2000). Broomrapes have evolved sophisticated systems for detecting the presence of host plants and coordinating their development with it (Joel and Losner-Goshen, 1994; Yoder, 1999; Bouwmeester et al., 2003). Following successful attachment, broomrape tissues adjacent to the host root grow into a bulbous structure called a tubercle (Kuijt, 1977). After approximately 4 weeks of growth, a floral meristem is produced, which emerges above ground to flower and disseminate seeds. Effective means for the control of broomrapes are few (Aly, 2007).

The best long-term strategy for controlling parasitic weeds may be through the identification and breeding of resistant crop genotypes (Cubero and Hernández, 1991; Ejeta et al., 1991). However, despite many years of work by plant breeders, only a handful of resistant crop cultivars are currently available.

Plants have evolved a variety of gene-silencing pathways mediated by small RNA sequences (siRNA), which are 21 or 24 nt in size. siRNA suppresses the expression of sequence-homologous genes at the transcriptional and post-transcriptional levels (Baulcombe, 2004; Ding, 2010). The production of hairpin RNA (hpRNA) in transgenic plants is a powerful tool for suppressing gene expression in plants (Wesley et al., 2001; Mansoor et al., 2006; Bandaranayake and Yoder, 2013) through a process known as post-transcriptional gene-silencing (PTGS).

Gene-silencing in plants has been shown to effectively control nematodes (Atkinson et al., 2003; Huang et al., 2006) and viruses (Prins et al., 2008; Leibman et al., 2015), and evidence is available to suggest a natural antiviral role for RNA silencing in vertebrates, fungi, worms, and flies (Fire et al., 1998; Denli and Hannon, 2003). RNAi strategies have also been tried for the control of parasitic plants such as *Triphysaria pusilla* (Benth.) T.I. Chuang and Heckard (Tomilov et al., 2008), *Striga hermonthica* (Delile) Benth. (de Framond et al., 2007; Ejeta and Gressel, 2007), *Phelipanche aegyptiaca* (Pers.) Pomel (Aly et al., 2009), and *Cuscuta pentagona* Engelm. (Alakonya et al., 2012).

Virus-induced gene-silencing (VIGS) is an RNA silencing-based technique used for the targeted down-regulation of a host gene, to allow the analysis of the function of that gene (Lu et al., 2003). It has also been used to silence a wide variety of genes in plants (Robertson, 2004). VIGS-derived dsRNA can be transferred from a host plant to herbivores (Kumar et al., 2012) and parasitic plants and suppress the expression of target genes (Aly et al., 2014). Using the VIGS technique, we have shown that transient knock-down of *Phelipanche aegyptiaca CCD7* and *CCD8* inhibits the development of parasite tubercles and the infestation process in tomato host plants (Aly et al., 2014).

The specific genes selected for silencing are genes that play critical roles in the life cycle of the parasite: 1-amino-cyclopropane-1-carboxylate synthase (*PaACS* synthase; accession no. AB219097) is a key regulatory enzyme in the ethylene biosynthetic pathway, which delays the flowering of the parasite growing on transgenic plants (Trusov and Botella, 2006). Mannose 6-phosphate reductase (*PaM6PR*; Aly et al., 2009) regulates mannitol content in *P. aegyptiaca*; mannitol is essential for the movement of water and nutrients from the host to the parasite (Everard et al., 1997; Delavault et al., 2002). Peroxidase (*PaPrx1*; accession no. AY692263) plays an important role in mediating the parasite's responses and signals during the early stages of infection (Keyes et al., 2001). It may also play a role in the infection process or in the development of the parasite, or even loosen the host cell wall (Foreman et al., 2003; Liszkay et al., 2004), thereby facilitating penetration.

We used two silencing strategies to degrade the RNA of three important *P. aegyptiaca* genes and demonstrated that levels of endogenous *PaACS* synthase and *PaM6PR* transcripts from *P. aegyptiaca* tubercles grown on transgenic or tobacco rattle virus (TRV)-mediated *Nicotiana benthamiana* plants were significantly reduced and significantly inhibited the development of the parasite.

MATERIALS AND METHODS

Selection of the Candidate Genes, Vector Construction and Generation of Transgenic Plants

Three *P. aegyptiaca* genes, *PaPrx1*, *PaM6PR*, and *PaACS*, were selected to be knocked down. The sequences of *PaPrx1* (AY692263), *PaM6PR* (Aly et al., 2009), and *PaACS* (AB219097) of *P. aegyptiaca* were fished out from NCBI and confirmed with the PPGP website.[1] The unique and non-homologous sequences of respective candidate genes were selected as described by Aly et al. (2009). RNA was isolated from *P. aegyptiaca* tubercles (Spectrum^{TM} Plant Total RNA Kit, STRN50, Sigma) and then used to prepare cDNA (Verso cDNA kit, AB-1453/A, Thermo) for the further amplification of selected sequences. The selected region of *PaPrx1* cDNA was amplified using the forward primer 5′-CGAGCTCCCAAGCAATTAAGTTTAGTG-3′ and the reverse primer 5′-GGGGTACCCCTCTCACGTGATATTGC-3′, flanking the *SacI* and *KpnI* sites, respectively, and cloned into pUC19. The selected region of the *PaM6PR* gene was amplified using the forward primer 5′-GGGGTACCTCCAAT GAGGATATGGAACTG-3′ and the reverse primer 5′-GCGTCGACGAGGTTGGAAGAGAACAATAC-3′, flanking the *KpnI* and *SalI* sites, respectively, and fused to the *PaPrx1* gene in the recombinant pUC19. The selected fragment of *PaACS* was amplified using the forward primer

[1] http://ppgp.huck.psu.edu

5′-GCGTCGACTTGATGACGATCGAGTGGCG-3′ and the reverse primer 5′-CCCAAGCTTATTTGCGGGCCAGCTGGAG-3′, flanking *SalI* and *HindIII*, respectively, and cloned in recombinant pUC19 containing both parts of the *PaPrx1* and *PaM6PR* genes. The recombinant clones containing the three fused genes were confirmed with restriction analysis and nucleotide sequencing. For the VIGS assay, the pma sequences (including the three segments of the parasite genes) were amplified using the forward primer 5′-GCGGCCGCTCTAGACCAAGCAATTAAGTTTAGTG-3′ and the reverse primer 5′-CCGCTCGAGGGATCCATTTGCGGGCCAGCTGG-3′, and then cloned into the pTRV2 vector at the *XbaI* and *BamHI* sites.

The above three fragment genes (799 nt) were also cloned in a hairpin configuration in a binary vector pBINPLUS under the control of the CaMV35S promoter, as described by Leibman et al. (2015). The transgenic tomato plants were generated using kanamycin as a selection marker (Barg et al., 1997; Leibman et al., 2015). T1 seeds were collected and sterilized with 70% ethanol for 1 min and 1% sodium hypochlorite and 0.01% Tween for 15–20 min, and then rinsed three times with sterilized distilled water. Ten seeds from each of the tomato T1 lines were grown on media (20 g sucrose, 4.4 g MS, pH 5.8, with 6 g phytagel) containing 100 μg/ml kanamycin. The petridishes containing the seeds were kept in a dark growth chamber for 3 days. After that period, the transgenic seedlings were transplanted into pots in a greenhouse.

RNA Isolation, RT-PCR, and qRT-PCR Analysis

Total RNA was extracted from 5 to 10 mm *P. aegyptiaca* tubercles grown on silenced and non-silenced *N. benthamiana* plants, and tubercles attached to transgenic tomato and control tomato plants. First-strand cDNA was synthesized using 1 μg of total RNA extracted from *P. aegyptiaca* tubercles. The quantitative reaction was performed using an ABI-Prism 7000 Real-Time PCR Detection System (Applied Biosystems) and SYBR Green Master Mix (Thermo-AB4162) according to the manufacturer's protocol. For real-time experiments, we used the following primers: for the *PaACS* gene, forward 5′-GGGCATGGTGGGTATTTGC-3′ and reverse 5′-TACTATGTGAGAATCTTGGGCTTGA 3′; for *PaM6PR*: forward 5′-CCAATGAGGATATGGAACTGTTGA-3′ and reverse 5′-CATGGGAGAGAAACTTATGCGAAAA-3′; and for *PaPrx1*: forward 5′ ATCCATCAACTTTGTTGCTGTGA-3′ and reverse 5′-ACGACATGTGCGAGAGTAGAATG-3′. Expression of the candidate target genes was normalized to the expression of the *Actin* gene using the forward primer 5′-ATGGGCCAGAAAGATGCATATGTT-3′ and the reverse primer 5′-GTGTGATGCCAAATTTTCTCCATGT-3′. Relative gene expression was calculated using the $2^{-\Delta\Delta Ct}$ method (Livak and Schmittgen, 2001). The qRT-PCR reaction conditions were as follows: 15 min activation at 95°C, followed by 40 cycles of 95°C for 10 s, 60°C for 15 s, and 70°C for 20 s. The qRT-PCR experiment was performed in biological and experimental duplicates using PCR Master Mix (cat.42-138; Apex^{TM}).

siRNA Analysis in Transgenic Roots and in VIGS-Assayed Plants

Total RNA was isolated from roots of transgenic tomato plants using an EZ-RNA II kit (Total RNA Isolation Kit, 20-410-100, Biological Industries). Expression of transgene siRNA were detected by northern blot analysis using a cDNA-transcript probe harboring the three target genes (^{32}P-labeled cDNA probe) according to a standard protocol (Aly et al., 2009).

VIGS Assay

Tobacco rattle virus-virus-induced gene-silencing experiments were performed with *N. benthamiana* seedlings that had 3–4 leaves. Agro-infiltration of tobacco leaves were performed as described by Bachan and Dinesh-Kumar (2012). In brief, the following plasmids: TRV-RNA1 (TRV), TRV-RNA2 (pTRV2:*PaPrx1-PaM6PR-PaACS*, abbreviated TRV:pma), and empty TRV were transformed with *Agrobacterium* strain EHA105 using standard protocols. Single colonies were inoculated for primary broth culture (5 ml), followed by secondary broth culture (50 ml) in the presence of suitable antibiotics. The colonies were then grown overnight at 28°C. The next day, 50 ml of cell culture was pelleted by centrifugation at 3000 rpm for 15 min. The recovered pellet was dissolved in infiltration medium (10 mM MES; 10 mM $MgCl_2$; 250 μ M acetosyringone in double-deionized water) adjusted to an OD of 1.0 (600 nm), and then incubated at room temperature for 3 h. *Agrobacterium* was introduced into the lower surface of the tobacco leaf with a 2.0 ml syringe. Just before infiltration, a culture of TRV and TRV:pma in a 1:1 ratio was prepared. RNA was isolated from TRV- and TRV:pma-infected leaves and roots 10 days after infiltration and then subjected to RT-PCR analysis. The expression analyses of TRV and TRV:pma were performed using the following primers: pTRV1 forward: 5′-CCTTTGAACGCGGTAGAACG-3′, pTRV1 reverse: 5′-TGCAGAGCAGGAACTCTATC-3′ and pTRV2 forward: 5′-TTACGGACGAGTGGACTTAG-3′ and pTRV2 reverse: 5′-CTATGGTAAGACAATGAGTCG-3′.

Evaluation of Plant Resistance to the Parasite

Phelipanche aegyptiaca seeds were collected from an infested tomato field in the Bet She'an Valley in eastern Israel. *N. benthamiana* plants were used as hosts for TRV infection. Host plants were transplanted into 2.0-L pots filled with soil (light-medium clay with 63% sand, 12% silt, and 22% clay) and grown in a greenhouse under natural light with an average 14 h of daylight and a temperature of 20 ± 6°C. The plants were watered and fertilized as needed. *N. benthamiana* seedlings were transplanted into pots containing soil infested with *P. aegyptiaca* seeds (20 mg), 7–10 days before agro-infiltration. Host roots from TRV-VIGS and control plants were collected 25–30 days after they were first exposed to the *P. aegyptiaca* seeds. *P. aegyptiaca* tubercles larger than 2 mm were counted and weighed and their RNA was then isolated.

The host resistance of transgenic tomato lines (T_1) was evaluated by challenging the host plants with the parasite seeds in

2.0-L pots. The roots were washed after the parasite inflorescences emerged above the ground. *P. aegyptiaca* tubercles larger than 2 mm were counted and weighed, and RNA was then isolated from those tubercles for the analysis of target gene expression. Host and parasite morphology and biomass were measured as described by Aly et al. (2009).

Peroxidase Assay

Peroxidase activity was evaluated using the Amplex® Red Hydrogen Peroxidase/Peroxidase Assay Kit (A22188, Molecular Probes). Briefly, 200 mg of the parasite tubercles attached to transgenic and non-transgenic host roots were collected and frozen in liquid nitrogen, and then homogenized in 1.2 ml of 0.2 M potassium phosphate buffer (pH 7.8). Samples were centrifuged at 15000 rpm for 20 min at 4°C. The supernatants were stored and the pellet was re-suspended again in 0.8 ml of the same buffer followed by centrifugation. Both supernatants were combined and stored on ice, and used to determine peroxidase activity.

RESULTS

P. aegyptiaca Target Gene Sequences

Based on the database of *P. aegyptiaca* ESTs from the Parasitic Plant Genome Project[2] (Westwood et al., 2011), PubMed-NCBI data sequences and an older database of *P. aegyptiaca* sequences (Aly et al., 2009), we identified and confirmed suitable DNA sequences (Supplementary Figure S1) from non-homologous regions of the three target genes that differ between *P. aegyptiaca*, tomato and *N. benthamiana*, to avoid silencing any host genes.

Silencing of *P. aegyptiaca* Target Genes via TRV Vectors

The selected target regions of *PaM6PR* (268bp), *PaACS* (299bp), and *PaPrx1* (232bp) from *P. aegyptiaca* (Supplementary Figure S1) were cloned in a transient expression system (TRV) vector (**Figure 1A**), as described by Liu et al. (2002). *N. benthamiana* plants were agro-infiltrated with the recombinant TRV2:pma and TRV (**Figure 1A**) according to the method described by Bachan and Dinesh-Kumar (2012). Accumulation of TRV and TRV:pma in roots and leaves of *N. benthamiana* plants was confirmed by RT-PCR (**Figure 1B**). The expression levels of the target gene mRNA in *P. aegyptiaca* grown on assayed *N. benthamiana* plants were evaluated using quantitative RT-PCR. This analysis showed that the transcript amounts of *PaACS* and *PaM6PR* were significantly reduced in the parasite tubercles growing on *N. benthamiana* plants infected with recombinant TRV as compared to *N. benthamiana* plants infected with TRV (**Figure 1C**). No significant suppression of the production of *PaPrx1* transcripts was observed in the parasite tubercles grown on *N. benthamiana* plants infected with recombinant TRV (**Figure 1C**).

[2] http://ppgp.huck.psu.edu/

Retardation of *P. aegyptiaca* Development on *N. benthamiana* Plants Infected with TRV:pma

Nicotiana benthamiana plants were assayed for resistance to *P. aegyptiaca* in pots in which infected plants were pre-challenged with the parasite seeds 10 days before agro-infiltration in the greenhouse (Aly et al., 2014). Parasite infection rates and the number and total weights of *P. aegyptiaca* tubercles larger than 2 mm were determined on TRV:pma and TRV control plants 2 weeks after agro-infiltration. TRV:pma-treated plants expressing the target sequences of *PaACS*, *PaM6PR*, and *PaPrx1*had significantly fewer parasite tubercles and the weight of those tubercles was also more than 50% lower among these plants, as compared to the control plants (**Figures 2A,B**). Growth of the parasite shoots also ceased (**Figure 2C**). Our data suggest that mobile siRNA might move from the host plant to the parasite tubercles and differentially affected the silencing of the target genes.

Since the efficacy of the trans-silencing of the target sequences in *P. aegyptiaca* was confirmed for at least two genes (*ACS* and *M6PR*) through the use of the transient VIGS strategy, we conducted experiments for stable transformation into tomato *Solanum lycopersicum* L.'MP-1'plants, to determine efficacy of this trans-silencing strategy in stable transgenic lines.

Characterization of Stable Transgenic Tomato Lines and Their Resistance to the Parasite

The binary pBINPLUS35:*pma* construct (**Figure 3A**) harboring fragments of *PaPrx1*, *PaM6PR*, and *PaACS* in a hairpin configuration was transformed into tomato *Solanum lycopersicum* 'MP-1'as described by Leibman et al. (2015). Several independent lines of transgenic tomato containing *pBINPLUS:pma* were developed through *Agrobacterium*-mediated transformation. Twenty-six independent transgenic tomato lines were generated and five lines (2, 17, 35, 45, and 59) were selected for use in further experiments based on transgene expression as determined by RT-PCR. PCR was used to confirm the presence of the transgene in the selected T1 transgenic lines (**Figure 3B**). A segregation ratio of close to 3:1 was noted for kanamycin resistance (data not shown), which may indicate the presence of a single locus in those transgenic lines. Expression levels of the transgene transcripts of lines 2, 17, 35, 45, and 59 (T1 progeny) were analyzed by RT-PCR, using target gene-specific primers (**Figure 3C**).

The transgene transcripts were detected only in the transgenic lines using specific primers (**Figure 3C**). Interestingly, the level of transgene transcript varied between the different target genes: *Prx1* $\geq$ *M6PR* $\geq$ *ACS*. This could be due to the orientation and position of the fragment in the inverted repeat construct, as was demonstrated by Wroblewski et al. (2014). In addition, the differences in transcript levels (**Figure 3C**) may reflect rapid processing of transgene dsRNA by DCLs to siRNA. So the low level of transcripts in line 59 may indicate a

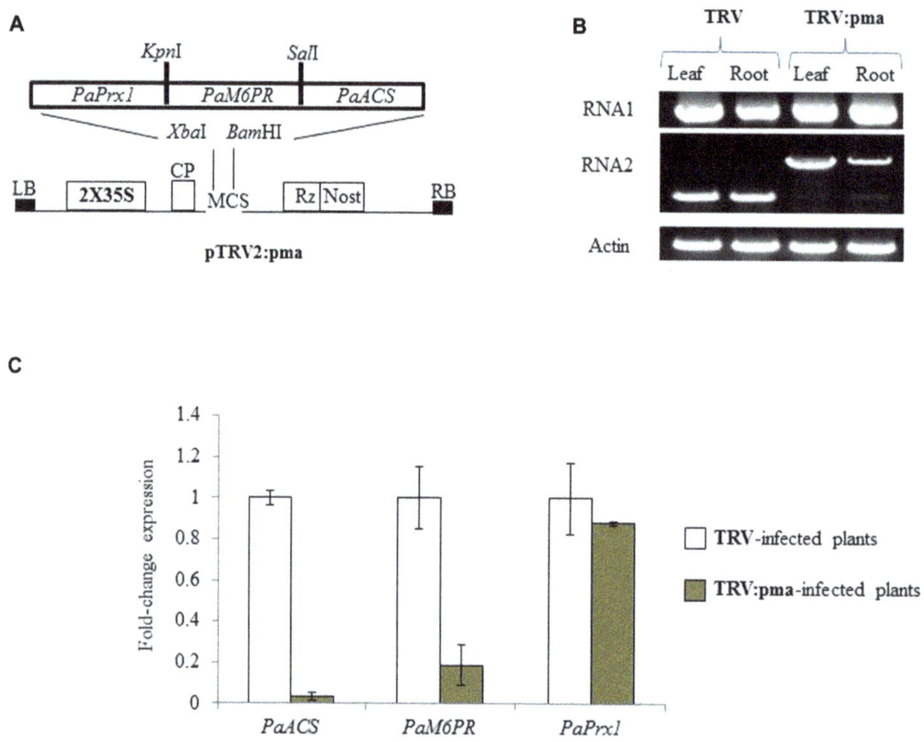

FIGURE 1 | Suppression of *PaACS*, *PaM6PR*, and *PaPrx1* mRNA in *Phelipanche aegyptiaca* via Tobacco rattle virus-virus-induced gene-silencing (TRV-VIGS) assayed in *Nicotiana benthamiana* plants. **(A)** Schematic representation of the construct cloned in the pTRV2 vector according to Liu et al. (2002). **(B)** Systemic infection of recombinant TRV:pma and the TRV control in leaves and roots of *N. benthamiana* plants. RT-PCR was used to assess the accumulation of TRV RNA1 and 2. The actin gene served as a control. **(C)** Quantification of *PaACS*, *PaM6PR*, and *PaPrx1* mRNA by qRT-PCR analysis was normalized to actin transcript levels in *P. aegyptiaca* tubercles attached to *N. benthamiana* plants that were infected with TRV and TRV:pma. All analyses were performed using three biological replicates. TRV-infected plants were calibrated to the value 1.

rapid processing to siRNA. This could explain why the low level of transcript in line 59 is associated with more efficient silencing.

To verify the transgene dsRNA processing by DICERs, we used northern blotting to analyze the accumulation of transgene siRNA in the roots of transgenic and non-transgenic lines. The accumulation of transgene siRNA was detected and confirmed in several lines, including lines 2, 17, and 59 (**Figure 3D**). The horticultural traits of the transgenic T1 tomato lines appeared normal and the plants were fertile under greenhouse conditions. No phenotypic differences were observed between these plants and the corresponding non-transformed MP-1 plants during the vegetative (**Figure 3E**) or reproductive growth stages (data not shown).

To determine whether transgene siRNA produced in the host would move into *P. aegyptiaca* and affect the accumulation of the parasite mRNA targets, we examined the expression levels of the target genes (*PaPrx1*, *PaM6PR*, and *PaACS*) in viable *P. aegyptiaca* tubercles. Our quantitative RT-PCR analysis showed that the level of endogenous target mRNA in the parasite tubercles was reduced relative to the levels in *P. aegyptiaca* tubercles grown on transgenic T1 tomato plants containing an empty vector (EV) or non-transgenic tomato plants (NT; **Figure 4A**).

Levels of *PaACS*, *PaM6PR*, and *PaPrx1* mRNA in *P. aegyptiaca* tubercles attached to plants of line 59 were significantly suppressed (more than 6-, 12- and 3-fold, respectively; **Figure 4A**). Significant mRNA suppression of *PaM6PR* and *PaPrx1* was observed among plants of line 17 and, among the plants of line 2, only *PaM6PR* mRNA was significantly suppressed (**Figure 4A**).

The resistance of the best candidate lines (2, 17, 45, and 59) to parasite development was evaluated in pot experiments. *P. aegyptiaca* infestation was examined in three separate experiments, which each included 10 biological replicates. To evaluate the resistance of the transgenic lines, we considered and counted only fresh and viable parasite tubercles. Our results indicate that the number of attached parasite tubercles was decreased significantly relative to the non-transgenic plants: 7-fold in line 59, 5-fold in line 17, and more than 2-fold in line 2 (**Figure 4B**). The fresh weights of parasite tubercles and shoots attached to lines 2, 17, and 59 were also significantly lower than those of the parasite tubercles and shoots attached to the control plants (**Figure 4C**). Dry weights of transgenic tomato shoots were significantly higher for lines 2, 17, and 59, as compared to the non-transgenic control plants (**Figure 4D**).

The resulting plants appeared normal and were fertile (**Figure 5A**). When grown in soil inoculated with *P. aegyptiaca*,

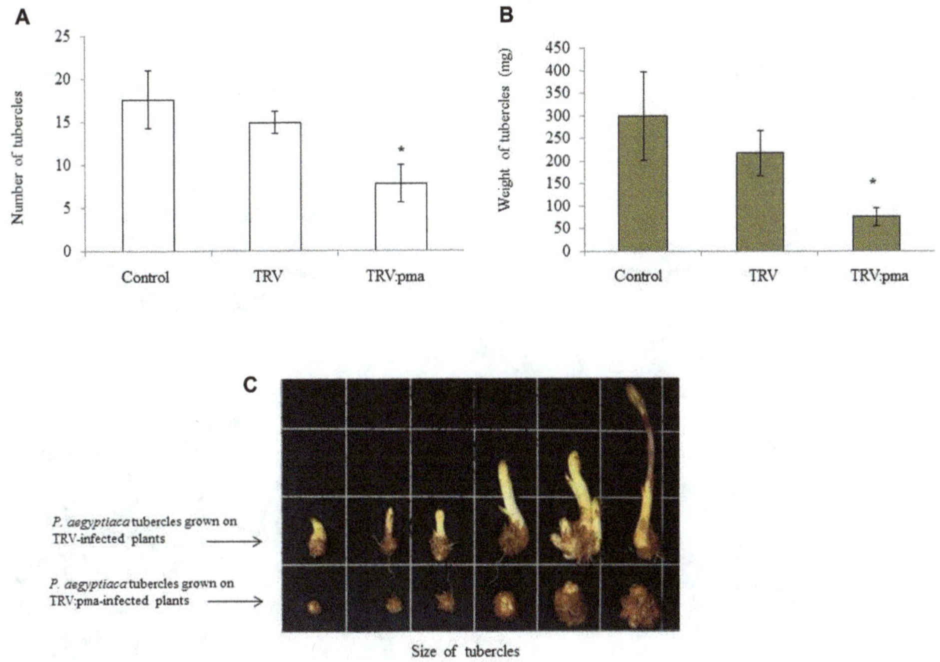

FIGURE 2 | Retardation of *P. aegyptiaca* development on VIGS-assayed *N. benthamiana* plants. The resistance of *N. benthamiana* plants to *P. aegyptiaca* was assayed by transplanting *N. benthamiana* seedlings into pots containing soil infested with *P. aegyptiaca* seeds (20 mg) 7–10 days before agro-infiltration. To evaluate host resistance to the parasite, host roots of rec-TRV- and TRV-treated plants were rinsed 25–30 days after they were challenged with *P. aegyptiaca* seeds. Tubercles larger than 2 mm (Diameter) were considered for analysis. The number of parasitic tubercles **(A)**, average weight of tubercles **(B)**, and representative tubercle growth of the parasitic plants attached to rec-TRV- and TRV-treated plants **(C)** were analyzed. Bars represent means of 10 replicates and vertical lines indicate SE values. Asterisks (*) indicate means different from that of the control and significant differences between empty vector (EV)-infiltrated plants and vector-containing target genes in the VIGS trials, as determined by Student's *t*-test, $\alpha = 0.05$.

transformed tomato lines 2, 17, 45, and 59 had significantly higher biomass accumulation than non-transgenic tomato lines (**Figure 5A**). Additionally, the transformed plants had higher proportions of necrotic and dead tubercles (**Figure 5B**), as compared to the non-transformed plants (**Figure 5C**). Specifically, the mean proportion of necrotic tubercles on non-transformed plants was 1%; whereas among the transgenic lines 2, 17, 45, and 59, the proportion ranged from 50 to 90%.

DISCUSSION

As described previously (Aly, 2007; Joel et al., 2007, 2013), parasitic weeds are difficult to control by conventional means due to their life style; they live in close association with the host roots and are concealed underground and out of sight until they have already inflicted irreversible damage. In this study, our hypothesis was based on our previous results showing partial silencing of a single gene (*M6PR*) in *P. aegyptiaca* (Aly et al., 2009). In order to increase the robustness of this resistance, we pyramided multiple hairpin sequences into single vector. We assumed that resistance to *P. aegyptiaca* in tomato would be improved by expressing dsRNA of multiple gene sequences involved with essential regulatory pathways in the parasite. We selected three genes that are important for the parasite's metabolism (*PaACS*, *PaM6PR*, and *PaPrx1*) for silencing. Suitable DNA fragments (Supplementary Figure S1) from non-homologous regions of the target genes that differ between *P. aegyptiaca* and tomato (to prevent silencing of endogenous genes in the host) were used.

To evaluate host resistance to parasitism by *P. aegyptiaca*, we used two different strategies to knock out three parasite target genes: VIGS and hairpin silencing. Using a VIGS system, we were able to knock-down two candidate genes, *PaACS* and *PaM6PR* (**Figure 1C**), but not *PaPrx1* (**Figure 1C**) in parasite tubercles attached to the host. However, in a stable transgenic system (hairpin-silencing strategy), significant suppression of *PaPrx1* transcripts was observed in the parasite tubercles attached to the roots of transgenic tomato plants of lines 17 and 59 (**Figure 4A**). The lacks of *PaPrx1* transcript suppression in the transient transformation system can be explained by the instability or inefficiencies of *PaPrx1*-siRNAs derived from TRV. Additionally, it might be that the siRNA itself was unable to reach the target gene due to its localization in the tissue.

It is also possible that the targeted *PaPrx1*genes is redundant with other members of their gene family compensating for the silenced genes. This assumption was confirmed, by a nucleotide blast of the selected *PaPrx1* region with the transcriptomic data of *P. aegyptiaca* ESTs from the Parasitic Plant Genome Project[3] at different developmental stages. Indeed, several members of

[3]http://ppgp.huck.psu.edu/

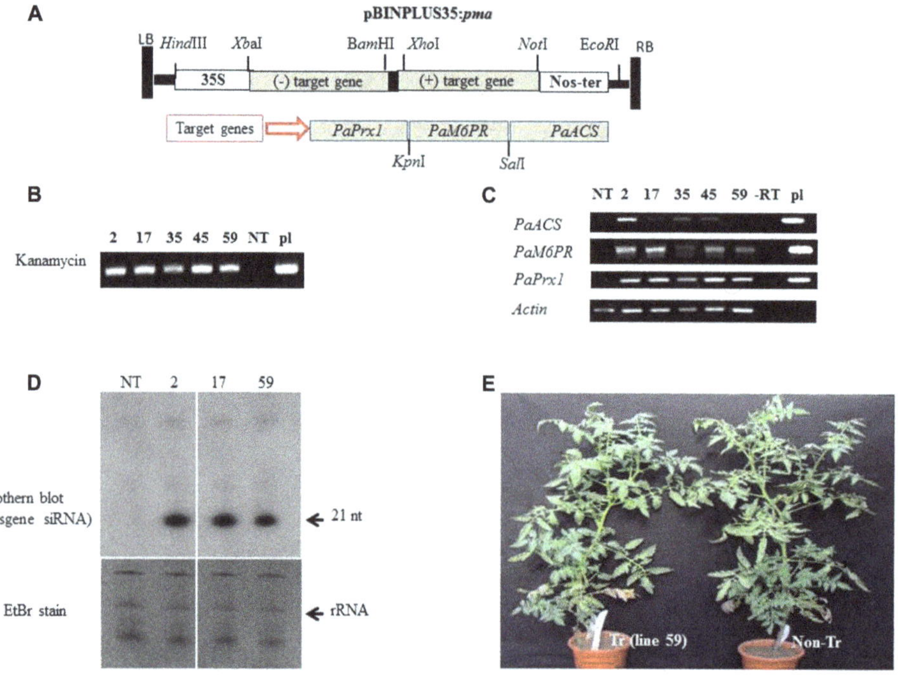

FIGURE 3 | Integration and expression of the *PaACS*, *PaM6PR,* and *PaPrx1* fragments in T1 transgenic tomato lines. **(A)** Schematic representation of the silencing construct pBINPLUS35S:*pma* binary vector harboring the target genes *PaACS*, *PaM6PR*, and *PaPrx1* in hairpin configuration. **(B)** The presence of the transgene in selected T1 lines (2, 17, 35, 45, and 59) was confirmed by PCR analysis of extracted DNA. Lanes NT and pl show the PCR products from the non-transgenic control plants and the pBINPLUS35S:*pma* binary vector, which served as a positive control. For RT-PCR analysis, total RNA was extracted from tomato roots and cDNA was then prepared using random hexamer primers. **(C)** Levels of the transgene transcripts were analyzed by RT-PCR of the self-pollinated progenies (T1) of the transgenic lines 2, 17, 35, 45, and 59. Expression of the actin gene was used as a control for the RT-PCR procedure, the construct pBINPLUS35S:*pma* (pl) served as a positive control and (–RT) served as a negative control. **(D)** Northern blot analysis of transgene-siRNA (t-siRNAs) accumulated in transgenic lines 2, 17, and 59. Non-transformed tomato (NT) served as a negative control. Approximately 30 μg of total RNA from each sample were separated on a 15% urea-PAGE gel and then transferred to a nylon NX membrane. Hybridization was performed with ^{32}P-labeled transcripts of the transgene clone. The gel was stained with ethidium bromide for RNA evaluation prior to transfer to nylon (EtBr stain). **(E)** Growth and appearance of the transgenic tomato plants (Tr) and non-transgenic (non-Tr) tomato plants in a greenhouse.

PaPrx1 were expressed during the early stages of host infection by the parasite (Supplementary Table S1).

We also measured peroxidase activity in tubercles attached to the roots of selected transgenic plants and found no significant suppression of peroxidase activity in either the VIGS system or the hairpin-silencing system (Supplementary Figure S2), with the exception of line 45, in which peroxidase activity was not correlated with the transcript level (**Figure 4A**). The observed peroxidase activity probably reflects the activity of multiple peroxidases rather than the targeted sequence of *PaPrx1*. Nevertheless, low-level suppression of the *PaPrx1* target gene did not affect the number or weight of the parasite tubercles that developed on the assayed plants treated with the silencing construct (TRV:pma sequences). The number and weight of the tubercles on these host plants were significantly lower than those observed for the control treatment (TRV), as a result of *PaACS* and *PaM6PR* silencing (**Figures 2A,B**). Additionally, the parasite tubercles that developed on the VIGS-assayed plants were small and necrotic and developed abnormally (**Figure 2C**). Accumulation of M6PR siRNA in transgenic tomatoes was shown to correlate with decreased levels of M6PR mRNA (Aly et al., 2009). A similar correlation was previously observed between the accumulation of siRNA and virus resistance (Bucher et al., 2006).

We assume that a silencing signal (i.e., mobile siRNA) travels a long distance (Molnar et al., 2011) and can move from host to parasite through haustoria and targeted the tubercles of the parasite genes. Such long-distance movement of mobile siRNA has also been observed between host plant tissue and the parasites *Triphysaria* (Tomilov et al., 2008) and *Phelipanche* (Aly et al., 2009). Although, TRV-VIGS strategy is based on transient expression and does not rely on transformation, it offers a tremendous advantage to analyze gene functions. Promising target genes identified by TRV-VIGS might be used for stable transformation. Therefore, for stably transformed tomato plants, constructs containing the selected sequences (*PaACS, PaM6PR,* and *PaPrx1*) from *P. aegyptiaca* were cloned into the pBINPLUS plasmid in hairpin configuration, as illustrated in **Figure 3A** and introduced into tomato [*S. lycopersicum* (Mill.)] plants. The presence of the transgenes in transgenic plants was verified by PCR and RT-PCR (**Figures 3B,C**). The accumulation of a large amount of siRNA in the transgenic host plants (2, 17, and 59; **Figure 3D**) could explain the significant reductions in the mRNA levels of the parasite tubercles grown on those transgenic plants.

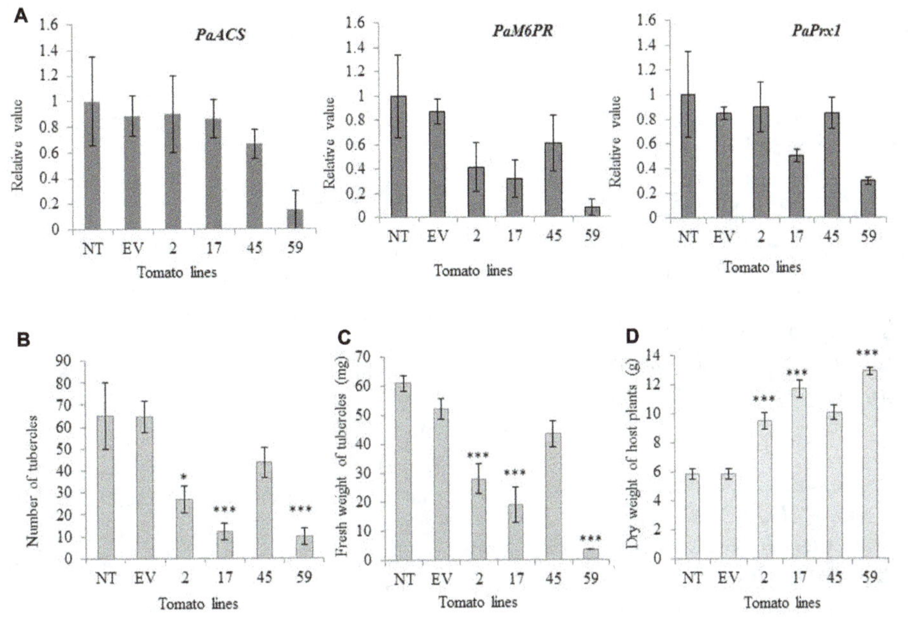

FIGURE 4 | mRNA levels of *PaACS, PaM6PR,* and *PaPrx1* in *P. aegyptiaca* tubercles and resistance of transgenic and non-transgenic lines to the parasite. **(A)** Quantification of *PaACS, PaM6PR,* and *PaPrx1* mRNA levels by qRT-PCR normalized to equal levels of actin transcripts in the underground tubercles of *P. aegyptiaca* including controls and transgenic tomato plants. Total RNA was extracted from 0.5 g of 3 to 5 pooled *P. aegyptiaca* tubercles grown on five transgenic T1 tomato plants (lines 2, 17, 45, and 59), a non-transgenic control plant (NT) and transgenic plants carrying an EV. Quantitative RT-PCR analysis was performed using primers specific for *PaACS, PaM6PR,* and *PaPrx1*. The data presented are relative values calculated following normalization to *P. aegyptiaca* actin with the $2^{-\Delta\Delta ct}$ program. The data are the means of three biological replicates. Bars represent the standard errors of three independent measurements. The graphs in panels **(B,C)** show the number and fresh weights of *P. aegyptiaca* tubercles attached to the transgenic and non-transgenic tomato plants in the greenhouse pot assay. *P. aegyptiaca* tubercles were collected from five transgenic T1 tomato plants (lines 2, 17, 45, and 59), a non-transgenic control plant (NT) and transgenic plants carrying an EV. Means ± SE were calculated based on 10 independent plants. For both experiments, * and *** indicate means different from NT and EV as determined by Student's *t*-test at $\alpha = 0.05$ and $\alpha = 0.001$, respectively. **(D)** Dry weights(g) of host plants were obtained as described by Hamamouch et al. (2005). Means ± SE were calculated based on 10 independent plants. For both experiments, *and *** indicate means significantly different from NT and EV as determined by Student's *t*-test at $\alpha = 0.05$ and $\alpha = 0.001$, respectively. (The weight in graph C and D represent individual average amount of each line).

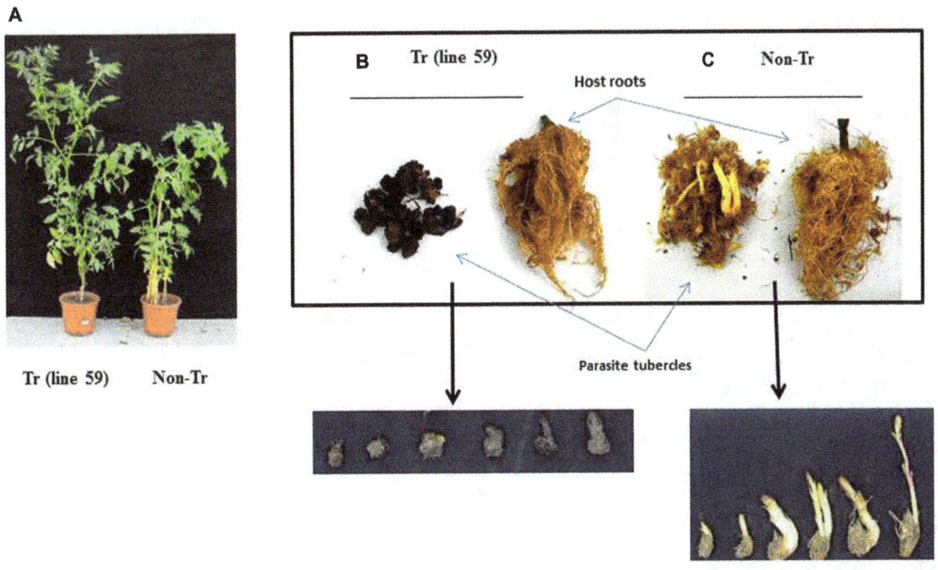

FIGURE 5 | Phenotypes of transgenic tomato plants and *P. aegyptiaca* tubercles grown in pot experiment in the greenhouse. **(A)** Growth and appearance of a representative transgenic tomato plant Tr (line 59) and a representative non-transgenic (non-Tr) plant. **(B)** Roots of transgenic line Tr (line 59) and the parasite tubercles originated from this line. **(C)** Roots of non-transgenic (non-Tr) and the parasite tubercles originated from this line.

Other transgenic lines (data not shown) had lower amounts of siRNA, reflecting lower levels of mRNA targeting of the parasite.

Silencing efficiency could be related to the amount of siRNA in the plant tissue or to the specific selected sequences of the target genes (Dunoyer et al., 2005). It is also possible that the efficiency of the haustorial connection varied between the different host lines. Silencing variability was shown for GUS activity in GUS-expressing *Triphysaria* that was grown on transgenic lettuce expressing GUS (Tomilov et al., 2008). In our study, the transgenic plants expressing the selected parasite genes were similar in appearance to non-transformed plants (**Figure 3E**), suggesting that the parasite target genes are not detrimental to the host.

Our results demonstrate differential silencing efficacy of the siRNA on target endogenous (*PaACS*, *PaM6PR*, and *PaPrx1*) mRNA from *P. aegyptiaca* tubercles grown on transgenic lines (2, 17, 45, and 59) (**Figure 4A**). For example, the quantity of *PaACS* transcript was significantly reduced in only in one line (line 59), the quantity of *PaM6PR* transcript was significantly reduced in three lines (2, 17, and 59), and the quantity of *PaPrx1* transcript was significantly reduced in two lines (17 and 59; **Figure 4A**).

Differences in the transcript levels or efficiency of mRNA silencing among the different transgenic lines could be related to the amount of siRNA in the host plant tissue or possibly to the efficiency of the haustorial connection to the host lines. In our previous study, accumulation of M6PR siRNA in transgenic tomatoes was shown to correlate with decreased levels of M6PR mRNA (Aly et al., 2009). Recently, the up-regulated expression of *SHOOT MERISTEMLESS-like* (*STM*) home box transcription factors was demonstrated during haustoria formation in *Cuscuta* (Alakonya et al., 2012). That study of transgenic tobacco expressing siRNA of *STM* specific to *Cuscuta* reported the reduced efficacy of dodder infection on transgenic tobacco plants and defects in haustorial connection, development, and establishment on the host (Alakonya et al., 2012).

In the current study, the transgenic lines (2, 17, and 59) had significantly fewer tubercles on their roots and the weight of those tubercles was also significantly reduced, as compared the control plants (**Figures 4B,C**). The transgenic plants (2, 17, 45, and 59) accumulated more biomass than both the non-transformed plants and the plants transformed with an empty vector in the presence of the parasite (**Figure 4D**). Furthermore, plants expressing the parasite target genes showed enhanced resistance to *P. aegyptiaca* as evidenced by abnormal parasite development and higher parasite mortality after attachment, as compared to non-transformed plants (**Figures 5A–C**). These results indicate that the resistance induced in lines 2, 17, and 59 through the use of hairpin silencing was considerable.

CONCLUSION

In light of the importance of parasitic weeds to world agriculture and the difficulty of obtaining resistance by conventional methods, we assume that genetic resistance based on the silencing of key metabolic genes in the parasite is now feasible. We used different experimental systems and demonstrated that the TRV-VIGS system can provide a rapid screening process for the silencing of potential candidate parasite genes. In addition, the results of our work involving a hairpin-silencing strategy showed that short interfering RNA molecules expressed in host plants affect gene expression in parasitic plants attached to host roots. However, in this context, further research will be required to identify more gene sequences critical to the growth of the parasite and to optimize the system for siRNA signal transmission from host to parasite for use with other promoter sequences.

AUTHOR CONTRIBUTIONS

RA conceived, planned, and supervised the work. ND performed the molecular work and transgenic analysis. HE analyzed the data. DL helped in siRNA analysis. DW contributed in tissue culture and tomato transformation. JA-N, SM, and ME contributed in data production. AG-O contributed in gene constructs.

FUNDING

This research was supported by research grant no. IS-4622-13 from BARD, the United States – Israel Binational Agricultural Research and Development Fund. Additional support was received from Chief Scientist of the Ministry of Agriculture and Rural Development (Israel), grant no.132-1646-13. Contribution from the Agricultural Research Organization, The Volcani Center, Bet-Dagan, Israel, no. 566/17.

ACKNOWLEDGMENT

ND is grateful to the PBC programmed of Israel Government for providing the Postdoctoral fellowship.

REFERENCES

Alakonya, A., Kumar, R., Koenig, D., Kimura, S., Townsley, B., Runo, S., et al. (2012). Interspecific RNA interference of *SHOOT MERISTEMLESS-like* disrupts *Cuscuta pentagona* plant parasitism. *Plant Cell* 24, 3153–3166. doi: 10.1105/tpc.112.099994

Aly, R. (2007). Conventional and biotechnological approaches for control of parasitic weeds. *In Vitro Cell. Dev. Biol. Plant* 43, 304–317. doi: 10.1007/s11627-007-9054-5

Aly, R. (2013). Trafficking of molecules between parasitic plants and their hosts. *Weed Res.* 53, 231–241. doi: 10.1111/wre.12025

Aly, R., Cholakh, H., Joel, D. M., Leibman, D., Steinitz, B., Zelcer, A., et al. (2009). Gene silencing of mannose 6-phosphate reductase in the parasitic weed *Orobanche aegyptiaca* through the production of homologous dsRNA sequences in the host plant. *Plant Biotechnol. J.* 7, 487–498. doi: 10.1111/j.1467-7652.2009.00418.x

Aly, R., Dubey, N. K., Yahyaa, M., Abu-Nassar, J., and Ibdah, M. (2014). Gene silencing of *CCD7* and *CCD8* in *Phelipanche aegyptiaca* by tobacco rattle virus system retarded the parasite development on the host. *Plant Signal. Behav.* 9:e29376. doi: 10.4161/psb.29376

Atkinson, H. J., Urwin, P. E., and McPherson, M. J. (2003). Engineering plants for nematode resistance. *Annu. Rev. Phytopathol.* 41, 615–639. doi: 10.1146/annurev.phyto.41.052002.095737

Bachan, S., and Dinesh-Kumar, S. (2012). "Tobacco rattle virus (TRV)-based virus-induced gene silencing," in *Antiviral Resistance in Plants: Methods in Molecular Biology*, Vol. 894, eds J. M. Watson and M.-B. Wang (New York City, NY: Humana Press), 83–92. doi: 10.1007/978-1-61779-882-5_6

Bandaranayake, P. C. G., and Yoder, J. I. (2013). Trans-specific gene silencing of acetyl-CoA carboxylase in a root-parasitic plant. *Mol. Plant Microbe Interact.* 26, 575–584. doi: 10.1094/MPMI-12-12-0297-R

Barg, R., Pilowsky, M., Shabtai, S., Carmi, N., Szechtman, A. D., Dedicova, B., et al. (1997). The TYLCV-tolerant tomato line MP-1 is characterized by superior transformation competence. *J. Exp. Bot.* 48, 1919–1923. doi: 10.1093/jxb/48.11.1919

Baulcombe, D. (2004). RNA silencing in plants. *Nature* 431, 356–363. doi: 10.1038/nature02874

Bouwmeester, H. J., Matusova, R., Zhongkui, S., and Beale, M. H. (2003). Secondary metabolite signalling in host–parasitic plant interactions. *Curr. Opin. Plant Biol.* 6, 358–364. doi: 10.1016/S1369-5266(03)00065-7

Bucher, E., Lohuis, D., van Poppel, P. M. J. A., Geerts-Dimitriadou, C., Goldbach, R., and Prins, M. (2006). Multiple virus resistance at a high frequency using a single transgene construct. *J. Gen. Virol.* 87, 3697–3701. doi: 10.1099/vir.0.82276-0

Cubero, J. I., and Hernández, L. (1991). Breeding faba bean (*Vicia faba* L.) for resistance to *Phelipanche crenata* Forsk. *Options Méditerr.* 10, 51–57.

de Framond, A., Rich, P. J., Mcmillan, J., and Ejeta, G. (2007). "Effects on Striga parasitism of transgenic maize armed with RNAi constructs targeting essential *S. asiatica* genes," in *Integrating New Technologies for Striga Control*, eds G. Ejeta and J. Gressel (Singapore: World Scientific Publishing Co.), 185–196.

Delavault, P., Simier, P., Thoiron, S., Véronési, C., Fer, A., and Thalouarn, P. (2002). Isolation of mannose 6-phosphate reductase cDNA, changes in enzyme activity and mannitol content in broomrape (*Orobanche ramosa*) parasitic on tomato roots. *Physiol. Plant.* 115, 48–55. doi: 10.1034/j.1399-3054.2002.1150105.x

Denli, A. M., and Hannon, G. J. (2003). RNAi: an ever-growing puzzle. *Trends Biochem. Sci.* 28, 196–201. doi: 10.1016/S0968-0004(03)00058-6

Ding, S.-W. (2010). RNA-based antiviral immunity. *Nat. Rev. Immunol.* 10, 632–644. doi: 10.1038/nri2824

Dunoyer, P., Himber, C., and Voinnet, O. (2005). DICER-LIKE 4 is required for RNA interference and produces the 21-nucleotide small interfering RNA component of the plant cell-to-cell silencing signal. *Nat. Genet.* 37, 1356–1360. doi: 10.1038/ng1675

Ejeta, G., Butler, L. G., Hess, D. E., and Vogler, R. K. (1991). "Parasitic Weeds," in *Proceedings of the 5th International Symposium of Parasitic Weeds*, eds J. K. Ransom, L. J. Musselman, A. D. Worsham, and C. Parker (Nairobi: CIMMYT), 539–544.

Ejeta, G., and Gressel, J. (2007). Integrating New Technologies for Striga Control: Towards Ending the Witch-Hunt. Singapore: World Scientific.

Everard, J. D., Cantini, C., Grumet, R., Plummer, J., and Loescher, W. H. (1997). Molecular cloning of mannose-6-phosphate reductase and its developmental expression in celery. *Plant Physiol.* 113, 1427–1435.

Fire, A., Xu, S., Montgomery, M. K., Kostas, S. A., Driver, S. E., and Mello, C. C. (1998). Potent and specific genetic interference by double-stranded RNA in *Caenorhabditis elegans*. *Nature* 391, 806–811. doi: 10.1038/35888

Foreman, J., Demidchik, V., Bothwell, J. H., Mylona, P., Miedema, H., Torres, M. A., et al. (2003). Reactive oxygen species produced by NADPH oxidase regulate plant cell growth. *Nature* 422, 442–446. doi: 10.1038/nature01485

Hamamouch, N., Westwood, J. H., Banner, I., Cramer, C. L., Gepstein, S., and Aly, R. (2005). A peptide from insects protects transgenic tobacco from a parasitic weed. *Transgenic Res.* 14, 227–236. doi: 10.1007/s11248-004-7546-1

Huang, G., Allen, R., Davis, E. L., Baum, T. J., and Hussey, R. S. (2006). Engineering broad root-knot resistance in transgenic plants by RNAi silencing of a conserved and essential root-knot nematode parasitism gene. *Proc. Natl. Acad. Sci. U.S.A.* 103, 14302–14306. doi: 10.1073/pnas.0604698103

Joel, D. M. (2000). The long-term approach to parasitic weed control: manipulation of specific developmental mechanisms of the parasite. *Crop Prot.* 19, 753–758. doi: 10.1016/S0261-2194(00)00100-9

Joel, D. M., Gressel, J., and Musselman, L. J. (2013). *Parasitic Orobanchaceae: Parasitic Mechanisms and Control Strategies.* Heidelberg: Springer Science & Business Media. doi: 10.1007/978-3-642-38146-1

Joel, D. M., Hershenhorn, J., Eizenberg, H., Aly, R., Ejeta, G., Rich, P. J., et al. (2007). "Biology and management of weedy root parasites," in *Horticultural Reviews*, Vol. 33, ed. J. Janick (London: John Wiley & Sons), 267. doi: 10.1002/9780470168011.ch4

Joel, D. M., and Losner-Goshen, D. (1994). "Early host-parasite interaction: models and observations of host root penetration by the haustorium of Orobanche," in *Proceedings of the 3rd International Workshop on Orobanche and Related Striga Research*, (Amsterdam: Royal Tropical Institute).

Keyes, W. J., Taylor, J. V., Apkarian, R. P., and Lynn, D. G. (2001). Dancing together. Social controls in parasitic plant development. *Plant Physiol.* 127, 1508–1512. doi: 10.1104/pp.010753

Kuijt, J. (1969). *The Biology of Parasitic Flowering Plants.* Berkeley, CA: University of California Press, 246.

Kuijt, J. (1977). Haustoria of phanerogamic parasites. *Annu. Rev. Phytopathol.* 15, 91–118. doi: 10.1146/annurev.py.15.090177.000515

Kumar, P., Pandit, S. S., and Baldwin, I. T. (2012). Tobacco rattle virus vector: a rapid and transient means of silencing *Manduca sexta* genes by plant mediated RNA interference. *PLOS ONE* 7:e31347. doi: 10.1371/journal.pone.0031347

Leibman, D., Prakash, S., Wolf, D., Zelcer, A., Anfoka, G., Haviv, S., et al. (2015). Immunity to tomato yellow leaf curl virus in transgenic tomato is associated with accumulation of transgene small RNA. *Arch. Virol.* 160, 2727–2739. doi: 10.1007/s00705-015-2551-7

Liszkay, A., van der Zalm, E., and Schopfer, P. (2004). Production of reactive oxygen intermediates (O_2-, H_2O_2, and OH) by maize roots and their role in wall loosening and elongation growth. *Plant Physiol.* 136, 3114–3123. doi: 10.1104/pp.104.044784

Liu, Y., Schiff, M., Marathe, R., and Dinesh-Kumar, S. P. (2002). Tobacco Rar1, EDS1 and NPR1/NIM1-like genes are required for N-mediated resistance to tobacco mosaic virus. *Plant J.* 30, 415–429. doi: 10.1046/j.1365-313X.2002.01297.x

Livak, K. J., and Schmittgen, T. D. (2001). Analysis of relative gene expression data using real-time quantitative PCR and the $2^{-\Delta\Delta C_T}$ method. *Methods* 25, 402–408. doi: 10.1006/meth.2001.1262

Lu, R., Martin-Hernandez, A. M., Peart, J. R., Malcuit, I., and Baulcombe, D. C. (2003). Virus-induced gene silencing in plants. *Methods* 30, 296–303. doi: 10.1016/S1046-2023(03)00037-9

Mansoor, S., Amin, I., Hussain, M., Zafar, Y., and Briddon, R. W. (2006). Engineering novel traits in plants through RNA interference. *Trends Plant Sci.* 11, 559–565. doi: 10.1016/j.tplants.2006.09.010

Molnar, A., Melnyk, C., and Baulcombe, D. C. (2011). Silencing signals in plants: a long journey for small RNAs. *Genome Biol.* 12:215. doi: 10.1186/gb-2010-11-12-219

Parker, C., and Riches, C. R. (1993). *Parasitic Weeds of the World: Biology and Control.* Wallingford: CAB International.

Prins, M., Laimer, M., Noris, E., Schubert, J., Wassenegger, M., and Tepfer, M. (2008). Strategies for antiviral resistance in transgenic plants. *Mol. Plant Pathol.* 9, 73–83. doi: 10.1111/j.1364-3703.2007.00447.x

Robertson, D. (2004). VIGS vectors for gene silencing: many targets, many tools. *Annu. Rev. Plant Biol.* 55, 495–519. doi: 10.1146/annurev.arplant.55.031903.141803

Tomilov, A. A., Tomilova, N. B., Wroblewski, T., Michelmore, R., and Yoder, J. I. (2008). Trans-specific gene silencing between host and parasitic plants. *Plant J.* 56, 389–397. doi: 10.1111/j.1365-313X.2008.03613.x

Trusov, Y., and Botella, J. R. (2006). Silencing of the ACC synthase gene *ACACS2* causes delayed flowering in pineapple [*Ananas comosus* (L.) Merr.]. *J. Exp. Bot.* 57, 3953–3960. doi: 10.1093/jxb/erl167

Wesley, S. V., Helliwell, C. A., Smith, N. A., Wang, M., Rouse, D. T., Liu, Q., et al. (2001). Construct design for efficient, effective and high-throughput gene silencing in plants. *Plant J.* 27, 581–590. doi: 10.1046/j.1365-313X.2001.01105.x

Westwood, J. H. (2000). Characterization of the orobanche–arabidopsis system for studying parasite–host interactions. *Weed Sci.* 48, 742–748. doi: 10.1186/1471-2105-15-S11-S13

Westwood, J. H., dePamphilis, C. W., Das, M., Fernández-Aparicio, M., Honaas, L. A., Timko, M. P., et al. (2011). The parasitic plant genome project: new tools for understanding the biology of orobanche and striga. *Weed Sci.* 60, 295–306. doi: 10.1614/WS-D-11-00113.1

Wroblewski, T., Matvienko, M., Piskurewicz, U., Xu, H., Martineau, B., Wong, J., et al. (2014). Distinctive profiles of small RNA couple inverted repeat-induced post-transcriptional gene silencing with endogenous RNA silencing pathways in *Arabidopsis*. *RNA* 20, 1987–1999. doi: 10.1261/rna.046532.114

Yoder, J. I. (1999). Parasitic plant responses to host plant signals: a model for subterranean plant–plant interactions. *Curr. Opin. Plant Biol.* 2, 65–70. doi: 10.1016/S1369-5266(99)80013-2

Conflict of Interest Statement: The authors declare that the research was conducted in the absence of any commercial or financial relationships that could be construed as a potential conflict of interest.

18

Recognition of *Orobanche cumana* Below-Ground Parasitism through Physiological and Hyper Spectral Measurements in Sunflower (*Helianthus annuus* L.)

Edited by:
Paul Christiaan Struik,
Wageningen University and Research,
Netherlands

Reviewed by:
Grama Nanjappa Dhanapal,
University of Agricultural Sciences,
Bangalore, India
Harro Bouwmeester,
University of Amsterdam, Netherlands
Aad Van Ast,
Wageningen University and Research,
Netherlands

***Correspondence:**
Jhonathan E. Ephrath
yoni@bgu.ac.il

Amnon Cochavi[1], Tal Rapaport[2], Tania Gendler[1], Arnon Karnieli[2], Hanan Eizenberg[3], Shimon Rachmilevitch[1] and Jhonathan E. Ephrath[1]*

[1] The French Associates Institute for Agriculture and Biotechnology of Drylands, The Jacob Blaustein Institutes for Desert Research, Ben-Gurion University of the Negev, Beer-Sheva, Israel, [2] The Remote Sensing Laboratory, The Swiss Institute for Dryland Environmental & Energy Research, The Jacob Blaustein Institutes for Desert Research, Ben Gurion University of the Negev, Beer-Sheva, Israel, [3] Department of Plant Pathology and Weed Research, Newe Ya'ar Research Center, Agricultural Research Organization, Volcani Center, Ramat Yishay, Israel

Broomrape (*Orobanche* and *Phelipanche* spp.) parasitism is a severe problem in many crops worldwide, including in the Mediterranean basin. Most of the damage occurs during the sub-soil developmental stage of the parasite, by the time the parasite emerges from the ground, damage to the crop has already been done. One feasible method for sensing early, below-ground parasitism is through physiological measurements, which provide preliminary indications of slight changes in plant vitality and productivity. However, a complete physiological field survey is slow, costly and requires skilled manpower. In recent decades, visible to-shortwave infrared (VIS-SWIR) hyperspectral tools have exhibited great potential for faster, cheaper, simpler and non-destructive tracking of physiological changes. The advantage of VIS-SWIR is even greater when narrow-band signatures are analyzed with an advanced statistical technique, like a partial least squares regression (PLS-R). The technique can pinpoint the most physiologically sensitive wavebands across an entire spectrum, even in the presence of high levels of noise and collinearity. The current study evaluated a method for early detection of *Orobanche cumana* parasitism in sunflower that combines plant physiology, hyperspectral readings and PLS-R. Seeds of susceptible and resistant *O. cumana* sunflower varieties were planted in infested (15 mg kg^{-1} seeds) and non-infested soil. The plants were examined weekly to detect any physiological or structural changes; the examinations were accompanied by hyperspectral readings. During the early stage of the parasitism, significant differences between infected and non-infected sunflower plants were found in the reflectance of near and shortwave infrared areas. Physiological measurements revealed no differences between treatments until *O. cumana* inflorescences emerged. However, levels of several macro- and microelements tended to decrease during the early stage of *O. cumana* parasitism. Analysis of leaf cross-sections revealed differences in range and in mesophyll structure as a result of different levels of nutrients in sunflower plants, manifesting the presence of *O. cumana* infections. The findings of an advanced PLS-R analysis emphasized the correlation between specific reflectance changes in the SWIR range and levels of various nutrients in sunflower plants. This work demonstrates potential for the early detection of *O. cumana* parasitism on sunflower roots using hyperspectral tools.

Keywords: broomrape, early attachment, mesophyll, PLS-R, minerals content

INTRODUCTION

For many years, the underground development of root parasitic plants has been a mystery for both researchers and farmers. Broomrape parasitism of host crop roots is generally recognized only when parasite inflorescences break through the soil surface, at which point irreversible damage has already been done to the crop (Mauromicale et al., 2008; Longo et al., 2010). Over time, we have learned more about the below-ground developmental stages of parasitic plants (Heide-Jørgensen, 2013) and that the germination of parasitic seeds is triggered by specific hormonal signals in the root exudates of host plants (Yoneyama et al., 2010). After recognition, the parasite seed germinates and the rootlet grows toward the host's roots. Finally, the parasite penetrates the host's roots and forms a vascular connection with the host (Yoneyama et al., 2013). The parasite then starts to exploit the host's water, mineral and energy resources, slowly accumulating biomass in preparation for its final stage of above-ground inflorescence (Westwood, 2013). In general, more severe infestations will lead to greater crop damage (Grenz et al., 2008).

Several methods have been developed over the years to reduce damage to sunflower (*Helianthus annuus* L.) caused by *Orobanche cumana* Waller, Sunflower Broomrape (Orobanchaceae family). Castejon-Munoz et al. (1993) showed that earlier sowing dates reduces parasite-induced damage to sunflowers. However, the utility of that method is limited, due to yield reductions caused by shifting sowing dates earlier into a period of cooler weather. Another approach for managing *O. cumana* parasitism is the development of resistant cultivars. However, such resistance is temporary and, after several growth cycles, a new strain of broomrape is usually able to parasitize the crop (Rubiales et al., 2009). A different approach to parasite management is the application of low rates of herbicides, which can reduce parasite damage in many crops (Eizenberg et al., 2012a; Cochavi et al., 2015). In practice, as the herbicide is transported from the foliage through the host vessels down to the parasite's tubercle, the parasite is injured (Eizenberg et al., 2012a). In this context, the rate and timing of herbicide applications to host foliage are crucial for optimal parasite management (Eizenberg et al., 2006); inaccurate herbicide application can lead to crop damage caused by the untreated parasite or too high doses of herbicide.

To predict the timing of the underground development of *O. cumana*, a thermal model was constructed (Eizenberg et al., 2012b). Using this model, a low rate of imazapic (an acetolactate synthase inhibitor) applied at the tubercle stage of parasite development provides highly effective control. Despite the efficacy of this herbicide, the thermal model predicts parasite development based on uniform spatial distribution of the parasite in the field, i.e., that essentially all host plants in the field are infected. Therefore, even non-parasitized plants are sprayed with herbicide, which can lead to unnecessary herbicide use and crop injury.

The use of physiological measurements can provide early evidence of parasite attachment to host roots. The advantage of physiological measurements lies in their ability to highlight changes in plant status before those changes become visible (Levitt, 1980). Nevertheless, to date, substantial broomrape-related physiological changes have been reported only after the parasite's inflorescence has already appeared above ground level (Barker et al., 1996; Alcántara et al., 2006), at which point herbicide applications are virtually ineffective in improving crop yield. Moreover, physiological measurements often require proficient manpower and intrusive instruments and may prove to be time- and cost-ineffective, due to biotic and abiotic variability in the field. A plausible solution to this latter issue is the use of hyperspectral remote-sensing instruments, which offer a faster, cheaper and non-destructive way to assess crop physiology (Mulla, 2013). Specifically, narrow-band visible (VIS, 350–700 nm) to shortwave infrared (SWIR, 1400–1700 nm) sensors are suitable for this purpose; they are sensitive enough to monitor slight changes in various biochemical properties (Blackburn, 2007).

Over the years, variations in numerous physiological properties have been spectrally monitored with such equipment, including changes in leaf water balance (Dzikiti et al., 2010), nutrient status (Menesatti et al., 2010), pigment content (Garrity et al., 2011), xanthophyll cycle activity and fluorescence (Rapaport et al., 2015) and more (Mulla, 2013). In the context of parasites, a previous study found changes in the reflectance of *O. cumana*-infected sunflower leaves illuminated by UV light (Ortiz-Bustos et al., 2016). Researchers found differences in the reflectance of wavelengths of 680 and 740 nm, which were related to changes in chlorophyll content and functioning.

Despite the potential advantages of hyperspectral signatures, narrow-band datasets tend to be highly noisy and therefore require the use of dimensionality-reduction techniques (Mariotto et al., 2013). Specifically, since only a few principal wavelengths of the entire VIS-SWIR spectrum are strongly correlated with physiological variables, whilst the rest are actually redundant – negatively affecting the derivation of indices – the use of optimal wavelength-selection processes is essential (Wold et al., 2001). This issue can be addressed by employing advanced statistical techniques, such as the partial least squares regression (PLS-R) method (Wold et al., 2001). That method was previously shown to be effective for highlighting the most physiologically sensitive wavebands and producing reliable spectral models (e.g., Rapaport et al., 2015). In practice, the PLS-R technique can be used to correlate collinear, noisy and distribution-free datasets, even when the number of predictors greatly exceeds the number of predictions – making it very suitable for dealing with hyperspectral signatures. Geladi and Kowalski (1986) demonstrated the superiority of targeted indices over traditional and broad-band spectral models.

This study focuses on the relationship between *O. cumana* parasitism and the reflectance of the host sunflower leaves. During the study, susceptible and resistant plants were examined, in order to evaluate the effect of *O. cumana* parasitism on different varieties of sunflowers. Physiological, anatomical, and chemical measurements of plant foliage were simultaneously taken. Statistical analysis, using commercially available software, was then applied to the measurements, to determine whether correlations exist between specific wavelengths and physiological, structural, and chemical changes in the leaf.

MATERIALS AND METHODS

Plant Material and Growth Conditions

Sunflower seeds (cv. D. Y. 3, which is *O. cumana*-susceptible, and cv. Emek 3, which is *O. cumana*-resistant; Sha'ar Ha'amakim Seeds, Sha'ar Ha'amakim, Israel) were sown. The seeds were planted in 4-L pots filled with infested or non-infested soil (Newe Ya'ar soil, Chromic Haploxererts, fine clayey, montmorillonitic, thermic, 55% clay, 25% silt and 20% sand, 2% organic matter and pH 7.2). Parasitic *O. cumana* seeds were brought from the Newe Ya'ar Research Center's collection. These seeds were collected in 2010 from Havat Gadash, Hahula Valley, Israel. The seeds were passed through a 300-mesh sieve and stored in the dark at 4°C. A germination test was performed under standard conditions at 25°C, with germination stimulated by the synthetic stimulant GR-24 (Yoneyama et al., 2013), which is commonly used for broomrape germination tests. The GR-24 was applied at a rate of 10 mg kg^{-1} after 12 days of pre-conditioning, resulting in a germination rate of 84%. Parasite seeds were mixed into the soil using a 50 L cement mixer, creating a concentration of 15 mg kg^{-1}. Over the course of the experiment, soil temperature was monitored using temperature data-loggers (UA-001-08 data logger, Onset, Co., Bourne, MA, United States) buried at a depth of 10 cm. The parasite's below-ground attachment was estimated using the thermal model developed by Eizenberg et al. (2012b). Over the course of the experiment, parasite development was also monitored visually by selecting pots, gently removing soil from the sunflower plants, washing their roots and examining any parasite development, as described by Eizenberg et al. (2006). Measurements were conducted as follows: the first measurements after two true leaves developed [23 days after planting (DAP)]; the second measurements after *O. cumana* tubercles were established on the host roots (31 DAP); the third measurement after 63% of *O. cumana* tubercles were established on the host roots (36 DAP); and the last measurement after *O. cumana* inflorescences emerged above the ground level (51 DAP). Pots were opened in order to validate *O. cumana* development according to the model. Each treatment contained 23 pots, with eight used for each of the measurements.

Plants were grown in a semi-controlled greenhouse (Midreshet Ben-Gurion, Ramat Hanegev, Israel) in which temperatures ranged between 20 and 30°C, with a radiation level of 800 μmol photons m^{-2} s^{-1} at midday. Plants were irrigated using 2 l h^{-1} drippers, for field capacity plus 10% in order to remove salts from the upper soil level. Run off water was monitored in order to evaluate the water required by the plants in the different treatments. Fertilizer (27:10:17 N-P-K) was applied through the drip-irrigation system at a rate of 1 g fertilizer/500 L water. Fertilizer application began 14 DAP, after the sunflower plants had begun to develop true leaves.

Spectral Measurements

Adaxial hyperspectral radiance was measured with a Pro-FR Field Spectrometer (Analytical Spectral Devices, Inc., Boulder, CO, United States) coupled to an 1800-12S Integrating Sphere (Li-Cor, Inc., Lincoln, NE, United States). The signatures were collected within the 350–2500 nm spectrum at a resolution of 1 nm and were normalized into reflectance units by frequently taking supplementary readings of $BaSO_4$ tablets. Measurements were taken from the youngest fully matured leaf, three measurements from each leaf and total of eight replicates, at midday. During the same period, physiological measurements were taken under clear, cloudless skies.

Physiological Measurements

Physiological measurements were taken starting after the formation of two true leaves (23 DAP); each of the measurements included eight replicates. All measurements were taken from the youngest fully matured leaf. Photosynthesis and transpiration rate, stomatal conductance, and vapor pressure deficit (VPD) were measured with a portable infra-red gas analyzer (IRGA) system (Li-6400, Li-Cor, Lincoln, NB, United States). Non-photochemical quenching (NPQ) was calculated based on the maximal fluorescence of the youngest fully matured leaf in the light (F_m') and dark-adapted (F_m, 30 min adaptation time) as shown in Eq. (1):

$$\frac{F_m - F_{m'}}{F_{m'}} \tag{1}$$

(Bilger and Bjorkman, 1990), using a Mini-PAM I (Walz, Effeltrich, Germany) photosynthesis yield analyzer. Relative water content was measured according to Smart and Bingham (1974); for each measurement, the youngest fully matured leaves were taken from four plants from each treatment. Chlorophyll content was measured as described by Bruinsma (1963): chlorophyll was extracted from leaf disks by soaking the disks overnight in dimethyl sulfoxide (DMSO), followed by spectrophotometer absorption measurements at wavelengths of 665 and 649 for chlorophyll *a* and *b* content evaluations, respectively.

Determination of Leaf Mineral Content

After physiological measurements were taken, the youngest fully matured leaf samples were taken for mineral analysis and oven-dried for 48 h in 65°C. Carbon, nitrogen, and hydrogen levels were measured using a FLASH 2000 CHNS/O

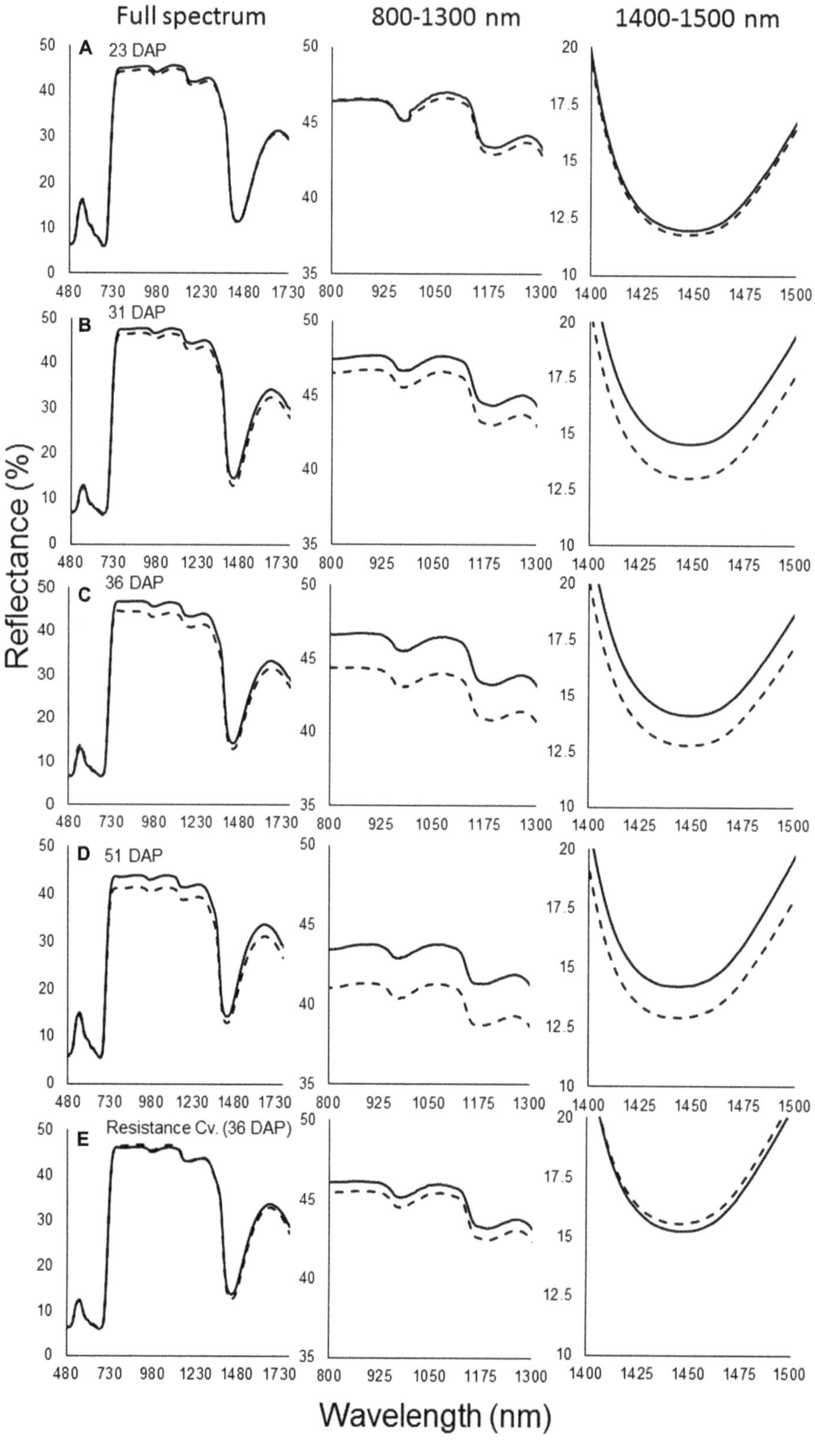

FIGURE 1 | Spectral signature of reflectance of non-infested (—) and infested (—) susceptible sunflower leaves. **(A)** Measurements taken at 23 days after planting (DAP), before *Orobanche cumana* tubercles became established; **(B)** 31 DAP, first attachment of *O. cumana*; **(C)** 36 DAP, 63% of all *O. cumana* attachments were established on the sunflower roots; **(D)** 51 DAP, *O. cumana* inflorescences emerged. **(E)** Resistance cultivar 36 days after planting. Lines represent means of eight plants; three measurements were made on each plant.

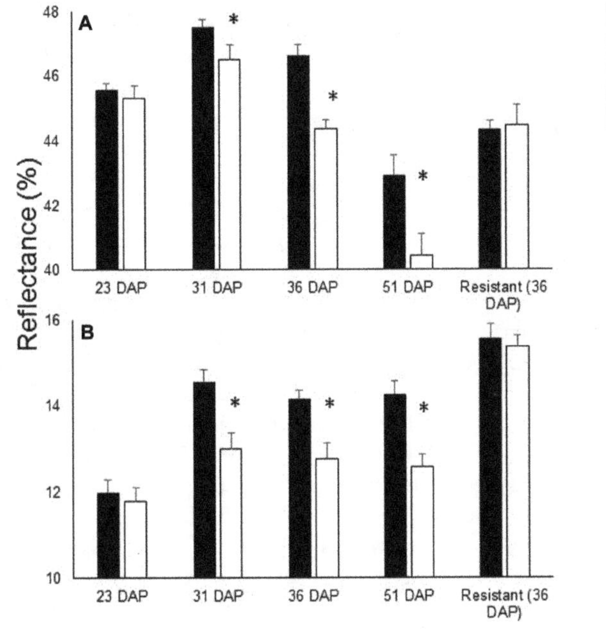

FIGURE 2 | Leaf reflectance of specific wavelengths in susceptible cultivar 23, 31, 36, and 51 DAP in addition to the reflectance of the resistant cultivar 36 DAP. **(A)** Average reflectance in the far-NIR (800–1300 nm) and **(B)** SWIR (1450 nm range). *O. cumana* attachments first observed at 31 DAP. There were eight replicates for each treatment. Bars refer to SE of the mean and asterisks indicate significant differences between treatments (0.05 ≥ α).

Analyzer (Thermo Fisher Scientific, Waltham, MA, United States). Additional mineral measurements were taken using Inductively Coupled Plasma Optical Emission Spectroscopy (ICP-OES, Varian 720 ES, Agilent Technologies, Santa Clara, CA, United States). Dry plant material was extracted by overnight ash in a 470°C oven, followed by an extraction using nitric acid. Samples were diluted with double-distilled water before measurements were taken. At each time point, mineral measurements involving five replications per treatment were taken.

Anatomic Analysis

Leaf tissue sections were taken from all treatments (susceptible and resistance cultivar, infected and non-infected plants) from the youngest fully matured leaf. Samples were taken from the widest part of the leaf without the central vein. Cross-sections were prepared as described by Rewald et al. (2012) and Rapaport et al. (2014). Digital images of the cross-sections were then taken with an Axio Imager A1 Light Microscope (Carl Zeiss Microscopy, Inc., TH, Germany) and 6 × 100 magnification images were produced using its AxioVision 4.6.3 software. Cross-sections were analyzed using Digimizer software (version 4.5.2, MedCalc Software, Ostend, Belgium). Measured parameters included tissue width and air space capacity. Air-space volume was defined as the proportion of the air spaces of the total leaf area. To evaluate air-space capacity in a uniform manner, analyzed leaf sections were chosen specifically for minimal presence of leaf vessels within a fixed area (300 μm × 300 μm square).

Statistical Analysis

The experiment was utilized a random design with eight replications and analyzed using one-way ANOVA. Means were compared with Tukey's HSD test, using JMP software (vers. 7, SAS, Inc., Cary, NC, United States). Correlations between physiological measurements and specific wavelength changes were calculated using PLS-R (PLS_Toolbox 7.03 software, Eigenvector Research, Inc., Manson, WA, United States) run in a MATLAB 7.12 environment (Mathworks, Inc., Natick, MA, United States), as explained in Rapaport et al. (2015). In short, (1) the independent (wavelengths) and dependent (physiology) variables were mean-centered during the pre-processing stage; (2) a PLS-R model was then built, and the contribution of all wavelengths was assessed using the variable importance in projection (VIP) statistic; (3) uncorrelated wavebands were then iteratively removed, to improve the model's accuracy; and (4)

TABLE 1 | Physiological measurements of infected and non-infected susceptible sunflower plants.

Treatment/DAP		23	31	36	51
Carbon assimilation (μmol m^{-2} s^{-1})	Control	11.29 ± 0.66	12.20 ± 0.74	14.53 ± 0.80	11.74 ± 1.94
	O. cumana infected	13.22 ± 0.84	11.27 ± 1.05	10.86 ± 1.48	9.76 ± 1.30
Stomatal conductance (mol m^{-2} s^{-1})	Control	0.40 ± 0.02	0.30 ± 0.02	0.41 ± 0.04	0.24 ± 0.06
	O. cumana infected	0.37 ± 0.04	0.23 ± 0.02	0.29 ± 0.06	0.19 ± 0.03
Transpiration (mmol m^{-2} s^{-1})	Control	6.74 ± 0.21	4.79 ± 0.34	6.56 ± 0.46	4.12 ± 1.00
	O. cumana infected	6.28 ± 0.40	3.8 ± 0.30	4.78 ± 0.76	3.38 ± 0.46
Non-photochemical quenching	Control	2.46 ± 0.22	2.32 ± 0.17	2.72 ± 0.26	2.16 ± 0.15
	O. cumana infected	2.12 ± 0.16	2.52 ± 0.86	2.63 ± 0.39	2.32 ± 0.20
Relative water content (%)	Control	75.61 ± 2.42	79.35 ± 5.01	86.84 ± 7.50	92.67 ± 3.83
	O. cumana infected	72.79 ± 1.41	79.31 ± 1.07	85.44 ± 4.30	91.33 ± 2.12
Chlorophyll content (mg l^{-1})	Control	0.03 ± 0.002	0.04 ± 0.004	0.04 ± 0.003	0.04 ± 0.002
	O. cumana infected	0.03 ± 0.001	0.04 ± 0.002	0.04 ± 0.001	0.03 ± 0.001

Orobanche cumana tubercles appear first in measurements done 31 DAP. Carbon assimilation, stomatal conductance, transpiration, and non-photochemical quenching include eight replications; relative water content, and chlorophyll content include four replications. Means comparison was done using one -way analysis of variance with Student's t-test (α < 0.05). The result represents the mean ± SE.

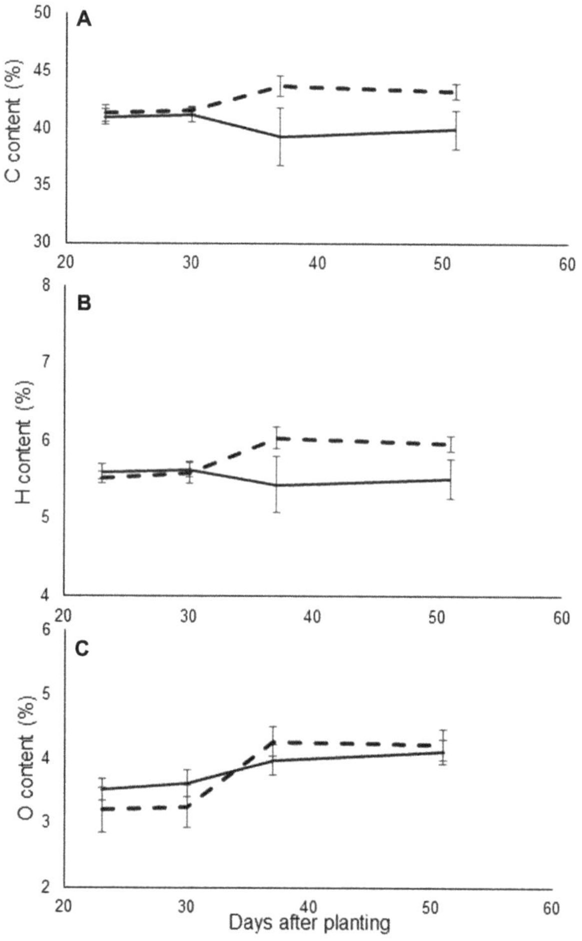

FIGURE 3 | (A) Carbon, **(B)** hydrogen, and **(C)** nitrogen content (% of total dry weight) of leaves in control (continuous line) and in *O. cumana*-infected (dashed line) susceptible sunflower plants. *O. cumana* infections appeared on plant roots 31 DAP. There were five replicates of each measurement. Bars represent standard errors of mean.

once no more areas for improvement were found, the robustness of the best model and of the most physiologically sensitive were assessed by considering the latent variable (LV) amount and the ability to explain the variation in the datasets. All the parameters were tested twice, once separately (physiological separately, minerals content, and anatomic cross-section separately) and after that in one experiment. Therefore only the second experiment is presented.

RESULTS

Spectral Measurements

The spectral signatures of the susceptible sunflower leaves indicated that *O. cumana* parasitism mainly affects reflectance in the far-NIR and SWIR wavebands. The first measurement was taken at 23 DAP and, at that time, no differences were seen between infested and non-infested pots (**Figure 1A**). At the next measurement (31 DAP), after *O. cumana* attachment appear on the host roots, differences in the reflectance between infected and uninfected susceptible plants were noted. These differences were pronounced in the far-NIR waveband, between 800 and 1300 nm. The reflectance level of the leaves of sunflower plants infected with *O. cumana* was found to be 1% lower in that range. Additional differences were found in the 1400–1500 nm range, wherein the reflectance of plants infected with *O. cumana* decreased by 1.5% (**Figure 1B**). The next measurement (36 DAP) conducted after 63% of the tubercle appear on the host roots according to the model, the reflectance differences between treatments were greater in the far-NIR range, but not in the SWIR range (~2% between treatments; **Figure 1C**). Furthermore, after the emergence of *O. cumana* inflorescences, the gap between treatments in the NIR area remained the same as they had been during the earlier measurements (**Figure 1D**).

During the experiment, no *O. cumana* development on the resistant cultivar roots was observed. Examination of the reflectance of the leaves of plants of the resistant cultivar revealed no differences between treatments throughout the experiment (i.e., whether in infested and non-infested soil) (**Figure 1E**). Statistical analysis indicated that the differences in the far-NIR and SWIR areas after *O. cumana* attachment are significant (**Figure 2**).

Physiological Measurements

Physiological measurements taken over the course of the experiment indicated no significant differences at the early attachment stage. No statistically significant differences were observed in the carbon assimilation, stomatal conductance, and transpiration rate of the parasitized plants compared with those of the control plants at any of the stages of parasite development (23–51 DAP). NPQ measurements also revealed no differences over the course of the experiment. Relative water content was similar in the infected and non-infected plants. Chlorophyll content did not show any significant differences until the *O. cumana* inflorescences emerged (**Table 1**). During the experiment, no physiological differences were found in the resistant cultivar between infested and non-infested pots.

Mineral Evaluations

Evaluations of leaf mineral content revealed differences in carbon and hydrogen values. Mineral contents were higher in the infected plants, as compared to the non-infected ones (**Figure 3**). However, no consistent pattern of differences in nitrogen content was observed among the different treatments.

The levels of other minerals were measured using ICP technology. Significant differences were found in the levels of the macro-elements potassium, phosphorus, magnesium, and sulfur between the infected and non-infected plants during the early stage of parasite development (**Figures 4A–D**). By contrast, the calcium content did not vary between the treatments (data not shown). Differences were noted in the levels of the microelements zinc and boron (**Figures 4E,F**, respectively). For all of these elements, the differences were more pronounced at the early attachment stage (starting at 31 DAP). The levels of other elements such as chlorine, sodium, iron and manganese

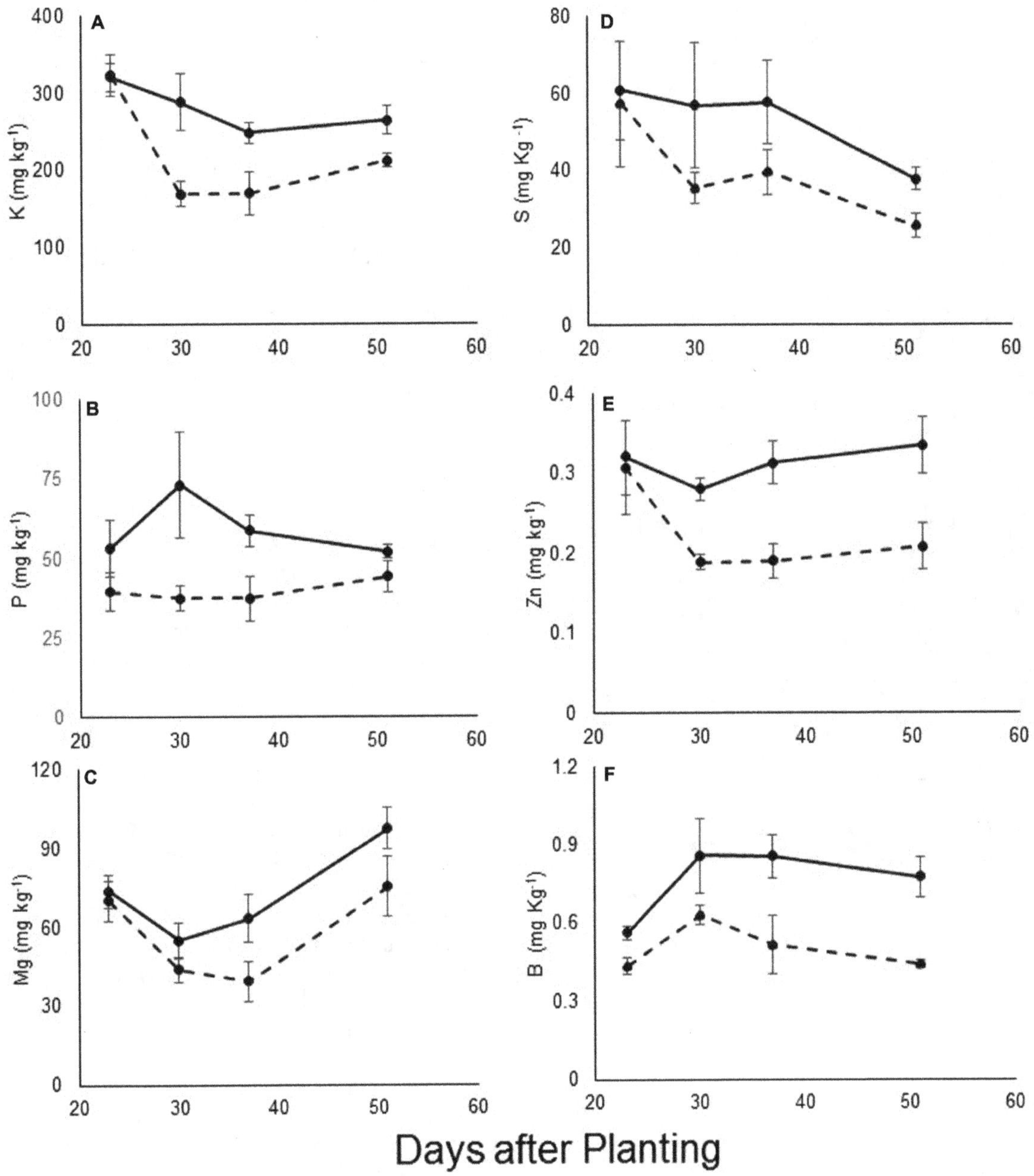

FIGURE 4 | (A) Phosphorus, **(B)** potassium, **(C)** magnesium, **(D)** sulfur, **(E)** zinc, and **(F)** boron content (mg kg^{-1}) of leaves of control (—) and *O. cumana*-infected (—) susceptible sunflower plants. *O. cumana* infections were first observed on plant roots 31 DAP. There were five replicates of each measurement. Bars represent standard errors of mean.

remained similar between treatments, and changed only after the emergence of the parasite (data not shown). No mineral content differences were found in the resistant cultivar between infested and non-infested pots at any time during the experiment.

Anatomical Analysis of Leaf Cross-Sections

On the first sampling date, no differences were found between the leaves of the control sunflower plants and those infected with *O. cumana*. In both treatments, relative air-space capacity, in both the spongy and the palisade mesophyll, was about 40%. However, a short time after the first *O. cumana* attachments appeared, differences were noted between treatments (**Figure 5**). The relative air-space capacity in the mesophyll of non-infected plants was about 20% on the following two sampling dates. By contrast, leaf cross-sections of *O. cumana*-infected plants exhibited air space capacity of about 30% (**Figure 5B**). Among the leaves of the resistant cultivar, no differences between infected and non-infected plants were

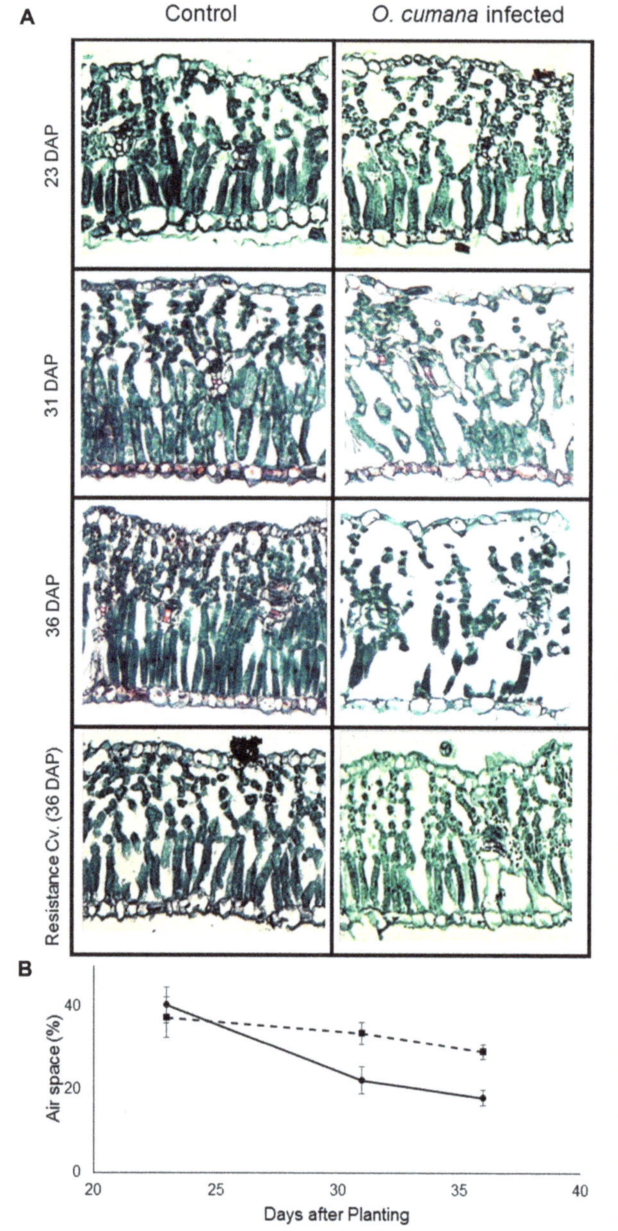

FIGURE 5 | (A) Sections of the youngest fully mature susceptible (23, 31, and 36 DAP) and resistant (36 DAP) sunflower leaves of control (left) and *O. cumana*-infested (right) sunflower plants. *O. cumana* tubercles were found starting from 31 DAP; no *O. cumana* attachment were found on the resistance cultivar roots. **(B)** Air spaces in non-infected (—) and *O. cumana*-infected (—) sunflower leaves. The relative amount of air space was calculated as the proportion of the total section area taken up by air space. There were four replicates for each treatment. Bars indicate standard errors of mean.

observed over the course of the experiment, therefore data not shown.

PLS-R Analysis

Statistical analysis revealed a strong correlation between the levels of different nutrients in the leaves and the reflectance of specific wavelengths (**Figure 6** and **Table 2**). VIP statistics revealed the shared importance of the wavelengths in both the VIS (primarily around 675 nm) and SWIR (mainly around 1400–1500 nm) wavebands. Specifically, element-correlated wavelengths were found at 685 nm for potassium; 700 and 1400 nm for phosphorus; 690, 1000, and 1450 nm for magnesium (**Figure 6A**); 670 and 1450 nm for sulfur; and 690 and 1400 nm for boron. However, zinc-correlated wavelength VIP scores were lower (**Figure 6B**).

Reflectance at the chosen wavelengths and the relative size of air spaces in the leaf cross-sections were highly correlated. Before the parasite attached, no significant differences in relative air-space capacity were noted between the treatments (**Figure 7A**). However, after the attachment of *O. cumana* to the sunflower roots, significant differences in air-space capacity were noted (**Figures 7B,C**).

DISCUSSION

Over the years, many options have been suggested for the management of root parasites. The timing for applying herbicides is crucial; they must be applied when the parasite is in its early developmental stage underground. However, the spatial distribution of parasites in the field are not considered when applying the herbicides. Hence, unnecessary herbicide applications are common. In this study, a non-destructive method detecting early parasitism was developed.

Hyperspectral measurements of leaves of infected and non-infected susceptible sunflower plants revealed that the first parasite attachments causes changes in the VIS-SWIR reflectance signature of sunflower leaves. Unlike other stress responses, which have been found to induce changes in the VIS range (350–700 nm; Rapaport et al., 2015), no changes in this range were noted in this study. Nonetheless, changes were detected in the NIR and SWIR ranges. Thus, the difference in the reflectance of the different treatments was greater on the following measurement time-points. These results are somewhat similar to those of Ortiz-Bustos et al. (2016), who demonstrated that differences in the leaf fluorescence of *O. cumana*-infected vs. non-infected sunflower plants are pronounced at 680 and 740 nm after UV light induction. By using UV-induced multicolor fluorescence imaging, the researcher found differences in the wavebands fluorescence. In this experiment, we measured the reflectance of a broad spectrum of wavelengths and noted differences at additional wavelengths in the non-visible light spectrum. The reflectance difference between treatments are thought to be related primarily to tissue structure and water absorption in the leaf (Carter, 1991; Rapaport et al., 2015).

Several physiological measurements were carried out over the course of this experiment, for early detection of the parasite. None of the measurements revealed any differences between treatments in the early stage of the parasitism. This finding is similar to those of previous studies (Barker et al., 1996; Hibberd et al., 1998), which demonstrated that the maintenance of plant functioning during parasite development is essential, due to the parasite's need for the energy and sugar produced by the host. Therefore, a decrease in plant functioning occurs only after the

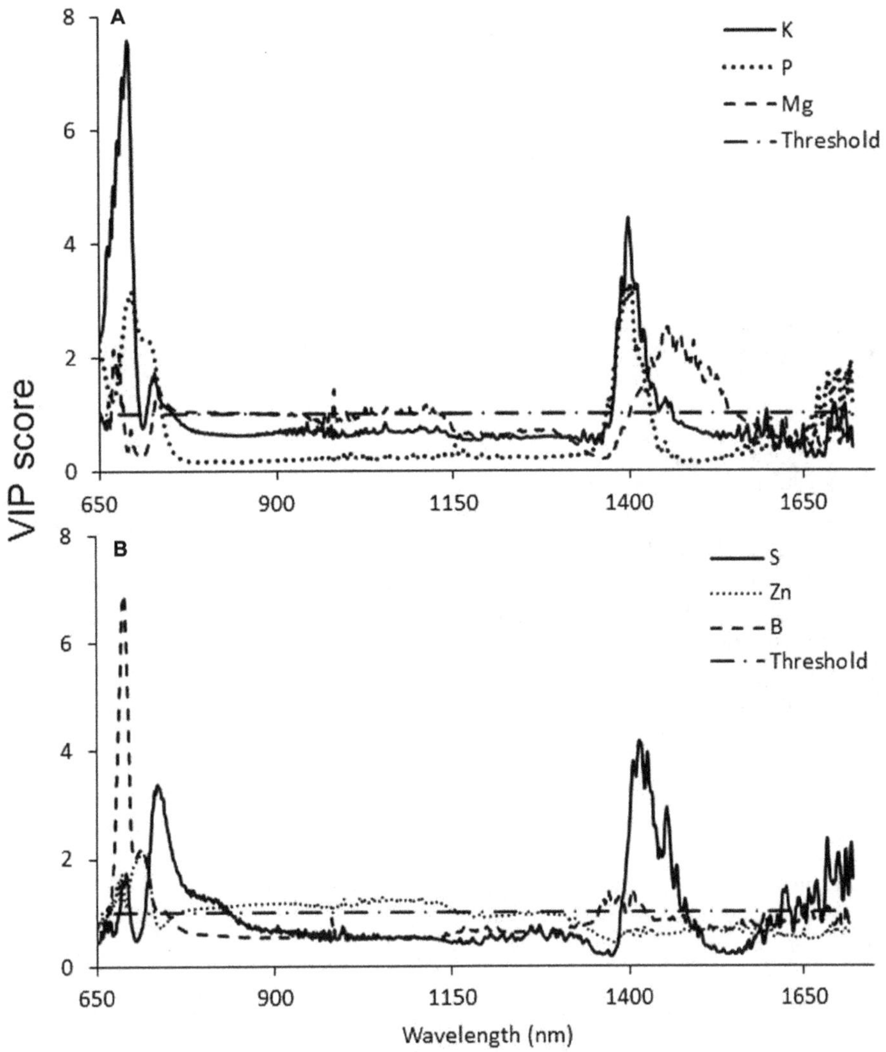

FIGURE 6 | Variable importance in projection (VIP) scores of potassium phosphorus and magnesium **(A)** and sulfur, zinc, and boron **(B)** content in susceptible sunflower leaves according to the PLS-R analysis. Values higher than the threshold (> 1) indicate a significant contribution of the specific wavelength to the overall variance.

parasite completes its life cycle. Although Ortiz-Bustos et al. (2016) reported inconsistent changes in chlorophyll content, in this study, no changes were seen in chlorophyll content. Moreover, no differences in the chlorophyll-related waveband (650–700 nm) were noted between the infected and non-infected plants.

As an obligate parasite, *O. cumana* gets all of its essential nutrients from its host. Furthermore, parasite development on host plant roots reduces the mineral content of the host plant (Westwood, 2013). In this experiment, reductions in the levels of several minerals (including essential macro- and microelements) were observed. The most significant reduction in the infected plants was found in levels of potassium, a very mobile element related to plant osmotic regulation. All of the other elements examined are known to be mobile or semi-mobile in the plant and can move rapidly to a strong sink. Carbon and hydrogen, however, showed an opposite pattern between treatments during the experiment; their levels increased (**Figure 3**). The relative increase in carbohydrate content can be explained by the increased demand for carbohydrates in the newly created sink, which leads to higher production of carbohydrates in the leaves (Nandula et al., 2000). Similar results were demonstrated in *Phelipanche aegyptiaca* parasitism of carrot and tomato (Nandula et al., 2000; Shilo et al., 2016). Another possible explanation is that a decrease in mineral content in the tissue leads to relative increases in the carbohydrate content. The results of this study contradict results of a previous study concerning the effect of *O. cumana* parasitism on mineral concentrations in sunflower leaves. However, that study dealt with *O. cumana* only after inflorescence emergence (Alcántara et al., 2006), while the current work focuses on earlier stages of the parasitism.

TABLE 2 | Evaluation of different elements partial least squares regression (PLS-R) models.

Element	Explained variance (%)	R^2	RMSEC	Q_3-Q_1	RPIQ
K	75.18	0.74	54.41	329.10	6.05
P	38.68	0.36	16.61	112.09	6.75
Mg	79.44	0.73	13.85	98.43	7.11
Zn	72.86	0.69	0.07	0.38	5.28
S	63.97	0.52	17.55	103.63	5.90
B	41.9	0.42	0.17	1.03	6.15

Each model was built from 40 observations (five replicates from two treatments over four time points), and include six latent variables. Q_3-Q_1, maximal element range between samples; RPIQ, (Q_3-Q_1)/RMSEC. Mineral contents are in mg Kg^{-1}.

The most noteworthy change found to be caused by *O. cumana* infection was the change in the relative amount of air space in the leaf mesophyll. This change in the mesophyll tissue was detected almost exactly at the predicted time for the first *O. cumana* attachment, according to Eizenberg et al. (2012b). Previously, changes in mesophyll tissue have been recorded in response to salinity stress and water deficit (Delfine et al., 1998; Rapaport et al., 2015). These studies demonstrated reduction in the mesophyll content and organization, caused by stress. Even earlier research demonstrated mesophyll disintegration caused by osmotic stress (Giles et al., 1974), stress known to induce abscisic acid metabolism in the plant (Zhu, 2002). Moreover, plant parasites are known to increase abscisic acid content in the host plant (Frost et al., 1997). This study demonstrates an instant relationship between *O. cumana* attachment and mesophyll disintegration. The observed damage to the mesophyll tissue can be explained by changes in mineral availability, primarily potassium, due to the new sink demands. Potassium deficit, due to it mobility, can lead to dramatic changes in the mesophyll structure (Zhao et al., 2001; Battie-Laclau et al., 2014).

Correlation between leaf mineral content and reflectance at specific wavelengths indicate that the changes in reflectance in the SWIR range are highly correlated with changes in levels of potassium, phosphorus, magnesium, and boron in the leaf. These related changes are known from previous work on mineral deficits in leaves (Al-Abbas et al., 1972). In addition, the statistical analysis revealed a high correlation between nutrient content and reflectance in the 650–750 nm range. However, no changes in reflectance related to *O. cumana* parasitism were observed within this range. The statistical analysis apparently revealed differences that were underway, but not yet manifested in a change in reflectance. Also, reflectance within this range is known to correlate closely with chlorophyll content and fluorescence (Ercoli et al., 1993) but no differences in chlorophyll content were noted in this study.

Changes in leaf structure and mesophyll structure, in particular, are known to be related to changes in far-NIR (800–1300 nm) reflectance (Johnson et al., 2005; Ollinger, 2011). Although PLS-R analysis revealed no differences in this area, the statistical analysis of the reflectance at these wavelengths revealed developing difference in this area after *O. cumana* attachment. The difference in reflectance between the *O. cumana*-infected and

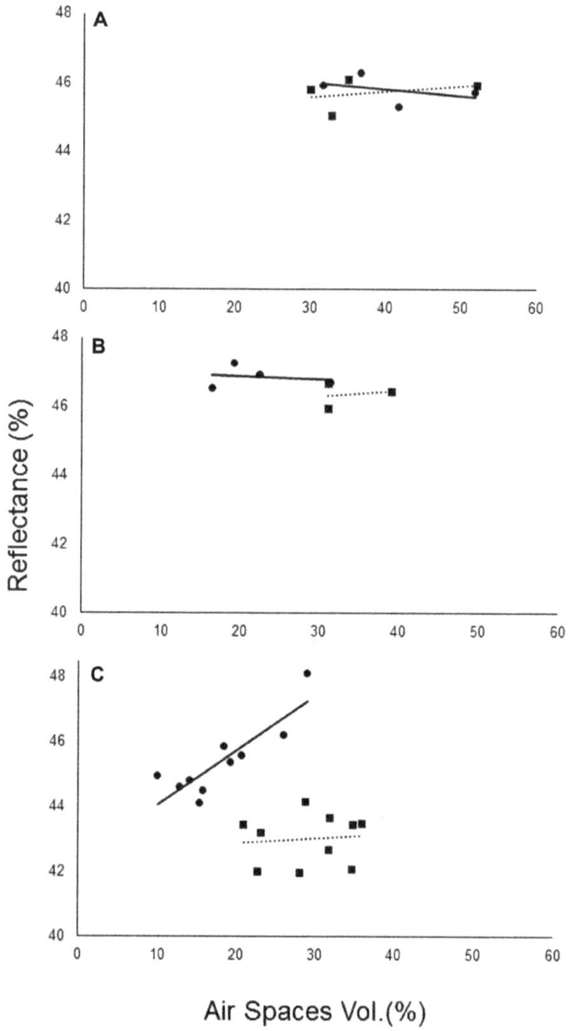

FIGURE 7 | Correlation between non-infected (—) and *O. cumana*-infected (—) air spaces in sunflower leaves and the average spectral reflectance at 800–1300 nm. **(A)** Before *O. cumana* attachment (23 DAP); **(B)** when *O. cumana* attachment was first observed (31 DAP); **(C)** 36 DAP in susceptible sunflower leaves.

non-infected control plants increased after attachment, from 1% after attachment to 2.5% after the emergence of the *O. cumana* inflorescences. The different treatments had a differential effect on the correlation between reflectance in the NIR range and leaf air space only after *O. cumana* attachment, indicating that reflectance within this range is indicative of *O. cumana* parasitism. Hence, use of these specific wavelengths can lead to good early detection of *O. cumana* in sunflower. Moreover, data from the control samples of resistant cultivar demonstrated that the presence of parasite seeds in the soil had no effect on leaf spectral, physiological and mineral content, nor on leaf mesophyll structure. Furthermore, the use of resistant variety support our assumption that the differences between the infected and non-infected sunflower are due to the parasitism and not from other physiological stresses.

CONCLUSION

Effective management of *O. cumana* requires the application of herbicide at the early stage of parasitism, at which point no external effects are visible. This study stresses two points. First, *O. cumana* causes rapid changes in leaf nutrient content, which lead to immediate changes in the structure of the leaf mesophyll of the host. Second, these changes can be observed using hyperspectral equipment. Combining the thermal model developed by Eizenberg et al. (2012b) with hyperspectral imaging should enable detection of the spatial distribution of the parasite at an early stage of parasitism. Future work should focus on how this integrated model can be applied in the field.

AUTHOR CONTRIBUTIONS

AC: performed greenhouse measurements, data analysis, and writing the manuscript. TR: performed greenhouse spectral measurements and PLS-R analysis. TG: performed anatomic cross-sections. AK: support in data analysis and writing. HE: head of the lab on parasitic weeds. SR: support in data analysis and writing. JE: support in data analysis and writing.

ACKNOWLEDGMENT

The authors wish to thank the Chief Scientist of the Israel Ministry of Agriculture and the ICA in Israel Foundation for partial support in this study.

REFERENCES

Al-Abbas, A. H., Barr, R., Hall, J. D., Crane, F. L., and Baumgardner, M. F. (1972). spectra of normal and nutrient-deficient maize leaves. *Agron. J.* 66, 16–20. doi: 10.2134/agronj1974.00021962006600010005x

Alcántara, E., Morales-garcía, M., and íaz-sánchez, J. D. (2006). Effects of broomrape parasitism on sunflower plants: growth, development and mineral nutrition. *J. Plant Nutr.* 29, 1199–1206. doi: 10.1080/01904160600767351

Barker, E. R., Press, M. C., Scholes, J. D., and Quick, W. P. (1996). Interactions between the parasitic angiosperm *Orobanche aegyptiaca* and its tomato host: growth and biomass allocation. *New Phytol.* 133, 637–642. doi: 10.1111/j.1469-8137.1996.tb01932.x

Battie-Laclau, P., Laclau, J., Beri, C., Mietton, L., Muniz, M. R. A., Arenque, B. C., et al. (2014). Photosynthetic and anatomical responses of *Eucalyptus grandis* leaves to potassium and sodium supply in a field experiment. *Plant Cell Environ.* 37, 70–81. doi: 10.1111/pce.12131

Bilger, W., and Bjorkman, O. (1990). Role of the xanthophyll cycle in photoprotection elucidated by measurements of light-induced absorbance changes, fluorescence and photosynthesis in leaves of *Hedera canariensis. Photosynth. Res.* 25, 173–185. doi: 10.1007/BF00033159

Blackburn, G. A. (2007). Hyperspectral remote sensing of plant pigments. *J. Exp. Bot.* 58, 855–867. doi: 10.1093/jxb/erl123

Bruinsma, J. (1963). The quantitative analysis of chlorophylls a and b in plant extracts. *Photochem. Photobiol.* 2, 241–249. doi: 10.1111/j.1751-1097.1963.tb08220.x

Carter, G. A. (1991). Primary and secondary effects of water content on the spectral reflectance of leaves. *Am. J. Bot.* 78, 916–924. doi: 10.2307/2445170

Castejon-Munoz, M., Romero-Munoz, F., and Garcia-Torres, L. (1993). Effect of planting date on broomrape (*Orobanche cernua* Loefi.) infections in sunflower (*Helianthus annuus* L.). *Weed Res.* 33, 171–176. doi: 10.1111/j.1365-3180.1993.tb01930.x

Cochavi, A., Achdari, G., Smirnov, Y., Rubin, B., and Eizenberg, H. (2015). Egyptian broomrape (*Phelipanche aegyptiaca*) management in carrot under field conditions. *Weed Technol.* 29, 519–528. doi: 10.1614/WT-D-14-00140.1

Delfine, S., Alvino, A., Zacchini, M., and Loreto, F. (1998). Consequences of salt stress on conductance to CO2 diffusion, Rubisco characteristics and anatomy of spinach leaves. *Aust. J. Plant Physiol.* 25, 395–402. doi: 10.1071/PP97161

Dzikiti, S., Verreynne, J. S., Stuckens, J., Strever, A., Verstraeten, W. W., Swennen, R., et al. (2010). Determining the water status of Satsuma mandarin trees [*Citrus Unshiu* Marcovitch] using spectral indices and by combining hyperspectral and physiological data. *Agric. For. Meteorol.* 150, 369–379. doi: 10.1016/j.agrformet.2009.12.005

Eizenberg, H., Aly, R., and Cohen, Y. (2012a). Technologies for smart chemical control of Broomrape (*Orobanche spp.* and *Phelipanche spp.*). *Weed Sci.* 60, 316–323. doi: 10.1614/WS-D-11-00120.1

Eizenberg, H., Hershenhorn, J., Achdari, G., and Ephrath, J. E. (2012b). A thermal time model for predicting parasitism of *Orobanche cumana* in irrigated sunflower — field validation. *F. Crops Res.* 137, 49–55. doi: 10.1016/j.fcr.2012.07.020

Eizenberg, H., Colquhoun, J. B., and Mallory-Smith, C. A. (2006). Imazamox application timing for small broomrape (*Orobanche minor*) control in red clover. *Weed Sci.* 54, 923–927. doi: 10.1614/WS-05-151R.1

Ercoli, L., Mariotti, M., Masoni, A., and Massantini, F. (1993). Relationship between nitrogen and chlorophyll content and spectral properties in maize leaves. *Eur. J. Agron.* 2, 113–117. doi: 10.1016/S1161-0301(14)80141-X

Frost, D. L., Gurney, A. L., Press, M. C., and Scholes, J. D. (1997). *Striga hermonthica* reduces photosynthesis in sorghum: the importance of stomatai limitations and a potential role for ABA? *Plant Cell Environ.* 20, 483–492. doi: 10.1046/j.1365-3040.1997.d01-87.x

Garrity, S. R., Eitel, J. U. H., and Vierling, L. A. (2011). Disentangling the relationships between plant pigments and the photochemical reflectance index reveals a new approach for remote estimation of carotenoid content. *Remote Sens. Environ.* 115, 628–635. doi: 10.1016/j.rse.2010.10.007

Geladi, P., and Kowalski, B. R. (1986). Partial least-squares regression: a tutorial. *Anal. Chim. Acta* 185, 1–17. doi: 10.1016/0003-2670(86)80028-9

Giles, K. L., Beardsell, M. F., and Cohen, D. (1974). Cellular and ultrastructural changes in mesophyll and bundle sheath cells of maize in response to water stress. *Plant Physiol.* 54, 208–212. doi: 10.1104/pp.54.2.208

Grenz, J. H., Istoc, V. A., Manschadi, A. M., and Sauerborn, J. (2008). Interactions of sunflower (*Helianthus annuus*) and sunflower broomrape (*Orobanche cumana*) as affected by sowing date, resource supply and infestation level. *F. Crops. Res.* 107, 170–179. doi: 10.1016/j.fcr.2008.02.003

Heide-Jørgensen, H. S. (2013). "Introduction: the parasitic syndrome in higher plants," in *Parasitic Orobanchaceae*, 1 Edn, eds D. M. Joel, J. Gressel, and L. J. Musselman (Heidelberg: Springer), 1–14.

Hibberd, J. M., Quick, W. P., Press, M. C., and Scholes, J. D. (1998). Can source - sink relations explain responses of tobacco to infection by the root holoparasitic angiosperm *Orobanche cernua*? *Plant Cell Environ.* 21, 333–340. doi: 10.1046/j.1365-3040.1998.00272.x

Johnson, D. M., Smith, W. K., Vogelmann, T. C., and Brodersen, C. R. (2005). Leaf architecture and direction of incident light influence mesophyll fluorescence profiles. *Am. J. Bot.* 92, 1425–1431. doi: 10.3732/ajb.92.9.1425

Levitt, J. (1980). *Responses of Plants to Environmental Stresses, Volume II: Water, Radiation, Salt, and Other Stresses.* London: Academic Press.

Longo, A. M. G., Monco, A. L., and Mauromicale, G. (2010). The effect of *Phelipanche ramosa* infection on the quality of tomato fruit. *Weed Res.* 50, 58–66. doi: 10.1111/j.1365-3180.2009.00752.x

Mariotto, I., Thenkabail, P. S., Huete, A., Slonecker, E. T., and Platonov, A. (2013). Hyperspectral versus multispectral crop-productivity modeling and type discrimination for the HyspIRI mission. *Remote Sens. Environ.* 139, 291–305. doi: 10.1016/j.rse.2013.08.002

Mauromicale, G., Lo Monaco, A., and Longo, A. M. G. (2008). Effect of branched broomrape (*Orobanche ramosa*) infection on the growth and photosynthesis of tomato. *Weed Sci.* 56, 574–581. doi: 10.1614/WS-07-147.1

Menesatti, P., Antonucci, F., Pallottino, F., Roccuzzo, G., Allegra, M., Stango, F., et al. (2010). Estimation of plant nutritional status by VIS - NIR spectrophotometric analysis on orange leaves [*Citrus sinensis* (L) Osbeck cv Tarocco]. *Biosyst. Eng.* 105, 448–454. doi: 10.1016/j.biosystemseng.2010.01.003
Mulla, D. J. (2013). Twenty five years of remote sensing in precision agriculture: key advances and remaining knowledge gaps. *Biosyst. Eng.* 114, 358–371. doi: 10.1016/j.biosystemseng.2012.08.009
Nandula, V. K., Foster, J. G., and Foy, C. L. (2000). Impact of egyptian broomrape (*Orobanche aegyptiaca* Pers.) parasitism on amino acid composition of carrot (*Daucus carota* L.). *J. Agric. Food Chem.* 48, 3930–3934. doi: 10.1021/jf991145w
Ollinger, S. V. (2011). Sources of variability in canopy reflectance and the convergent properties of plants. *New Phytol.* 189, 375–394. doi: 10.1111/j.1469-8137.2010.03536.x
Ortiz-Bustos, C. M., Pérez-Bueno, M. L., Barón, M., and Molinero-Ruiz, L. (2016). Fluorescence imaging in the red and far-red region during growth of sunflower plantlets. diagnosis of the early infection by the parasite *Orobanche cumana*. *Front. Plant Sci.* 7:884. doi: 10.3389/fpls.2016.00884
Rapaport, T., Hochberg, U., Rachmilevitch, S., and Karnieli, A. (2014). The effect of differential growth rates across plants on spectral predictions of physiological parameters. *PLoS One* 9:e88930. doi: 10.1371/journal.pone.0088930
Rapaport, T., Hochberg, U., Shoshany, M., Karnieli, A., and Rachmilevitch, S. (2015). Combining leaf physiology, hyperspectral imaging and partial least squares-regression (PLS-R) for grapevine water status assessment. *ISPRS J. Photogramm. Remote Sens.* 109, 88–97. doi: 10.1016/j.isprsjprs.2015.09.003
Rewald, B., Raveh, E., Gendler, T., Ephrath, J. E., and Rachmilevitch, S. (2012). Phenotypic plasticity and water flux rates of citrus root orders under salinity. *J. Exp. Bot.* 63, 2717–2727. doi: 10.1093/jxb/err457
Rubiales, D., Verkleij, J., Vurro, M., Murdoch, A. J., and Joel, D. M. (2009). Parasitic plant management in sustainable agriculture. *Weed Res.* 49, 1–5. doi: 10.1111/j.1365-3180.2009.00741.x
Shilo, T., Zygier, L., Rubin, B., Wolf, S., and Eizenberg, H. (2016). Mechanism of glyphosate control of *Phelipanche aegyptiaca*. *Planta* 244, 1095–1107. doi: 10.1007/s00425-016-2565-8
Smart, R. E., and Bingham, G. E. (1974). Rapid estimates of relative water content. *Plant Physiol.* 53, 258–260. doi: 10.1104/pp.53.2.258
Westwood, J. H. (2013). "The physiology of the established parasite-host association," in *Parasitic Orobanchaceae*, 1 Edn, eds D. M. Joel, J. Gressel, and L. J. Musselman (Heidelberg: Springer).
Wold, S., Sjostrom, M., and Eriksson, L. (2001). PLS-regression: a basic tool of chemometrics. *Chemom. Intell. Lab. Syst.* 58, 109–130. doi: 10.1016/S0169-7439(01)00155-1
Yoneyama, K., Awad, A. A., Xie, X., Yoneyama, K., and Takeuchi, Y. (2010). Strigolactones as germination stimulants for root parasitic plants. *Plant Cell Physiol.* 51, 1095–1103. doi: 10.1093/pcp/pcq055
Yoneyama, K., Ruyter-Spira, C., and Bouwmeester, H. J. (2013). "Induction of germination," in *Parasitic Orobanchaceae*, 1 Edn, eds D. M. Joel, J. Gressel, and L. J. Musselman (Heidelberg: Springer), 167–194. doi: 10.1007/978-3-642-38146-1_10
Zhao, D., Costerhuis, D. M., and Bednarz, C. W. (2001). Influence of potassium deficiency on photosynthesis, chlorophyll content, and chloroplast ultrastructure of cotton plants. *Photosynthetica* 39, 103–109. doi: 10.1023/A:1012404204910
Zhu, J. K. (2002). Salt and drought stress signal transduction in plants. *Annu. Rev. Plant Biol.* 53, 247–273. doi: 10.1146/annurev.arplant.53.091401.143329

Conflict of Interest Statement: The authors declare that the research was conducted in the absence of any commercial or financial relationships that could be construed as a potential conflict of interest.

The reviewer AVA and handling Editor declared their shared affiliation, and the handling Editor states that the process met the standards of a fair and objective review.

19

Sunflower Resistance to Broomrape (*Orobanche cumana*) is Controlled by Specific QTLs for Different Parasitism Stages

Edited by:
Monica Fernandez-Aparicio,
Institut National de la Recherche
Agronomique, France

Reviewed by:
Estefanía Carrillo,
Instituto Nacional de Investigaciones
Agropecuarias, Ecuador
Sara Fondevilla,
Institute for Sustainable
Agriculture-Consejo Superior
de Investigaciones Científicas, Spain
Belén Román,
Instituto de Investigación y Formación
Agraria y Pesquera, Spain

***Correspondence:**
Stéphane Muños
stephane.munos@toulouse.inra.fr

*Johann Louarn[1], Marie-Claude Boniface[1], Nicolas Pouilly[1], Leonardo Velasco[2], Begoña Pérez-Vich[2], Patrick Vincourt[1] and Stéphane Muños[1]**

[1] LIPM, Université de Toulouse, INRA, CNRS, Castanet-Tolosan, France, [2] Instituto de Agricultura Sostenible-Consejo Superior de Investigaciones Científicas, Cordoba, Spain

Orobanche cumana (sunflower broomrape) is an obligatory and non-photosynthetic root parasitic plant that specifically infects the sunflower. It is located in Europe and in Asia, where it can cause yield losses of over 80%. More aggressive races have evolved, mainly around the Black Sea, and broomrape can rapidly spread to new areas. Breeding for resistance seems to be the most efficient and sustainable approach to control broomrape infestation. In our study, we used a population of 101 recombinant inbred lines (RILs), derived from a cross between the two lines HA89 and LR1 (a line derived from an interspecific cross with *Helianthus debilis*). Rhizotrons, pots and field experiments were used to characterize all RILs for their resistance to *O. cumana* race F parasitism at three post vascular connection life stages: (i) early attachment of the parasite to the sunflower roots, (ii) young tubercle and (iii) shoot emergence. In addition, RIL resistance to race G at young tubercle development stage was evaluated in pots. The entire population was genotyped, and QTLs were mapped. Different QTLs were identified for each race (F from Spain and G from Turkey) and for the three stages of broomrape development. The results indicate that there are several quantitative resistance mechanisms controlling the infection by *O. cumana* that can be used in sunflower breeding.

Keywords: broomrape, sunflower, resistance, QTL mapping, *Orobanche cumana*, plant–plant interaction, candidate genes, parasitic weeds

INTRODUCTION

The parasitic weed *Orobanche cumana* (sunflower broomrape) is an obligatory and non-photosynthetic root parasitic plant of the sunflower (*Helianthus annuus* L.) and is a substantial threat in Europe, especially in countries around the Black Sea and in Spain (Molinero-Ruiz et al., 2013). *O. cumana* has a negative effect on sunflower development. The infected plants are smaller, the sunflower head diameter is reduced and up to 80% of yield losses are observed (Alcántara et al., 2006; Duca, 2015).

Unlike other weedy *Orobanche* species, for which genetic resistance in the host is of quantitative nature (horizontal), genetic resistance to *O. cumana* in the sunflower is in most cases qualitative or vertical (Fernández-Martínez et al., 2015). For this reason, *O. cumana* populations are commonly classified into physiological races (Vranceanu et al., 1980) that periodically surpass all the available resistance sources. Eight races of *O. cumana*, A through H, have been reported thus far, with races F, G, and H commonly reported in several countries (Kaya, 2014). Different mechanisms have been described that might determine the rapid emergence of new races of *O. cumana* including recombination and increase of genetic diversity, mutation and selection within specific gene pools, or gene flow between wild and weedy *O. cumana* populations (Pineda-Martos et al., 2013, 2014).

Several methods of broomrape control are available with more or less efficiency. Different crop management solutions can be used: soil solarization (Mauromicale et al., 2005), biological control (Thomas et al., 1998; Louarn et al., 2012) and the use of herbicides such as imidazolinone combined with herbicide-tolerant sunflower hybrids (Tan et al., 2005). However, breeding for genetic resistance remains the most efficient method. The first introgression of genetic resistance to broomrape on the sunflower was conducted in the former USSR (Pustovoit, 1966). Genetic resistance has been characterized in wild *Helianthus* spp., and introgression of the resistance gene from interspecific crosses has been reported (Jan and Fernandez-Martínez, 2002; Velasco et al., 2007). Even if neither quantitative trait loci (QTLs) nor major genes were mapped for the resistance to *O. cumana* race G, Velasco et al. (2012) showed that the resistance (from *H. debilis* subsp. *tardiflorus*) to the race G of *O. cumana* was dominant and controlled by a single locus in their population. Several major *Or* resistance genes controlling the resistance to specific *O. cumana* races have been used in breeding programs (Fernández-Martínez et al., 2008). However, there are only two reports for the molecular genetic mapping of resistance loci. The first one concerns the *Or5* gene conferring resistance to race E (Lu et al., 2000; Tang et al., 2003; Pérez-Vich et al., 2004). The second report details the mapping of QTLs for resistance to *O. cumana* race F (Pérez-Vich et al., 2004). Six QTLs controlling the number of *O. cumana* emergences in the field have been detected on five linkage groups (LG). No genes have been cloned, and the molecular mechanisms involved in the resistance mechanisms remain unknown.

The life cycle of broomrape is composed of several steps from seed germination to plant flowering and seed production (Gibot-Leclerc et al., 2012; Yang et al., 2015). These steps can be roughly classified into four stages (**Figure 1**). During stage 1, the germination of the *O. cumana* seeds is induced by the host. Germination is one of the most studied steps of the broomrape life cycle. The molecules secreted by the host root system play a major role in the induction of broomrape germination (Fernández-Aparicio et al., 2009a). Two main types of molecules exuded by sunflower roots are known to induce *O. cumana* seeds germination: strigolactones and sesquiterpene lactones (Cook et al., 1966; Joel et al., 2011; Yoneyama et al., 2011; Raupp and Spring, 2013). Germination is followed by stage 2, in which the fixation of the parasite to the sunflower root, root penetration and establishment of vascular connections between the parasite and the host is achieved by means of the haustorium developed at the tip of *O. cumana* radicle (Hassan et al., 2004). After vascular connection, broomrape begins to derive phloemic flow acting as a strong nutrient sink. During stage 3, nutrient storage organ called tubercle develop quickly at the attachment point from which an underground shoot will eventually develop (Aly et al., 2009). The last stage is the above ground stage 4, which begins with the emergence of *O. cumana* from the ground and ends with flowering and seed production and dispersal. During the broomrape life cycle, several resistance mechanisms operating at the pre-attachment, pre-haustorial, or post-haustorial stages of the parasite-host interaction have been reported (Fernández-Martínez et al., 2015). These include mechanisms acting at the pre-attached stage such as decreased strigolactone exudation by host roots (Jamil et al., 2011; Fernández-Aparicio et al., 2014), or mechanisms such as cell wall deposition, vessel occlusion, broomrape cellular disorganization occurring during host invasion leading to incompatible attachments or tubercle necrosis such as that observed in the sunflower resistant line LR1 (Labrousse, 2001).

Single major resistance genes permit an efficient resistance to diseases. However, genetic resistance based on major dominant genes shows weak sustainability. Breeding for sustainable resistance needs to combine QTLs and major genes (Lindhout, 2002; Palloix et al., 2009; Brun et al., 2010). Additionally, significant gains can be made through breeding approaches informed by increasing understanding of the physiology of the parasitic plant association (Pérez-Vich et al., 2013). Accordingly, the objective of this study was to determine the genetic architecture of quantitative resistance in a RIL population combined with a candidate gene approach based both on race F field phenotyping and race F and G laboratory screening for resistance mechanisms.

MATERIALS AND METHODS

Sunflower Lines and *Orobanche cumana* Races

A population of 101 recombinant inbred lines (RILs) (F_8) was obtained by single-seed descent from a cross between two parental lines HA89 and LR1. LR1 is an inbred line derived from an interspecific cross involving the wild species *Helianthus debilis* (ecotype 215 in the INRA collection) (Labrousse, 2001). The sunflower line 2603 was used as a susceptible control in each experiment (Labrousse, 2001). All sunflower lines are part of the French genetic resources collection maintained by INRA (crb.tournesol@toulouse.inra.fr). The same population was previously described by Labrousse et al. (2004) for the resistance to *O. cumana* race E. A large variability of response was observed in the population for their capacity to induce the germination of *O. cumana* seeds, to control the development of tubercles and to induce tubercle necrosis. Using *O. cumana* race E, the authors found that LR1sunflower line disables the vascular connection leading to necrosis of 100% established tubercles (Labrousse, 2001). HA89 was found to be more

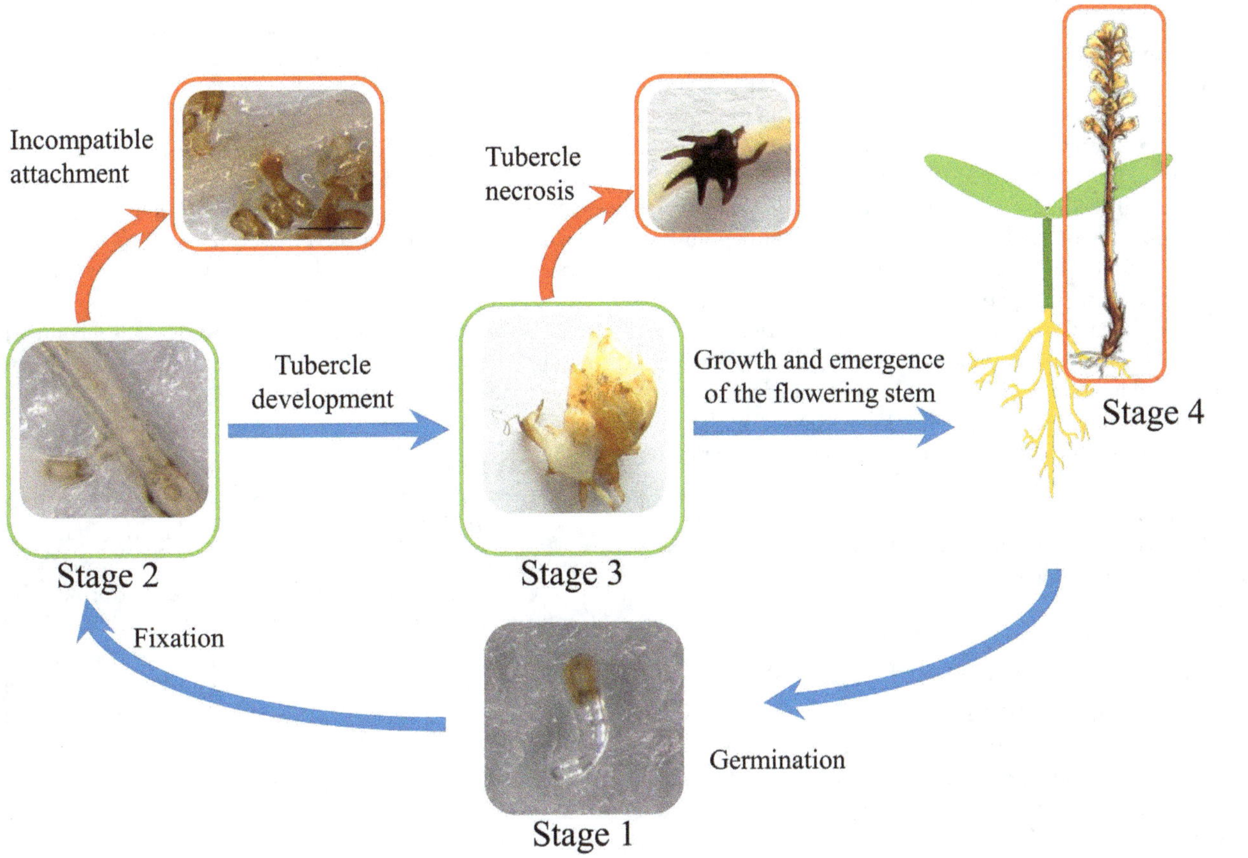

FIGURE 1 | Life cycle of *O. cumana*. Stage 1 corresponds to broomrape seed germination induced by sunflower root exudates. During stage 2, *O. cumana* radicle attaches, invades and connects the vascular system of the sunflower root by means of haustorium. Once *O. cumana* is attached and the nutrient diversion towards the parasite is established, *O. cumana* develops a tubercle (stage 3). During the stage 4, the stem emerges through the soil surface with the subsequent onset of flowering. The red arrows show putative resistance mechanism to *O. cumana* at stage 2 (incompatible attachment) and at stage 3 (tubercle necrosis).

susceptible to *O. cumana* race E than LR1, inducing necrosis only in 60% of the established tubercles but allowing further development of the remaining parasites (Labrousse et al., 2004).

The RIL population was characterized for plant height in Toulouse (France) in an *O. cumana*-free experimental field. The plants were sown in 4 m long rows of 10 plants. Plant height was recorded by measuring the main stem height in five individual plants (in the center of the row) at flowering stage during 2015 growing season.

Orobanche cumana seeds, race F, were collected in Marchena (Province of Sevilla, Spain) in 2012. During field experiments in 2014 and 2015, the *O. cumana* population parasitized sunflower line NR5 (carrying the *Or5* gene conferring resistance to race E) but not the P-96 line (conferring resistance to *O. cumana* race F), indicating that this population corresponded to race F (data not shown). *O. cumana* seeds race G were collected in 2000 in an experimental field located at Çeşmekolu (Kirklareli Province, Turkey). This *O. cumana* population was confirmed as race G because it parasitized sunflower lines K-96, P-96 and R-96 (carrying resistance to race F) (Fernández-Martínez et al., 2004).

Rhizotron Experiments: Evaluation of Resistance to Race F Early Post-vascular Development

Orobanche cumana seeds were surface-sterilized with 2.6% NaClO for 5 min and were rinsed thoroughly with sterile water. Race F of *O. cumana* seeds were spread on moistened sterile glass fiber filter paper (GF/A, Whatman) and incubated for 7 days in the dark at 22°C to allow *O. cumana* seed conditioning, a process used in broomrape experiments to break seed dormancy and promote seed sensitivity to molecules inducing germination secreted by host roots (Lechat et al., 2012). The sunflower seeds were first germinated in moistened filter paper for 3 days and then transferred to sand for four additional days. Subsequently, the rhizotron system was used to enable the interaction between 7-days conditioned *O. cumana* seeds and the roots of 7-days old sunflower plants. Briefly, a rhizotron is made of two PVC glasses (12 cm × 12 cm) confining the sunflower roots and the broomrape seeds on a glass fiber filter. Sterilized blocks of rockwool (Grodan, ROCKWOOL) of 1cm of thickness kept the glass fiber filter wet where the root system was deposited. Rhizotrons were placed in a growth chamber under a 16 h photoperiod and a constant temperature of 22°C during day and night periods. Plants were watered daily with half strength Long Ashton Nutrient Solution (LANS, Hewitt, 1966). A balanced incomplete-block design was performed to study the attachment stage. Seventeen blocks of 30 different RILs, permitting the phenotyping of each RIL five times, and the two parental lines HA89 and LR1 were designed using the R package DBI. After 2 weeks of growth, the numbers of compatible attachments (CA) and incompatible attachments (IA) were counted under a stereomicroscope (S6D, Leica). Attachments (stage 2 in **Figure 1**) were considered as compatible when the radicle of *O. cumana* was observed penetrating the host root, characterized by a slight swelling of the *O. cumana* radicle. Attachments were considered as incompatible when the radicle of *O. cumana* had initiated host root penetration but were stopped by host defenses mechanisms. The *O. cumana* radicle was not connected to the vascular system of the host leading to the death of the parasite. In addition, darkening of host cells at the penetration point was observed. The rate of IA was calculated as the percentage of IA out the total of attached *O. cumana* radicles (IA and CA).

Pot Experiments: Evaluation of Resistance to Race F and G at Young Tubercle Development

The evaluation of each RIL for resistance to *O. cumana* races F and G (stage 3 in **Figure 1**) was performed according to Louarn et al. (2012). Sunflower seeds were first germinated in moistened filter paper for 3 days. The substrate (charred clay for race F and a mixture of sand and vermiculite for race G) was inoculated with 60 mg of *O. cumana* seeds per liter of substrate and was moistened with water allowing the *Orobanche* seeds to undergo conditioning during 7 days. Then, 3 days pre-germinated sunflower seeds were sown in the pot (volume of 70 ml) and kept in a growth chamber under a 16 h photoperiod and a constant temperature of 22°C during day and night periods. Plants were watered daily with LANS (~10 ml/plant). After 5 weeks of growth, the level of infection by *O. cumana* was determined for each RIL by counting the number of healthy and necrotic tubercles on 5 (race F) or 9 (race G) sunflower plants from 5 or 3 independent experiments, respectively. The rate of necrotic tubercle was calculated as the percentage of necrotic tubercle out of the total number of tubercles (healthy and necrotic). The susceptible line 2603 and the race F resistant line P-96 were used as controls.

Field Evaluation of *O. cumana* Race F Emergence

The RILs were evaluated together with the parental lines for *O. cumana* race F resistance under artificial inoculation in field conditions in the spring and summer of 2014 and 2015 (stage 4 in **Figure 1**). Sunflower seeds were germinated during 2 days in moistened filter paper and subsequently transferred to small pots (7 cm × 7 cm × 7 cm) containing a mixture of sand and peat (1:1, v:v). Previously, soil (~180 g) was carefully mixed with 50 mg of *O. cumana* seeds to obtain a homogeneously infested substrate. The plants were watered by hand as needed and kept in a growth chamber for 15–20 days (time necessary for transplanting all plants to the field) at 25°C/20°C (day/night) with a 14 h photoperiod for incubation. They were then transplanted to a field plot at the experimental farm of the Institute for Sustainable Agriculture (CSIC, Córdoba, Southern Spain) in which only race F experiments have been conducted since 1999. The assays were transplanted between the 31st of March and the 2nd of April 2014 and between the 25th of March and the 27th of March 2015. The evaluation consisted of three replicates of 15 plants by row for each experiment. The plants were set 33 cm apart in 5 m row with 1 m separation between rows. The race F susceptible lines 2603 and NR5, and the resistant line P-96 were used as controls. The number of emerged broomrapes per sunflower plant was counted at the time of sunflower maturity.

Genotyping

Genomic DNA for the 101 RILs and the two parental lines was extracted using the Kit DNeasy Plant Mini Kit (Qiagen©). The DNA concentration was adjusted to 10 ng.μl^{-1} in water.

The hybridization experiments were performed by the Gentyane platform (Plateforme Gentyane, UMR INRA/UBP 1095 Génétique Diversité et Ecophysiologie des Céréales, 5 chemin de Beaulieu - 63039 Clermont-Ferrand, France) on a GeneTitan® (Affymetrix) according to the manufacturer's instructions. The AXIOM array was built using a set of 586 985 SNPs in the frame of the SUNRISE project[1]. Genotypic data were obtained with the software Axiom Analysis Suite[2].

The genotypic data were filtered, and SNPs were selected according to the following criteria: (a) the three replicates of the two parental lines were homozygous and consistent, (b) HA89 and LR1 were polymorphic, (c) the allelic frequency in the population was between 40 and 60%, (d) the missing data in the 101 RILs did not exceed 5%, and (e) the number of the heterozygous data in the whole population did not exceed 5%. We obtained a set of 21 201 SNPs for the genetic mapping. The genetic map was obtained using CarthaGène software (de Givry et al., 2005). After merging the markers ("mrkmerges" function), the LGs were obtained with the "group 0.2 3" function. The genetic distance between markers was calculated using the "lkh 1 −1" function and the genetic maps were obtained with the "bestprintd" function. The resulting genetic map, containing 951 markers obtained from 21 201 heterozygous SNPs, is shown in Supplementary Figure S1. The sequences of the markers mapped are included in Supplementary File S1.

QTL Mapping

Quantitative trait loci detection was performed with MCQTL (Jourjon et al., 2005) with a threshold corresponding to a Type I error rate of 1% at the genome-wide level, as determined after 3 000 replications of the resampling process for each trait. Biomercator (version 4.2) was used for the visualization of the different QTLs (Arcade et al., 2004).

[1] http://www.sunrise-project.fr/en/

[2] http://www.affymetrix.com

Statistical Analysis

The relationship between the rate of IA, the rate of necrotic tubercles, the number of healthy tubercles and the number of broomrape shoot emergences in field in 2014 and 2015 were studied by a PCA analysis. PCA was performed with the package FactoMineR using R software (version 3.1.2) under the Rcmdr environment (version 2.1-7). The Pearson correlation test was performed to determine the R coefficient between the different traits.

In Silico Mapping of Candidate Genes

The candidate gene sequences were obtained from Letousey et al. (2007), Radwan et al. (2008), and Li and Timko (2009). These sequences were used to search for the putative orthologous cDNAs in a sunflower transcriptomic database by BLAST[3]. The sunflower sequences were then used for *in silico* mapping by BLAST alignment to a genetic map obtained from the international consortium of sunflower genome sequencing and composed of 454 contigs[4]. The BLAST results were selected according to the following criteria: (i) identity between the query and the subject sequences greater than 95% and (ii) length of the alignment greater than 400 nt. The colinearity between the two genetic maps was determined by a BLAST alignment of the AXIOM markers linked to the QTLs on the 454 contig database.

RESULTS

Phenotypic Evaluation of the RIL Population

Evaluation of the Resistance to Race F at the Stage 2 and 3 of Broomrape Development

The first physical interaction between sunflower and broomrape occurs underground after seed germination when *O. cumana* fixes to the root system of the host (stage 2, **Figure 1**). This stage was evaluated in rhizotrons, and two different phenotypes were observed. When a sunflower line was susceptible to *O. cumana* race F at this stage, a CA allowed the parasite to attach and connect with the vascular system of the host root (**Figure 2A**). When a sunflower line was resistant, IA was observed. The IA was characterized by a browning of the interaction zone between the parasite and the sunflower root (**Figure 2B**). The two parental lines showed a significant difference for rate of IA at stage 2 (**Figure 2C**). Indeed, minimal IA was observed in the parental line HA89 (3.51%) compared to LR1, which produced a high rate of IA (46.51%). In the segregating population, RILs showed a continuous quantitative resistance profile from susceptible (IA $\leq$ 10%) to resistant (IA $\geq$ 50%). Thus, 14 lines showed a high rate of IA (IA $\geq$ 50%), and a low rate of IA (IA $\leq$ 10%) was observed for 19 lines (**Figure 2C**).

Following successful attachment and vascular connection during stage 2, *O. cumana* develops a storage organ called tubercle (stage 3, **Figure 1**). The RIL population was phenotyped in pots for stage 3 development of *O. cumana* race F, by counting the number of healthy and necrotic tubercles attached to each sunflower line (**Figures 3A,B**). The necrosis of the tubercles was characterized by browning and death of the parasite (**Figure 3A**). In contrast, the healthy tubercles remained yellow and became larger, allowing the next stages of development of *O. cumana* (**Figure 3B**). Although the size of the tubercles was not systematically measured, the healthy tubercles observed after 5 weeks were approximately 1 cm in diameter, the bigger tubercles showing a differentiated bud (**Figure 3A**).

The number of healthy tubercles observed in the two parental lines HA89 and LR1 was similarly low: 3.4 and 2.4 tubercles/plant on average, respectively. Despite the low number of *O. cumana* tubercles observed in the two parental lines, the segregating population allowed a wider range of tubercle development (from 0 to 18 tubercles/plant, **Figure 3C**). Some lines exhibited a high resistance level, with less than 1 healthy tubercle in average, similar to the level of resistance in the resistant line P-96. Highly susceptible RILs were also found with a susceptible reaction similar or higher to sunflower line 2603 (12.85 tubercles/plant on average) (**Figure 3C**).

Besides the absolute number of healthy *O. cumana* tubercles per plant, the rate of necrotic tubercles out of the total tubercles developed was also calculated as a possible discriminating mechanism of resistance (**Figure 3D**). Negligible values of necrosis of tubercle were observed in *O. cumana* attached to either parental lines HA89 and LR1 or in the susceptible control line 2603. As mentioned above, *O. cumana* was unable to develop beyond stage 2 in the resistant control P-96 and in consequence no tubercles of any kind (healthy or necrotic) were observed. Most of the RILs induced negligible or low level of necrotic tubercles (<30%), except for 3 RILs that exhibited an average of necrosis higher than 50% (**Figure 3D**).

Evaluation of the Resistance to Race G at Stage 3 of Broomrape Development

Race G of *O. cumana* is found only in countries around the Black Sea. The population was not evaluated in field, for precautionary quarantine reasons; therefore the resistance level at stage 3 of tubercle development was measured in a confined growth chamber. For race G, the number of healthy tubercles developing on the susceptible line 2603 was 10.88 tubercles per plant on average and was 7.22 and 8.44 healthy tubercles/plant on average, for HA89 and LR1 respectively. However, the number of *O. cumana* tubercles on the roots of parental lines in race G was more than twice that in race F. The level of infestation was also higher in the RILs population when challenged to race G. Despite no fully resistant RIL (no tubercle development) was observed, a wider range of responses was observed in the RIL population when compared with the parental lines (from 1.14 to 16.55 tubercles/plant, Supplementary Figure S2A). Necrosis of tubercle was not observed except in 8 RILs with an average of tubercle necrosis higher than 20%. (Supplementary Figure S2B). Interestingly, two of these RILs also induced significant necrosis in tubercles when challenged with *O. cumana* race F. There was no correlation for number of healthy tubercles in the sunflower

[3]https://www.heliagene.org/HaT13l/

[4]http://www.sunflowergenome.org/

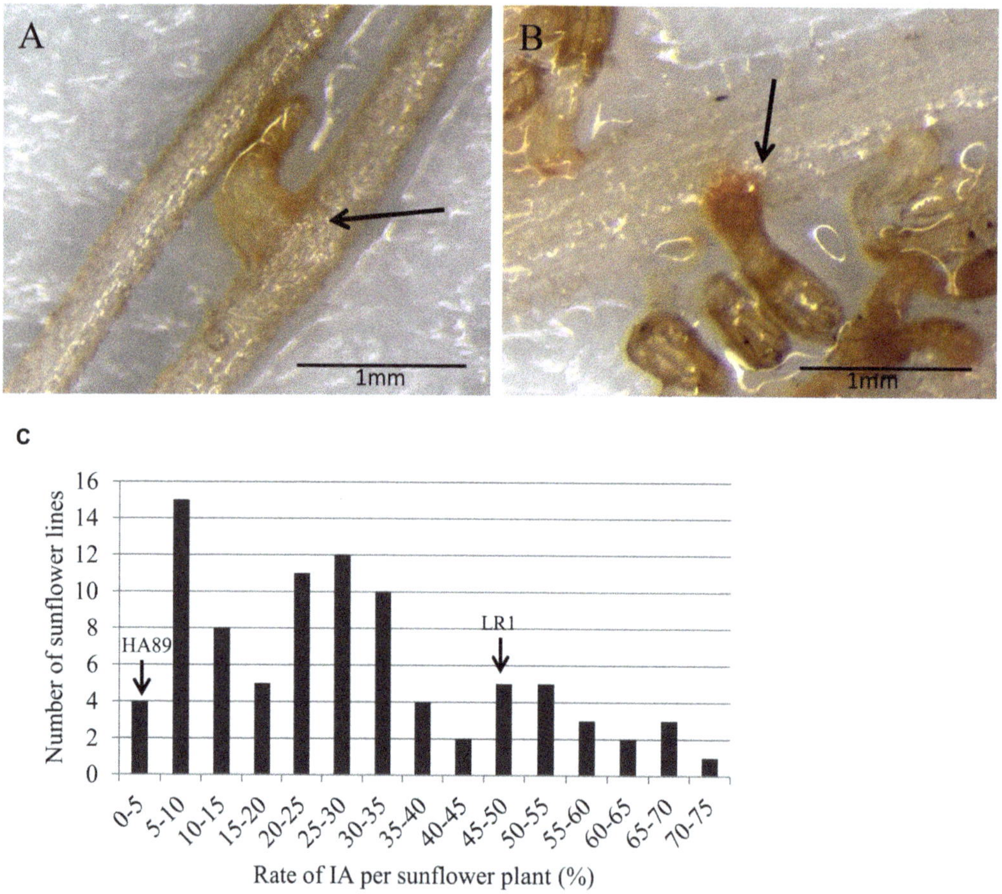

FIGURE 2 | Description of the resistance mechanism inhibiting *O. cumana* development at stage 2 in the RIL population HA89xLR1. The photos show **(A)** compatible attachment and **(B)** incompatible attachment between *O. cumana* and sunflower roots. The arrows indicate the attachment area between the radicle of *O. cumana* race F and the sunflower roots. **(C)** Distribution of the rate of IA counted 2 weeks after infection by *O. cumana* race F. The data are calculated from at least 3 replicates for each sunflower line. Replicates with less than 5 attachments (IA + CA) were discarded. The two parental lines HA89 and LR1 showed 3.5 and 46.5% of IA, respectively.

population between the screenings made with race F and race G (Supplementary Figure S2C).

Resistance to Race F in the Field

Phenotyping of the RIL population in the field was performed in 2014 and 2015. In both years, the two parental lines showed a similar resistance profile to *O. cumana* race F (**Figure 4**). A large number of resistant lines were observed in both years (12 and 10 RILs showing less than 1 broomrape emergence per sunflower plant in 2014 and 2015 respectively). In 2015 (**Figure 4B**), the overall intensity of the attack was higher than in 2014 (**Figure 4A**), with a consistent two times (1.94) increase *O. cumana* emergence average per sunflower plant across the entire RIL population allowing to establish a high correlation between experimental years (**Figure 4C**).

Plant size can be related with intensity of *Orobanche* emergence in the field masking genetic resistance. Vigorous hosts can increase the chances of parasitic encounter through bigger root systems or, once *Orobanche* is attached, they can provide better nutritive supply, therefore increasing the rate of parasitic emergence. In order to discover possible associations between sunflower vigor and the levels of *O. cumana* emergence observed in Cordoba, the plant height of the RIL population was characterized in *O. cumana*-free field conditions and its variability (Supplementary Figure S3) was mapped to one QTL on LG 5 (Supplementary Figure S5). The variability observed for sunflower plant height was not significantly related with the level of *O. cumana* emergence in sunflower (Supplementary Figure S4).

Correlation between the Resistance Mechanisms at Stage 2, 3, and 4 of Broomrape Development

PCA analysis was performed to determine the relationship between the different traits measured for race F of *O. cumana* (**Figure 5**). The first axis of the PCA explained more than 50% of the variability, distinguishing between healthy tubercle development in pots and shoot emergence in field on the basis of the rate of IA and the necrosis rate. This PCA analysis also highlights the correlation between the field observations in 2014 and 2015, as shown in **Figure 4C**, as well as the

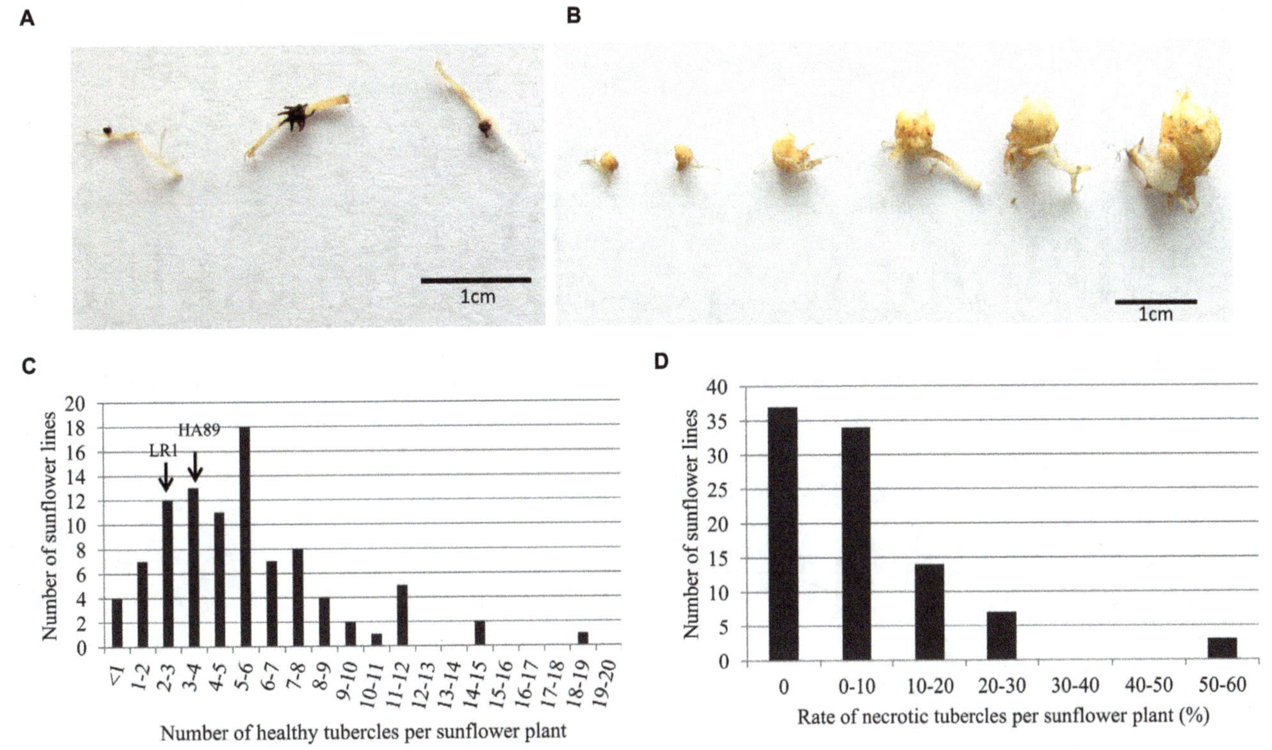

FIGURE 3 | Evaluation of the resistance level to *O. cumana* race F in the RILs population HA89xLR1 at stage 3. Photos illustrate the different developmental stages observed for either **(A)** necrotic tubercle or **(B)** healthy tubercle after 5 weeks post infection. **(C)** Distribution of the number of healthy tubercle in the RILs population HA89xLR1 5 weeks after infection. The two parental lines HA89 and LR1 show 3.4 and 2.4 healthy tubercle per plant, respectively **(D)** Distribution of the rate of necrotic tubercle in the RILs population HA89xLR1 5 weeks after infection. No necrotic tubercles were observed for the two parental lines. Value represents the number of RILs for each range of the number of healthy tubercle or rate of necrotic tubercle per sunflower plant. Five independent experiments were performed with one single plant per RIL in each experiment.

correlation between susceptibility in the field and the number of healthy tubercles. Interestingly, the rate of IA and the necrosis of the tubercles were negatively correlated to the number of healthy broomrape tubercles and the number of *O. cumana* field emergences. When resistance data to race G was added, no correlation was observed in the RIL population for number of healthy tubercles (data not shown). This result is consistent with the data from Supplementary Figure S1C.

Finally, the different RILs were distributed according to the different axes of the PCA (**Figure 6**). The parental lines HA89 and LR1 and the control lines P-96 and 2603 were added to the figure as supplementary individuals. The population was divided into three distinct groups. The two smaller groups were composed of 7 and 8 RILs and were grouped with the resistant line P-96 and the susceptible line 2603, respectively. The majority of the RILs (86) were grouped together with the two parental lines HA89 and LR1, exhibiting a partial resistance profile to race F of *O. cumana*.

QTL Mapping

The 101 RILs were genotyped using a high-throughput genotyping tool. The genotypic data were used to obtain a genetic map of 1795.8 cM for the 17 LGs of the sunflower genome. Twenty one thousand two hundred and one markers were mapped and located on 951 different genetic bins, with a mean distance between bins of 1.9 cM. Combined with the phenotypic data, the genotypic data were used for mapping the QTLs. A total of 17 QTLs were found to control resistance to *O. cumana* race F and G (**Figure 7**). These QTLs were localized on 9 LGs. None of the QTLs controlling the different resistant traits for race F were mapped in the same region, and only one QTL controlling the resistance to race F and G colocalized on LG 9.

From the 13 QTLs, we identified two QTLs with a strong effect. The first, on LG16, controlled the healthy tubercles of race G of *O. cumana*. The second controlled the number of emergences in the field and was mapped on LG13. This QTL was identified in both years 2014 and 2015 and was the only one controlling resistance in the field.

The coordinate of the first axis of the PCA was used to perform QTL analysis. Four different QTLs were identified. Two of these QTLs localized with QTLs on LG04 and LG11, controlling the number of healthy tubercles for *O. cumana* race F and the rate of necrotic tubercles for *O. cumana* race G, respectively. The other QTLs were also found on LG04 (**Figure 7**).

Analysis of the additive effect showed that the genetic resistance to race F of *O. cumana* at 2 QTLs is coming from LR1 and at 4 QTLs from HA89. The QTLs controlling the number of emergences, E-2014-F-13.1 and E-2015-F-13.1, explain between 14 and 31% of the variability and the resistant alleles are carried

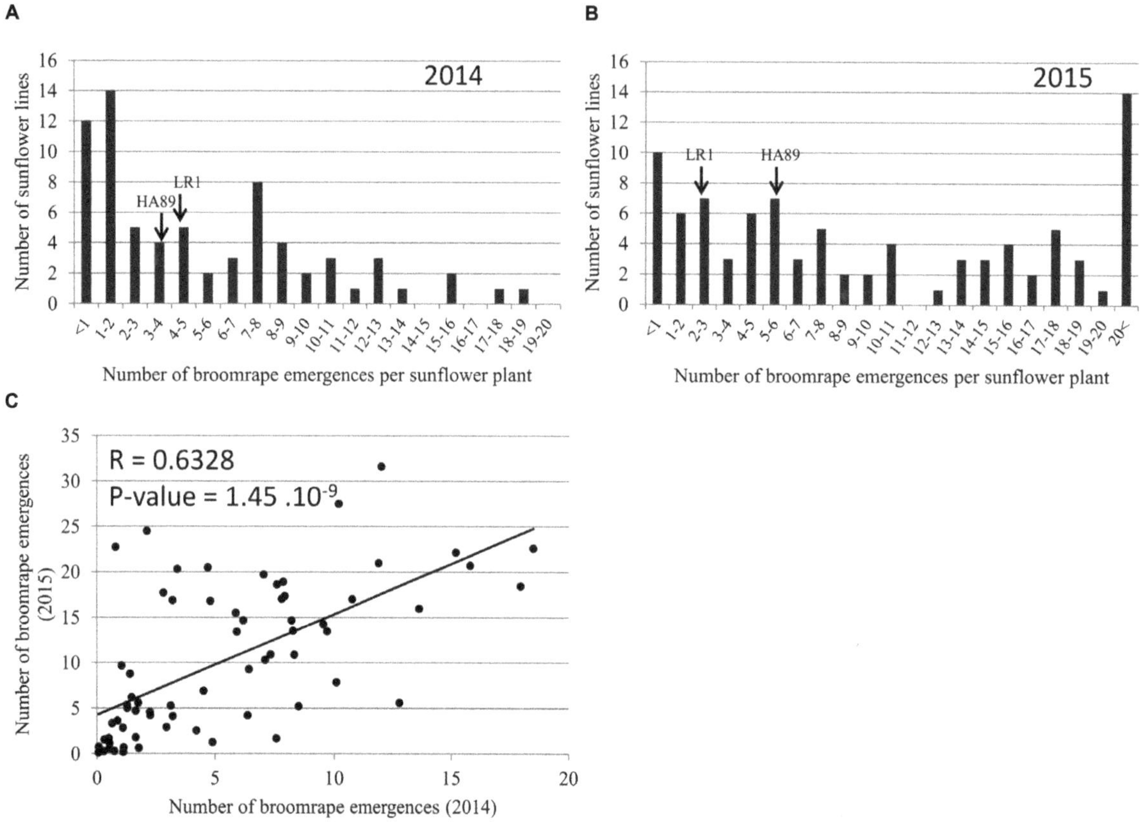

FIGURE 4 | Characterization of the resistance level in field at the stage 4 on the RILs population HA89xLR1. Distribution of the number of *O. cumana* emerged shoots per sunflower plant in 2014 **(A)** and in 2015 **(B)**. The data represent the mean of three replicates of 15 sunflower plants in field (Cordoba, Spain). **(C)** Relationship between the number of *O. cumana* emerged shoots in 2014 and 2015.

by the parental line HA89 (**Figure 7B**). IA is induced by a locus from the parental line LR1 (**Figure 7B**). For the control of race G, the two most important QTLs identified HT-G-16.1 and HT-G-16.2, were found on the same LG and are very close to each other. These two QTLs have 2 distinct parental origins, HT-G-16.1 is from LR1 and HT-G-16.2 is from HA89 (**Figure 7B**).

Candidate Gene Mapping

We attempted a candidate gene approach, based on previous functional results on the plant–parasite interaction. Nucleotide-binding site leucine-rich repeat (NBS-LRR) proteins play an important role in plant resistance to pathogens (McHale et al., 2006). Radwan et al. (2008) have identified several NBS-LRR homologs in the sunflower. Based on genetic map positions, only one QTL, "IA-F-15.1" on the LG15, colocalized with one NBS-LRR gene identified by Radwan et al. (2008). This NBS-LRR gene was coded by EF559379.1 cDNA.

Li and Timko (2009) were the first to identify gene-for-gene resistance to parasitic plants, and they have identified RSG3-301 and its predicted coiled-coil-nucleotide-binding-site-leucine-rich repeat (CC-NBS-LRR). BlastX analysis of this protein was performed on the sunflower genome to identify its homolog, and the 20 best hits were mapped on the sunflower genome. Interestingly, only three of these homologs colocalized with QTLs identified to control the resistance to *O. cumana*. These homologs colocalized with QTLs controlling broomrape field emergence (on the LG13, HaT13l034464), controlling the capacity to induce IA (on the LG15, HaT13l008311) and controlling the induction of necrotic tubercle on LG17 (HaT13l008327).

Letousey et al. (2007) have identified HaGSL1 to be induced during incompatible interactions between *O. cumana* and the sunflower root of LR1. HaGSL1 was mapped on the LG09 of the sunflower genome, but it was not localized under the QTLs found on this LG.

DISCUSSION

QTL Mapping

In this study, few phenotypic differences were observed between the two parental lines HA89 and LR1 for races F and G, as previously observed by Labrousse et al. (2004) for race E. Interestingly, the IA rate observed for *O. cumana* race F, was significantly different between HA89 (3.51%) and LR1 (46.51%). This difference was not observed at stage 3 of *O. cumana* development. It would be interesting to evaluate the two parental

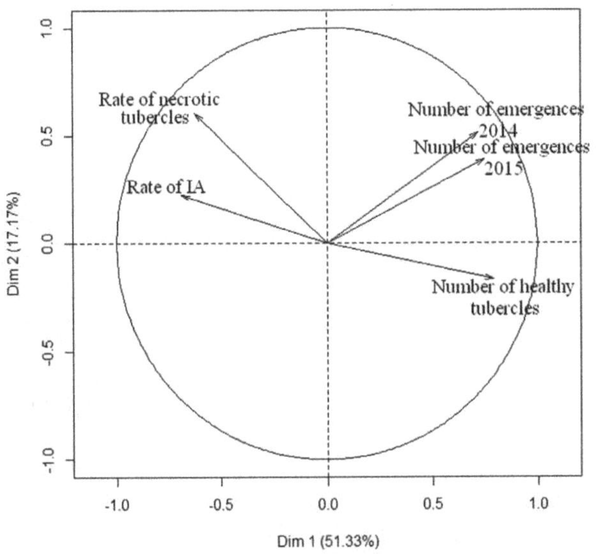

FIGURE 5 | Relationship between the different sunflower resistance traits to *O. cumana* race F. PCA was performed with all phenotypic data from the 101 RILs. The two first axis of the PCA explain 68.5% of the variability.

lines in a kinetic of *O. cumana* development between 2 to 5 weeks in order to determine the importance of this resistance mechanism. The RILs population was not suitable to map major genes because the two parental lines do not exhibit a clear discriminating resistance profiles at stage 3 and at stage 4 of *O. cumana* development. The parental lines could also be monomorphic for loci controlling some traits that will not allow the detection of these QTLs in our study. However, some recombinant lines obtained from the cross between HA89 and LR1 exhibited a higher level of resistance or susceptibility than the two parental lines, which indicates complex multigenic control of the resistance that can be increased by particular allele combinations for several loci. Velasco et al. (2012) showed that the resistance carried by *H. debilis subsp tardiflorus* was monogenic; indicating that the use of some *H. debilis* accessions in breeding programs could be an additional way to improve resistance to *O. cumana*.

The expression of resistance is complex because it arises from a combination of several resistance mechanisms acting at different steps of broomrape development. We mapped QTLs controlling some of these steps. Even though some traits were correlated (**Figure 5**), we never observed QTLs that colocalized in the same genetic region (**Figure 7**). We identified only one QTL explaining more than 30% of the phenotypic variability in the entire population. Almost all QTLs explained between 9 and 30% of the phenotypic variation of the traits. This result suggests that many minor quantitative loci are involved in the expression of the traits and cannot be detected in our genetic design. The first report identifying QTLs for broomrape resistance in sunflower was provided by Pérez-Vich et al. (2004), who mapped 7 QTLs controlling the resistance to *O. cumana* race F and found that each of them had a contribution (R^2) varying from 11 to 38%, similarly to those observed in this study. However, the early stages of the interaction (**Figure 1**) were not taken into account in this previous study. We identified a stable and strong QTL on LG13 for resistance in the field, and our results suggest that the final expression of the resistance in the field is not correlated with one specific mechanism but is due to a combination of resistance mechanisms acting at the early stages of broomrape development.

Race-Specific Quantitative Resistance

The number of healthy tubercles at stage 3 was the trait best correlated with the number of emergences in the field for race F of *O. cumana* (**Figure 5**). The resistance to race G of *O. cumana* was only measured in controlled conditions at stage 3. Compared to race F, twice more tubercles were observed in the RIL population using *O. cumana* race G. These experiments were performed in same controlled conditions (nutrition, light, temperature and hygrometry). Then, the differences found between the 2 races cannot be explained by environmental effects but were due to the higher virulence of race G. No correlation was observed between the number of broomrape tubercles found for race F and G in the RIL population (Supplementary Figure S2C**)**. As expected, the QTL analysis showed that only two QTLs for the resistance of both races colocalized on LG09 (**Figure 7**). The race-specific resistance to broomrape has previously been reported for quantitative loci (Pérez-Vich et al., 2004) and for resistance control by major genes (Vranceanu et al., 1980). The genetic control of the race-specific resistance in the pathosystem *H. annuus/O. cumana* is similar to the one found for downy mildew in the *H. annuus/P. halstedii* pathosystem (Vear et al., 2007). It would be interesting to further explore the genetic control of the race-specific resistance by dynamic phenotyping of the interaction from the germination of the broomrape seeds to the final steps of parasite development. Such experiments could identify the important steps involved in the interaction for the different races.

New Phenotyping Tools Need to be Developed

In our study, the induction of iIAs during the early stages of development played an important role in the resistance mechanisms. However, the attachment is a small and difficult structure to observe. New phenotyping tools are necessary for a better characterization of the interaction between the sunflower roots and *O. cumana*. Furthermore, increasing the depth (qualitative) and the throughput (quantitative) of phenotyping would enrich the quality of the genetic analysis. Such phenotyping methods have successfully been used to describe the development of the roots. These rapid and efficient tools have been one of the major challenges of the decade. Several recent reports have helped to improve the characterization of the root architecture (Iyer-Pascuzzi et al., 2010; Clark et al., 2013; Slovak et al., 2014). Despite the difficulty in observing the attachment, the IA shows a browning at the point of attachment between the host and the parasite. These browning

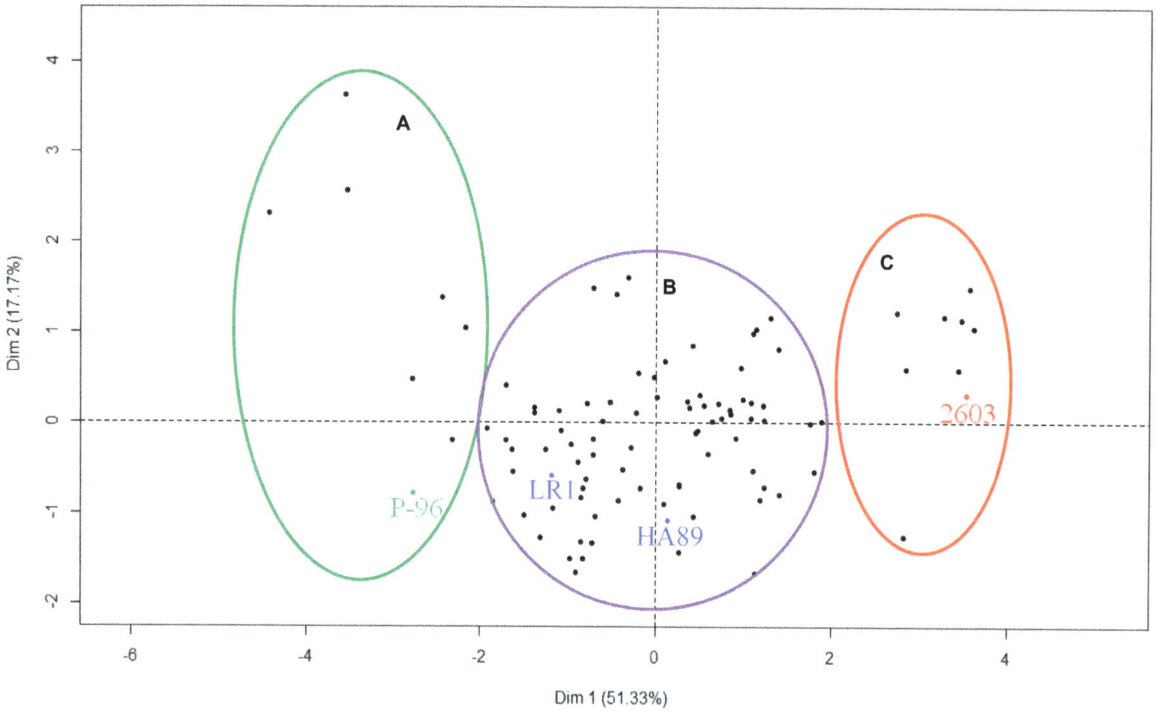

FIGURE 6 | Phenotypic diversity within the complete RIL population for *O. cumana* race F. Three groups were found according to the first dimension of the PCA. They correspond to resistant lines **(A)**, susceptible lines **(C)** and partially resistant lines **(B)**. The four control lines HA89, LR1, P-96 and 2603 have been added as additional samples to be shown on the figure but they have not been taken into account in the PCA analysis.

contrasts with the sunflower root could be used to detect IA in high definition pictures. Furthermore, the development of high throughput phenotyping tools will allow for the kinetic analysis of the infection, which could detect new resistance mechanisms.

Effect of the Environment and Sunflower Plant Height on Resistance

The same field in Cordoba (Spain) was used to perform two experiments in 2014 and 2015. We observed a correlation between data obtained for the whole population during both years (**Figure 4C**; Supplementary Figure S4). However, the number of *O. cumana* emergences in 2015 was approximately twice in average the number of emergences in 2014. Temperature has been shown to have an effect on *O. cumana* development. Sukno et al. (2001) and Eizenberg (2003) have found that a moderate increase of the temperature had a positive effect on the intensity of sunflower infection by broomrape. The increase of the infection level between both years could be due to the increased temperature in Cordoba in 2015 compared to 2014. We obtained the climatic data for both years[5] between March and June. In March 2015, an increase of the mean temperature of 1°C was observed compared to 2014. After transplanting, the higher maximal temperature observed during the first week in 2015 (24.5°C) than in 2014 (17.9°C) could improve the attachment of the parasite to the host root and could explain the differences in the number of emergences between both years. In April, the mean temperature was similar, but the mean temperatures in May and June 2015 were approximately 2°C higher than in 2014. The effect of the temperature on the level of infestation did not affect the genetic control of the resistance. A stable QTL was mapped on LG13 with data from 2014 and 2015 (**Figure 7**). Besides temperature, other environmental factors could influence *O. cumana* development across seasons. For the field experiments, we used an artificial inoculation in controlled conditions during the first 20 days of the sunflower life. New studies should be conducted to take into account the soil management, rain, nutritive input, weeds and soil microflora composition, to provide a complete understanding of the dynamics of the parasitic interaction and to identify the key environmental factors affecting the development of *O. cumana* and thus indirectly the level of resistance in the sunflower.

It has been observed that the host plant vigor was correlated with *Orobanche* shoot emergences in several pathosystems (Fernández-Aparicio et al., 2009b, 2012a; Fondevilla et al., 2010). The same RIL population was evaluated in 2015 in Toulouse (France) for plant height (as estimate of sunflower vigor). Plant height in the RIL population showed a normal distribution (Supplementary Figure S3). No correlation between plant height and the number of broomrape shoot emergences in field season 2014 and 2015 was observed (Supplementary Figure S4). We mapped a QTL controlling plant height to LG05, where no

[5]http://power.larc.nasa.gov

A

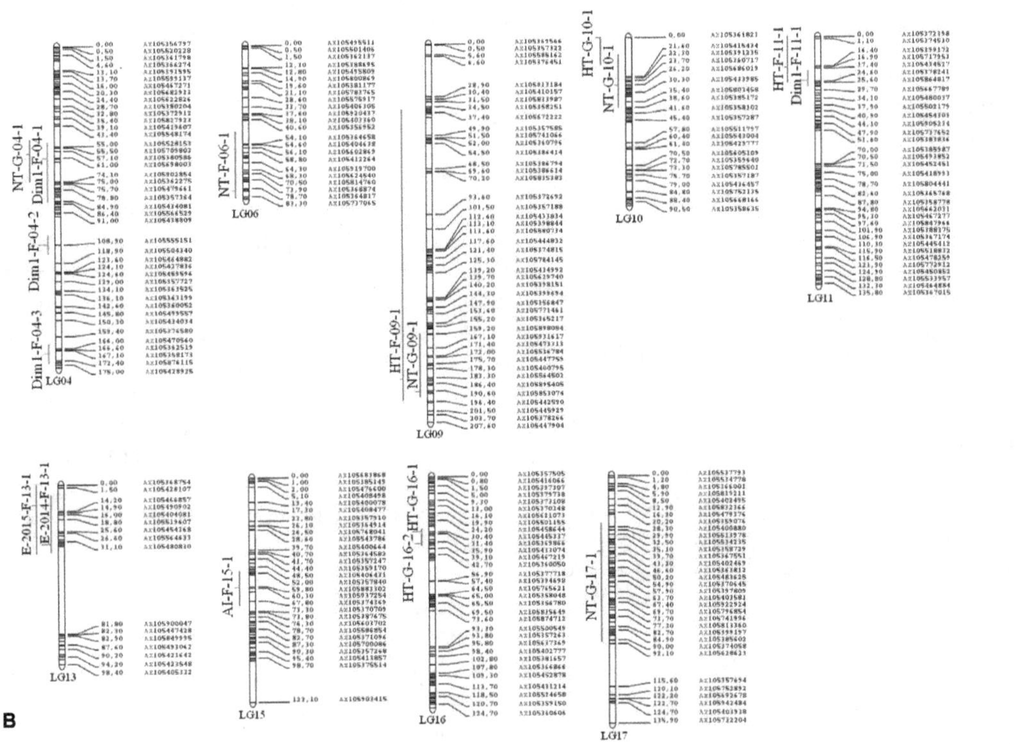

B

	Trait	QTL	LG	Position (cM)	Interval support	LOD	R^2	Additive effect	Markers flanking the QTL
Race F	IA	IA-F-15.1	15	60.3	51.18 - 70.73	3.32	0.13	0.97	AX105937254/AX105872010
	Healthy Tubercle	HT-F-09.1	09	184.8	36.10 - 196.50	2.56	0.09	0.45	AX105576408/AX105693984
		HT-F-11.1	11	17.4	0 - 25.69	3.33	0.12	0.54	AX105434527/AX105378241
	Necrotic Tubercle	NT-F-06.1	06	80.2	47.67 - 84.50	3.39	0.12	-0.08	AX105372900/AX105737065
	Emergence (field 2014)	E-2014-F-13.1	13	33	6.16 - 37.90	6.45	0.31	1.33	AX105372031/AX105370341
	Emergence (field 2015)	E-2015-F-13.1	13	7.5	0 - 38.25	3.70	0.14	1.64	AX105428107/AX105466857
	ACP-Dim1	Dim1-F-04-1	04	61.5	52.68 - 85.68	4.48	0.18	0.35	AX105471636/AX105491160
	ACP-Dim1	Dim1-F-04-2	04	108.9	103.73 - 114.51	7.66	0.30	-0.51	AX105555151/AX105504340
	ACP-Dim1	Dim1-F-04-3	04	169.1	163.83 - 174.65	4.78	0.19	0.33	AX105358173/AX105876115
	ACP-Dim1	Dim1-F-11-1	11	24.6	19.00 - 28.08	4.55	0.18	0.29	AX105378241/AX105864817
Race G	Healthy tubercle	HT-G-10.1	10	0	0 - 15.72	3.90	0.15	0.47	AX105361821/AX105415434
		HT-G-16.1	16	30.4	28.97 - 30.76	8.09	0.30	-1.25	AX105445337/AX105369866
		HT-G-16.2	16	37	31.40 - 38.38	5.53	0.21	0.98	AX105680427/AX105725318
	Necrotic Tubercle	NT-G-04.1	04	61.5	51.47 - 74.69	2.89	0.11	-1.46	AX105380586/AX105698003
		NT-G-09.1	09	191.8	159.17 - 194.30	3.77	0.14	-1.71	AX105426024/AX105437681
		NT-G-10.1	10	23.7	9.69 - 39.38	3.00	0.11	-1.48	AX105365035/AX105520142
		NT-G-17.1	17	46.6	26.31 - 91.60	3.32	0.13	-1.53	AX105896867/AX105391615

FIGURE 7 | Quantitative trait loci (QTL) mapping of the different quantitative resistance traits to *O. cumana* race F and G. (A) The genetic map used to detect QTLs covered the 17LGs (951 genetic bins and 1795.8 cM lengths). QTLs are indicated at the left of each Linkage Group. **(B)** Summary of the different QTL found for the resistance to *O. cumana* race F and G. QTL name nomenclature is: trait-broomrape race-linkage group-QTL number on the linkage group. Additive effect, (–) be equivalent to a decrease of the value due to LR1 alleles. The traits used for QTL mapping were: IA, ratio of incompatible attachments (IA); Emergence, the number of emerged shoot broomrape per sunflower plant under field conditions in 2014 and 2015; Healthy Tubercle, the number of healthy tubercles; Necrotic Tubercle, ratio of necrotic tubercles; ACP-Dim1, the coordinates of the first axis from the PCA analysis. QTL detection was performed with MCQTL (Jourjon et al., 2005) with a threshold corresponding to a Type I error rate of 1%.

resistance QTLs was mapped (Supplementary Figure S5). These clues indicate that the resistance to *O. cumana* race F and plant height are not linked in our RIL population HA89xLR1.

Candidate Genes Located in the QTL Regions

For breeding programs, the stable QTL found on LG13, which controls the number of emergences in the field is the one that could be most rapidly used. Pérez-Vich et al. (2004) have mapped a QTL controlling resistance to race F in the same genetic region using a different mapping population (P-21 × P-96). In our conditions, this QTL explains 15–30% of the overall phenotypic variability, which is similar to the effect of the QTL reported by Pérez-Vich et al. (2004). Furthermore, we identified a cDNA (HaT13l034464), showing homology with a gene coding for a CC-NBS-LRR protein described by Li and Timko (2009) that were located in the interval support of this QTL. NBS-LRR proteins play important roles in gene-for-gene resistance (McHale et al., 2006).

Two NBS-LRR genes (GenBank accession number EF559379.1 and HaT13l008311) were identified in the interval support of a QTL mapped to LG15. One of these genes encodes for a NBS-LRR protein (HaT13l008311) showing homology to RSG3-301 that controls the incompatible interaction between *Striga gesnerioides* and cowpea (*Vigna unguiculata*) (Li and Timko, 2009). These authors found that RSG3-301 induces an early resistance mechanism and acts on the attachment zone between the roots of *Vigna unguiculata* and *S. gesnerioides*. In our study, a similar phenotype (IA) was observed and evaluated for all RILs of the population (**Figure 2**), and one QTL was found to control this resistance mechanism on LG15. The attachment of the broomrape radicle to the sunflower root is an essential step for the parasite to establish the interaction to redirect sunflower assimilates for its growth. However, we show that the sunflower has a resistance mechanism to block this interaction. In some lines, there is a rapidly incompatible interaction mechanism. This response is characterized by browning at the point of attachment between the host and the parasite. A similar incompatible phenotype of the *Orobanche* penetration process has been described in several crop species including *Vicia* spp. response to *O. aegyptiaca* (Goldwasser et al., 2000), and the response of rice (Gurney et al., 2006) or *Sorghum* (Mohamed et al., 2003) to *Striga* spp. Further cytological or biochemical experiments would be needed to investigate the nature of the incompatible phenotype observed in sunflower resistant RILs. Several defense mechanisms could underlay the incompatible phenotype including hypersensitive response (Morel and Dangl, 1997), a rapid and strong resistance reaction usually mediated by specific recognition of pathogen-derived effectors by the cognate resistance protein in the host during a gene-for-gene interaction. To date, the *Orobanche*-encoded molecular cues that determine the resistant phenotype in the host are unknown. Identification of candidate genes by gene expression comparison between compatible and incompatible interactions could provide a better understanding of the interaction mechanism.

With a full sequence of the sunflower genome, we could identify other candidate genes. For all of the candidate genes that we have identified, more experiments are needed, and it would be interesting to test their functional roles in resistance by reverse genetics using EMS mutants (Sabetta et al., 2011; Kumar et al., 2013) or to measure their expression level in different genotypes or during the different steps of the interaction.

CONCLUSION

For the first time, the early stages of the *O. cumana/H. annuus* pathosystem were used to map QTLs. No pleiotropic QTLs were found and, these QTLs controlled specific developmental stages. To complete the overall panel of possible resistance mechanisms, it is important to characterize the induction of *O. cumana* seed germination by sunflower root exudates and the induction of the haustorium at the genetic level. It was found that the low capacity to induce broomrape seed germination by the host was a good way to control broomrape in field (Jamil et al., 2011; Fernández-Aparicio et al., 2012b). However, Labrousse et al. (2004) have not found a correlation between the capacity to induce broomrape seed germination and the level of infection in the sunflower.

The emergence of new virulent races has frequently been observed, and new race-specific resistance loci need to be identified. Even though the introgression of major resistance genes is an easy and quick solution for breeding, more sustainable resistance has to be developed. In other pathosystems, the additive effect of minor QTLs has improved resistance to broomrape (Fondevilla et al., 2010; Gutiérrez et al., 2013). Breeding methods that integrate QTLs are a good way to improve the sustainability of sunflower resistance to broomrape. Our results show that several quantitative resistance loci can be used. However, the diversity of broomrape populations needs to be detailed, and the emergence of new races and new geographical infected areas must be monitored.

No resistance genes have yet been cloned, and the molecular mechanism underlying the resistance to *O. cumana* is poorly described. One of the main future goals will be to better understand the interaction between the sunflower and *O. cumana*.

AUTHOR CONTRIBUTIONS

M-CB and NP contributed to the production of the data. LV and JL contributed to the production and to the analysis of the data. BP-V, PV, and SM contributed to the analysis of the data and to the coordination of the project.

FUNDING

Promosol (17, rue du Louvre F-75001 Paris, France) financially supported this work.

ACKNOWLEDGMENTS

We gratefully thank Jessica Issakof for her help in performing phenotyping experiments. The Gentyane genomic platform (Plateforme Gentyane, UMR INRA/UBP 1095 Génétique Diversité et Ecophysiologie des Céréales, 5 chemin de Beaulieu – 63039 Clermont-Ferrand, France) performed the AXIOM genotyping experiments. Biogemma (Biogemma, 6 Chemin de Panedautes, 31700 Mondonville, France) provided *O. cumana* (race F) seeds. The AXIOM array has been developed in the frame of the SUNRISE project (ANR-11-BTBR-0005, http://www.sunrise-project.fr). We thank the three reviewers of the manuscript and MF-A(associate editor) for their comments and their corrections that helped to improve our article.

REFERENCES

Alcántara, E., Morales-García, M., and Díaz-Sánchez, J. (2006). Effects of broomrape parasitism on sunflower plants: growth, development, and mineral nutrition. *J. Plant Nutr.* 29, 1199–1206. doi: 10.1080/01904160600767351

Aly, R., Cholakh, H., Joel, D. M., Leibman, D., Steinitz, B., Zelcer, A., et al. (2009). Gene silencing of mannose 6-phosphate reductase in the parasitic weed *Orobanche aegyptiaca* through the production of homologous dsRNA sequences in the host plant. *Plant Biotechnol. J.* 7, 487–498. doi: 10.1111/j.1467-7652.2009.00418.x

Arcade, A., Labourdette, A., Falque, M., Mangin, B., Chardon, F., Charcosset, A., et al. (2004). BioMercator: integrating genetic maps and QTL towards discovery of candidate genes. *Bioinformatics* 20, 2324–2326. doi: 10.1093/bioinformatics/bth230

Brun, H., Chèvre, A.-M., Fitt, B. D., Powers, S., Besnard, A.-L., Ermel, M., et al. (2010). Quantitative resistance increases the durability of qualitative resistance to *Leptosphaeria maculans* in *Brassica napus*. *New Phytol.* 185, 285–299. doi: 10.1111/j.1469-8137.2009.03049.x

Clark, R. T., Famoso, A. N., Zhao, K., Shaff, J. E., Craft, E. J., Bustamante, C. D., et al. (2013). High-throughput two-dimensional root system phenotyping platform facilitates genetic analysis of root growth and development: root phenotyping platform. *Plant Cell Environ.* 36, 454–466. doi: 10.1111/j.1365-3040.2012.02587.x

Cook, C. E., Whichard, L. P., Turner, B., Wall, M. E., and Egley, G. H. (1966). Germination of witchweed (*Striga lutea* lour.): isolation and properties of a potent stimulant. *Science* 154, 1189–1190. doi: 10.1126/science.154.3753.1189

de Givry, S., Bouchez, M., Chabrier, P., Milan, D., and Schiex, T. (2005). CARHTA GENE: multipopulation integrated genetic and radiation hybrid mapping. *Bioinformatics* 21, 1703–1704. doi: 10.1093/bioinformatics/bti222

Duca, M. (2015). Historical aspects of sunflower researches in the republic of moldova. *Helia* 38, 79–92. doi: 10.1515/helia-2014-0028

Eizenberg, H. (2003). Resistance to broomrape (*Orobanche* spp.) in sunflower (*Helianthus annuus* L.) is temperature dependent. *J. Exp. Bot.* 54, 1305–1311. doi: 10.1093/jxb/erg129

Fernández-Aparicio, M., Flores, F., and Rubiales, D. (2009a). Recognition of root exudates by seeds of broomrape (*Orobanche* and *Phelipanche*) species. *Ann. Bot.* 103, 423–431. doi: 10.1093/aob/mcn236

Fernández-Aparicio, M., Flores, F., and Rubiales, D. (2009b). Field response of *Lathyrus cicera* germplasm to crenate broomrape (*Orobanche crenata*). *Field Crops Res.* 113, 321–327. doi: 10.1016/j.fcr.2009.06.009

Fernández-Aparicio, M., Flores, F., and Rubiales, D. (2012a). Escape and true resistance to crenate broomrape (*Orobanche crenata* Forsk.) in grass pea (*Lathyrus sativus* L.) germplasm. *Field Crops Res.* 125, 92–97. doi: 10.1016/j.fcr.2011.09.003

Fernández-Aparicio, M., Kisugi, T., Xie, X., Rubiales, D., and Yoneyama, K. (2014). Low strigolactone root exudation: a novel mechanism of broomrape (*Orobanche* and *Phelipanche* spp.) resistance available for faba bean breeding. *J. Agric. Food Chem.* 62, 7063–7071. doi: 10.1021/jf5027235

Fernández-Aparicio, M., Moral, A., Kharrat, M., and Rubiales, D. (2012b). Resistance against broomrapes (*Orobanche* and *Phelipanche* spp.) in faba bean (*Vicia faba*) based in low induction of broomrape seed germination. *Euphytica* 186, 897–905. doi: 10.1007/s10681-012-0686-0

Fernández-Martínez, J. M., Domínguez, J., and Pérez-Vich, B. (2008). Update on breeding for resistance to sunflower broomrape. *Helia* 31, 73–84. doi: 10.2298/HEL0848073F

Fernández-Martínez, J. M., Pérez-Vich, B., Akhtouch, B., Velasco, L., Muñoz-Ruz, J., Melero-Vara, J. M., et al. (2004). Registration of four sunflower germplasms resistant to race F of broomrape. *Crop Sci.* 44, 1033–1034. doi: 10.2135/cropsci2004.1033

Fernández-Martínez, J. M., Pérez-Vich, B., and Velasco, L. (2015). "Sunflower broomrape (*Orobanche cumana* Wallr.)," in *Sunflower Oilseed Chemistry, Production, Processing and Utilization*, eds E. Martínez-Force, N. T. Dunford, and J. J. Salas (Champaign IL: AOCS Press), 129–156.

Fondevilla, S., Fernández-Aparicio, M., Satovic, Z., Emeran, A. A., Torres, A. M., Moreno, M. T., et al. (2010). Identification of quantitative trait loci for specific mechanisms of resistance to *Orobanche crenata* Forsk. in pea (*Pisum sativum* L.). *Mol. Breed.* 25, 259–272. doi: 10.1007/s11032-009-9330-7

Gibot-Leclerc, S., Sallé, G., Reboud, X., and Moreau, D. (2012). What are the traits of *Phelipanche ramosa* (L.) Pomel that contribute to the success of its biological cycle on its host *Brassica napus* L.? *Flora* 207, 512–521. doi: 10.1016/j.flora.2012.06.011

Goldwasser, Y., Plakhine, D., Kleifeld, Y., Zamski, E., and Rubin, B. (2000). The differential susceptibility of Vetch (*Vicia* spp.) to *Orobanche aegyptiaca*: anatomical studies. *Ann. Bot.* 85, 257–262. doi: 10.1006/anbo.1999.1029

Gurney, A. L., Slate, J., Press, M. C., and Scholes, J. D. (2006). A novel form of resistance in rice to the angiosperm parasite *Striga hermonthica*. *New Phytol.* 169, 199–208. doi: 10.1111/j.1469-8137.2005.01560.x

Gutiérrez, N., Palomino, C., Satovic, Z., Ruiz-Rodríguez, M. D., Vitale, S., Gutiérrez, M. V., et al. (2013). QTLs for *Orobanche* spp. resistance in faba bean: identification and validation across different environments. *Mol. Breed.* 32, 1–14. doi: 10.1007/s11032-013-9920-2

Hassan, E. A., El-Akkad, S. S., Moustafa, S. M., and El-Awadi, M. E. (2004). Histochemical aspects of penetration and vascular connection of broomrape haustoria in the host root, and the possible implication of phenylpropanoids. *Int. J. Agric. Biol.* 6, 430–434.

Hewitt, J. (1966). *Sand and Water Culture Methods Used in the Study of Plant Nutrition*. Farnham Royal: Commonwealth Agricultural Bureaux.

Iyer-Pascuzzi, A. S., Symonova, O., Mileyko, Y., Hao, Y., Belcher, H., Harer, J., et al. (2010). Imaging and analysis platform for automatic phenotyping and trait ranking of plant root systems. *Plant Physiol.* 152, 1148–1157. doi: 10.1104/pp.109.150748

Jamil, M., Charnikhova, T., Houshyani, B., Ast, A., and Bouwmeester, H. J. (2011). Genetic variation in strigolactone production and tillering in rice and its effect on *Striga hermonthica* infection. *Planta* 235, 473–484. doi: 10.1007/s00425-011-1520-y

Jan, C. C., and Fernandez-Martínez, J. M. (2002). Interspecific hybridization, gene transfer, and the development of resistance to the broomrape race F in Spain. *Helia* 25, 123–136. doi: 10.2298/hel0236123j

Joel, D. M., Chaudhuri, S. K., Plakhine, D., Ziadna, H., and Steffens, J. C. (2011). Dehydrocostus lactone is exuded from sunflower roots and stimulates germination of the root parasite *Orobanche cumana*. *Phytochemistry* 72, 624–634. doi: 10.1016/j.phytochem.2011.01.037

Jourjon, M.-F., Jasson, S., Marcel, J., Ngom, B., and Mangin, B. (2005). MCQTL: multi-allelic QTL mapping in multi-cross design. *Bioinformatics* 21, 128–130. doi: 10.1093/bioinformatics/bth481

Kaya, Y. (2014). "Current situation of sunflower broomrape around the world," in *Proceedings of the 3rd International Symposium on Broomrape Sunflower Córdoba Spain 3–6 June 2014* (Córdoba: International Sunflower Association), 9–18.

Kumar, A. P., Boualem, A., Bhattacharya, A., Parikh, S., Desai, N., Zambelli, A., et al. (2013). SMART–sunflower mutant population and reverse genetic tool for crop improvement. *BMC Plant Biol.* 13:38. doi: 10.1186/1471-2229-13-38

Labrousse, P. (2001). Several mechanisms are involved in resistance of *Helianthus* to *Orobanche cumana* Wallr. *Ann. Bot.* 88, 859–868. doi: 10.1006/anbo.2001.1520

Labrousse, P., Arnaud, M. C., Griveau, Y., Fer, A., and Thalouarn, P. (2004). Analysis of resistance criteria of sunflower recombined inbred lines against *Orobanche cumana* Wallr. *Crop Prot.* 23, 407–413. doi: 10.1016/j.cropro.2003.09.013

Lechat, M.-M., Pouvreau, J.-B., Peron, T., Gauthier, M., Montiel, G., Veronesi, C., et al. (2012). PrCYP707A1, an ABA catabolic gene, is a key component of *Phelipanche ramosa* seed germination in response to the strigolactone analogue GR24. *J. Exp. Bot.* 63, 5311–5322. doi: 10.1093/jxb/ers189

Letousey, P., De Zélicourt, A., Vieira Dos Santos, C., Thoiron, S., Monteau, F., Simier, P., et al. (2007). Molecular analysis of resistance mechanisms to *Orobanche cumana* in sunflower. *Plant Pathol.* 56, 536–546. doi: 10.1111/j.1365-3059.2007.01575.x

Li, J., and Timko, M. P. (2009). Gene-for-gene resistance in *Striga*-cowpea associations. *Science* 325, 1094–1094. doi: 10.1126/science.1174754

Lindhout, P. (2002). The perspectives of polygenic resistance in breeding for durable disease resistance. *Euphytica* 124, 217–226. doi: 10.1023/A:1015686601404

Louarn, J., Carbonne, F., Delavault, P., Bécard, G., and Rochange, S. (2012). Reduced germination of *Orobanche cumana* seeds in the presence of arbuscular mycorrhizal fungi or their exudates. *PLoS ONE* 7:e49273. doi: 10.1371/journal.pone.0049273

Lu, Y. H., Melero-Vara, J. M., Garcia-Tejada, J. A., and Blanchard, P. (2000). Development of SCAR markers linked to the gene Or5 conferring resistance to broomrape (*Orobanche cumana* Wallr.) in sunflower. *Theor. Appl. Genet.* 100, 625–632. doi: 10.1007/s001220050083

Mauromicale, G., Monaco, A. L., Longo, A. M. G., and Restuccia, A. (2005). Soil solarization, a nonchemical method to control branched broomrape (*Orobanche ramosa*) and improve the yield of greenhouse tomato. *Weed Sci.* 53, 877–883. doi: 10.1614/WS-05-023R1.1

McHale, L., Tan, X., Koehl, P., and Michelmore, R. W. (2006). Plant NBS-LRR proteins: adaptable guards. *Genome Biol.* 7, 212. doi: 10.1186/gb-2006-7-4-212

Mohamed, A., Ellicott, A., Housley, T. L., and Ejeta, G. (2003). Hypersensitive response to *Striga* infection in *Sorghum*. *Crop Sci.* 43, 1320–1324. doi: 10.2135/cropsci2003.1320

Molinero-Ruiz, L., García-Carneros, A. B., Collado-Romero, M., Raranciuc, S., Domínguez, J., and Melero-Vara, J. M. (2013). Pathogenic and molecular diversity in highly virulent populations of the parasitic weed *Orobanche cumana* (sunflower broomrape) from Europe. *Weed Res.* 54, 87–96. doi: 10.1111/wre.12056

Morel, J.-B., and Dangl, J. L. (1997). The hypersensitive response and the induction of cell death in plants. *Cell Death Differ.* 4, 671–683. doi: 10.1038/sj.cdd.4400309

Palloix, A., Ayme, V., and Moury, B. (2009). Durability of plant major resistance genes to pathogens depends on the genetic background, experimental evidence and consequences for breeding strategies. *New Phytol.* 183, 190–199. doi: 10.1111/j.1469-8137.2009.02827.x

Pérez-Vich, B., Akhtouch, B., Knapp, S. J., Leon, A. J., Velasco, L., Fernandez-Martínez, J. M., et al. (2004). Quantitative trait loci for broomrape (*Orobanche cumana* Wallr.) resistance in sunflower. *Theor. Appl. Genet.* 109, 92–102. doi: 10.1007/s00122-004-1599-7

Pérez-Vich, B., Velasco, L., Rich, P. J., and Ejeta, G. (2013). "Marker-assisted and physiology-based breeding for resistance to root parasitic orobanchaceae," in *Parasitic Orobanchaceae*, eds D. M. Joel, J. Gressel, and L. J. Musselman (Heidelberg: Springer), 369–391.

Pineda-Martos, R., Pujadas-Salvà, A. J., Fernández-Martínez, J. M., Stoyanov, K., Velasco, L., and Pérez-Vich, B. (2014). The genetic structure of wild *Orobanche cumana* Wallr. (Orobanchaceae) populations in Eastern Bulgaria reflects introgressions from weedy populations. *Sci. World J.* 2014, 1–15. doi: 10.1155/2014/150432

Pineda-Martos, R., Velasco, L., Fernández-Escobar, J., Fernández-Martínez, J. M., and Pérez-Vich, B. (2013). Genetic diversity of *Orobanche cumana* populations from Spain assessed using SSR markers. *Weed Res.* 53, 279–289. doi: 10.1111/wre.12022

Pustovoit, V. (1966). *Selection, Seed Culture and Some Agrotechnical Problems of Sunflower* (transl. from the Russian in 1976). Delhi: Indian National Scientific Documentation Centre.

Radwan, O., Gandhi, S., Heesacker, A., Whitaker, B., Taylor, C., Plocik, A., et al. (2008). Genetic diversity and genomic distribution of homologs encoding NBS-LRR disease resistance proteins in sunflower. *Mol. Genet. Genomics* 280, 111–125. doi: 10.1007/s00438-008-0346-1

Raupp, F. M., and Spring, O. (2013). New sesquiterpene lactones from sunflower root exudate as germination stimulants for *Orobanche cumana*. *J. Agric. Food Chem.* 61, 10481–10487. doi: 10.1021/jf402392e

Sabetta, W., Alba, V., Blanco, A., and Montemurro, C. (2011). sunTILL: a TILLING resource for gene function analysis in sunflower. *Plant Methods* 7, 20. doi: 10.1186/1746-4811-7-20

Slovak, R., Goschl, C., Su, X., Shimotani, K., Shiina, T., and Busch, W. (2014). A scalable open-source pipeline for large-scale root phenotyping of *Arabidopsis*. *Plant Cell* 26, 2390–2403. doi: 10.1105/tpc.114.124032

Sukno, S., Fernández-Martínez, J. M., and Melero-Vara, J. M. (2001). Temperature effects on the disease reactions of sunflower to infection by *Orobanche cumana*. *Plant Dis.* 85, 553–556. doi: 10.1094/PDIS.2001.85.5.553

Tan, S., Evans, R. R., Dahmer, M. L., Singh, B. K., and Shaner, D. L. (2005). Imidazolinone-tolerant crops: history, current status and future. *Pest Manag. Sci.* 61, 246–257. doi: 10.1002/ps.993

Tang, S., Heesacker, A., Kishore, V. K., Fernandez, A., Sadik, E. S., Cole, G., et al. (2003). Genetic mapping of the Or5 gene for resistance to *orobanche* race E in sunflower. *Crop Sci.* 43, 1021–1028. doi: 10.2135/cropsci2003.1021

Thomas, H., Sauerborn, J., Müller-Stöver, D., Ziegler, A., Bedi, J. S., and Kroschel, J. (1998). The potential of *Fusarium oxysporum* f.sp. orthoceras as a biological control agent for *Orobanche cumana* in sunflower. *Biol. Control* 13, 41–48. doi: 10.1006/bcon.1998.0642

Vear, F., Serre, F., Roche, P., Walser, P., and Tourvieille de Labrouhe, D. (2007). Recent research on downy mildew resistance useful for breeding industrialuse sunflowers. *Helia* 30, 45–54. doi: 10.2298/HEL0746045V

Velasco, L., Pérez-Vich, B., Jan, C. C., and Fernández-Martínez, J. M. (2007). Inheritance of resistance to broomrape (*Orobanche cumana* Wallr.) race F in a sunflower line derived from wild sunflower species. *Plant Breed.* 126, 67–71. doi: 10.1111/j.1439-0523.2006.01278.x

Velasco, L., Pérez-Vich, B., Yassein, A. A. M., Jan, C.-C., and Fernández-Martínez, J. M. (2012). Inheritance of resistance to sunflower broomrape (*Orobanche cumana* Wallr.) in an interspecific cross between *Helianthus annuus* and *Helianthus debilis* subsp. tardiflorus. *Plant Breed.* 131, 220–221. doi: 10.1111/j.1439-0523.2011.01915.x

Vranceanu, A. V., Tudor, V. A., Stoenescu, F. M., and Pirvu, N. (1980). "Virulence group of *Orobanche cumana* Wallr., differential hosts and resistance sources and genes in sunflower," in *Proceedings of the Ninth International Sunflower Conference*, Torremolinos, 71–82.

Yang, Z., Wafula, E. K., Honaas, L. A., Zhang, H., Das, M., Fernandez-Aparicio, M., et al. (2015). Comparative transcriptome analyses reveal core parasitism genes and suggest gene duplication and repurposing as sources of structural novelty. *Mol. Biol. Evol.* 32, 767–790. doi: 10.1093/molbev/msu343

Yoneyama, K., Xie, X., Kisugi, T., Nomura, T., Sekimoto, H., Yokota, T., et al. (2011). Characterization of strigolactones exuded by Asteraceae plants. *Plant Growth Regul.* 65, 495–504. doi: 10.1007/s10725-011-9620-z

Conflict of Interest Statement: The authors declare that the research was conducted in the absence of any commercial or financial relationships that could be construed as a potential conflict of interest.

20

Secondary Effects of Glyphosate Action in *Phelipanche aegyptiaca*: Inhibition of Solute Transport from the Host Plant to the Parasite

Edited by:
Richard S. Winder,
Natural Resources Canada, Canada

Reviewed by:
Stephen Oscar Duke,
United States Department of
Agriculture, USA
Nacer Bellaloui,
United States Department of
Agriculture-ARS, USA

***Correspondence:**
Tal Shilo
talshilo44@gmail.com

Tal Shilo[1,2], Baruch Rubin[2], Dina Plakhine[1], Shira Gal[3], Rachel Amir[4], Yael Hacham[4], Shmuel Wolf[2] and Hanan Eizenberg[1]*

[1] Department of Plant Pathology and Weed Research, Agricultural Research Organization, Newe Ya'ar Research Center, Ramat Yishay, Israel, [2] The Robert H. Smith Faculty of Agriculture, Food and Environment, The Robert H. Smith Institute of Plant Sciences and Genetics, The Hebrew University of Jerusalem, Rehovot, Israel, [3] Department of Entomology, Agricultural Research Organization, Newe Ya'ar Research Center, Ramat Yishay, Israel, [4] Migal Galilee Technology Center, Kiryat Shmona, Israel

It is currently held that glyphosate efficiently controls the obligate holoparasite *Phelipanche aegyptiaca* (Egyptian broomrape) by inhibiting its endogenous shikimate pathway, thereby causing a deficiency in aromatic amino acids (AAA). While there is no argument regarding the shikimate pathway being the primary site of the herbicide's action, the fact that the parasite receives a constant supply of nutrients, including proteins and amino acids, from the host does not fit with an AAA deficiency. This apparent contradiction implies that glyphosate mechanism of action in *P. aegyptiaca* is probably more complex and does not end with the inhibition of the AAA biosynthetic pathway alone. A possible explanation would lie in a limitation of the translocation of solutes from the host as a secondary effect. We examined the following hypotheses: (a) glyphosate does not affects *P. aegyptiaca* during its independent phase and (b) glyphosate has a secondary effect on the ability of *P. aegyptiaca* to attract nutrients, limiting the translocation to the parasite. By using a glyphosate-resistant host plant expressing the "phloem-mobile" green fluorescent protein (GFP), it was shown that glyphosate interacts specifically with *P. aegyptiaca*, initiating a deceleration of GFP translocation to the parasite within 24 h of treatment. Additionally, changes in the entire sugars profile (together with that of other metabolites) of *P. aegyptiaca* were induced by glyphosate. In addition, glyphosate did not impair germination or seedling development of *P. aegyptiaca* but begun to exert its action only after the parasite has established a connection to the host vascular system and became exposed to the herbicide. Our findings thus indicate that glyphosate does indeed have a secondary effect in *P. aegyptiaca*, probably as a consequence of its primary target inhibition—via inhibition of the translocation of phloem-mobile solutes to the parasite, as was simulated by the mobile GFP. The observed disruption in the metabolism of major sugars that are abundant in *P. aegyptiaca* within 48 h after glyphosate treatment provides a possible explanation for this inhibition of translocation and might reflect a critical secondary effect of the herbicide's primary action that results in loss of the parasite's superior sink for solutes.

Keywords: Egyptian broomrape, *Orobanche*, germination, tomato, phloem, translocation, sugars

INTRODUCTION

The obligate holoparasite Egyptian broomrape, *Phelipanche aegyptiaca* (Orobanchaceae, formerly known as *Orobanche aegyptiaca*), parasitizes dicotyledonous plants and causes severe yield loss in numerous vegetable crops (Eizenberg et al., 2012, 2013). The life cycle of *P. aegyptiaca*—like that of other members of the genus *Phelipanche*—can be roughly divided into two main phases (Joel, 2000): a short independent phase, starting from seed conditioning and continuing through germination to the establishment of a vascular connection to the host root, and a parasitic phase, which lasts throughout the rest of the parasite lifecycle. During the latter phase, the parasite depends entirely on the host plant for water, carbohydrates and nutrients (Heide-Jørgensen, 2013). The parasite attaches to the roots of its host by a haustorium, whose role is initially attachment to and penetration into the host root tissue (Joel and Losner-Goshen, 1994) and, later, the establishment of both xylem and phloem connections between the parasite and its host (Dörr and Kollmann, 1995; Joel, 2013). The haustorium has some functional similarities both to roots, in that it absorbs water and minerals, and to sink organs in that it attracts carbon molecules (Westwood, 2013). In the host-parasite system, organic carbon, nitrogen and some minerals are transported primarily via the phloem (Hibberd et al., 1999; Hibberd and Jeschke, 2001). Translocation to the parasite of small molecules, such as sugars, herbicides and possibly amino acids, also takes place, as has been previously described in some members of the genus *Orobanche* (Aber et al., 1983; Arjona-Berral et al., 1990; Nandula et al., 1999; Diaz-Sanchez et al., 2002). In addition, it is known that macromolecules are also transported in the host-parasite system. For example, viral DNA and RNA were transferred from infected tomato (*Solanum lycopersicum*) and tobacco (*Nicotiana tabacum*) plants to *P. aegyptiaca* (Gal-On et al., 2009), and movement of green fluorescent protein (GFP) was demonstrated between parasitized transgenic tomato plants, expressing GFP under the *At*SUC2 promoter (Imlau et al., 1999), and *P. aegyptiaca* (Aly et al., 2011).

The ability of plant sink tissues, such as fruits and roots, to import carbon is generally attributed (as was first posited by Ernst Münch, 1930) to the differential pressure caused by a gradient of sugars along the source-sink path and their consumption in the sink (Wang et al., 1993; Knoblauch et al., 2016). In *P. aegyptiaca*, it is, however, still not clear exactly what determines the parasite's sink strength and what allows it to overcome the sinks in the host plant. Some lines of evidence indicate that by accumulating high levels of mineral ions, sugars and sugar alcohols, holoparasites of the Orobanchaceae maintain a low water potential, which facilitates the uptake of water and nutrients, as has been observed in some algae and fungi (Harloff and Wegmann, 1993; Westwood, 2013; Hacham et al., 2016). The cleavage of the translocated sucrose to glucose and fructose by invertase (Aber et al., 1983; Fernández-Aparicio et al., 2016) is thought to contribute further to elevating the osmotic pressure and thereby to provide the driving force for translocation of assimilates to the parasite (Delavault et al., 2002; Abbes et al., 2009b). Another factor contributing to elevation of the osmotic pressure in *P. aegyptiaca* and some of its close relatives is believed to be high levels of the sugar alcohol mannitol (Aly et al., 2009). The entire mannitol biosynthetic pathway has been documented in *P. ramosa* and *O. crenata* (Harloff and Wegmann, 1993), as has the activity of the key enzyme mannose-6-phosphate reductase (M6PR; EC 1.1.1.224) in *P. ramosa* (Delavault et al., 2002). It seems that this pathway is of importance to parasite survival, especially since silencing of the parasite's *M6PR* gene, by using a siRNA construct, resulted in a reduction of mannitol levels concomitant with increased mortality of *P. aegyptiaca* (Aly et al., 2009).

Despite the considerable amount of work that has been devoted to *P. aegyptiaca*, this parasitic plant still poses a challenge in weed management, particularly because only a few herbicides are available for its control (Eizenberg et al., 2013). Among these, the non-selective herbicide glyphosate can efficiently control *P. aegyptiaca* (Joel et al., 1995), even in low doses (Jacobsohn and Kelman, 1980; Cochavi et al., 2015). Glyphosate inhibits the enzyme 5-enolpyruvyl shikimate-3-phosphate synthase (EPSPS; EC 2.5.1.19) (Steinrücken and Amrhein, 1980), a key enzyme in the shikimate pathway and in the aromatic amino acids (AAA) biosynthesis pathway. After absorption into the plant foliage, glyphosate is translocated mainly into the symplast and accumulates mostly in sink tissues (Gougler and Geiger, 1981; Schulz et al., 1990). Tubercles of *P. aegyptiaca* and of other close relatives in the genus have been shown to accumulate high amounts of [^{14}C]glyphosate within 24 h after application of the herbicide to host plants (Arjona-Berral et al., 1990; Jain and Foy, 1997; Diaz-Sanchez et al., 2002). Furthermore, accumulation of shikimate has been detected in *P. aegyptiaca* as early as 10 h after treatment (HAT) with glyphosate (Shilo et al., 2016), indicating the presence of the herbicide in the parasite tissue. In addition to its movement in the plant, some evidence indicates that glyphosate is exuded from the roots of treated plants and also acts in the rhizosphere (Coupland and Caseley, 1979; Kremer et al., 2005; Laitinen et al., 2007). While it is clear that glyphosate reaches and acts in *P. aegyptiaca* tubercles, the fate of the parasite seeds and seedlings upon exposure to glyphosate in the host rhizosphere, before attachment occurs, remains unknown. While there is no argument that glyphosate inhibits specifically its target enzyme—EPSPS, the mode of action of glyphosate is thought to be more complex than the inhibition of the AAA biosynthetic pathway alone. The importance of the shikimate pathway in the metabolism of primary and secondary compounds implies that its inhibition might influence a variety of processes in the plant (Tzin and Galili, 2010). Each one of these processes, once inhibited, could potentially lead to death of the plant. On the one hand, in some autotrophic plants symptoms resulting from the application of glyphosate tend to develop slowly, in keeping with a deficiency of AAA and disruption of protein biosynthesis (Gresshoff, 1979; Duke and Powles, 2008). On the other hand, several secondary effects of glyphosate action have been described, including: (1) a decrease in the rate of photosynthesis and in concentrations of photosynthates in source tissues (Servaites et al., 1987; Geiger and Bestman, 1990; Fuchs et al., 2002); (2) inhibition of the biosynthesis of phenolic compounds, resulting in impaired defense mechanisms (Lévesque and Rahe, 1992); (3) interference with auxin levels, leading—in glyphosate resistant-cotton—to re-organization of the anther tissue and resultant male sterility (Yasuor et al., 2006); (4) injury to glyphosate-resistant-soybean possibly as a result of phytotoxicity of aminomethylphosphonic acid (AMPA), a metabolite of glyphosate (Reddy et al., 2004); (5) leaching of glyphosate along with other metabolites into the soil, resulting in improved pathogen success (Kremer et al., 2005); (6) limitation of water availability to the shoots (Fuchs et al., 2002); and (7) disruption of carbon metabolism, leading to limitation of transport within the plant, either of glyphosate itself (Geiger et al., 1999) or of carbon compounds (Geiger and Bestman, 1990; Geiger et al., 1999; Orcaray et al., 2010).

The inhibition of EPSPS causes a loss of feedback inhibition of 3-deoxy-D-arabinoheptulosonate 7-phosphate synthase (DAHPS; EC 2.5.1.54), an enzyme of the shikimate pathway, which leads to an extremely high accumulation of shikimate (Duke, 1988). Since the shikimate pathway metabolizes about 30% of the assimilated carbon in photosynthetic plants (Maeda and Dudareva, 2012), it is reasonable to assume that related metabolic pathways may be affected within hours after the inhibition of EPSPS, due to the impairment of carbon metabolism (Servaites et al., 1987; Siehl, 1997; Orcaray et al., 2012). These changes, in turn, might lead to changes in source-sink relations and limit the translocation of solutes, as seen in autotrophic plants and also in a study of *Cuscuta campestris* parasitism, in which the translocation of [^{14}C]sucrose and GFP in the phloem were limited by glyphosate application to the host plants (Nadler-Hassar et al., 2004).

In *P. aegyptiaca*, levels of AAAs were significantly decreased within 48 h after application of glyphosate to a tomato host, which is indicative both of the presence of an active shikimate pathway in the parasite and implies of the dependence of the parasite on self-production of AAAs (Shilo et al., 2016). This finding of a metabolite deficiency in *P. aegyptiaca* is something of an enigma in light of its ability to strongly attract nutrients, including proteins (Aly et al., 2011) and free amino acids (Aber et al., 1983; Gaudin et al., 2014), from the host. We thus posit that glyphosate acts in *P. aegyptiaca* not only by creating a deficiency in AAA, but also through disruption of pathways that are associated with the shikimate pathway, causing the inhibition of transport of solutes from the host to the parasite.

The objective of the current study was thus to examine aspects of *P. aegyptiaca* control by glyphosate that are yet to be resolved, as follows: (a) during the independent life cycle phase, to examine the effect of the herbicide on seed germination, on the viability of the seedlings and the ability of the seedlings to establish the vital connection to the hosts' vascular system, and (b) during the parasitism phase, to examine the possibility of a secondary effect of glyphosate's primary mechanism acting through the inhibition of translocation of solutes from the host to the parasite.

MATERIALS AND METHODS

Plant Material

Tomato plants (*Solanum lycopersicum* Mill., var. UC-82, kindly provided by "Hishtil," Israel) were used as the glyphosate-sensitive host (GST). The tomato germplasm 1232 (F_4 self-cross, Zygier, 2005) was used as a glyphosate-resistant host (GRT). A transgenic tomato expressing GFP in its companion cells under the *At*SUC2 promoter (Imlau et al., 1999) was used as a GFP host sensitive to glyphosate (GGST). In these plants, GFP is secreted into the phloem system and is transported throughout the plant along the source-to-sink pathway (Imlau et al., 1999; Nadler-Hassar et al., 2004; Birschwilks et al., 2006; Aly et al., 2011). A GFP host resistant to glyphosate (GGRT) was prepared by crossing GGST plants with GRT plants. The F_1 hybrids (GGRT) were resistant to glyphosate and expressed GFP in the phloem.

Phelipanche aegyptiaca Pers. seeds were collected in a commercial processing tomato field in the Hula valley in Northern Israel and were surface sterilized by soaking in 70% ethanol for 3 min and then in 1% sodium hypochlorite for 10 min, followed by washing four times in sterile water, before being dried and stored at 4°C in the dark.

Germination of *P. aegyptiaca*

Seeds of *P. aegyptiaca* were spread in petri dishes on GF/A filter paper (Whatman, GE healthcare, UK) discs soaked in sterile distilled water and left for pre-conditioning for 7 days in an incubator (21.5°C constant). Thereafter, a solution of the strigolactone-analog GR24 (30 µl; 10^{-5} M) was added to the filter paper discs. Glyphosate (30 µl) was applied to the petri dishes as described below, either at the beginning of the pre-conditioning period (for the herbicide to be taken up during imbibition) or 2 days after addition of GR24 (after the germination process had commenced). Germination was documented 7 days after addition of GR24.

Experimental System of Host-Parasite Association

Hosts and parasites were grown hydroponically in polyethylene bags (PEBs) in a growth chamber [20°C/25°C (night/day); 14 h light, 150–200 µE m^{-2} s^{-1}] according to Parker and Dixon (1983), as modified by Eizenberg et al. (2003). Briefly, 4-week-old tomato seedlings were thoroughly washed free of soil, re-rooted and transferred to the center of a PEB containing half-strength Hoagland solution (Hoagland and Arnon, 1950) on GF/A filter paper. PEBs were kept inside a dark box to exclude light from the root zone. For sugars analysis, GRT were used as host plants. With the planting of GRT plants in the PEB, sterile *P. aegyptiaca* seeds were spread on the GF/A paper. GR24 solution (4 ml; 5 ppm) was injected into the PEB 10 days after planting.

For GFP accumulation analysis, GGST and GGRT were used as host plants. Sterile *P. aegyptiaca* seeds were pre-conditioned in a petri dish as described above. Twenty four hours following the addition of GR24 (125 µl; 10^{-5} M), the seeds were transferred to the PEB in which they were placed close to randomly selected host roots at a distance not greater than 0.5 mm (**Figure 1A**). Germination of *P. aegyptiaca* occurred uniformly 3 days after the application of GR24, when the seeds were already in the PEBs; that day was recorded as the day of germination according to which all treatments were applied.

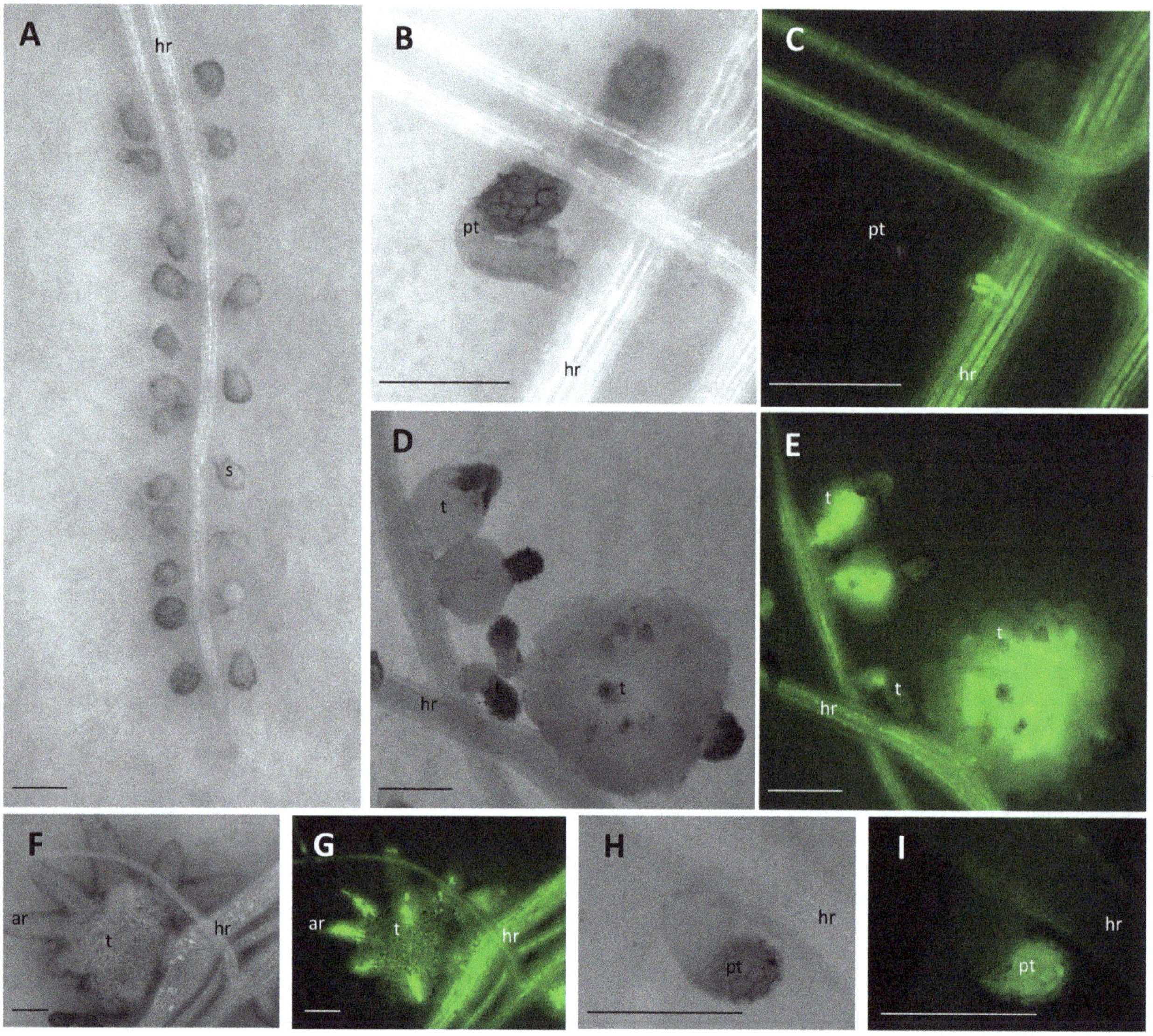

FIGURE 1 | Parasitism of *P. aegyptiaca* on the "GFP-mobile" host. (A) Seeds of *P. aegyptiaca* were placed up to 0.5 mm from the root. **(B,C)** The parasite endophyte displaying fluorescence after forming a primary phloem connection. **(D,E)** Tubercles of *P. aegyptiaca* accumulating GFP. **(F,G)** "Spider" tubercle translocating GFP to its adventitious roots through the parasite phloem. **(H,I)** Attachment of *P. aegyptiaca* seed to a wild-type host. Notice the auto-fluorescence of the seed coat and the lack of fluorescence in the root of the host or in the tubercle. Images **(C,E,G,I)** were captured under GFP3 filter. hr, host root; s, seedling; pt, primary tubercle; t, tubercle; ar, adventitious root. Bar length represents 0.5 mm.

Glyphosate Application

For the germination analysis, glyphosate (Roundup®, 360 g ae l^{-1}, Monsanto, USA) was applied in one of three doses: 18, 180, or 1,800 mg ae l^{-1}. The highest dose was chosen to simulate an equivalent dose to 360 g ae ha^{-1} (in a spraying volume of 2,001). For the GFP accumulation analysis, glyphosate was applied to the tomato plants grown in PEB either on the day of *P. aegyptiaca* germination or 7 days after germination (DAG) at a dose of 18 μg per plant on three leaves in 4 drops of 5 μl each. For the sugars analysis, glyphosate was similarly applied 5 weeks after planting in PEB.

GFP Accumulation Analysis

GFP accumulation was monitored *in situ* in *P. aegyptiaca* parasitizing GGST and GGRT hosts from germination onwards for a period of 14 DAG or in a higher resolution experiment at smaller intervals for a period of 48 h. As soon as *P. aegyptiaca* penetrated the vascular system of a "GFP-mobile" host and connected to its phloem elements, intensified fluorescence was visible. In very early stages of the parasitism, the fluorescence was observed in the endophyte itself (**Figures 1B,C**), and later, in the forming tubercle (**Figures 1D,E**), including in the parasitic phloem system, once formed (**Figures 1F,G**). Although, auto-fluorescence of the seed coat of *P. aegyptiaca* was also detected

(**Figures 1H,I**), it did not interfere with the observation and measurement of fluorescence in the tubercle itself. Images were captured using a fluorescent stereo microscope (Leica M205FA, Leica Microsystems GmbH, Germany), under a GFP3 filter (excitation: 450–490 nm, Leica Microsystems GmbH, Germany). The number of fluorescent tubercles was recorded as a measure of the progress of connection to the host phloem, and images were taken for subsequent measurement of the tubercles' area and fluorescence intensity by using ImageJ software (US National Institutes of Health, Bethesda, MD, USA, Schneider et al., 2012) according to Burgess et al. (2010) and Gavet and Pines (2010). Fluorescence intensity was quantified with the aim to assess the extent of translocation of the phloem solutes to the parasite, according to the following equation:

$$FI = ID - (TA \times MFB) \quad (1)$$

where *FI*—the intensity of fluorescence, *ID*—the integrated density, *TA*—the area of the tubercle and *MFB*—the mean fluorescence of the background. Fluorescence intensity was also measured in untreated GST plants to eliminate the possibility of auto-fluorescence of the tissues.

Sugars Analysis

GRT hosts and parasites were grown in PEB and treated with glyphosate as described above. Tubercles and host roots (100 mg fresh weight) were sampled 48 HAT, frozen immediately in liquid nitrogen, and stored at −80°C until analysis.

Soluble carbohydrates were extracted from the above tissues and quantified according to Lisec et al. (2006) and Kazachkova et al. (2013) with minor modifications. Frozen samples were ground in 1 ml of extraction buffer containing HPLC-grade methanol:chloroform:distilled water (50:20:20 v:v:v). Norleucine (6.6 μl; 2 mg/ml) was added as an internal quantitative standard for the polar phase. Samples were then centrifuged (20,800 *g*) for 10 min at 4°C, and supernatants were transferred to new tubes. Distilled water (300 μl) and chloroform (300 μl) were added, and the vials were vortexed thoroughly and centrifuged (20,800 *g*) for 5 min at 4°C. Aliquots (300 μl) of the polar upper phase were transferred to new tubes and then dried in a vacuum concentrator (Centrivap concentrator, Labconco, Kansas city, MO, USA) overnight without heating and stored at −80°C until derivatization. For derivatization of the polar phase, frozen samples were dried again in vacuum concentrator for 30 min. To each tube, 40 μl of methoxyamine hydrochloride (20 mg/ml in pyridine) were added, and the mixture was shaken for 2 h at 37°C. Thereafter, 100 μl of MSTFA (N-methyl-N (trimethylsilyl)-trifluoroacetamide) and 7 μl of alkane standard mixture were added, and the mixture was shaken again for 30 min at 37°C. After a short spin, samples were transferred to GCMS auto-sampler vials.

Analysis was performed on a GCMS system (Agilent 7890A, Agilent Technologies, Inc., Santa Clara, CA, USA) coupled with a mass selective detector (Agilent 5975c, Agilent Technologies, Inc., Santa Clara, CA, USA) using a capillary VF-5ms column (Agilent Technologies, Inc., Santa Clara, CA, USA). Chromatograms and mass spectra were evaluated using MSD Chemstation (Agilent Technologies, Inc., Santa Clara, CA, USA). Sugars were identified and annotated in comparison with standards and with libraries of the National Institute of Standards and Technology (NIST) and the Max-Planck Institute for Plant Physiology Golm Metabolome Database (GMD). The peak area of each sugar metabolite was normalized according to the norleucine internal standard.

Statistical Analysis

Statistical analysis was carried out as described by Onofri et al. (2010). The experiments were arranged in a fully randomized design, with 3–6 replications as indicated in the figures. ANOVA and mean comparison were calculated by Student's *t*-test ($\alpha < 0.05$) using JMP (version 5.1; SAS Inc., NC, USA). Non-linear regressions were computed by SigmaPlot (version 11.01; SPSS Inc., Chicago, IL, USA) and were used to describe the number of attachments and the accumulation of GFP in *P. aegyptiaca* and the response of these factors to glyphosate treatment according to the following equations:

$$\text{Sigmoid, 3 parameters:} \quad y = \frac{a}{1 + e^{-\left(\frac{x - x_0}{b}\right)}} \quad (2)$$

where *x*—the time in DAG, *y*—the number of fluorescent tubercles or the intensity of fluorescence (as indicated in **Tables 1**, **2**), *a*—the upper asymptote, X_0—the inflection point, representing the time at 50% of the amplitude, and *b*—the slope at the inflection point.

$$\text{Gaussian, 3 parameters:} \quad y = ae^{\left[-0.5\left(\frac{x - x_0}{b}\right)^2\right]} \quad (3)$$

where *x*—the time in DAG, *y*—the number of fluorescent tubercles or the intensity of fluorescence (as indicated in **Tables 1**, **2**), *a*—the amplitude, X_0—the time at the peak point, and *b* controls the spread of the curve.

Partial least square discriminant analysis (PLS-DA) was computed for the peak area measurements of the sugars using Metaboanalyst 3.0 (http://www.metaboanalyst.ca/) (Xia and Wishart, 2011; Xia et al., 2015) with auto scaling (mean-centered and divided by the standard deviation of each variable).

RESULTS

The Independent Stage of *P. aegyptiaca*

Glyphosate Has No Effect on Germination and Seedling Development of *P. aegyptiaca*

The fate of *P. aegyptiaca* seeds and seedlings exposed to glyphosate in the soil solution is not known, but it is relevant to the understanding whether glyphosate also prevents success of this parasite during its independent stage of life. In a petri dish experiment, it was found that under low doses of glyphosate (18 and 180 mg ae l^{-1}) imbibed during the pre-conditioning period (**Figure 2A**) the portion of germinated glyphosate-treated seeds (for both doses) remained similar to that in the untreated group according to comparison of the

TABLE 1 | Regression coefficients and statistical analysis for Figure 4.

Figure	y Variable	Treatment	Regression		Parameters			Regression	
					Coefficient	SE	*P*	*RMSE*	*P*
Figure 4B	No. of fluorescent tubercles (% of max)	Untreated	Sigmoid (Equation 2)	a	98.78	1.42	<0.0001	3.62	<0.0001
				b	1.13	0.09	<0.0001		
				X_0	5.00	0.10	<0.0001		
		Glyphosate	Gaussian (Equation 3)	a	96.70	3.38	<0.0001	6.84	<0.0001
				b	2.12	0.08	<0.0001		
				X_0	7.64	0.10	<0.0001		
Figure 4C	Tubercle area (mm^2)	Untreated	Sigmoid (Equation 2)	a	2.01	0.32	0.0002	0.08	<0.0001
				b	2.32	0.41	0.0005		
				X_0	10.86	0.88	<0.0001		
		Glyphosate	Sigmoid (Equation 2)	a	0.07	0.004	<0.0001	0.009	0.0011
				b	1.68	0.63	0.0259		
				X_0	1.47	0.63	0.0465		
Figure 4D	Fluorescence intensity (arbitrary units)	Untreated	Sigmoid (Equation 2)	a	150.62	10.48	<0.0001	15.35	<0.0001
				b	1.19	0.25	0.0003		
				X_0	8.49	0.32	<0.0001		
		Glyphosate	Gaussian (Equation 3)	a	2.69	0.40	<0.0001	0.85	0.0151
				b	3.16	0.65	0.0003		
				X_0	7.02	0.55	<0.0001		

Glyphosate was applied pre-attachment on the day of germination (0 DAG).

TABLE 2 | Regression coefficients and statistical analysis for Figure 5.

Figure	y Variable	Treatment	Regression		Parameters			Regression	
					Coefficient	SE	*P*	*RMSE*	*P*
Figure 5B	No. of fluorescent tubercles (% of max)	Untreated	Sigmoid (Equation 2)	a	98.78	1.42	<0.0001	3.62	<0.0001
				b	1.13	0.09	<0.0001		
				X_0	5.00	0.10	<0.0001		
		Glyphosate	Gaussian (Equation 3)	a	103.61	4.54	<0.0001	9.55	<0.0001
				b	2.67	0.14	<0.0001		
				X_0	8.19	0.14	<0.0001		
Figure 5C	Tubercle area (mm^2)	Untreated	Sigmoid (Equation 2)	a	2.01	0.32	0.0002	0.08	<0.0001
				b	2.32	0.41	0.0005		
				X_0	10.86	0.88	<0.0001		
		Glyphosate	Sigmoid (Equation 2)	a	0.40	0.03	<0.0001	0.05	<0.0001
				b	1.63	0.56	0.0183		
				X_0	5.13	0.58	<0.0001		
Figure 5D	Fluorescence intensity (arbitrary units)	Untreated	Sigmoid (Equation 2)	a	150.62	10.48	<0.0001	15.35	<0.0001
				b	1.19	0.25	0.0003		
				X_0	8.49	0.32	<0.0001		
		Glyphosate	Gaussian (Equation 3)	a	39.40	2.90	<0.0001	6.85	<0.0001
				b	2.87	0.29	<0.0001		
				X_0	9.11	0.27	<0.0001		

Glyphosate was applied post-attachment, a week after germination (7 DAG).

means ($P < 0.0001$). When glyphosate was applied during the germination process (**Figure 2B**), the portion of germinated seeds was significantly higher for both doses, reaching 67% or 71% respectively, compared with 52% in the untreated group. For untreated seeds or seeds treated with low-dose glyphosate, similar portions of undeveloped seedlings were observed (**Figure 2**). These seedlings apparently started the process of germination but did not develop further, remaining spiky and sometimes necrotic. These results indicate that exposure of *P. aegyptiaca* during the independent growth stage to doses of glyphosate that may persist in the soil and are common in *P. aegyptiaca* management practice will basically not inhibit seed germination or the establishment of seedlings. In contrast, for the high dose of 1,800 mg ae l^{-1} glyphosate, the germination of *P. aegyptiaca* was almost completely inhibited, even though GR24 was added at the end of the pre-conditioning period. For both application times—at the beginning of the preconditioning period and 2 days after GR24 addition—the sum of ungerminated seeds plus nonviable seedlings totaled to more than 99 and 98%, respectively (**Figures 2A,B**).

Glyphosate Has No Effect on Attachment of *P. aegyptiaca* to the Host Root

To examine the effect of glyphosate on the ability of *P. aegyptiaca* to attach to the host root, an *in-situ* experimental system was constructed to allow tracking of the process of attachment to the host vascular system by using GFP as a marker for the start of the translocation of solutes.

The count of fluorescent tubercles showed that connection to the phloem started as early as 2 DAG, while most of the connections occurred between 4 and 8 DAG (**Figures 3**, **4B**). This pattern was similar for *P. aegyptiaca* parasitizing both GGST and GGRT hosts (*F*-test for regressions), and therefore the data for the two hosts were pooled (**Figure 4B**). When glyphosate was applied during the independent phase (pre-attachment), *P. aegyptiaca* seedlings were able to penetrate into the phloem similarly to the untreated group (**Figure 4A**), and first connections were similarly visible 2 DAG (**Figure 4B**). For both treated and untreated groups, most of the connections were established up to 7 DAG, but from that day onwards the number of fluorescent tubercles decreased in the treated groups (**Figure 4B**). A loss of GFP fluorescence and darkening occurred for each specific tubercle 3–4 days after the connection to the host (**Figure 4A** vs. **Figure 3**). These results indicate that *P. aegyptiaca* seedlings have the ability to attach normally, but once attached, their development is impaired by the herbicide. *P. aegyptiaca* responded similarly to glyphosate on both sensitive and resistant hosts (*F*-test for the regressions) and thus, here too, the data for the two hosts were pooled to create a single non-linear peak regression (**Figure 4B**). To normalize the number of attachments and established tubercles, the number of fluorescent tubercles was computed as percent of the maximal number (**Figure 4B**). However, it is important to note that *P. aegyptiaca* did successfully establish a phloem connection in the treated group to an extent similar to that for the untreated parasites (data not shown).

The Parasitic Stage of *P. aegyptiaca*

Glyphosate Hinders Normal Development of the Young Tubercle

Once connected to the vascular system of the host, *P. aegyptiaca* enters the parasitic life phase. In the early stages of the parasitism, the tubercles of *P. aegyptiaca* developed and increased in size (area) exponentially (**Figure 4C**). Additionally, proper development of the tubercle was indicated by its ability to attract solutes from the host phloem, as demonstrated by the level of fluorescence intensity, indicating accumulation of GFP (**Figures 3**, **4D**). Accumulation of GFP started in the untreated tubercles at 4 DAG, while massive accumulation took place 7–10 DAG. The pattern of GFP transport in the phloem followed a sigmoidal trend and did not differ significantly between the two hosts (*F*-test) and therefore the data for the two hosts were pooled (**Figure 4D**). The maximal values of the measured fluorescent intensity—represented in the upper asymptote (**Figure 4D**; parameter *a* in **Table 1**)—reflect the drawback inherent in the non-destructive method of monitoring fluorescence when the tubercle tissue becomes more differentiated and less transparent. As a result, the measurement time is limited to the first week or so after attachment. In later developmental stages, when the parasite's phloem is formed and endogenous translocation of GFP is visible, the parasite tissues are even less transparent (**Figure 1G**).

Glyphosate that was applied during the independent phase charged the host plant with the herbicide, and the glyphosate therefore reached the parasite immediately after the connection had been established. As opposed to the observations on the untreated tubercles, in treated plants, *P. aegyptiaca* was unable to develop beyond the initial tubercle structure that was prevalent in the first 2–3 days after attachment (**Figures 4A,C**). In parallel, there was no substantial accumulation of GFP (**Figure 4D**), reflecting immediate injury to the tubercle tissue and inability of the tubercle to develop further.

Glyphosate Inhibits GFP Translocation to the Parasite

During the parasitic phase, *P. aegyptiaca* relies on a continuous supply of nutrients from the host plant that is crucial for its growth and development. Therefore, it was important to clarify whether glyphosate has a secondary effect in inhibiting the translocation of solutes to the parasite and over what time frame. Despite the continuous translocation of GFP in GGRT roots—visible by its fluorescence (**Figure 5A**)—a darkening of the tubercles and cessation of GFP transport were observed for all the tubercles 3–4 days after "post-attachment" application of glyphosate (**Figure 5A** vs. **Figure 3**). This phenomenon was accompanied by the inability of the tubercles to develop and increase in area, as they did in the untreated control (**Figure 5C**). During the first 2 days after treatment, the fluorescence in the tubercles continued to increase, indicating a continuous accumulation of GFP, and maximal fluorescence was observed 9 DAG (**Figure 5D**). Nevertheless, this increase took place at lower rates compared with the accumulation rate in the untreated control, and the difference could be observed as early as 8 DAG, 24 HAT with glyphosate. This time frame was confirmed in another experiment with more frequent observations revealing

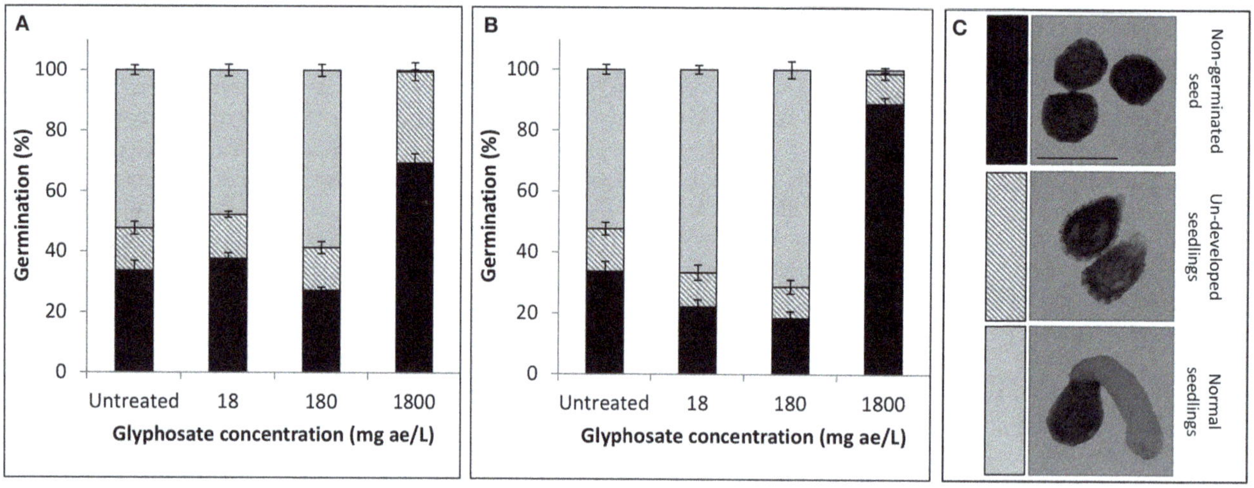

FIGURE 2 | Germination rate (%) of *P. aegyptiaca* seeds after treatment with glyphosate (A) at the beginning of the pre-conditioning period or **(B)** 2 days after addition of GR24. Key—**(C)** Classification of seeds and seedlings corresponding to bar colors, as follows: non-germinated seeds (black); undeveloped seedlings (stripes); normal seedlings (gray). Error bars represent standard error ($n = 4$). Bar length represents 0.5 mm.

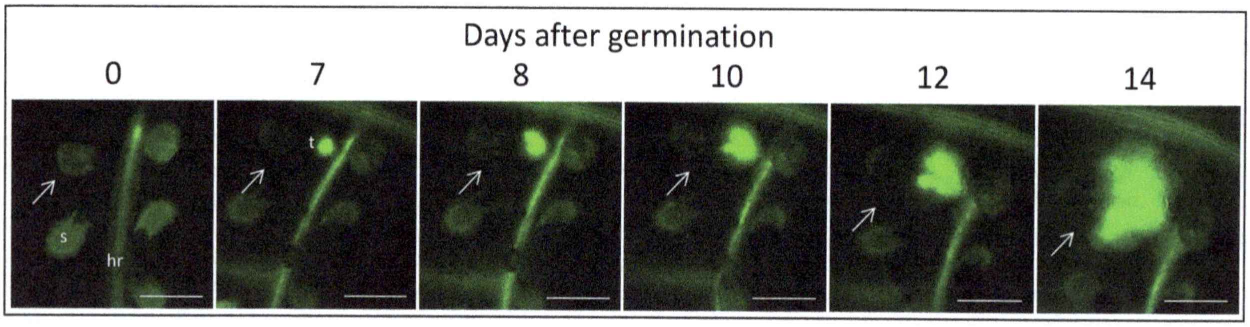

FIGURE 3 | Time course of attachment to untreated GGRT plants, captured under a GFP3 filter. hr, host root; s, seedling; t, tubercle; Arrows indicate attachments. Bar length represents 0.5 mm.

that the levels of fluorescence intensity were stabilized as early as 24 HAT (data not shown). The results from both experiments indicate a deceleration or a complete inhibition of GFP movement to the parasite tubercles within a day after glyphosate application. In addition, it is important to note that the inhibition described above was similar for GGST and GGRT hosts, thereby demonstrating the specific influence of glyphosate solely on the parasite tissues.

Glyphosate Alters the Sugars Profile of *P. aegyptiaca*

It is generally held that the reliance of *P. aegyptiaca* on the metabolism of certain carbohydrates allows it to attract more solutes from the vascular system of the host. In this part of the study, we sought an explanation for the inhibition of solutes translocation by glyphosate by examining changes in the sugars profile of *P. aegyptiaca*. To provide a graphic summary of the metabolomics data clustering patterns resulting from glyphosate treatment, a PLS-DA was computed. Profiles of sugars in *P. aegyptiaca* and GRT roots are presented in the score plots given in **Figure 6**. The analysis revealed a distinction between glyphosate-treated and untreated plants that clearly reflects the effect of glyphosate on the profile of the sugars metabolites in the parasite (**Figure 6A**). Although, there is no complete separation between the two groups, clustering could be observed. It appears that the sugars profile was less profoundly affected by the treatment in comparison with the total metabolites (data not shown), that included also the amino acids and other acids that participate in amino acids biosynthesis (Shilo et al., 2016). Naturally, the metabolites that are directly involved in the shikimate pathway and in amino acids biosynthetic pathways were separated into two distinct clusters (data not shown). These results strengthen our hypothesis that the herbicidal action of glyphosate includes alterations to the sugars profile in the parasite.

In the host roots, glyphosate treatment did not substantially change the metabolic profiles of sugars (**Figure 6B**). Parasitism, however, significantly altered this profile, as is apparent from their separate clustering (light blue, **Figure 6B**). It seems that the treatment of the parasitized host with glyphosate somewhat "pushed" the metabolic profile in the direction of the

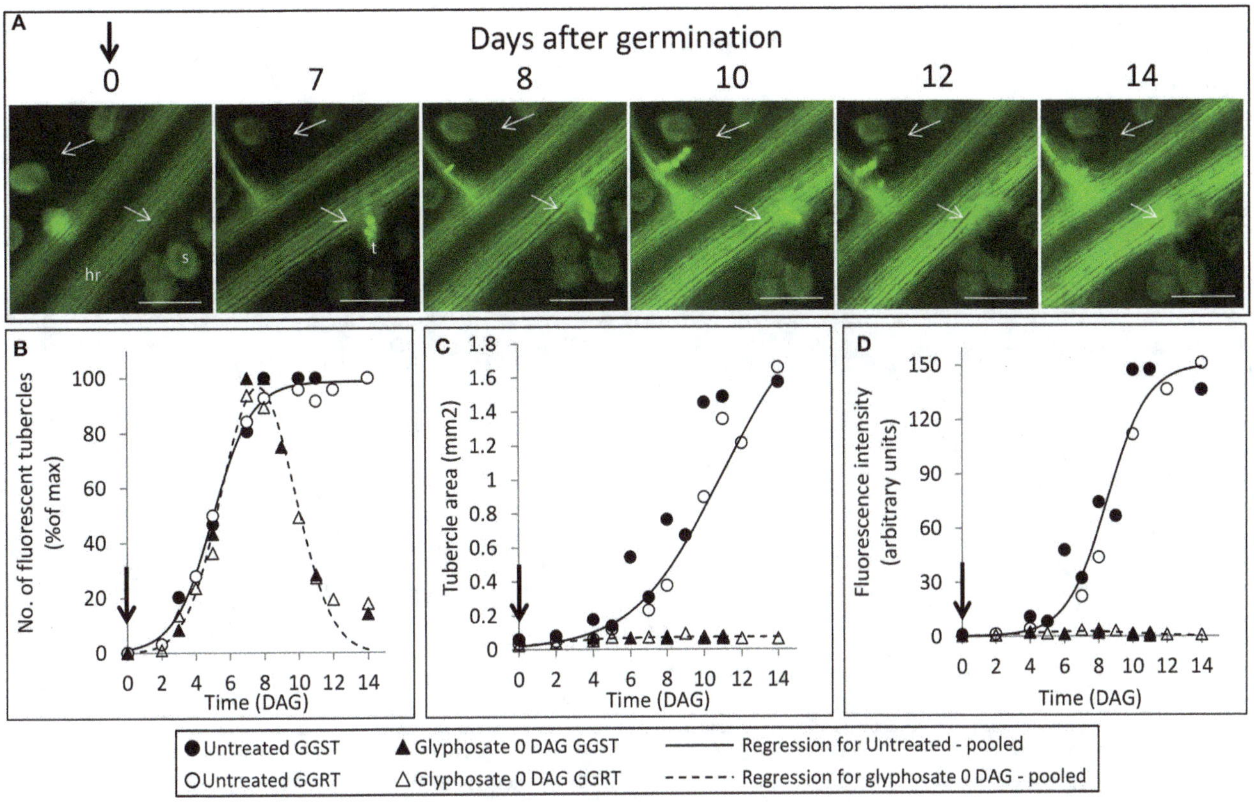

FIGURE 4 | Attachment and development of *P. aegyptiaca* following treatment with glyphosate at 0 DAG (pre-attachment). (A) Time course of attachment to GGRT plants following treatment with glyphosate, captured under a GFP3 filter. hr, host root; s, seedling; t, tubercle; Arrows indicate attachments. Bar length represents 0.5 mm. **(B)** Number of active attachments (represented by the number of fluorescent tubercles), **(C)** area of the tubercle, and **(D)** accumulation of GFP in the tubercles (represented by the intensity of fluorescence) on GGST hosts (filled symbols) and on GGRT hosts (open symbols) following treatment with glyphosate at 0 DAG (triangles) or without treatment (circles). For non-linear regressions, data from GGST and GGRT hosts was subjected to the *F*-test and combined. DAG—days after germination. Regression parameters are presented in **Table 1**. ($n = 6$). Vertical arrows indicate the time of glyphosate application (0 DAG).

non-parasitized clusters, probably as a result of elimination of the parasites.

The 15 soluble carbohydrates that were most abundant in *P. aegyptiaca* tubercles and in GRT roots are presented in **Figures 7, 8**. Mannitol and sucrose were found to be the most abundant sugars in the tubercles of *P. aegyptiaca* (**Figures 7A,B**). As implied in the score plot, treatment with glyphosate created a shift in most of the sugars that were detected in the parasite tubercles 48 HAT. Levels of the hexoses, fructose and glucose, were 3- and 4-fold lower than those detected in untreated control by 48 HAT, respectively (**Figures 7E,G**). In contrast, levels of other sugars, such as mannitol, glucose-6-P and fructose-6-P, exhibited an increase of approximately 2.5-fold at 48 HAT as compared with the untreated tubercles (**Figures 7A,J,L**). Sucrose levels did not differ significantly from the untreated control (**Figure 7B**).

In contrast to the findings in the tubercles, the levels of the sugars that were detected in the roots of the GRT host plants were not significantly altered by glyphosate treatment (**Figure 8**), confirming the strong resistance of the GRT host to glyphosate and indicating a specific influence of the herbicide on carbon metabolism in the parasite. Nonetheless, almost all sugars in the host root exhibited substantial increases in response to *P. aegyptiaca* parasitism. The most abundant sugars in the roots of parasitized GRT plants were sucrose, inositol-*myo*, fructose, mannitol, and galactinol (**Figures 8A–E**). Mannitol levels were almost undetectable in non-parasitized roots but soared to a level 20-fold higher than that in parasitized roots (**Figure 8D**). Sucrose levels increased by about 5-fold in response to the parasitism (**Figure 8A**), while inositol-*myo* and fructose increased by almost 7-fold when compared with non-parasitized roots (**Figures 8B,C**). Some sugars in parasitized roots, i.e., trehalose and arabitol, fell after treatment with glyphosate, probably as a result of the elimination of *P. aegyptiaca* (**Figures 8F,O**). Mannose and xylose were also detected in the roots of GRT plants but at lower levels and therefore are not presented in **Figure 8**.

Some of the sugars that were detected in *P. aegyptiaca* were notably elevated in comparison with levels in the adjacent host roots, indicating their accumulation or biosynthesis in the parasite. Among these sugars, mannitol was present in very high levels, 12-fold higher than those in the parasitized host roots (**Figures 7A, 8D**). Fructose and glucose levels were also higher

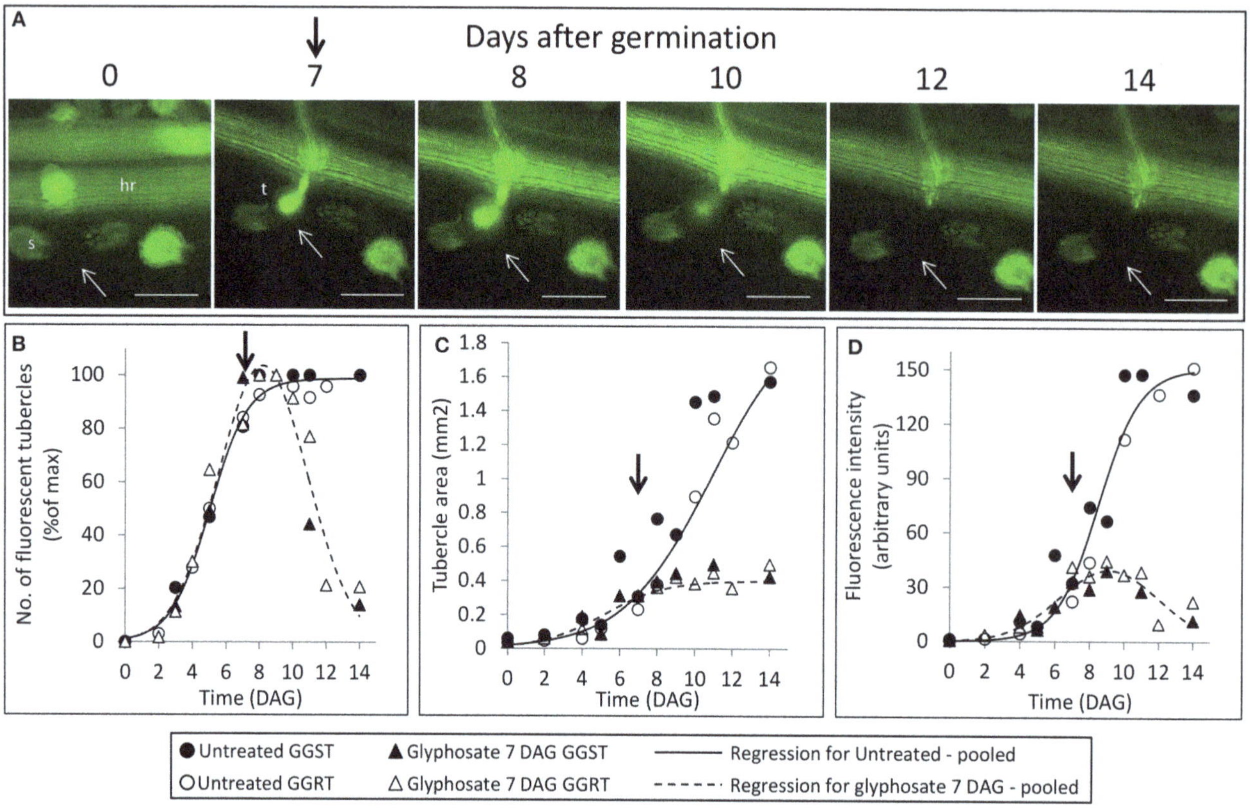

FIGURE 5 | Attachment and development of *P. aegyptiaca* following treatment with glyphosate at 7 DAG (post-attachment). (A) Time course of attachment to GGRT following treatment with glyphosate, captured under a GFP3 filter. hr, host root; s, seedling; t, tubercle; Arrows indicate attachments. Bar length represents 0.5 mm. **(B)** Number of active attachments (represented by the number fluorescent tubercles), **(C)** area of the tubercle, and **(D)** accumulation of GFP in the tubercles (represented by the intensity of fluorescence) on GGST hosts (filled symbols) and on GGRT hosts (open symbols) following treatment with glyphosate at 7 DAG (triangles) or without treatment (circles). For non-linear regressions, data from GGST and GGRT hosts was subjected to *F*-test and combined. DAG—days after germination. Regression parameters are presented in **Table 2**. ($n = 6$). Vertical arrows indicate the time of glyphosate application (7 DAG).

in the tubercles than in roots of the host, with an almost 2-fold difference (**Figures 7E,G, 8C,G**). Other sugars that were also found to be substantially elevated in comparison with levels in the host roots were galactinol (**Figures 7C, 8E**) and galactose (**Figures 7F, 8K**), being10- and 42-fold higher, respectively. It is important to note that sucrose was not among the accumulating sugars in *P. aegyptiaca* tubercles, and its levels were largely unchanged compared to those in the adjacent host roots.

DISCUSSION

In the present study we demonstrated the secondary effects of glyphosate in controlling the holoparasite *P. aegyptiaca* in both of its life cycle phases—independent and parasitic.

Effect of Glyphosate on *P. aegyptiaca* during the Independent Phase

Glyphosate is usually applied to the plant foliage, from where it moves to the roots and sink tissues, mainly via the phloem (Gougler and Geiger, 1981; Schulz et al., 1990). In several species, it has been shown that glyphosate is exuded from the roots to the rhizosphere (Gomes et al., 2014). For example, glyphosate exudation was recorded from the roots of both soybean (*Glycine max*) and glyphosate-resistant soybean as early as 2 days after treatment (Kremer et al., 2005). In quinoa (*Chenopodium quinoa*), a residue of 4% of the applied glyphosate was detected in the adjacent soil 8 days after treatment (Laitinen et al., 2007). A similar phenomenon has been described in herbicides of the imidazolinones, whose exudation to the rhizosphere has been exploited as a weed control strategy against parasitic weeds (Kanampiu et al., 2002; Colquhoun et al., 2006; Eizenberg et al., 2013).

When simulating the effect of glyphosate on the parasite during the independent phase, we found that glyphosate did not inhibit the germination of *P. aegyptiaca* seeds or impair the potency of the seedlings (**Figure 2**), with the exception of the highest examined concentration of glyphosate (1,800 mg ae l^{-1}) for which almost absolute inhibition of germination and seedling viability were observed (**Figure 2**). This dose is, however, unlikely to be found in the soil solution under *P. aegyptiaca* management practices (Cochavi et al., 2015, 2016). In addition, even if higher doses of glyphosate are applied, the seed is likely to encounter only an extremely small fraction of the initial herbicide dose in the soil because of the herbicide interaction

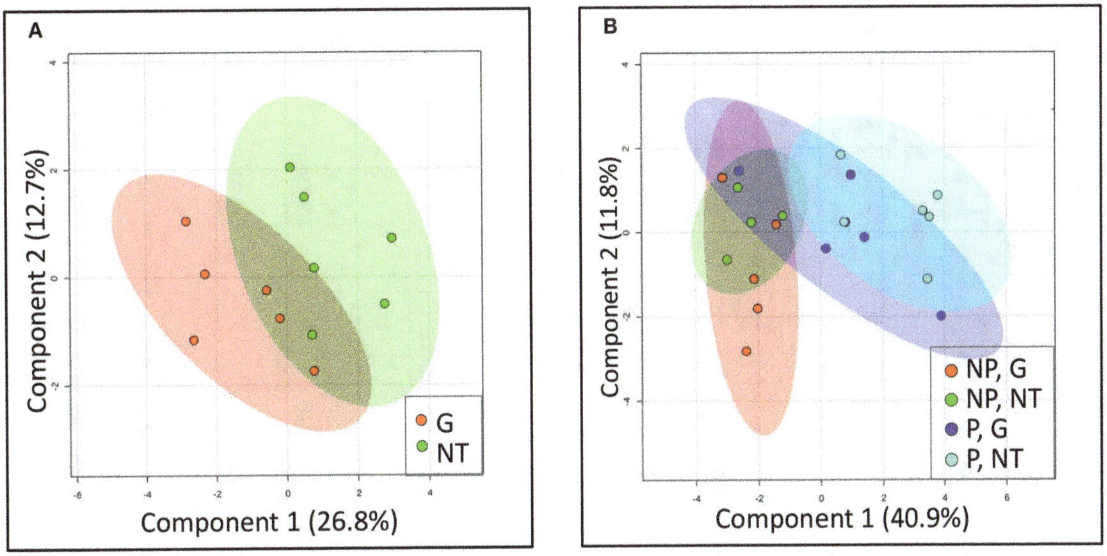

FIGURE 6 | Clustering of sugars profile using a partial least square–discriminant analysis (PLS-DA) model. Models were constructed for **(A)** *P. aegyptiaca* under treatment with glyphosate (G) or without treatment (NT) and for **(B)** GRT roots comparing parasitized (P) or non-parasitized (NP) roots under treatment with glyphosate (G) or without treatment (NT). Each plotted point is a projection of the samples data on a two-dimensional plot. Colored area represents 95% confidence region. Co-variance explained by the two primary components is indicated in parentheses.

with and adsorption to soil particles (Borggaard and Gimsing, 2008). The phenomenon of inhibition of germination by high concentrations of different substances has also been described by Whitney and Carsten (1981), who observed negative effects on germination and radicle elongation under high concentrations of germination stimulants. In those cases, the radicle may not even emerge from within the seed coat, making it appear as if the seeds have not germinated at all (Whitney and Carsten, 1981; Joel and Bar, 2013).

Lower doses of glyphosate actually had a positive effect on germination when the herbicide was applied after the germination process had begun (**Figure 2B**). The phenomenon of hormesis to herbicides is well documented and was reported in different plants in relation to glyphosate (Belz and Duke, 2014). In some *Orobanche* and *Phelipanche* species, similar findings have been reported for some growth regulators, such as herbicides of the group of phytoene synthase inhibitors, which had an additive effect in enhancing germination of the parasite when applied together with a germination stimulant (Chae et al., 2004; Song et al., 2005, 2006).

Seedlings of *P. aegyptiaca* were able to attach to both GGST and GGRT hosts roots even as long as 7 days after treatment with glyphosate (**Figures 4A,B**). This result further supports our hypothesis that glyphosate does not harm *P. aegyptiaca* seeds or seedlings during the independent phase of growth and that its herbicide action begins to take effect only after the parasite has established a connection to the vascular system of the host and begins to draw in glyphosate.

The establishment of a phloem connection by *P. aegyptiaca* to its host is a crucial step in the parasite survival, representing the transition from an independent organism incapable of long-term existence to a parasitic lifestyle (Joel, 2000). The progress of this developmental phase can be described by a non-linear regression (**Figure 4B**), indicating the important time points of the first established attachments and the range during which most connections take place in a population of *P. aegyptiaca* seeds.

In the GGST and GGRT plants, darkening (loss of fluorescence) of the glyphosate-treated attachments appeared 3–4 days after their connection to the host phloem (**Figures 4A,B**) as a result of translocation of the herbicide and the beginning of the mortality processes. Yet, it is reasonable to assume that injury to the parasite had begun to occur even earlier, possibly within hours after treatment, as the impairment of tubercle development was indeed visible at that time (**Figures 4A,C**). It has previously been shown, with fluorescein diacetate staining, that loss of viability in even larger tubercles of *P. aegyptiaca* was evident a few hours after treatment with glyphosate (Shilo et al., 2016). The early injury was specific to the haustorium region, which constituted most of the parasitic tissues in the young tubercle, as was the case in this study. In addition, the fluorescence in the tubercles might not exactly parallel the viability of the tissues because some time might be required for the GFP to be degraded or vacated to other tissues that will have become stronger sinks.

Effect of Glyphosate on *P. aegyptiaca* during the Parasitic Phase

As mentioned above, previous studies have shown that glyphosate translocates mainly in the phloem and readily reaches the sink tissues. For example, movement of [^{14}C]glyphosate to *O. crenata* parasitizing garden peas (*Pisum sativum*) or faba beans (*Vicia faba*) or to *O. cumana* parasitizing sunflower (*Helianthus annuus*) plants reached a maximum 3 days after application (Arjona-Berral et al., 1990; Jain and Foy, 1997; Diaz-Sanchez et al., 2002). However, measurements of shikimate

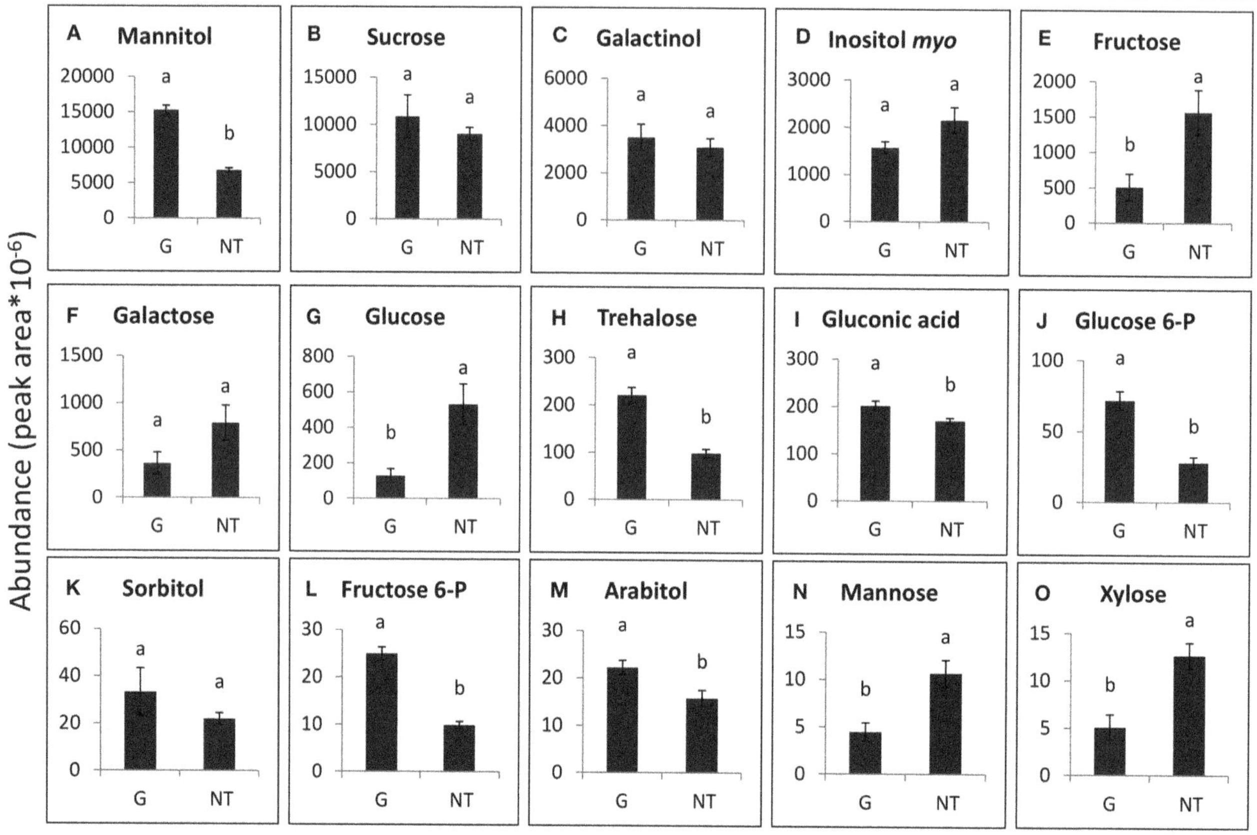

FIGURE 7 | Levels of the major sugars that were detected in *P. aegyptiaca* tubercles 48 h after treatment (HAT) with glyphosate (G) or untreated (NT). Graphs are arranged in descending abundance. Error bars represent standard error ($n = 3$). ANOVA was computed according to Student's t-test ($P < 0.05$). **(A)** Mannitol, **(B)** Sucrose, **(C)** Galactinol, **(D)** Inositol *myo*, **(E)** Fructose, **(F)** Galactose, **(G)** Glucose, **(H)** Trehalose, **(I)** Gluconic acid, **(J)** Glucose 6-P, **(K)** Sorbitol, **(L)** Fructose 6-P, **(M)** Arabitol, **(N)** Mannose, and **(O)** Xylose.

accumulation showed that the herbicide had already reached the tubercles a few hours after the application (Shilo et al., 2016). As we hypothesized, in our study of the tomato–*P. aegyptiaca* association, glyphosate had a rapid effect on translocation, limiting the parasite's ability to attract solutes via the phloem. According to the stabilization in fluorescence levels of GFP in the tubercles, it was evident that at least a deceleration, if not complete inhibition, of movement of GFP occurred within 24 HAT (**Figure 5D**). Due to this lack of synchronization between GFP fluorescence and the actual cessation of translocation of the protein, it is difficult to pinpoint exactly when the full inhibition of translocation occurred. Nonetheless, despite the macromolecular size of GFP, the movement of this protein in "GFP-mobile" plants is considered to provide a good simulation of solute translocation via the phloem and partitioning among different sink tissues (Imlau et al., 1999; Nadler-Hassar et al., 2004; Birschwilks et al., 2006). Thus, even a deceleration in GFP movement will provide an indication of general inhibition of solute translocation from the phloem. For example, translocation of GFP and [^{14}C]sucrose to the shoot parasite *Cuscuta campestris* parasitizing tobacco plants was inhibited within 3 days of glyphosate application (Nadler-Hassar et al., 2004).

GGRT plants, which have the parental background of GRT plants, are not affected—symptomatically or biochemically—by the currently used dosages of glyphosate, as is apparent from their stable profiles of amino acids and other organic acids, including shikimate (Shilo et al., 2016). In addition, GGRT plants continued to distribute GFP throughout their vascular system, with their roots remaining fluorescent a week and even 2 weeks after glyphosate application (**Figures 4A, 5A**, respectively). These findings mean that the GGRT host continues to distribute solutes despite the treatment with glyphosate, as opposed to the GGST host that sustain an injury of its own. Nonetheless, the translocation pattern of GFP and its inhibition after treatment with glyphosate were similar in both glyphosate-sensitive and glyphosate-resistant hosts parasitized by *P. aegyptiaca*. The importance of this finding lies in its contribution to differentiating between the effects of glyphosate on the parasite and those on the host plants, indicating that it is the parasite that is independently responsible for the extent of translocation in the host-parasite system.

In addition, it was found that glyphosate created a shift in the sugars profile of *P. aegyptiaca*, measured at 48 HAT (**Figures 6A, 7**) but had little effect on the sugars profile in the

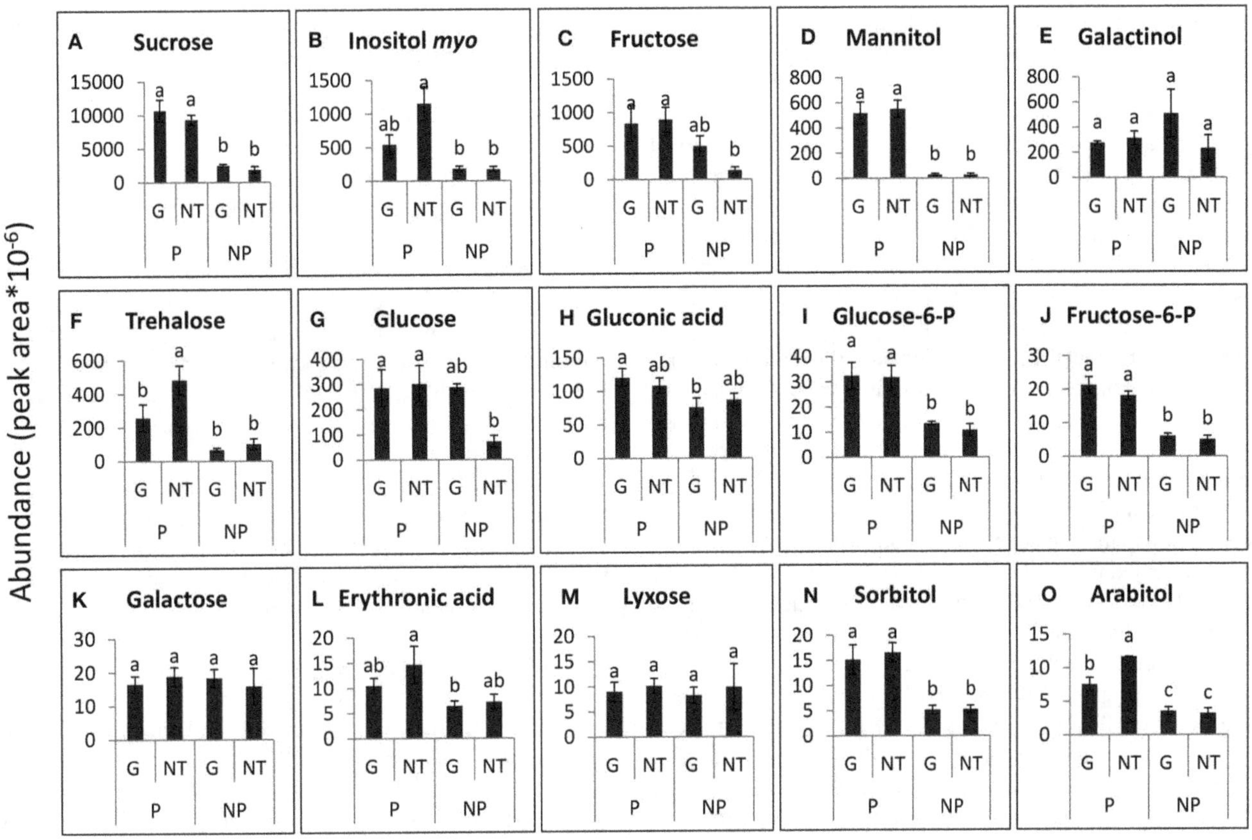

FIGURE 8 | Levels of the major sugars that were detected in the roots of GRT parasitized by *P. aegyptiaca* (P) or not parasitized (NP), 48 h after treatment (HAT) with glyphosate (G) or untreated (NT). Graphs are arranged in descending abundance. Error bars represent standard error ($n = 3$). ANOVA was computed according to Student's *t*-test ($P < 0.05$). **(A)** Sucrose, **(B)** Inositol *myo*, **(C)** Fructose, **(D)** Mannitol, **(E)** Galactinol, **(F)** Trehalose, **(G)** Glucose, **(H)** Gluconic acid, **(I)** Glucose-6-P, **(J)** Fructose-6-P, **(K)** Galactose, **(L)** Erythronic acid, **(M)** Lyxose, **(N)** Sorbitol, and **(O)** Arabitol.

host roots, with or without parasite (**Figures 6B, 8**). This finding implies that the changes in the sugars profile are not a result of inhibition of the translocation of sugars but are more likely to be the result of disruption of metabolism in the parasite tissues. Apparently, metabolites participating in the pathways closest to the shikimate pathway will be affected before the sugars. About 30% of the assimilated carbon in plants enter the shikimate pathway (Maeda and Dudareva, 2012). If EPSPS is inhibited, a large portion of this carbon accumulates in the form of shikimate, preventing the metabolism of carbon assimilates by other important pathways. It was previously found that the accumulation of shikimate in the parasite in the tomato–*P. aegyptiaca* system started at 10 HAT (Shilo et al., 2016), but it probably takes time for the parasite to accumulate the substantial levels of shikimate that will force alterations to other important carbon metabolic pathways. In *P. aegyptiaca*, some carbon metabolic pathways, e.g., that of mannitol, are unique to this family of root parasites and are thought to be of crucial importance to its ability to attract solutes and successfully compete with the sinks of the host (Westwood, 2013). Therefore, an alteration of the carbon metabolism in *P. aegyptiaca* might provide a possible explanation for the weakening of its sink strength, actually preventing it from maintaining the driving force necessary for the attraction of solutes. It has previously been reported that glyphosate also interrupts, within hours of application, carbon metabolism in other plant species and in other organisms. For instance, glyphosate disrupted C_3 cycle metabolism in source leaves of sugar beet plants a few hours after application, inducing limited translocation throughout the plant (Geiger et al., 1999). In velvetleaf (*Abutilon theophrasti*), glyphosate had a more destructive effect on sink tissues because of its rapid movement to the root system (Fuchs et al., 2002). In both leaves and roots of pea plants, changes in levels of sugars were observed in response to glyphosate application, as manifested by an accumulation of sucrose, hexose and starch 1 day after treatment (Orcaray et al., 2012). In addition, changes in levels of transcripts related to the glycolysis process were reported in soybean as early as four HAT with glyphosate (Zhu et al., 2008). Nonetheless, based on our results, we are not able to attribute the alterations in sugars profile to a specific event occurring as a result of the inhibition of EPSPS and the shikimate pathway. Further investigation is required to determine the exact pathways in *P. aegyptiaca* that are influenced following EPSPS inhibition.

Tubercles of *P. aegyptiaca* displayed elevated amounts of mannitol (**Figure 7A**) and alterations in all other sugars (namely, fructose, glucose, mannose, glucose-6-P and fructose-6-P) that participate in mannitol biosynthesis 48 HAT with glyphosate (**Figure 7**; Harloff and Wegmann, 1993). Mannitol is known to serve as an osmoregulant in algae and fungi, and, as mentioned above, its biosynthetic pathway is crucial to the *P. aegyptiaca* and its holoparasite relatives (Harloff and Wegmann, 1993; Aly et al., 2009). In addition, the levels of fructose and glucose—the products of sucrose cleavage that enable its continuous transport (Draie et al., 2011; Fernández-Aparicio et al., 2016)—were significantly decreased in the tubercles in response to glyphosate treatment (**Figure 7**). In contrast, the levels of sucrose in *P. aegyptiaca* tubercles were similar to its levels in the adjacent tomato roots (**Figures 7B, 8A**). This finding was surprising, since one would expect that the inhibition of translocation would be reflected in reduced levels of translocated sucrose. However, it is possible that simultaneous synthesis of sucrose might take place, as becomes apparent from the reduction of its products. has been documented in nodules of nodulated lupine (*Lupinus albus*) plants treated with glyphosate in which there was elevated activity of sucrose synthase concomitant with increased sucrose levels and a reduction in starch production (De María et al., 2006).

The fact that glyphosate induced little effect on the profile of sugars in GRT roots (**Figures 6E, 8**) further supports our supposition of the high resistance of the GRT host to glyphosate. On the other hand, the parasitism *per se* had a strong effect on the amounts of sugars that were abundant in the host roots (**Figures 6E, 8**). In other cases of root parasitism, metabolomics changes were stimulated in the tissues of the host plant. For example, in the root hemi-parasite *Striga hermontica* parasitizing sorghum (*Sorghum bicolor*) and in *O. foetida* parasitizing faba beans certain changes in the metabolomic profiles were observed in the xylem sap and leaf phloem of these plants, respectively (Pageau et al., 2003; Abbes et al., 2009a). In contrast, Hacham et al. (2016) found that parasitism of *P. aegyptiaca* had no profound effect on metabolic profile of the tomato host, inducing elevation of only 8 of a total of 59 metabolites that were detected in its roots, although some of those metabolites were carbohydrates. It is important to remember that sometimes even different phenological stages, let alone different experimental systems and different environmental conditions, could affect the metabolomics profile and the translocation profile in the host-parasite system (Westwood, 2013) and might provide an explanation for the above differences.

CONCLUSIONS

Holoparasites differ from autotrophic plants in that they depend on photosynthesis and carbon assimilation of another organism. Nonetheless, it seems that *P. aegyptiaca* does rely on its own carbon metabolism to provide the driving force for the attraction of solutes. In the current study, it was shown that glyphosate has a secondary mechanism of action in *P. aegyptiaca*, acting via inhibition of the translocation of large molecules (e.g., GFP) and probably of other phloem solutes as well. Although, the detailed chain of events is not yet clear, it appears that glyphosate's alteration of the parasite's sugars profile after EPSPS inhibition might provide an explanation for the decreased ability of *P. aegyptiaca* to attract solutes via the phloem. In addition, it was shown that glyphosate affects *P. aegyptiaca* only post attachment to the host root and does not seem to affect parasite seed germination and establishment on the host root.

AUTHOR CONTRIBUTIONS

TS, BR, SW, and HE conceived and designed research. TS conducted experiments. DP, SG, and YH assisted with experiments execution. RA assisted with data analysis. TS analyzed data and wrote the manuscript. All authors read and approved the manuscript.

ACKNOWLEDGMENTS

The authors wish to thank the Office of the Chief Scientist, Israel Ministry of Agriculture and The Plant Production and Marketing Board for funding this project (Grant no. 132-1499-11). The authors also would like to thank Jacklin Abu-Nassar, Evgeny Smirnov, Guy Achdari, and Hammam Ziadna for the technical assistance in performing the experiments. The first author (TS) is the recipient of scholarships from the Teomim Foundation, the Ministry of Agriculture and Rural Development, the Regional Jezreel Valley R&D, the Goldwasser-Zeisman Foundation and the Sam Hamburg Foundation.

REFERENCES

Abbes, Z., Kharrat, M., Delavault, P., Chaïbi, W., and Simier, P. (2009a). Nitrogen and carbon relationships between the parasitic weed *Orobanche foetida* and susceptible and tolerant faba bean lines. *Plant Physiol. Biochem.* 47, 153–159. doi: 10.1016/j.plaphy.2008.10.004

Abbes, Z., Kharrat, M., Delavault, P., Chaïbi, W., and Simier, P. (2009b). Osmoregulation and nutritional relationships between *Orobanche foetida* and faba bean. *Plant Signal. Behav.* 4, 336–338. doi: 10.4161/psb.4.4.8192

Aber, M., Fer, A., and Salle, G. (1983). Transfer of organic substances from the host plant *Vicia faba* to the parasite *Orobanche crenata*. *Z. Pflanzenphysiol.* 112, 297–308. doi: 10.1016/S0044-328X(83)80047-6

Aly, R., Cholakh, H., Joel, D. M., Leibman, D., Steinitz, B., Zelcer, A., et al. (2009). Gene silencing of mannose 6-phosphate reductase in the parasitic weed *Orobanche aegyptiaca* through the production of homologous dsRNA sequences in the host plant. *Plant Biotechnol. J.* 7, 487–498. doi: 10.1111/j.1467-7652.2009.00418.x

Aly, R., Hamamouch, N., Abu-Nassar, J., Wolf, S., Joel, D. M., Eizenberg, H., et al. (2011). Movement of protein and macromolecules between host plants and the parasitic weed *Phelipanche aegyptiaca* Pers. *Plant Cell Rep.* 30, 2233–2241. doi: 10.1007/s00299-011-1128-5

Arjona-Berral, A., Mesa-Garcia, J., and Garcia-Torres, L. (1990). Distribution of ^{14}C-glyphosate in legumes parasitized by *Orobanche crenata*. *Weed Res.* 30, 53–60. doi: 10.1111/j.1365-3180.1990.tb01687.x

Belz, R. G., and Duke, S. O. (2014). Herbicides and plant hormesis. *Pest Manage. Sci.* 70, 698–707. doi: 10.1002/ps.3726

Birschwilks, M., Haupt, S., Hofius, D., and Neumann, S. (2006). Transfer of phloem-mobile substances from the host plants to the holoparasite *Cuscuta* sp. *J. Exp. Bot.* 57, 911–921. doi: 10.1093/jxb/erj076

Borggaard, O. K., and Gimsing, A. L. (2008). Fate of glyphosate in soil and the possibility of leaching to ground and surface waters: a review. *Pest Manage. Sci.* 64, 441–456. doi: 10.1002/ps.1512

Burgess, A., Vigneron, S., Brioudes, E., Labbé, J. C., Lorca, T., and Castro, A. (2010). Loss of human Greatwall results in G2 arrest and multiple mitotic defects due to deregulation of the cyclin B-Cdc2/PP2A balance. *Proc. Natl. Acad. Sci. U.S.A.* 107, 12564–12569. doi: 10.1073/pnas.0914191107

Chae, S. H., Yoneyama, K., Takeuchi, Y., and Joel, D. M. (2004). Fluridone and norflurazon, carotenoid-biosynthesis inhibitors, promote seed conditioning and germination of the holoparasite *Orobanche minor*. *Physiol. Plant.* 120, 328–337. doi: 10.1111/j.0031-9317.2004.0243.x

Cochavi, A., Achdari, G., Smirnov, E., Rubin, B., and Eizenberg, H. (2015). Egyptian broomrape (*Phelipanche aegyptiaca*) management in carrot under field conditions. *Weed Technol.* 29, 519–528. doi: 10.1614/WT-D-14-00140.1

Cochavi, A., Rubin, B., Smirnov, E., Achdari, G., and Eizenberg, H. (2016). Factors affecting Egyptian broomrape (*Orobanche aegyptiaca*) control in carrot. *Weed Sci.* 64, 321–330. doi: 10.1614/WS-D-15-00123.1

Colquhoun, J. B., Eizenberg, H., and Mallory-Smith, C. A. (2006). Herbicide placement site affects small broomrape (*Orobanche minor*) control in red clover. *Weed Technol.* 20, 356–360. doi: 10.1614/WT-04-327R2.1

Coupland, D., and Caseley, J. C. (1979). Presence of ^{14}C activity in root exudates and guttation fluid from *Agropyron repens* treated with ^{14}C-labelled glyphosate. *New Phytol.* 83, 17–22. doi: 10.1111/j.1469-8137.1979.tb00721.x

Delavault, P., Simier, P., Thoiron, S., Véronési, C., Fer, A., and Thalouarn, P. (2002). Isolation of mannose 6-phosphate reductase cDNA, changes in enzyme activity and mannitol content in broomrape (*Orobanche ramosa*) parasitic on tomato roots. *Physiol. Plant.* 115, 48–55. doi: 10.1034/j.1399-3054.2002.1150105.x

De María, N., Becerril, J. M., García-Plazaola, J. I., Hernández, A., De Felipe, M. R., and Fernández-Pascual, M. (2006). New insights on glyphosate mode of action in nodular metabolism: role of shikimate accumulation. *J. Agric. Food Chem.* 54, 2621–2628. doi: 10.1021/jf058166c

Diaz-Sanchez, J., Lopez-Martinez, N., Lopez-Granados, F., De Prado, R., and Garcia-Torres, L. (2002). Absorption, translocation, and fate of herbicides in *Orobanche cumana*-sunflower system. *Pestic. Biochem. Physiol.* 74, 9–15. doi: 10.1016/S0048-3575(02)00119-0

Dörr, I., and Kollmann, R. (1995). Symplasmic sieve element continuity between *Orobanche* and its host. *Bot. Acta* 108, 47–55. doi: 10.1111/j.1438-8677.1995.tb00830.x

Draie, R., Péron, T., Pouvreau, J.-B., Véronési, C., Jégou, S., Delavault, P., et al. (2011). Invertases involved in the development of the parasitic plant *Phelipanche ramosa*: characterization of the dominant soluble acid isoform, PrSAI1. *Mol. Plant Pathol.* 12, 638–652. doi: 10.1111/j.1364-3703.2010.00702.x

Duke, S. O. (1988). "Glyphosate," in *Herbicides Chemistry, Degradation, and Mode of Action*, eds P. C. Kearney and D. D. Kaufman (New York, NY: Marcel Dekker Inc.), 1–72.

Duke, S. O., and Powles, S. B. (2008). Glyphosate: a once-in-a-century herbicide. *Pest Manage. Sci.* 64, 319–325. doi: 10.1002/ps.1518

Eizenberg, H., Aly, R., and Cohen, Y. (2012). Technologies for smart chemical control of broomrape (*Orobanche* spp. and *Phelipanche* spp.). *Weed Sci.* 60, 316–323. doi: 10.1614/WS-D-11-00120.1

Eizenberg, H., Hershenhorn, J., Ephrath, J. H., and Kanampiu, F. (2013). "Chemical control," in *Parasitic Orobanchaceae: Parasitic Mechanisms and Control Strategies*, eds D. M. Joel, J. Gressel, and L. J. Musselman (Berlin; Heidelberg: Springer-Verlag), 415–428.

Eizenberg, H., Plakhine, D., Hershenhorn, J., Kleifeld, Y., and Rubin, B. (2003). Resistance to broomrape (*Orobanche* spp.) in sunflower (*Helianthus annuus* L.) is temperature dependent. *J. Exp. Bot.* 54, 1305–1311. doi: 10.1093/jxb/erg129

Fernández-Aparicio, M., Reboud, X., and Gibot-Leclerc, S. (2016). Broomrape weeds. underground mechanisms of parasitism and associated strategies for their control: a review. *Front. Plant Sci.* 7:135. doi: 10.3389/fpls.2016.00135

Fuchs, M. A., Geiger, D. R., Reynolds, T. L., and Bourque, J. E. (2002). Mechanisms of glyphosate toxicity in velvetleaf (*Abutilon theophrasti* medikus). *Pestic. Biochem. Physiol.* 74, 27–39. doi: 10.1016/S0048-3575(02)00118-9

Gal-On, A., Naglis, A., Leibman, D., Ziadna, H., Kathiravan, K., Papayiannis, L., et al. (2009). Broomrape can acquire viruses from its hosts. *Phytopathology* 99, 1321–1329. doi: 10.1094/PHYTO-99-11-1321

Gaudin, Z., Cerveau, D., Marnet, N., Bouchereau, A., Delavault, P., Simier, P., et al. (2014). Robust method for investigating nitrogen metabolism of ^{15}N labeled amino acids using AccQ•Tag ultra performance liquid chromatography-photodiode array-electrospray ionization-mass spectrometry: application to a parasitic plant–plant interaction. *Anal. Chem.* 86, 1138–1145. doi: 10.1021/ac403067w

Gavet, O., and Pines, J. (2010). Progressive activation of cyclinB1-Cdk1 coordinates entry to mitosis. *Dev. Cell* 18, 533–543. doi: 10.1016/j.devcel.2010.02.013

Geiger, D. R., and Bestman, H. D. (1990). Self-limitation of herbicide mobility by phytotoxic action. *Weed Sci.* 38, 324–329.

Geiger, D. R., Shieh, W. J., and Fuchs, M. A. (1999). Causes of self-limited translocation of glyphosate in *Beta vulgaris* plants. *Pestic. Biochem. Physiol.* 64, 124–133. doi: 10.1006/pest.1999.2419

Gomes, M. P., Smedbol, E., Chalifour, A., Hénault-Ethier, L., Labrecque, M., Lepage, L., et al. (2014). Alteration of plant physiology by glyphosate and its by-product aminomethylphosphonic acid: an overview. *J. Exp. Bot.* 65, 4691–4703. doi: 10.1093/jxb/eru269

Gougler, J. A., and Geiger, D. R. (1981). Uptake and distribution of N-phosphonomethylglycine in sugar beet plants. *Plant Physiol.* 68, 668–672. doi: 10.1104/pp.68.3.668

Gresshoff, P. M. (1979). Growth inhibition by glyphosate and reversal of its action by phenylalanine and tyrosine. *Funct. Plant Biol.* 6, 177–185. doi: 10.1071/pp9790177

Hacham, Y., Hershenhorn, J., Dor, E., and Amir, R. (2016). Primary metabolic profiling of Egyptian broomrape (*Phelipanche aegyptiaca*) compared to its host tomato roots. *J. Plant Physiol.* 205, 11–19. doi: 10.1016/j.jplph.2016.08.005

Harloff, H. J., and Wegmann, K. (1993). Evidence for a mannitol cycle in *Orobanche ramosa* and *Orobanche crenata*. *J. Plant Physiol.* 141, 513–520. doi: 10.1016/S0176-1617(11)80449-9

Heide-Jørgensen, H. S. (2013). "Introduction: the parasitic syndrome in higher plants," in *Parasitic Orobanchaceae: Parasitic Mechanisms and Control Strategies*, eds D. M. Joel, J. Gressel, and L. J. Musselman (Berlin; Heidelberg: Springer-Verlag), 1–18.

Hibberd, J. M., and Jeschke, W. D. (2001). Solute flux into parasitic plants. *J. Exp. Bot.* 52, 2043–2049. doi: 10.1093/jexbot/52.363.2043

Hibberd, J. M., Quick, W. P., Press, M. C., Scholes, J. D., and Jeschke, W. D. (1999). Solute fluxes from tobacco to the parasitic angiosperm *Orobanche cernua* and the influence of infection on host carbon and nitrogen relations. *Plant Cell Environ.* 22, 937–947. doi: 10.1046/j.1365-3040.1999.00462.x

Hoagland, D. R., and Arnon, D. I. (1950). The water-culture method for growing plants without soil. *California Agric. Exp. Station Circ.* 347, 1–32.

Imlau, A., Truernit, E., and Sauer, N. (1999). Cell-to-cell and long-distance trafficking of the green fluorescent protein in the phloem and symplastic unloading of the protein into sink tissues. *Plant Cell* 11, 309–322. doi: 10.1105/tpc.11.3.309

Jacobsohn, R., and Kelman, Y. (1980). Effectiveness of glyphosate in broomrape (*Orobanche* spp.) control in four crops. *Weed Sci.* 28, 692–699.

Jain, R., and Foy, C. L. (1997). Translocation and metabolism of glyphosate in Egyptian broomrape (*Orobanche aegyptiaca*) infested tomato (*Lycopersicon esculentum*) plants. *Quarterly* 25, 1–7.

Joel, D. M. (2000). The long-term approach to parasitic weeds control: manipulation of specific developmental mechanisms of the parasite. *Crop Protect.* 19, 753–758. doi: 10.1016/S0261-2194(00)00100-9

Joel, D. M. (2013). "Functional structure of the mature haustorium," in *Parasitic Orobanchaceae: Parasitic Mechanisms and Control Strategies*, eds D. M. Joel, J. Gressel, and L. J. Musselman (Berlin; Heidelberg: Springer-Verlag), 25–60.

Joel, D. M., and Bar, H. (2013). "The seed and the seedling," in *Parasitic Orobanchaceae: Parasitic Mechanisms and Control Strategies*, eds D. M. Joel, J. Gressel, and L. J. Musselman (Berlin; Heidelberg: Springer-Verlag), 147–166.

Joel, D. M., Kleifeld, Y., Losner-Goshen, D., Herzlinger, G., and Gressel, J. (1995). Transgenic crops against parasites. *Nature* 374, 220–221. doi: 10.1038/374220a0

Joel, D. M., and Losner-Goshen, D. (1994). The attachment organ of the parasitic angiosperms *Orobanche cumana* and *O. aegyptiaca* and its development. *Can. J. Bot.* 72, 564–574. doi: 10.1139/b94-075

Kanampiu, F. K., Ransom, J. K., Friesen, D., and Gressel, J. (2002). Imazapyr and pyrithiobac movement in soil and from maize seed coats to control *Striga* in legume intercropping. *Crop Protect.* 21, 611–619. doi: 10.1016/S0261-2194(01)00151-X

Kazachkova, Y., Batushansky, A., Cisneros, A., Tel-Zur, N., Fait, A., and Barak, S. (2013). Growth platform-dependent and -independent phenotypic and metabolic responses of *Arabidopsis* and its halophytic relative, *Eutrema salsugineum*, to salt stress. *Plant Physiol.* 162, 1583–1598. doi: 10.1104/pp.113.217844

Knoblauch, M., Knoblauch, J., Mullendore, D. L., Savage, J. A., Babst, B. A., Beecher, S. D., et al. (2016). Testing the Münch hypothesis of long distance phloem transport in plants. *eLife* 5:e15341. doi: 10.7554/eLife.15341

Kremer, R., Means, N., and Kim, S. (2005). Glyphosate affects soybean root exudation and rhizosphere micro-organisms. *Int. J. Environ. Anal. Chem.* 85, 1165–1174. doi: 10.1080/03067310500273146

Laitinen, P., Rämö, S., and Siimes, K. (2007). Glyphosate translocation from plants to soil – does this constitute a significant proportion of residues in soil? *Plant Soil* 300, 51–60. doi: 10.1007/s11104-007-9387-1

Lévesque, C. A., and Rahe, J. E. (1992). Herbicide interactions with fungal root pathogens, with special reference to glyphosate. *Annu. Rev. Phytopathol.* 30, 579–602. doi: 10.1146/annurev.py.30.090192.003051

Lisec, J., Schauer, N., Kopka, J., Willmitzer, L., and Fernie, A. R. (2006). Gas chromatography mass spectrometry-based metabolite profiling in plants. *Nat. Protoc.* 1, 387–396. doi: 10.1038/nprot.2006.59

Maeda, H., and Dudareva, N. (2012). The shikimate pathway and aromatic amino acid biosynthesis in plants. *Annu. Rev. Plant Biol.* 63, 73–105. doi: 10.1146/annurev-arplant-042811-105439

Münch, E. (1930). *Die Stoffbewegungen in der Pflanze*. Jena: Gustav Fischer Verlag.

Nadler-Hassar, T., Goldshmidt, A., Rubin, B., and Wolf, S. (2004). Glyphosate inhibits the translocation of green fluorescent protein and sucrose from a transgenic tobacco host to *Cuscuta campestris* Yunk. *Planta* 219, 790–796. doi: 10.1007/s00425-004-1288-4

Nandula, V. K., Foy, C. L., and Orcutt, D. M. (1999). Glyphosate for *Orobanche aegyptiaca* control in *Vicia sativa* and *Brassica napus*. *Weed Sci.* 47, 486–491.

Onofri, A., Carbonell, E. A., Piepho, H. P., Mortimer, A. M., and Cousens, R. D. (2010). Current statistical issues in Weed Research. *Weed Res.* 50, 5–24. doi: 10.1111/j.1365-3180.2009.00758.x

Orcaray, L., Igal, M., Marino, D., Zabalza, A., and Royuela, M. (2010). The possible role of quinate in the mode of action of glyphosate and acetolactate synthase inhibitors. *Pest Manage. Sci.* 66, 262–269. doi: 10.1002/ps.1868

Orcaray, L., Zulet, A., Zabalza, A., and Royuela, M. (2012). Impairment of carbon metabolism induced by the herbicide glyphosate. *J. Plant Physiol.* 169, 27–33. doi: 10.1016/j.jplph.2011.08.009

Pageau, K., Simier, P., Le Bizec, B., Robins, R. J., and Fer, A. (2003). Characterization of nitrogen relationships between *Sorghum bicolor* and the root-hemiparasitic angiosperm *Striga hermonthica* (Del.) Benth. using K15NO3 as isotopic tracer. *J. Exp. Bot.* 54, 789–799. doi: 10.1093/jxb/erg081

Parker, C., and Dixon, N. (1983). The use of polyethylene bags in the culture and study of *Striga*spp. and other organisms on crop roots. *Ann. Appl. Biol.* 103, 485–488. doi: 10.1111/j.1744-7348.1983.tb02787.x

Reddy, K. N., Rimando, A. M., and Duke, S. O. (2004). Aminomethylphosphonic acid, a metabolite of glyphosate, causes injury in glyphosate-treated, glyphosate-resistant soybean. *J. Agric. Food Chem.* 52, 5139–5143. doi: 10.1021/jf049605v

Schneider, C. A., Rasband, W. S., and Eliceiri, K. W. (2012). NIH Image to ImageJ: 25 years of image analysis. *Nat. Methods* 9, 671–675. doi: 10.1038/nmeth.2089

Schulz, A., Münder, T., Holländer-Czytko, H., and Amrhein, N. (1990). Glyphosate transport and early effects on shikimate metabolism and its compartmentation in sink leaves of tomato and spinach plants. *Z. Naturforsch. C Bio. Sci.* 45, 529.

Servaites, J. C., Tucci, M. A., and Geiger, D. R. (1987). Glyphosate effects on carbon assimilation, ribulose bisphosphate carboxylase activity, and metabolite levels in sugar beet leaves. *Plant Physiol.* 85, 370–374. doi: 10.1104/pp.85.2.370

Shilo, T., Zygier, L., Rubin, B., Wolf, S., and Eizenberg, H. (2016). Mechanism of glyphosate control of *Phelipanche aegyptiaca Planta* 244, 1095–1107. doi: 10.1007/s00425-016-2565-8

Siehl, D. L. (1997). "Inhibitors of EPSP synthase, glutamine synthetase, and histidine synthesis," in *Herbicide Activity: Toxicology, Biochemistry and Molecular Biology*, eds M. R. Roe, J. D. Burton, and R. J. Kuhr (Amsterdam: IOS Press), 37–68.

Song, W. J., Zhou, W. J., Jin, Z. L., Cao, D. D., Joel, D. M., Takeuchi, Y., et al. (2005). Germination response of *Orobanche* seeds subjected to conditioning temperature, water potential and growth regulator treatments. *Weed Res.* 45, 467–476. doi: 10.1111/j.1365-3180.2005.00477.x

Song, W. J., Zhou, W. J., Jin, Z. L., Zhang, D., Yoneyama, K., Takeuchi, Y., et al. (2006). Growth regulators restore germination of *Orobanche* seeds that are conditioned under water stress and suboptimal temperature. *Aust. J. Agric. Res.* 57, 1195–1201. doi: 10.1071/AR06131

Steinrücken, H. C., and Amrhein, N. (1980). The herbicide glyphosate is a potent inhibitor of 5-enolpyruvylshikimic acid-3-phosphate synthase. *Biochem. Biophys. Res. Commun.* 94, 1207–1212. doi: 10.1016/0006-291X(80)90547-1

Tzin, V., and Galili, G. (2010). New insights into the shikimate and aromatic amino acids biosynthesis pathways in plants. *Mol. Plant* 3, 956–972. doi: 10.1093/mp/ssq048

Wang, F., Sanz, A., Brenner, M. L., and Smith, A. (1993). Sucrose synthase, starch accumulation, and tomato fruit sink strength. *Plant Physiol.* 101, 321–327. doi: 10.1104/pp.101.1.321

Westwood, J. H. (2013). "The physiology of the estsblished parasite-host association," in *Parasitic Orobanchaceae: Parasitic Mechanisms and Control Strategies*, eds D. M. Joel, J. Gressel, and L. J. Musselman. (Berlin; Heidelberg: Springer-Verlag), 87–114.

Whitney, P. J., and Carsten, C. (1981). Chemotropic response of broomrape radicles to host root exudates. *Ann. Bot.* 48, 919–921.

Xia, J., Sinelnikov, I. V., Han, B., and Wishart, D. S. (2015). MetaboAnalyst 3.0—making metabolomics more meaningful. *Nucleic Acids Res.* 43, W251–W257. doi: 10.1093/nar/gkv380

Xia, J., and Wishart, D. S. (2011). Web-based inference of biological patterns, functions and pathways from metabolomic data using MetaboAnalyst. *Nat. Protoc.* 6, 743–760. doi: 10.1038/nprot.2011.319

Yasuor, H., Abu-Abied, M., Belausov, E., Madmony, A., Sadot, E., Riov, J., et al. (2006). Glyphosate-induced anther indehiscence in cotton is partially temperature dependent and involves cytoskeleton and secondary wall modifications and auxin accumulation. *Plant Physiol.* 141, 1306–1315. doi: 10.1104/pp.106.081943

Zhu, J., Patzoldt, W. L., Shealy, R. T., Vodkin, L. O., Clough, S. J., and Tranel, P. J. (2008). Transcriptome response to glyphosate in sensitive and resistant soybean. *J. Agric. Food Chem.* 56, 6355–6363. doi: 10.1021/jf801254e

Zygier, L. (2005). Effect of *Glyphosate Absorption* and *Translocation* on Egyptian *Broomrape* (Orobanche aegyptiacaPers.) *Control* in *Tomato* (Lycopersicon esculentum Mill.). Master's thesis, The Hebrew University of Jerusalem, Rehovot.

Conflict of Interest Statement: The authors declare that the research was conducted in the absence of any commercial or financial relationships that could be construed as a potential conflict of interest.

Permissions

All chapters in this book were first published in APWR, by Frontiers Media SA; hereby published with permission under the Creative Commons Attribution License or equivalent. Every chapter published in this book has been scrutinized by our experts. Their significance has been extensively debated. The topics covered herein carry significant findings which will fuel the growth of the discipline. They may even be implemented as practical applications or may be referred to as a beginning point for another development.

The contributors of this book come from diverse backgrounds, making this book a truly international effort. This book will bring forth new frontiers with its revolutionizing research information and detailed analysis of the nascent developments around the world.

We would like to thank all the contributing authors for lending their expertise to make the book truly unique. They have played a crucial role in the development of this book. Without their invaluable contributions this book wouldn't have been possible. They have made vital efforts to compile up to date information on the varied aspects of this subject to make this book a valuable addition to the collection of many professionals and students.

This book was conceptualized with the vision of imparting up-to-date information and advanced data in this field. To ensure the same, a matchless editorial board was set up. Every individual on the board went through rigorous rounds of assessment to prove their worth. After which they invested a large part of their time researching and compiling the most relevant data for our readers.

The editorial board has been involved in producing this book since its inception. They have spent rigorous hours researching and exploring the diverse topics which have resulted in the successful publishing of this book. They have passed on their knowledge of decades through this book. To expedite this challenging task, the publisher supported the team at every step. A small team of assistant editors was also appointed to further simplify the editing procedure and attain best results for the readers.

Apart from the editorial board, the designing team has also invested a significant amount of their time in understanding the subject and creating the most relevant covers. They scrutinized every image to scout for the most suitable representation of the subject and create an appropriate cover for the book.

The publishing team has been an ardent support to the editorial, designing and production team. Their endless efforts to recruit the best for this project, has resulted in the accomplishment of this book. They are a veteran in the field of academics and their pool of knowledge is as vast as their experience in printing. Their expertise and guidance has proved useful at every step. Their uncompromising quality standards have made this book an exceptional effort. Their encouragement from time to time has been an inspiration for everyone.

The publisher and the editorial board hope that this book will prove to be a valuable piece of knowledge for researchers, students, practitioners and scholars across the globe.

List of Contributors

Alberto Martín-Sanz
Pioneer Hi-Bred Agro Servicios Spain SL, Sevilla, Spain

Jebri Malek, José M. Fernández-Martínez, Begoña Pérez-Vich and Leonardo Velasco
Institute for Sustainable Agriculture, Consejo Superior de Investigaciones Científicas, Córdoba, Spain

Henry S. Nzioki
Kenya Agriculture and Livestock Research Organization, Machakos, Kenya

Florence Oyosi
Liberty Initiators Network, Maseno, Kenya

Cindy E. Morris
Plant Pathology, INRA-PACA, Avignon, France
Department of Plant Sciences and Plant Pathology, Montana State University, Bozeman, MT, USA

Eylul Kaya, Alice L. Pilgeram and David C. Sands
Department of Plant Sciences and Plant Pathology, Montana State University, Bozeman, MT, USA

Claire S. Baker
Biotech Investments, Bozeman, MT, USA

Ariel Venezian
Efal Agri Ltd., Netanya, Israel

Evgenia Dor, Guy Achdari, Dina Plakhine, Evgeny Smirnov and Joseph Hershenhorn
Department of Plant Phytopathology and Weed Research, Newe Ya'ar Research Center, Agricultural Research Organization, Ramat Yishai, Israel

Carmen M. Ortiz-Bustos and Leire Molinero-Ruiz
Department of Crop Protection, Institute for Sustainable Agriculture - Spanish National Research Council, Cordoba, Spain

María L. Pérez-Bueno and Matilde Barón
Estación Experimental del Zaidín, Spanish National Research Council, Granada, Spain

Xavier Reboud
INRA, UMR 1347 Agroécologie, Dijon, France

Stephanie Gibot-Leclerc
Agrosup-Dijon, UMR 1347 Agroécologie, Dijon, France

Hiroaki Samejima
Graduate School of Agricultural Science, Kobe University, Kobe, Japan
International Cooperation Center for Agricultural Education, Nagoya University, Nagoya, Japan

Abdel G. Babiker
College of Agricultural Studies, Sudan University of Science and Technology, Khartoum North, Sudan

Ahmed Mustafa
Gezira Research Station, Agricultural Research Corporation, Wad Madani, Sudan

Yukihiro Sugimoto
Graduate School of Agricultural Science, Kobe University, Kobe, Japan

Alejandro Pérez-de-Luque
Área de Mejora y Biotecnología, IFAPA, Centro Alameda del Obispo, Córdoba, Spain

Fernando Flores
Departamento de Ciencias Agroforestales, ETSI La Rábida, Universidad de Huelva, Huelva, Spain

Diego Rubiales
Institute for Sustainable Agriculture, Consejo Superior de Investigaciones Científicas (CSIC), Córdoba, Spain

Amnon Cochavi
Department of Phytopathology and Weed Research, Newe Ya'ar Research Center, Agricultural Research Organization, Ramat Yishay, Israel
R. H. Smith Institute of Plant Sciences and Genetics in Agriculture, Faculty of Agricultural, Food and Environmental Sciences, The Hebrew University of Jerusalem, Rehovot, Israel

Baruch Rubin
R. H. Smith Institute of Plant Sciences and Genetics in Agriculture, Faculty of Agricultural, Food and Environmental Sciences, The Hebrew University of Jerusalem, Rehovot, Israel

Hanan Eizenberg
Department of Phytopathology and Weed Research, Newe Ya'ar Research Center, Agricultural Research Organization, Ramat Yishay, Israel

Mónica Fernández-Aparicio
Agroécologie, AgroSup Dijon, INRA, Université Bourgogne Franche-Comté, Dijon, France
CSIC, Institute for Sustainable Agriculture, Córdoba, Spain

Alexandre Bernard and Laurent Falchetto
INRA, UE0115 Domaine Expérimental d'Epoisses, Breteniére, France

Pascal Marget
Agroécologie, AgroSup Dijon, INRA, Université Bourgogne Franche-Comté, Dijon, France
INRA, UE0115 Domaine Expérimental d'Epoisses, Breteniére, France

Bruno Chauvel, Christian Steinberg, Stephanie Gibot-Leclerc, David A. Bohan and Xavier Reboud
Agroécologie, AgroSup Dijon, INRA, Université Bourgogne Franche-Comté, Dijon, France

Cindy E. Morris
INRA, UR0407 Pathologie Végétale, Montfavet, France

Angela Boari and Maurizio Vurro
CNR, Institute of Sciences of Food Production, Bari, Italy

Lilach Iasur Kruh
Department of Weed Science, Newe Ya'ar Research Center, Agricultural Research Organization, Ramat Yishay, Israel
Department of Biotechnology Engineering, ORT Braude College, Karmiel, Israel

Tamar Lahav and Shiri Freilich
The Institute of Plant Sciences, Newe Ya'ar Research Center, Agricultural Research Organization, Ramat Yishay, Israel

Jacline Abu-Nassar, Guy Achdari and Radi Aly
Department of Weed Science, Newe Ya'ar Research Center, Agricultural Research Organization, Ramat Yishay, Israel

Raghda Salami
Department of Biotechnology Engineering, ORT Braude College, Karmiel, Israel

Mathieu Rolland and Aurélie Dupuy
GEVES, Beaucouzé, France

Aude Pelleray and Philippe Delavault
Laboratoire de Biologie et Pathologie Végétales, Université de Nantes, Nantes, France

Hammam Ziadne
Department of Plant Pathology and Weed Research, Newe Ya'ar Research Center, Agricultural Research Organization, Ramat Yishai, Israel

Ludmila Tsechansky and Ellen R. Graber
Institute of Soil, Water and Environmental Sciences, Agricultural Research Organization, The Volcani Center, Beit Dagan, Israel

Mónica Fernández-Aparicio
Institute for Sustainable Agriculture, Consejo Superior de Investigaciones Científicas, Córdoba, Spain
INRA, UMR1347 Agroécologie, Dijon, France

Fernando Flores
Escuela Técnica Superior de Ingeniería - Universidad de Huelva, Palos de la Frontera, Spain

Diego Rubiales
Institute for Sustainable Agriculture, Consejo Superior de Investigaciones Científicas, Córdoba, Spain

Luíza Teixeira-Costa and Gregório C. T. Ceccantini
Group of Ecological Wood Anatomy and Parasitic Plant Biology, Department of Botany, Institute of Biosciences, University of São Paulo, São Paulo, Brazil

Delphine Moreau, Stéphanie Gibot-Leclerc, Annette Girardin, Olivia Pointurier, Carole Reibel, Florence Strbik and Nathalie Colbach
Agroécologie, AgroSup Dijon, INRA, Université de Bourgogne Franche-Comté, Dijon, France

Noam Nativ, Yael Hacham and Rachel Amir
Migal Galilee Technology Center, Kiryat Shmona, Israel
Biotechnology Department, Tel-Hai College, Upper Galilee, Israel

Neeraj K. Dubey, Jackline Abu-Nassar and Sally Marzouk
Department of Plant Pathology and Weed Research, Newe Ya'ar Research Center, Agricultural Research Organization, Volcani Center, Ramat Yishay, Israel

Diana Leibman and Amit Gal-On
Department of Plant Pathology and Weed Research, Agricultural Research Organization, Volcani Center, Rishon LeZion, Israel

Dalia Wolf and Menahem Edelstein
Department of Plant Science, Agricultural Research Organization, Volcani Center, Rishon LeZion, Israel

Amnon Cochavi, Tania Gendler, Shimon Rachmilevitch and Jhonathan E. Ephrath
The French Associates Institute for Agriculture and Biotechnology of Drylands, The Jacob Blaustein Institutes for Desert Research, Ben-Gurion University of the Negev, Beer-Sheva, Israel

Tal Rapaport and Arnon Karnieli
The Remote Sensing Laboratory, The Swiss Institute for Dryland Environmental & Energy Research, The Jacob Blaustein Institutes for Desert Research, Ben Gurion University of the Negev, Beer-Sheva, Israel

Johann Louarn, Marie-Claude Boniface, Nicolas Pouilly, Patrick Vincourt and Stéphane Muños
LIPM, Université de Toulouse, INRA, CNRS, Castanet-Tolosan, France

Tal Shilo
Department of Plant Pathology and Weed Research, Agricultural Research Organization, Newe Ya'ar Research Center, Ramat Yishay, Israel
The Robert H. Smith Faculty of Agriculture, Food and Environment, The Robert H. Smith Institute of Plant Sciences and Genetics, The Hebrew University of Jerusalem, Rehovot, Israel

Shmuel Wolf
The Robert H. Smith Faculty of Agriculture, Food and Environment, The Robert H. Smith Institute of Plant Sciences and Genetics, The Hebrew University of Jerusalem, Rehovot, Israel

Shira Gal
Department of Entomology, Agricultural Research Organization, Newe Ya'ar Research Center, Ramat Yishay, Israel

Index